AF388923

Animal Nutrition and Productions

Animal Nutrition and Productions

Editor

Daniel Simeanu

MDPI • Basel • Beijing • Wuhan • Barcelona • Belgrade • Manchester • Tokyo • Cluj • Tianjin

Editor
Daniel Simeanu
Ion Ionescu de la Brad
University of Life Sciences Iasi
Iasi, Romania

Editorial Office
MDPI
St. Alban-Anlage 66
4052 Basel, Switzerland

This is a reprint of articles from the Special Issue published online in the open access journal *Agriculture* (ISSN 2077-0472) (available at: https://www.mdpi.com/journal/agriculture/special_issues/animal_nutrition_productions).

For citation purposes, cite each article independently as indicated on the article page online and as indicated below:

LastName, A.A.; LastName, B.B.; LastName, C.C. Article Title. *Journal Name* **Year**, *Volume Number*, Page Range.

ISBN 978-3-0365-8010-4 (Hbk)
ISBN 978-3-0365-8011-1 (PDF)

Contents

About the Editor

Daniel Simeanu

Daniel Simeanu, Ph.D., University hab. Professor, Iași University of Life Sciences, Romania, acquired all the degrees in higher education in Romania, from a preparatory degree in 1999 to a tenured professor in 2021. He defended his doctoral thesis in 2003 and his habilitation thesis in 2021 to become a Ph.D. supervisor. Research areas: nutrition and feeding; influence of feeding on animal production quality. Publishing activity: 193 papers in national and international journals; 46 in Web of Science journals. He has managed four research projects and participated as a member of 25 grant teams. He has published five books as the sole or first author and co-authored 15 university books and handbooks. He was awarded two prizes, by the Romanian Academy (2004) and by the EUROINVENT Exhibition of Creativity and Innovation (2019) for two referential books in the field of animal sciences engineering (Treaty of Aviculture - coord. by Vacaru-Opris, I. and Animal Productions – Simeanu D., Dolis, M.G.)

Editorial

Animal Nutrition and Productions

Daniel Simeanu [1] and Răzvan-Mihail Radu-Rusu [2,*]

[1] Department of Control, Expertise and Services, Faculty of Food and Animal Sciences, "Ion Ionescu de la Brad" University of Life Sciences, 8 Mihail Sadoveanu Alley, 700489 Iasi, Romania; dsimeanu@uaiasi.ro

[2] Department of Animal Resources and Technology, Faculty of Food and Animal Sciences, "Ion Ionescu de la Brad" University of Life Sciences, 8 Mihail Sadoveanu Alley, 700489 Iasi, Romania

* Correspondence: radurazvan@uaiasi.ro

Citation: Simeanu, D.; Radu-Rusu, R.-M. Animal Nutrition and Productions. *Agriculture* 2023, 13, 943. https://doi.org/10.3390/agriculture13050943

Received: 21 April 2023
Accepted: 23 April 2023
Published: 25 April 2023

Animal nutrition and feeding, along with other sciences (reproduction, genetics, hygiene, management, etc.), play a major role in improving animal production. In this vein, it is unanimously accepted that animal nutrition and feeding do not create high-potential animals, but allow them reach their productive potential. It is unthinkable to breed, artificially select and genetically improve animals in inadequate feeding conditions; in the same way, the normal reproduction of animals relies, to a decisive extent, on a rational diet based on the latest scientific achievements.

The importance of nutrition in animal production is also highlighted by the weight of the cost of feed (over 50% of the costs of animal products). The attention that animal breeders pay to nutrition and feeding determines, to a large extent, the profitability of their livestock.

Animal feeding has a major influence on the quality of products of animal origin, and on the proximate composition of different components of the body (proteins, lipids, water), the sensory (colour, taste, smell) and biochemical–functional (such as the proportion of different fatty acids) properties and even on their technological qualities (in milk, meat, eggs, etc.).

Animals are biological transformers of dietary matter and energy into high-quality raw matters (meat, eggs, milk and honey) for human food, but also into raw materials (e.g., wool and leather) for human clothing and accessories. Thus, animal nutrition is a fundamental topic for all farm animal productions as well as for the health and wellbeing of companion animals.

Animal products have been essential components of human food for many centuries and in many cultures. In developed countries, the share of products of animal origin in food is important; thus, for example in the USA [1], the typical daily ration for humans provides approx. 2100 kcal, to which animal products contribute 47% and vegetable products 53%.

Moreover, at the European level, a direct relationship between infant mortality and milk production has been highlighted (the increase of 100 kg of milk production/inhabitant causes a reduction in infant mortality of 2.20/1000 inhabitants). Additionally, a direct link has been established between the increase in milk consumption and the decrease in alcoholism.

The global population was 7.4 billion in 2016 and is projected to reach 9.6 billion in 2050. With the increase in global population and the per capita human consumption of meat, milk and eggs, it is expected that the demand for animal protein and other animal products will increase by 70% globally by 2050 [2].

It is also important to mention that recent evidence suggests that a major reduction in animal production would produce both important ecological disturbances and strong economic instability worldwide. However, the increased consumption of food of animal origin requires a larger area of land to be used to feed one person (approx. 0.37 ha, compared to only 0.09–0.1 ha in the case of vegetarian food). According to the biological law of cycles and trophic levels, the ingested organic matter is found in the consumer's body in a ratio of 1/10; therefore, in herbivores, from a consumption of 10 kg of organic substance, only 1 kg of organic substance is deposited in the body.

Due to the continuous growth of the population and other causes, the area of land cultivated with cereals (the main source of food for non-ruminant animals), expressed in relation to the number of inhabitants, has decreased and continues to decrease.

Increasing animal productivity, possibly primarily through genetic improvements associated with appropriate feeding and a more efficient use of feed, ensures a better use of agricultural land, allowing a reduction in animal flocks without decreasing total production.

This Special Issue, "Animal Nutrition and Productions", comprises a total of 20 articles, of which one is a review, from a total of 139 researchers: 49 from Romania, 34 from Brazil, 13 from the USA, 11 from Tunisia, 8 each from Italy and Mexico, 7 from Saudi Arabia, 3 from Egypt and 1 each from Spain, Canada, Jordan, Germany, France and the Netherlands.

The published articles can be classified in nine research directions, including the influence of nutrition on the quantitative and qualitative production of milk in different species (buffaloes, camels); the influence of nutrition on the quantitative and qualitative production of meat from different species (rabbit, swordfish, European catfish), as well as the changes obtained after preservation (chicken, turkey and duck); the influence of nutrition on reproductive function; the digestibility of some animal feed sources; the influence of nutrition on egg production (hen or quail); microbiological evaluation of combined feeds intended for broiler chickens; the assessment and monitoring of pollutants from animal farms; the quality of honey production; and a phylogenetic analysis of endangered animal breeds.

The first category includes two articles that study the impact of the quality of different feed sources on milk production in camels and buffaloes. Thus, Abdelrahman M.M. et al. [3] conducted a study on the nutritional value of camel milk obtained from animals that were fed during the winter with a total mixed ration (TMR) together with alfalfa hay. Thirty-seven lactating multiparous camels were chosen for the study. The onset period of the study was mid-lactation and the geographical area of this research was Al-Kharj in Saudi Arabia. Females were divided into two groups, one fed exclusively alfalfa hay ad libitum (C), and group two (T) received an additional mixture (TMR) consisting of barley, wheat, palm kernel cake, soybean husks, vitamins and minerals at an amount of 4 kg/head/day. Milk and blood samples were collected from the studied animals in the middle of the winter season. A significantly ($p < 0.05$) higher concentration of protein and organic matter was found in the milk of group T camels. In addition, the levels of Mg, Co, Fe and Zn in milk increased significantly ($p < 0.05$) in group T compared to group C. The concentration of minerals Ca, P, Mg, Cu, I, Se, Zn and Cd in the blood of camels from group T was significantly ($p < 0.05$) higher than that of group C. Blood serum levels from the youth of group T were significantly ($p < 0.05$) higher for all minerals compared to the values recorded in the comparison group (C), the exception was iodine. In youth, there were significant signals and correlations between Co and Mn and most of the other minerals investigated. In conclusion, supplementation with TMR in the T group of camels during lactation, in the winter season, can be recommended because it leads to improvements in the chemical composition and mineral profile of camel milk.

In another case study, Serrapica F. et al. [4] analysed the effect of raising buffaloes outside traditional breeding areas in Italy. Thus, in their study, researchers from Italy chose 20 pregnant buffaloes (*Bubalus bubalis*) that were moved from the traditional breeding lot area, on the southern coast of Italy, to a farm in the interior of the peninsula located in a hilly area. After the buffaloes calved, data were collected monthly, throughout lactation, on the productive and reproductive performance of the studied animals. The applied feeding regime was also studied throughout 4–6 months of lactation to evaluate the in vivo digestibility of locally produced forages (maize silage compared to hay) and their effects of on the milk production. At the same time, the sensory properties of the mozzarella cheese obtained from the milk of the studied animals were also analysed.

In terms of the applied feeding, no significant differences were found; however, a comparison of the results obtained with the values recorded in the previous lactation revealed major differences. Milk production decreased by 37.2%, and the protein content of milk was reduced by 6.1%. At the same time, the fat content of milk improved (+4.5%). Severe deteriorations in reproductive indicators, such as decreases in gestation rate (-13.3%) and an increase in calving interval (+26.9%), were reported. The duration of lactation was less than the standard value of 270 days. Artificial insemination failure and lower environmental temperatures are likely the origin of these results, while the body condition score at the end of the observation period and data from the feeding study do not indicate feeding errors. Overall, the distinctive reproductive characteristics of buffalo, the lower environmental temperatures and the specificity of the mozzarella cheese production process are the main issues that farmers should consider when they want to expand buffalo farming outside the traditional growing areas of Italy.

The second category, the largest one, with five articles, includes studies carried out on fish (*Polyodon spathula*, *Silurus glanis*), hares and rabbits, lamb and different types of poultry meat preserved by smoking.

The first study includes data on the qualitative and nutritional evaluation of the meat production of paddlefish sturgeon of different ages, raised and fed under the conditions of some fish farms in Romania. The authors, coordinated by Simeanu D. [5], contribute detailed knowledge on the nutritional properties of the flesh of second- and third-summer paddlefish sturgeon, but also more specifically of the different muscle groups. The study included analyses of the chemical composition and cholesterol and collagen content of fillets obtained from two- or three-year-old *Polyodon spathula* sturgeons reared and fed in polyculture conditions with carp (*Cyprinus carpio*) and Asian cyprinids (*Ctenopharyngodon idella*, *Hypophthalmichthys molitrix* and *Aristichthys nobilis*). The nutritional analysis was even more detailed, as it calculated the nutritional value, the profile of acids and amino acids, the sanogenic indices and the biological value of proteins for the epaxial and hypaxial muscle groups. The chemical analysis of the fillets, related to age, indicated slightly higher values for three-summer-old compared to two-summer-old sturgeon fish: +5.32% dry matter, +0.89% protein, +41.21% fat, +10.94% gross energy, +2.94% collagen; in contrast, for water, minerals and W/P ratio the values were lower by 1.52%, 10.08% and 2.29%, respectively. The nutritional evaluation of *Polyodon spathula* sturgeon meat highlighted that the fats of these fish are of good quality with a significant presence of PUFA (about 22% of total fatty acids) and with good sanogenic indices values (PI = 7.01–8.77; AI = 0.57; TI = 0.38–0.39; hFA = 33.01–41.34; h/H = 1.9). Additionally, the proteins of these fish are of good quality for youth and adult consumers (EAAI (%) = 156.11; BV = 158.46; NI (%) = 28.30) and good enough for children (EAAI (%) = 96.41; BV) = 93.39; NI (%) = 17.45). The authors recommend that paddlefish sturgeons should be capitalised when they are three summers old, when they have the best nutritional quality.

In another study concerning fish meat, a group of researchers from USV Iași led by Simeanu C. [6] carried out research to evaluate the quantitative and qualitative meat of European catfish (*Silurus glanis*) from two different environments: aquaculture, where it benefited from a feeding specific to farms of this type, and in the Prut river in Romania, where it fed naturally. The samples were grouped into two groups: AG—fish from aquaculture and RG—fish from a natural environment, the Prut river. The catfish were measured, and biometric and conformational indices were calculated. The best values were found in fish from aquaculture. The Fulton coefficient reached 0.82 in RG and 0.91% in AG catfish. The fleshy index was 19.58% in AG aquaculture fish and 20.79% in wild fish, indicating better productivity capabilities in AG catfish. The yield at slaughter was calculated, and the qualitative analysis of the resulting fillets was carried out at different moments of refrigeration conservation (0–15 days). In the AG samples, water content decreased by 8.87%, protein by 27.66% and lipid by 29.58%. In the case of RG samples, the loss reached 8.59% for water, 25.16% for protein and 29% for lipids. The fatty acid profile was also studied and sanogenic indices were calculated. The studied European catfish presented good levels of PUFA (31–35%), and the AI reached 0.35–0.41 while the TI varied between 0.22 and 0.27. In conclusion, the origin of the fish and especially the refrigeration duration influence the proximate composition of the meat and its nutritional value.

A nutritional and technological perspective on rabbit (*Oryctolagus cuniculus*) and hare (*Lepus europaeus Pallas*) meat is presented by Frunză G. et al. [7]. In this study, the authors analysed meat from 50 Flemish Giant farmed rabbits and 50 hunted hares from three regions of the carcasses; the *Longissimus dorsi* (LD), *Semimembranosus* (SM) and *Triceps brachii* (TB) muscles were harvested. The proximate composition and the fatty acid profile were evaluated, then the energy content and lipid sanogenic indices (polyunsaturated, atherogenic, thrombogenic, hypocholeroserolemic/hypercholesterolemic ratio and the nutritional index—NVI) were calculated. To highlight the technological qualities of the meat, the pH values at 24 and 48 h after slaughter, the cooking loss (CL) and the water-holding capacity (WHC) were assessed. The gross energy was higher in TB samples from rabbit compared to hare due to higher fat accumulations ($p < 0.001$). The pH value was higher for TB muscles in both categories of meat, WHC was higher in hare ($p < 0.001$) and CL was higher in rabbit ($p < 0.001$). PI values were 6.72 in hare and 4.59 in rabbit. AI reached 0.78 in hare and 0.73 in rabbit. TI was calculated at 0.66 in hare and 0.39 in rabbit. The h/H ratio reached 3.57 in hare and 1.97 in Giant Flemish rabbit, while the NVI was 1.48 in hare and 1.34 in rabbit. Meat from both species is nutritionally valuable to human consumers, with values better than other farmed mammals or other wild species of mammals or birds. The study highlighted that hare meat is healthier than rabbit meat because it has lower fat content, less energy and contains better quality lipids.

The quantitative and qualitative aspects of meat from lambs fed Tifton 85 grass (*Clynodon* sp.) in different proportions were dealt with by the team led by Corrêa Y. [8]. The authors aimed to determine the optimal ratio between fibrous and concentrated feedstuffs on the quantitative and qualitative properties of Santa Ines lamb meat. Diets were composed of Tifton 85 grass hay (*Cynodon* sp.) and concentrate mixtures (soybean meal, corn meal, urea and mineral mix) and consisted of five forage/concentrate ratios of 88:12 (C12), 69:31 (C31), 50:50 (C50), 31:69 (C69) and 12:88 (C88). After 63 days, animals were slaughtered and the carcasses, dressed yield and meat physicochemical properties were evaluated. The high proportion of concentrates in the lamb diet provided a higher intake of dry matter, better feed conversion, well-conformed carcasses, good slaughter yield and higher meat fat content. The addition of 50% concentrate improved carcass conformation and the physicochemical parameters of the meat in a similar way to the 88% concentrate diet but with leaner meat, meeting the requirements of the current consumer market.

Another article published in this Special Issue, written by Coroian C. et al. [9], deals with the presence of polycyclic aromatic hydrocarbons (PAHs) in traditionally smoked poultry. The attention of researchers has been drawn to the fact that increasing attention is being paid to the sensory, nutritional and sanogenic properties of meat in general and of poultry meat in particular. They evaluated how a traditional meat preservation method—hot smoking with natural wood smoke—adds certain polycyclic aromatic hydrocarbons (PAHs) to chicken, duck and turkey. A one-day versus two-day smoking period and three types of smoking wood (plum, cherry and beech) showed that the highest concentrations of PAHs were present in duck meat, regardless of smoking duration or type of wood. An overall higher PAH concentration was quantified when beech wood was used, followed by cherry and plum wood. Fluorene associated with beech wood gave the highest values for day 1 and day 2, followed by duck meat and turkey meat, respectively. Highly significant differences ($p < 0.001$) were usually observed for duck meat compared to chicken and turkey meat, but absolute values for anthracene, phenanthrene or fluoranthene were also easily observed. As expected, smoking for two days contributed to higher concentrations of PAHs in the meat.

In the third category, there are three articles. The first addresses the interesting topic of the effect of supplementation with *Saccharomyces cerevisiae* on the reproductive performance of rams fed a wheat-straw-based ration. The experiment led by Ben Saïd S. et al. [10]

comprised 14 Queue Fine de l'Ouest rams aged between 3 and 4 years with an average body weight of 54.7 ± 2.03 kg and an average body condition score of 3.5 ± 0.5. The study was carried out throughout a period of 80 days, during the breeding season. The rams were divided into two homogeneous groups (n = 7) and were maintained in individual floor pens. The control group was fed a daily diet containing 1 kg of wheat straw and 700 g of concentrates, and the experimental group received the same diet but supplemented with 10 g of *S. cerevisiae*/head/day. Although the supplementation did not significantly influence ram body weight, sperm volume and concentration, dry matter intake, protein digestibility and nitrogen balance, the authors found that the addition of *S. cerevisiae* improved ($p < 0.05$) the digestibility of dry matter by 7.3%, of organic matter by 11.9% and of crude fibre by 24%. In addition, the degree of sperm motility increased in the case of the group that received yeast compared to the control (3.7 ± 0.24 vs. 1.9 ± 0.27, $p < 0.05$). There was also a decrease in the total number of dead and abnormal spermatozoa for the yeast-fed group as opposed to the control group (9.28 ± 0.95 vs. 26.8 ± 3.85% and 25.5 ± 3.33%, respectively, vs. 59.2 ± 2.78%; $p < 0.05$). Therefore, the addition of *S. cerevisiae* to the diets of rams during the breeding season could improve nutrient digestibility and reproductive performance.

The influence of nutrition on the reproductive performance of sheep was also approached by Nechifor I. et al. [11], who studied females from the Karakul de Botoșani breed. Two groups, L1 (control) and L2 (experimental), were formed of adult females aged between two and six years. Group L2 received supplemental feeding 25 days before mating. Improvements in body condition and significant increases in live weight occurred up to the time of mating in ewes receiving the booster feed (L2) ($p < 0.01$ vs. L1). The number of lambs at calving was influenced by body condition score (BCS), especially in ewes with a BCS of 2. The total number of lambs weaned by ewes with a BCS of 2 was different compared to that of ewes with a BCS of 2.5 or 3.0. All results showed that supplementary feeding applied to ewes before mating influenced their reproductive and economic performance, and increased the live weight of lambs at weaning in group L2 ($p < 0.01$ in lambs from ewes with BCS 2 and $p < 0.001$ in lambs from ewes with BCS 2.5 to 3.5). Stimulative feeding applied to ewes positively influenced their reproductive performance and generated better economic results (23–27% better live weight in lambs from supplemented ewes with only 1.3% additional feed costs), thus supporting its applicability in farm conditions.

Reproductive performance influenced by animal feeding was also studied by the research team led by Edmunds C.E. [12], who studied the effect of increasing manganese intake from an organic source on the reproductive performance of sows. Sows (n = 39; 231 ± 8 kg) were randomly assigned to one of three dietary Mn inclusion levels (CON: 0 ppm Mn; PRO20: 20 ppm Mn; PRO40: 40 ppm Mn). Groups PRO20 and PRO40 were initiated at breeding and continued through two parturitions. Sows were grouped by parity within each farrowing group. Data were analysed as a completely randomized block design using the MIXED procedure of SAS with diet as a fixed effect and group as a random effect. Feed intake for lactation increased in PRO20 sows compared to CON and PRO40 sows ($p < 0.05$). PRO20 and PRO40 sows farrowed piglets that achieved an improved average daily gain from birth to weaning (CON 214 g/d; PRO20 237 g/d; 220 g/d; $p < 0.05$) compared to CON sows. Milk fat content was lower in PRO20 (5.5%) and PRO40 (6.1%; $p < 0.05$) compared to CON (7.8%) sows, possibly due to increased milk demand for suckling. The introduction of supplementary Mn during two cycles of gestation and lactation resulted in the improved birth weight and growth rate of piglets until weaning.

Another direction of research addressed in this Special Issue is the digestibility of some feed sources for animals. Thus, the group of authors led by Ammar H. [13] studied the nutritional value of *Ajuga iva* (*A. iva*) harvested from three distinct altitude regions in Tunisia (Dougga, Mograne and Nabeul). Proximate composition, phenolic concentration, gas production and in vitro dry matter digestibility were assessed. The highest concentrations of neutral detergent fibres and acid detergent fibres were found in *A. iva* cultivated in Nabeul. In contrast, the highest concentration of crude protein was observed in plants from Mograne and the lowest one ($p < 0.01$) was measured in Dougga samples. In addition, cultivation regions affected the concentrations of free radical scavenging activity, total flavonoids and total polyphenols ($p < 0.01$). The highest free radical scavenging activity was observed in *A. iva* grown in Dougga and Mograne. The highest gas production rate ($p < 0.05$) and lag time were observed in *A. iva* grown in the Mograne and Nabeul regions. Dry matter digestibility differed between regions and analytical methods. The highest ($p < 0.01$) DM degradability was observed in plants harvested from Mograne and Dougga, while the lowest value ($p < 0.01$) was recorded for those from Nabeul. Additionally, metabolizable energy (ME) and protein digestibility values were higher in *A. iva* harvested from Mograne compared to those collected from the other areas. In conclusion, the nutritional value of *A. iva* differed between regions and care should be taken when developing recommendations for the use of *A. iva* as feed, including the adoption of season- and region-specific feeding strategies.

In another study dealing with the digestibility of some feedstuffs, Zúñiga-Serrano A. et al. [14] carried out an extensive bibliographic study on the antimicrobial and digestive effects of *Yucca schidigera* extracts on animal production and their consequences for the environment. Plant extracts have been used throughout time in traditional medicine, mainly for their antimicrobial activity and medicinal effects. Plant-derived products contain secondary metabolites that inhibit pathogenic microbial growth, similar to conventional drugs. These secondary metabolites can improve animal health and production in a more natural or organic way and help reduce the use of pharmacological drugs in animal feed, which is a major concern for emerging microbial resistance. Plant secondary metabolites can be cost-effective while improving the production efficiency of ruminants, non-ruminants and fish. Among the plant-derived products, *Yucca schidigera* extract (YSE) contains steroidal saponins as the main active component. YSE has multiple biological effects, including the inhibition of some pathogenic bacteria, protozoa and nematodes. YSE is used to control odours and ammonia and consistently improve poultry production. In pigs, the results are still inconclusive. In ruminants, YSE acts against protozoa, has selective action against bacteria and reduces archaea populations; all these effects are reflected in a reduction in polluting gas emissions, mainly methane, although they are not observed in all fuel conditions. These effects of YSE are discussed in this review. YSE has potential as a natural feed additive for sustainable animal production while contributing to the mitigation of pollutant gas emissions.

The final study that falls within the research area of feed digestibility in farm animals is that of Zanine A. et al. [15], who investigated the effect of cotton lint consumption on feed intake, nutrient digestibility, nitrogen balance and blood parameters in rams. Twenty rams weighing 30.2 ± 3.7 kg and aged 12 ± 1.3 months were studied and distributed in a completely randomized design with four treatments and five replications. The diets consist of 50% forages and 50% concentrates. Treatments consisted of replacing dietary corn with cotton lint at levels of 0, 70, 140 and 210 g/kg dry matter (DM). The feed intake of the rams was determined during the last three days of the experimental period. The linear effect decrease ($p < 0.05$) was observed for the nutritional fraction intake; however, neutral detergent fibre (NDF) intake and plasma urea-N were not affected ($p > 0.05$) by cotton consumption levels. The apparent digestibility of DM, crude protein, fat and non-nitrogen extractives were affected ($p < 0.05$), except for total carbohydrate. There was a low effect ($p < 0.05$) on the efficiency of SM and NDF intake. Nitrogen balance (g/day) and glucose levels (mg/dL) were reduced following the administration of cotton in the diets. Up to 70 g/kg of cotton fluff (lint) can be added to ram feed without adversely affecting DM intake and without altering digestibility, nitrogen balance, plasma urea and glucose concentrations; however, negative effects can be recorded when using larger amounts of cotton fluff, as a decrease in dry matter intake has been noted.

The category dealing with egg production includes two papers. In one of the articles, Usturoi A. et al. [16] studied the correlations between the health status of laying hens and their productivity. The study was conducted on the ISA Brown hybrid, monitored over a period of 25–55 weeks. The groups were represented by fowl raised differently in halls equipped with aviary or improved cage batteries, under real production conditions. The blood biochemical traits of hens, as well as quantitative indicators of egg production, were studied. A cumulative production of 199.24 eggs/week/head was achieved in birds reared in the aviary compared to 199.98 from the improved cage batteries, which led to an average laying intensity of 91.82% and 92.17%, respectively. Flock casualties reached 4.14% (aviary) and 2.98% (cages). Daily average feed intake reached 122.20 g feed/head in the aviary and 115.87 g feed/head in cages, while the feed conversion ratio was 133.09 g feed/egg in aviary versus 125.69 in cages. The aviary rearing system has proven to provide optimal conditions for the expression of the natural behaviours of fowl, and it had a positive impact on metabolic functions, resulting in good health and high productive levels, comparable to those of birds exploited in cages.

The second article by Carvalho L.C. et al. [17] highlights a very interesting topic, and the authors provide a substantial contribution to the determination of the optimal amino acid ratio for Japanese quail based on the quality of egg production. A completely randomized design was used, with 120 Japanese quails, 12 treatments and 10 replicates per treatment. The treatments consisted of a basal balanced protein (BP) and the 11 combined feed diets that were obtained by reducing BP by 40%, using a specific test for Lys, Met + Cys, Thr, Trp, Arg, Gly + Ser, Val, Ile, Leu, His and Phr + Tyr. The study lasted 25 days. At the end, egg weight (EW), albumen height, albumen diameter, albumen index, yolk height, yolk diameter, yolk index, Haugh units, mineral shell weight (ESW) and egg shell participation percentage were measured. The ideal ratio was calculated when a statistical difference was detected using Dunnett's test. Only the variables EW and ESW were different in relation with BP. The ideal amino acid ratios considering Lys as 100 for EW and ESW were Met + Cys 82 and 83, Thr 60 and 68, Trp 18 and 21, Arg 109 and 112, Gly + Ser 99 and 102, Val 77 and 87, Ile 61 and 67, Leu 155 and 141, His 34 and 37, and Phe + Try 134 and 133, respectively. Applying the reduction method with a 40% limitation of dietary amino acid depletion allowed the EW and ESW variables to be sensitive for all amino acids tested. Thus, it was possible to simultaneously establish the ideal profile of essential amino acids in the diet, focusing on target traits (which in this study were EW and ESW).

The microbiological evaluation of chicken broiler feed was handled by a group of researchers from USV Iași led by Lăpușneanu D. [18], who conducted a case study in a combined feed factory in Romania. The study was carried out between 2019 and 2020 and 334 samples of raw materials and 601 samples of combined feed were collected and analysed. The raw materials (corn, wheat, soybean meal and sunflower) were tested for the presence of yeasts and moulds with the following results: 1.3×10^3, 9.5×10^2, 6.4×10^2 and $7,4 \times 10^2$ cfu/g in 2019 and 1.5×10^2, 1.0×10^3, 5.2×10^2 and 7.1×10^2 cfu/g in 2020. Combined feed samples (starter, grower and finisher) revealed the following mean values: 5.9×10^2, 4.2×10^2 and 4.2×10^2 cfu/g in 2019 and 5.3×10^2, 6.5×10^2 and 5.8×10^2 cfu/g in 2020. Potentially toxigenic fungi of the genera *Aspergillus*, *Penicillium* and *Fusarium* occurred most frequently in all samples. For raw materials, the highest numbers of positive samples for *Aspergillus* were recorded in both years: 66.6% in 2019 and 100% in 2020 for maize, 50% in 2019 and 75% in 2020 for wheat, 76% in 2019 and 87.5% in 2020 for soybean meal and 71.4% in 2019 and 100% in 2020 for sunflower meal. In starter combined feeds, *Aspergillus* predominated in 2019 (46.6%), while in 2020, *Penicillium* and *Cladosporium* were identified in most samples (50%); in grower and finisher feeds, *Aspergillus* was predominantly identified in 2019 (60% and 72.2% of the samples, respectively) and in 2020 (61.5% and 46.6%, respectively). All bacteriological analyses to identify possible contamination with *Salmonella spp.*, *E. coli* and *Clostridium perfringens* produced negative results. Based on the results obtained in this study, it could be concluded that the monitoring and analysis of microbiological hazards in a feed factory are necessary to be able to prevent contaminations of this type because they can have a direct impact on feed and food safety.

Another side of obtaining animal products is its polluting potential. This topic was a concern of researchers from USAMV Bucharest, led by Popa R.A. [19], who carried out a comparative assessment of the resulting air pollutant dynamics in a dairy farm using an LoT platform. The study is based on the necessity of the awareness and permanent monitoring of sources of pollution to ensure the efficient management of the farm. The authors conducted a case study of air pollutants in a cattle farm in different seasons (winter and summer) and the assessed the correlation between their variations and microclimate parameters. In this study, a further comparison was carried out between values estimated using the EMEP methodology (European Monitoring and Evaluation Programme, 2019) for air pollutant emissions and values measured by sensors in a hybrid farm decision support platform. Interactions between microclimate and pollutant emissions were found that can provide a model for farm activities, which is useful as managing tool for farmers. Starting from the fact that the estimates using the EMEP methodology do not consider the natural and artificial ventilation of the stable, as they are used for a global assessment of atmospheric air quality, the authors recommend the use of sensors and alert systems that analyse pollutant concentrations in real time to ensure animal welfare.

Another type of study included in this Special Issue was related to the quality of honey. Albu A. et al. [20] analysed the phenolic and flavonoid contents and also the raw chemical composition of Romanian monofloral honey. In this study, 28 samples (acacia, linden, rapeseed, sunflower and mint) were analysed. Pearson's test revealed a strong positive correlation between total phenolic content and total flavonoids ($r = 0.76$) and color intensity ($r = 0.72$). For total flavonoid content, correlations were strongly positive with colour intensity ($r = 0.81$), ash content ($r = 0.76$) and electrical conductivity ($r = 0.73$). The relevant levels of polyphenols and flavonoids identified in the types of honey analysed demonstrate its antioxidant potential, with essential nutritional and sanogenic characteristics in human nutrition.

The second article dealing with the quality of honey offers us a profile of several varieties of monofloral honey from Romania. Thus, Pop I.M. et al. [21] evaluated 7 monofloral honey varieties (linden, acacia, rapeseed, sunflower, mint, raspberry and chestnut) from a physical-chemical point of view (moisture, specific gravity, pH, free acidity, ash, electrical conductivity, total phenols and total flavonoid content, K, Ca, Mg, Na and P). The quality parameters that were investigated fell within the recommended limits established by the regulated standards. Sample analyses indicated the presence of antioxidants such as total phenols content (TPC) (17.9–73.2 mg GAE/100 g) and total flavonoids content (TFC) (0.84–4.81 mg QE/100 g) and high amounts of K (101–1462 mg kg^{-1}), Ca (58.3–167.5 mg kg^{-1}), Mg (24.8–330.6 mg kg^{-1}), Na (94.5–233.3 mg kg^{-1}) and P (34.1–137.2 mg kg^{-1}). Pearson's correlations between certain parameters (such as colour $\times$ TFC, colour $\times$ Mg, colour $\times$ P, electrical conductivity $\times$ ash, mm Pfund $\times$ TFC, TPC $\times$ TFC, K $\times$ Ash, P $\times$ Mg) along with principal component analysis, hierarchical clusters analysis and ANOVA statistics reveal three main factors that explain the variability of the data set and could be attributed to stability: minerals and colour/antioxidant contributions. FTIR spectra confirmed the authenticity of all monofloral honeys. The experimental data confirmed the influence of environmental elements (soil, water, air) on the composition of honey and highlighted the quality of honey as a complete food and therapeutic product.

For the end of this editorial, we have saved an article that presented a phylogenetic analysis of the genetic diversity of the Romanian cattle breed named Steppe Grey, a breed that is on the way to extinction. Since 2000, FAO has been drawing attention to the significant decline in the Podolian cattle population group, which also includes the Romanian Steppe Grey. Currently, this breed is on the verge of extinction, counting less than 100 heads throughout the country. Due to its qualities of resilience, adaptability and increased resistance to diseases and severe climatic conditions, the Steppe Grey is considered a valuable genetic pool for improving livestock production. The study carried out by Davidescu M.A. et al. [22] aimed to quantify the genetic diversity of a population of 32 cattle from northeastern Romania (historical province of Moldova) by analysing two mitochondrial markers, Cytochrome B and Loop D, that proved to be relevant for studies of genetic diversity and phylogeny. The results obtained, based on statistical data analysis derived from nucleotide sequencing analysis software (DnaSP, SeaView, MegaX, PopArt, etc.), demonstrated that the breed belonged to the ancestral haplogroup P′QT, with direct ancestry from *Bos taurus primigenius*. Within this haplogroup, five cattle were identified that could be used in artificial selection and mating, with the aim of preserving valuable genetic resources, improving the biological resilience of other cattle breeds against various adverse environmental factors and protecting biodiversity.

This Special Issue of *Agriculture*, focused on animal nutrition and productions, includes articles that are diversified in approach and represent some of the latest preoccupations of specialised research, aiming at continuously developing knowledge in the targeted fields. The authors of this editorial are pleased to have completed this Special Issue, underlining the valuable contributions of researchers from all over the world, many of which are young people representing the future of research in their respective fields. We hope that Series II will be as successful and will give us the opportunity to write a second editorial.

Author Contributions: Conceptualization, D.S. and R.-M.R.-R.; investigation, D.S. and R.-M.R.-R.; writing—original draft preparation, D.S.; writing—review and editing, R.-M.R.-R.; visualization, D.S. and R.-M.R.-R.; All authors have read and agreed to the published version of the manuscript.

Acknowledgments: We would like to sincerely thank all authors who submitted papers to the Special Issue of *Agriculture* entitled "Animal Nutrition and Productions", to the reviewers of these papers for their constructive comments and thoughtful suggestions, and the editorial staff of *Agriculture*.

Conflicts of Interest: The authors declare no conflict of interest.

References

1. Pond, W.G.; Church, D.B.; Pond, K.R.; Schoknecht, P.A. *Basic Animal Nutrition and Feeding*, 5th ed.; Jhon Wiley & Sons, Inc.: New York, NY, USA, 2004; pp. 1–15.
2. Guoyao, W. *Principles of Animal Nutrition*; CRC Press: Boca Raton, FL, USA; Taylor & Francis Group: Boca Raton, FL, USA, 2018; pp. XXIII–XXIV.
3. Abdelrahman, M.M.; Alhidary, I.A.; Matar, A.M.; Alobre, M.M.; Alharthi, A.S.; Faye, B.; Aljumaah, R.S. Effect of Total Mixed Ratio (TMR) Supplementation on Milk Nutritive Value and Mineral Status of Female Camels and Their Calves (*Camelus dromedarius*) Raised under Semi Intensive System during Winter. *Agriculture* **2022**, *12*, 1855. [CrossRef]
4. Serrapica, F.; Masucci, F.; De Rosa, G.; Braghieri, A.; Sarubbi, F.; Garofalo, F.; Grasso, F.; Di Francia, A. Moving Buffalo Farming beyond Traditional Areas: Performances of Animals, and Quality of Mozzarella and Forages. *Agriculture* **2022**, *12*, 1219. [CrossRef]
5. Simeanu, D.; Radu-Rusu, R.-M.; Mintas, O.S.; Simeanu, C. Qualitative and Nutritional Evaluation of Paddlefish (*Polyodon spathula*) Meat Production. *Agriculture* **2022**, *12*, 1965. [CrossRef]
6. Simeanu, C.; Măgdici, E.; Păsărin, B.; Avarvarei, B.-V.; Simeanu, D. Quantitative and Qualitative Assessment of European Catfish (*Silurus glanis*) Flesh. *Agriculture* **2022**, *12*, 2144. [CrossRef]
7. Frunză, G.; Murariu, O.C.; Ciobanu, M.-M.; Radu-Rusu, R.-M.; Simeanu, D.; Boișteanu, P.-C. Meat Quality in Rabbit (*Oryctolagus cuniculus*) and Hare (*Lepus europaeus Pallas*)—A Nutritional and Technological Perspective. *Agriculture* **2023**, *13*, 126. [CrossRef]
8. Corrêa, Y.; Santos, E.; Oliveira, J.; Carvalho, G.; Pinto, L.; Pereira, D.; Assis, D.; Cruz, G.; Panosso, N.; Perazzo, A.; et al. Diets Composed of Tifton 85 Grass Hay (*Cynodon* sp.) and Concentrate on the Quantitative and Qualitative Traits of Carcass and Meat from Lambs. *Agriculture* **2022**, *12*, 752. [CrossRef]
9. Coroian, C.O.; Coroian, A.; Becze, A.; Longodor, A.; Mastan, O.; Radu-Rusu, R.-M. Polycyclic Aromatic Hydrocarbons (PAHs) Occurrence in Traditionally Smoked Chicken, Turkey and Duck Meat. *Agriculture* **2023**, *13*, 57. [CrossRef]
10. Ben Saïd, S.; Jabri, J.; Amiri, S.; Aroua, M.; Najjar, A.; Khaldi, S.; Maalaoui, Z.; Kammoun, M.; Mahouachi, M. Effect of *Saccharomyces cerevisiae* Supplementation on Reproductive Performance and Ruminal Digestibility of Queue Fine de l'Ouest Adult Rams Fed a Wheat Straw-Based Diet. *Agriculture* **2022**, *12*, 1268. [CrossRef]
11. Nechifor, I.; Florea, M.A.; Radu-Rusu, R.-M.; Pascal, C. Influence of Supplemental Feeding on Body Condition Score and Reproductive Performance Dynamics in Botosani Karakul Sheep. *Agriculture* **2022**, *12*, 2006. [CrossRef]
12. Edmunds, C.E.; Cornelison, A.S.; Farmer, C.; Rapp, C.; Ryman, V.E.; Schweer, W.P.; Wilson, M.E.; Dove, C.R. The Effect of Increasing Dietary Manganese from an Organic Source on the Reproductive Performance of Sows. *Agriculture* **2022**, *12*, 2168. [CrossRef]
13. Ammar, H.; Kholif, A.E.; Soltan, Y.A.; Almadani, M.I.; Soufan, W.; Morsy, A.S.; Ouerghemmi, S.; Chahine, M.; de Haro Marti, M.E.; Hassan, S.; et al. Nutritive Value of *Ajuga iva* as a Pastoral Plant for Ruminants: Plant Phytochemicals and In Vitro Gas Production and Digestibility. *Agriculture* **2022**, *12*, 1199. [CrossRef]
14. Zúñiga-Serrano, A.; Barrios-García, H.B.; Anderson, R.C.; Hume, M.E.; Ruiz-Albarrán, M.; Bautista-Martínez, Y.; Sánchez-Guerra, N.A.; Vázquez-Villanueva, J.; Infante-Rodríguez, F.; Salinas-Chavira, J. Antimicrobial and Digestive Effects of *Yucca schidigera* Extracts Related to Production and Environment Implications of Ruminant and Non-Ruminant Animals: A Review. *Agriculture* **2022**, *12*, 1198. [CrossRef]

15. Zanine, A.; Castro, W.; Ferreira, D.; Sousa, A.; Parente, H.; Parente, M.; Santos, E.; Geron, L.; Lima, A.G.; Ribeiro, M.; et al. The Effect of Cotton Lint from Agribusiness in Diets on Intake, Digestibility, Nitrogen Balance, Blood Metabolites and Ingestive Behaviour of Rams. *Agriculture* **2022**, *12*, 1262. [CrossRef]
16. Usturoi, A.; Usturoi, M.-G.; Avarvarei, B.-V.; Pânzaru, C.; Simeanu, C.; Usturoi, M.-I.; Spătaru, M.; Radu-Rusu, R.-M.; Doliş, M.-G.; Simeanu, D. Research Regarding Correlation between the Assured Health State for Laying Hens and Their Productivity. *Agriculture* **2023**, *13*, 86. [CrossRef]
17. Carvalho, L.C.; Malheiros, D.; Lima, M.B.; Mani, T.S.A.; Pavanini, J.A.; Malheiros, R.D.; Silva, E.P. Determination of the Optimal Dietary Amino Acid Ratio Based on Egg Quality for Japanese Quail Breeder. *Agriculture* **2023**, *13*, 173. [CrossRef]
18. Lăpuşneanu, D.M.; Simeanu, D.; Radu-Rusu, C.-G.; Zaharia, R.; Pop, I.M. Microbiological Assessment of Broiler Compound Feed Production as Part of the Food Chain—A Case Study in a Romanian Feed Mill. *Agriculture* **2023**, *13*, 107. [CrossRef]
19. Popa, R.A.; Popa, D.C.; Pogurschi, E.N.; Vidu, L.; Marin, M.P.; Tudorache, M.; Suciu, G.; Bălănescu, M.; Burlacu, S.; Budulacu, R.; et al. Comparative Evaluation of the Dynamics of Animal Husbandry Air Pollutant Emissions Using an IoT Platform for Farms. *Agriculture* **2023**, *13*, 25. [CrossRef]
20. Albu, A.; Radu-Rusu, R.-M.; Simeanu, D.; Radu-Rusu, C.-G.; Pop, I.M. Phenolic and Total Flavonoid Contents and Physicochemical Traits of Romanian Monofloral Honeys. *Agriculture* **2022**, *12*, 1378. [CrossRef]
21. Pop, I.M.; Simeanu, D.; Cucu-Man, S.-M.; Pui, A.; Albu, A. Quality Profile of Several Monofloral Romanian Honeys. *Agriculture* **2023**, *13*, 75. [CrossRef]
22. Davidescu, M.-A.; Simeanu, D.; Gorgan, D.-L.; Ciorpac, M.; Creanga, S. Analysis of Phylogeny and Genetic Diversity of Endangered Romanian Grey Steppe Cattle Breed, a Reservoir of Valuable Genes to Preserve Biodiversity. *Agriculture* **2022**, *12*, 2059. [CrossRef]

Article

Effect of Total Mixed Ratio (TMR) Supplementation on Milk Nutritive Value and Mineral Status of Female Camels and Their Calves (*Camelus dromedarius*) Raised under Semi Intensive System during Winter

Mutassim M. Abdelrahman [1,*], Ibrahim A. Alhidary [1], Abdulkareem M. Matar [1], Mohsen M. Alobre [1], Abdulrahman S. Alharthi [1], Bernard Faye [2,3] and Riyadh S. Aljumaah [1]

[1] Department of Animal Production, College of Food and Agriculture Sciences, King Saud University, P.O. Box 2460, Riyadh 11451, Saudi Arabia
[2] CIRAD-ES, UMR SELMET, Campus International de Baillarguet, 34398 Montpellier, France
[3] Faculty of Biology and Biotechnology, Department of Biotechnology, Kazakh National University Al-Farabi, Almaty 050013, Kazakhstan
* Correspondence: amutassim@ksu.edu.sa

Abstract: This study was conducted to investigate the nutritional values of female camels' milk and the minerals status, for them and their calves, when fed a total mixed ration (TMR) beside alfalfa hay during winter. Thirty-seven lactating multiparous female camels and their calves were selected at mid-lactation in the Al-Kharj region. Group one was fed only alfalfa hay ad libitum (C) and group two was supplemented with a total mixed ratio (TMR; 4 kg/head/day), primarily containing a mix of barley grain, wheat feed, palm kernel cake, soya hulls, vitamins and minerals. Milk and blood samples were collected in the middle of the winter season and analyzed for minerals using ICP-OES. A significant ($p < 0.05$) higher concentration was observed for protein and inorganic matter in milk from female camels supplemented with TMR in the T group. In addition, Mg, Co, Fe and Zn levels in milk significantly ($p < 0.05$) increased compared with the control group (C). Blood serum concentrations of Ca, P, Mg, Cu, I, Se, Zn and Cd minerals in female camels were significantly ($p < 0.05$) higher in the T group compared to the C group. Blood serum of the calves in the T group was significantly ($p < 0.05$) at higher levels for all minerals than in the control group (C); except iodine. Furthermore, significant correlations were reported between Co and Mn elements with most other minerals under investigation. In conclusion, TMR supplementation in the T group of female camels during lactation in the winter season is highly recommended since it improves the milk composition and mineral profile.

Keywords: female camels; milk; minerals; heavy metals; winter; total mixed ration

Citation: Abdelrahman, M.M.; Alhidary, I.A.; Matar, A.M.; Alobre, M.M.; Alharthi, A.S.; Faye, B.; Aljumaah, R.S. Effect of Total Mixed Ratio (TMR) Supplementation on Milk Nutritive Value and Mineral Status of Female Camels and Their Calves (*Camelus dromedarius*) Raised under Semi Intensive System during Winter. *Agriculture* **2022**, *12*, 1855. https://doi.org/10.3390/agriculture12111855

Academic Editor: Daniel Simeanu

Received: 18 September 2022
Accepted: 1 November 2022
Published: 4 November 2022

Publisher's Note: MDPI stays neutral with regard to jurisdictional claims in published maps and institutional affiliations.

1. Introduction

The ecosystem in the Middle East and Africa can be primarily classified as arid or semi-arid, with approximately 70% of their lands receiving less than 100 mm of rainfall per year. Traditionally livestock grazing on these lands has been a nomadic system. The inadequate rangeland forage has forced nomadic people to shift to other feed resources and consequently change their production system from nomadic to primarily settled, semi-intensive systems [1]. The main consequence for the feeding behavior of camels in this context is the change from a highly diversified diet (with important variability in nutritive value and grazed ecosystems) to a very monotonous diet (typically alfalfa + more or less barley + more or less concentrate [2].

Camels (*Camelus dromedarius*) are important dairy animals in semi-arid and arid nomadic and other communities as a source of high-quality animal protein, fatty acids, vitamins and minerals contained in milk [3]. Camel production systems are now shifting toward the semi-intensive approach that depends mainly on feed supplements to meet nutrient requirements. Feed supplements consisting of alfalfa hay, Rhodes grass, barley, wheat bran, crop by-products and, rarely, seasonal grazing pasture are the main source of nutrients for camels which do not cover the nutrient requirements including trace minerals, especially during lactation [4]. In camels, features such as low growth rate, seasonal breeding, long gestation period, abortion, low milk production, diseases and the high mortality rate of pre-weaning newborns and dams appear to be the major constraints on the improvement of their productivity and general performance [5]. However, camels are well known for their unique anatomy and physiology, which helps them to be suited for harsh environments, poor feeding and water scarcity. However, the problem with this feature is the unknown nutritional subclinical deficiencies during different seasons, which markedly reduce camel performance and profitability. Many scientific studies have demonstrated alterations in the absorption and metabolism of glucose, protein, lipids and minerals and effects on liver biological function in heat or cold-stressed subjects [6–8].

Seasons are well documented to affect grazing ruminants' health and productivity since there is significant variation in feed availability and nutritive values. Moreover, the animal requirements differ according to seasons, such as summer and winter. Winter is a very stressful season for camels, especially lactating ones, affecting milk yield and composition and so pre-weaned newborn camels' health and growth rate. Dwyer et al. [9] reported the mortality rate of livestock raised under an extensive system as follows: 9% of beef cattle, 15% of sheep, 20% of goat and 30% of camels. The causes of high mortality rate in camels may be caused by poor dam nutrition and late colostrum taking, which impairs immunity transfer and, consequently, neonatal health and growth [10].

Trace and macro minerals are crucial for animal health and productivity, including camels, especially when they become a limiting factor of the diet [11]. They play a pivotal role in many physiological activities, and their deficiency causes various pathological problems and metabolic defects [12]. The levels of nutrients and trace minerals intake are known to affect the reproductive ability of male or female camels [13,14]. The trace minerals such as selenium (Se), copper (Cu), zinc (Zn), manganese (Mn), cobalt (Co), iron (Fe), iodine (I) and molybdenum (Mo) are involved in normal growth and productivity of livestock [15]. Infertility, non-infectious abortion, anemia and metabolic diseases are some of the main clinical signs of deficiencies and abnormalities [16,17]. A few scientific studies have shown evidence of camels' sensitivity to trace and macro mineral disorders as a result of either deficiency or toxicity in the same way as other ruminants [18]). Faye et al. [18,19], and Zong-Ping et al. [20] have reported several incidences of clinical mineral deficiencies in camels being underestimated because signs of subclinical deficiencies may remain undetected for long periods. Regarding toxicity, evidence is even rarer. For example, some cases of selenosis were described [21], as well as fluorosis [22].

Most of these studies in semi-arid and arid areas fail to consider the main factors that affect the mineral and nutritional status of camels, such as seasons, age, breed, sex, physiological status, and management risk factors, especially those linked to the feeding system [23]. Such information is essential to establish a solid background for developing supplementation programs during different seasons to improve camels' reproductive and production efficiency and, consequently, milk and meat quality and quantity. The results indicated a lower concentration of some trace minerals in blood serum and tissue compared with normal levels in camels, significantly varying by breed [24]. Alhidary et al. [25] found that supplementing trace minerals to camels improves the average daily gain, feed efficiency, liver mineral reserve, and immune response. The objective of this study was to investigate the effect of total mixed ratio supplementation to one-humped female camels' calves on milk composition and minerals status during the winter season compared to female camels fed only alfalfa hay raised under semi-intensive systems.

2. Materials and Methods

2.1. Experimental Site

This study was performed in the Al-Kharj agricultural region, located southeast of Riyadh province, Saudi Arabia (23°59′ N 47°09′ E–24°22′ N 47°06′ E). The climate in this area is typically arid, hot in the summer and cold in winter, with annual rain precipitation of 132 mm. Winter starts in mid-September to the end of January, the coldest in December. Moreover, the maximum temperature is 18 °C and the lowest is 5 °C.

2.2. Samples Collection and Management

Blood samples were taken with sterilized vacutainer tubes from the jugular vein of 37 multiparous female camels (at mid-lactation; ±2 months) and their calves, raised under a semi-intensive system in Al-Kharj, Riyadh, with a different feeding protocol: Group one (C) twenty female camels fed alfalfa *ad libitium* and group two (T) 17 female camels were fed Total Mix Ration (TMR; 4 kg/head/day) beside the alfalfa hay ad libitum. The chemical composition of the TMR is presented in Table 1. Blood samples were centrifuged (2000 for 10 min) to obtain serum. The serum was treated immediately with 10% trichloroacetic acid (TCA) (1-part serum: 4 parts TCA) and centrifuged (1500× *g* for 10 min). The supernatants were collected and stored at −20 °C until mineral analysis was performed. At the same time, milk samples were collected from each female camel and prepared for composition and mineral analysis.

Table 1. Composition and calculated nutrient content of the experimental diet.

Ingredients, %	Treatments	
	Alfalfa	TMR
Barley grain	-	30.45
Wheat feed	-	26.00
Wheat Bran	-	5.00
Palm kernel cake	-	17.55
Soya Hulls	-	13.40
Salt	-	1.00
Limestone	-	2.50
Molasses	-	3.00
Acid buffer	-	1.00
Commercial Premix	-	0.10
Calculated analysis		
Dry matter, %	91.3	90.8
Crude protein, %	14.1	13.24
Crude fiber, %	28.7	12.72
ME, Mcal/kg	16.8	2.79
Calcium (%)	378.86	617.27
Phosphorous (%)	1956.6	330.58
Iron (µg/g)	0.185	4.5
Copper (µg/g)	8.58	12.84
Zinc (µg/g)	12.5	15.65
Manganese (µg/g)	2064.6	2148.6
Selenium (µg/g)	1.47	17.51

Milk samples were immediately pipetted to prevent settling prior to removing the sample. Milk composition (protein %, fat %, lactose % and total solid %) was analyzed after collection using Milko-Scan FT6000 (Foss, Hillerød, Denmark). For trace metals analysis; a 0.5000 ± 0.001 g milk sample was weighed in an acid-washed Teflon™ vessel. Then, 1 mL HNO_3 (65% Riedel-de Haen, Hanover, Germany), 1 mL HCl (36% Avonchem, Macclesfield, UK), 1 mL H_2O_2 (30% *w/v* Avonchem, UK) and 1 mL deionized H_2O (Milli-Q quality) were added to the sample before loading on the digestion units. The samples were digested according to pre-set temperature program recommended by [26]. The digested samples were diluted in a 25 mL volumetric flask using 0.1 normality HCl and mixed very well. Subsamples (5 mL) were taken in sterilized tubes for mineral analysis.

2.3. Mineral Analysis

Determinations of major, trace and heavy metals were determined using ICP-OES equipped with a Meinhard Nebulizer type A2. Argon (purity higher than 99.999% supplied by AH group (Dammam, Saudi Arabia) was used to sustain plasma and as a carrier gas. The operating conditions employed for the ICP-OES determination were 1300 W RF power, 15 L min^{-1} plasma flow, 0.2 L min^{-1} auxiliary flow, 0.8 L min^{-1} nebulizer flow, and 1.5 mL min^{-1} sample uptake rate. The axial and radial view was used for metals determination, while two-point background correction and three replicates were used to measure the analytical signal, with the processing mode being the peak area. The emission intensities were obtained for the most sensitive lines free of spectral interference. The calibration standards were prepared by diluting the stock multi-elemental standard solution (1000 mg L^{-1}) in 0.5% (v/v) nitric acid. The calibration curves for all elements were in the range of 1.0 ng mL^{-1} to 1.0 μg mL^{-1} (1–1000 ppb).

2.4. Statistical Analysis

Data were analyzed by using a completely randomized design, through ANOVA (variance analysis), general linear model procedure (Proc GLM) of SAS (v. 9.4, SAS Institute Inc., Cary, NC, USA). The statistical model was $Y = \mu + T_i + \varepsilon_{ij}$, where: Y = minerals concentration in different tissues, blood serum, and milk (P, Mg, Co, Cu, Fe, I, Mn, Se, Cd, Pb); T_i = the effect of the C or T diets; and ε_{ij} = random error. The variation coefficients (R^2 and VC) were considered to determine the models' validity. A correlation was performed to explain the relationship between the minerals in serum, blood and milk of female camels and their calves. Differences among means were compared through LSD (least statistical differences) considering $p < 0.05$.

3. Results

3.1. Female Camel Nutrition and Milk Composition

Protein content was significantly ($p < 0.05$) higher for female camels' milk in the T group supplemented with TMR (2.71%) than female camels' milk in group C fed only alfalfa hay (2.13%). In addition, the content of inorganic matter was significantly ($p < 0.05$) higher in the female camels' milk fed with TMR and alfalfa hay (0.79%) than in the C group (0.65%) fed only alfalfa. Fat and lactose content did not statistically differ ($p > 0.05$) between female camels' milk in group C (2.29 % and 4.48 %, respectively), and group T (2.49 % and 4.22 %, respectively), as reported in Table 2.

Table 2. Camel milk's nutritive value raised under semi-intensive system during winter.

Nutrients, (%)	Group-C ($n = 17$)	Group-T ($n = 20$)	SEM	p Value
protein	2.13 [b]	2.71 [a]	0.18	0.051
Fat	2.29	2.49	0.21	0.66
Lactose	4.48	4.22	0.15	0.40
Total solid	9.45	9.81	0.44	0.70
SNF	7.07	7.39	0.29	0.58
Inorganic matter	0.65 [b]	0.79 [a]	0.02	0.04

[a,b] Within a column, means without a common superscript are significantly different. Group-C: fed only alfalfa hay ad libitum during the winter season in a semi-intensive production system. Group-T: fed alfalfa hay ad libitum and 4 kg TMR/head/day during the winter season in a semi-intensive production system. p-value: less than 0.05 considered significantly different. SEM: Standard error of means.

3.2. Female Camel Nutrition and Minerals Content in Milk

The effects of feeding management on mineral concentrations in the milk and serum of female camels and their calves are shown in Tables 3–5. On average, the rank of the trace minerals levels, from higher to lower, in the milk of the control group were as follows: iodine (46.15 µg/mL), followed by iron (27.77 µg/mL), copper (14.06 µg/mL), selenium (3.90 µg/mL) and zinc (2.05 µg/mL). Other minerals, such as cadmium (0.045 µg/mL), were present at the lowest concentration. Moreover, the highest value for macro minerals was reported for Ca, followed by P, and the lowest was Mg for both groups. The same trend was reported in the treated group with TMR. The content of Mg, Co, Fe, and Zn in female camels' milk (group T) fed TMR, and alfalfa hay in winter was significantly ($p < 0.05$) higher than in female camels' milk (group C) fed only alfalfa hay (Table 3). The serum concentration of macro minerals Ca and P were significantly ($p < 0.05$) higher in the female camel of the T group than in the C group. In contrast, the serum concentration of Mg in female camels in the C group that were fed alfalfa hay was significantly ($p < 0.05$) higher than in female camels in the T group. On the other hand, serum concentrations of trace minerals such as Cu and Zn were significantly ($p < 0.05$) higher in female camels in the T group while the serum concentrations of selenium were significantly ($p < 0.05$) higher in female camels of group C compared to T group. In addition, the serum concentration of the heavy metal, Cd was significantly ($p < 0.05$) higher in the serum of female camel in the C group compared to the T group (Table 4).

A significantly ($p > 0.05$) higher concentration of macro minerals (Ca, P and Mg) was observed in calves' blood serum that suckled from female camels supplemented with TMR. Similarly, the trace minerals (Co, Cu, Fe, Mg, Se and Zn) and heavy metals (Ca and Cd) were significantly ($p < 0.05$) higher in concentration in the serum of calves that suckled from female camels from group T than the blood serum of calves suckling female camels' milk from the control group (Table 5). The two groups observed no significant differences in iodine concentrations between serum female camels and calves (Table 5).

Table 3. Effect of concentrate supplementation to female camels beside alfalfa on element concentration of milk during winter.

Mineral	Group-C (n = 17)	Group-T (n = 20)	SEM	p-Value
	Macro minerals, mg/dL			
Calcium	214.80	201.9	1.64	0.71
Phosphorus	130.38	130.52	8.57	0.99
Magnesium	27.84 [b]	29.36 [a]	2.41	0.053
	Trace minerals, µg/mL			
Cobalt	0.311 [b]	0.536 [a]	0.14	0.04
Copper	14.06	15.48	1.41	0.34
Iron	27.77 [b]	32.99 [a]	1.52	0.03
Iodine	46.15	56.61	10.24	0.64
Manganese	1.954	2.669	0.33	0.29
Selenium	3.909	3.706	0.34	0.78
Zinc	2.058 [b]	2.694 [a]	0.41	0.054
	Heavy metals, µg/mL			
Cadmium	0.045	0.051	0.01	0.57
Lead	0.580	0.555	0.16	0.95

[a,b] Within a column, means without a common superscript are significantly different. Group-C: fed only alfalfa hay ad libitum during the winter season in a semi-intensive production system. Group-T: fed alfalfa hay ad libitum and 4 kg TMR/head/day during the winter season in a semi-intensive production system. p-value: less or equal to 0.05 considered significantly different. SEM: Standard error of means.

Table 4. Element concentrations (ug/mL) in the blood serum of female camels raised in a semi-intensive system during winter.

Mineral	Group-C (*n* = 17)	Group-T (*n* = 20)	SEM	*p* Value
	Macro minerals, µg/mL			
Calcium	231.3 [b]	257.1 [a]	0.083	0.03
Phosphorus	139.77 [b]	219.87 [a]	16.88	0.01
Magnesium	35.62 [a]	30.571 [b]	1.186	0.02
	Trace minerals, µg/mL			
Cobalt	0.117	0.118	0.001	0.98
Copper	1.616 [b]	2.279 [a]	0.32	0.053
Iron	0.535	0.585	0.03	0.48
Iodine	0.053	0.092	0.002	0.16
Manganese	0.875	0.789	0.01	0.54
Selenium	1.253 [a]	0.962 [b]	0.03	0.050
Zinc	0.127 [b]	0.159 [a]	0.01	0.052
	Heavy metals, µg/mL			
Cadmium	0.024 [a]	0.023 [b]	0.0003	0.01
Lead	0.006	0.005	0.0008	0.87

[a,b] Within a column, means without a common superscript are significantly different. Group-C: fed only alfalfa hay ad libitum during the winter season in a semi-intensive production system. Group-T: fed alfalfa hay ad libitum and 4 kg TMR/head/day during the winter season in a semi-intensive production system. *p*-value: less or equal to 0.05 is considered to be significantly different. SEM: Standard error of means.

Table 5. Elements concentration (ug/mL) in blood serum of calves s raise in a semi-intensive system during winters.

Mineral	Group-C (*n* = 17)	Group-T (*n* = 20)	SEM	*p*-Value
	Macro minerals, µg/mL			
Calcium	293.0 [b]	326.1 [a]	0.080	0.04
Phosphorus	282.4 [b]	384.1 [a]	11.73	0.04
Magnesium	31.90 [b]	39.82 [a]	1.964	0.03
	Trace minerals, µg/mL			
Cobalt	0.01 [b]	0.02 [a]	0.004	0.04
Copper	1.46 [b]	2.82 [a]	0.41	0.052
Iron	0.55 [b]	0.76 [a]	0.04	0.01
Iodine	0.11	0.12	0.01	0.64
Manganese	0.08 [b]	0.09 [a]	0.003	0.051
Selenium	1.08 [b]	1.38 [a]	0.07	0.02
Zinc	0.18 [b]	0.20 [a]	0.008	0.055
	Heavy metals, µg/mL			
Cadmium	0.023 [b]	0.027 [a]	0.0001	0.002
Lead	0.005 [b]	0.009 [a]	0.0003	0.06

[a,b] Within a column, means without a common superscript are significantly different. Group-C: fed only alfalfa hay ad libitum during the winter season in a semi-intensive production system. Group-T: fed alfalfa hay ad libitum and 4 kg TMR/head/day during the winter season in a semi-intensive production system. *p*-value: less or equal to 0.05 considered significantly different. SEM: Standard error of means.

3.3. Correlation between Minerals Concentrations

Mn and Co concentrations in the blood serum of female camel and milk and calf serum were significantly ($p < 0.05$) correlated with most other minerals under investigation. In the C group, the Mn in the serum of female camels and their calves showed significant ($p < 0.05$) positive correlations with Mn, Cu, Fe, and Zn. On the other hand, Co showed significant ($p < 0.05$) positive correlations with Cu and Se and significant ($p < 0.05$) negative correlations with Mn and Fe within the C group (Table 6). In the T group, Mn in the serum of female camels and their calves showed a significant ($p < 0.05$) positive correlation with Mn, Co, Se and Zn, while it showed a significant ($p < 0.05$) negative correlation with Fe. In addition, the mineral Co showed significant ($p < 0.05$) positive correlations with Mn, Co, Cu, Se, and Zn in group T (Table 6).

Table 6. Correlation coefficient of some trace elements for female camels and their calves blood serum.

Mineral	Mn [NB]	Cu [NB]	Fe [NB]	I [NB]	Co [NB]	Se [NB]	Zn [NB]
Group-C							
Mn [D]	0.990 **	0.946 *	0.962 *	NC	−0.945 *	NC	0.901 *
Co [D]	−0.945 *	0.903 **	−0.962 *	NC	0.813 *	0.914 *	NC
Group-T							
Mn [D]	0.835 *	NC	−0.967 *	NC	0.989 **	0.874 **	0.768 *
Co [D]	0.893 *	0.946 **	NC	NC	0.958 **	0.822 *	0.903 *

* = $p < 0.05$; ** = $p < 0.01$; NC = No correlation; D = female camels; NB = Newborn.

The correlation coefficients of minerals Mn and Co with minerals in the blood serum and milk of female camel are presented in Table 7. In group C, Mn showed significant ($p < 0.05$) positive correlations with minerals Fe, I, Se and Zn while, Co showed significant ($p < 0.05$) positive correlations with minerals Mn, Fe, I and Co. In addition, the results showed that Mn showed a significant ($p < 0.05$) positive correlation with minerals Mn, Cu, Se and Zn minerals in group T (Table 7).

Table 7. Correlation coefficient of some trace elements for female camels' blood serum and milk.

Mineral	Mn [M]	Cu [M]	Fe [M]	I [M]	Co [M]	Se [M]	Zn [M]
			Group-C				
Mn [D]	NC	NC	0.961 **	0.964 *	NC	0.864 **	0.871 *
Co [D]	0.990 **	NC	0.973 **	0.961 *	0.788 *	NC	NC
			Group-T				
Mn [D]	0.936 **	0.988 *	−0.898 ***	NC	NC	0.966 ***	0.875 **
Co [D]	0.932 **	NC	NC	NC	NC	0.768 *	0.945 **

* = $p < 0.05$; ** = $p < 0.01$; *** = $p < 0.001$; NC = No correlation; D = female camels; M = Milk.

The correlation coefficient of Mn and Co with other minerals in the blood serum of calves and female camel milk is presented in Table 8. Mn showed significant ($p < 0.05$) positive correlations with Fe, I, and Se, while Co was significantly ($p < 0.05$) positively correlated with Fe and negatively with Cu in the C group. In addition, Mn showed significant ($p < 0.05$) positive correlations with Fe, I, Co and Zn, and significant ($p < 0.05$) negative correlation with Se. On the other hand, Co showed significantly ($p < 0.05$) positive correlations with Mn, Fe, I and Zn, and negative ones with Se in the T group (Table 8).

Table 8. Correlation coefficient of some trace elements for calves' blood serum and milk.

Mineral	Mn [M]	Cu [M]	Fe [M]	I [M]	Co [M]	Se [M]	Zn [M]
			Group-C				
Mn [NB]	NC	NC	0.871 *	0.810 *	NC	0.756 *	NC
Co [NB]	NC	−0.891 *	0.903 **	NC	0.988 **	NC	NC
			Group-T				
Mn [NB]	NC	NC	0.837 **	0.967 *	*0.875 **	−0.880 *	0.785 ***
Co [NB]	0.965 **	NC	0.837 **	0.938 *	0.817 *	−0.889 *	0.856 ***

* = $p < 0.05$; ** = $p < 0.01$; *** = $p < 0.001$; NC = No correlation; NB = newborn calves; M = Milk.

4. Discussion

In Saudi Arabia, there is little information on mineral bioavailability so, the practice of mineral supplementing is little or non-existent; therefore, it is possible that dietary imbalance is associated with mineral deficiencies. For dietary mineral imbalances and their interrelationship, elements were determined in the blood serum and milk of female camels with their calves, during winter. The current study aimed to estimate the micro and macro minerals and heavy metals in the blood serum and milk of female camels and their calves as indicators for determining the effect of supplementing TMR to female camels on their newborns' mineral status during winter. So, for trace minerals to be essential, they must be part of biologically active molecules and affect animal productivity and health [11].

The female camel milk in the current study has low fat, protein, and total solid content compared to Chinese Bactrian camel milk [27] and dromedary milk from Pakistan [28]. Still, it is similar to dromedary milk [29,30] from different Saudi camel breeds [31] and Moroccan camels [32]. In their meta-analysis, Konuspayeva et al. [33] showed higher concentrations of chemical parameters in Bactrian and Asian dromedary compared to dromedary from Arabian Peninsula and Africa. Unlike other dairy products, camel milk's composition and mineral content is mainly influenced by its water content [33]. Additionally, camel milk is influenced by other factors such as feeding conditions, camel breed, stage of lactation, age, and the number of calves [34]. In the present study, the results showed a significant effect of feeding TMR on the protein content of female camels' milk, which could be due to the protein content of the diet. In addition, the content of inorganic matter in the milk of female camels fed TMR was also significantly higher by (14%) compared to the C group. Milk yield was not recorded in the present study, which could explain some results regarding differences in milk composition between groups.

Dietary mineral bioavailability has not been well studied in camels and described in the literature, but it is well documented in other ruminant animals. Micronutrients such as Zn, Fe, Cu, Mn, Co, Mo, and Cr are required in minute amounts for homeostasis and optimal body function, but excessive intake can have negative health consequences [35]. Thus, our expectations regarding mineral metabolism in camels may differ from those of other ruminants because of their digestive process, feeding behavior and other physiological characteristics. In the present study, it was reported that the main mineral content (Ca, P, and Mg) of female camels' milk was higher compared to the milk of Chinese Bactrian camels [27] and the milk of dromedary camels in Saudi Arabia [31,35]. However, the Mg content of female camel milk was lower compared to the data reported for camels in Saudi Arabia [35] and Somali, and Tur-kana camel breeds [36], which were performed in many studies with large variations among them. The zinc concentration in the female camel milk was similar to the value observed by Dell'Orto, Cattaneo, Beretta and Baldi [36] The reason for these large variations could be due to the analytical methods used or factors not considered. In this study, the trace mineral content in female camels' milk was significantly higher than in cow's milk [37]. In contrast to previous studies in other countries, these values differ. The Fe and Cu content (2.98 and 0.14 mg/100 g) were close to the results of three dromedary camels reported by Al-Wabel et al. [38] The Cu content was slightly lower than the value (0.22 mg/100 mL) reported by Raziq et al. [39] for camels from mountainous Baluchistan, Pakistan, but significantly higher than the Fe and Cu content (0.23 and 0.061 mg/100 g) reported by Soliman [40]. The high Fe, Cu, and P contents of camel milk, as well as the fact that camels are fed dry plants, contribute to camel milk being very salty and high in chloride [41]. In the current study, female camels fed TMR had higher concentrations of Co, Fe and Zn in their milk than female camels from the C group. According to the result of this study and previous studies, the overall means differ, which may be due to several different factors, such as different physiological conditions (e.g., stages of lactation), management, feeding, and statistical methodology. The effect of trace elements in the diet on acid-base status is minimal because they are absorbed in such small amounts [42].

The data obtained in this study showed that the minerals Ca, P, Cu, and Zn were in significantly higher concentrations in the serum of female camel-fed TMR than in those fed only alfalfa hay. Interestingly, the serum of calves that suckled from female camel-fed TMR had the highest levels of macro (Ca, P and Mg) and trace mineral (Co, Cu, Fe, Mn,

Se and Zn) than that of other calves. On the contrary, Dell'Orto [36] reported that feeding supplements to lactating camels had no effect on calf plasma Ca, P, Mg, Zn and Fe levels. The average plasma phosphorus concentration ranges from 4 to 9 mg/100 mL, depending on the species and age of the animal, and P levels are more influenced by diet than Ca levels [16]. Goff [42] found that serum P levels exceed 2.0 mmol/L and are related to 1,25-dihydroxyvitamin D, which cannot be converted from 25-hydroxyvitamin D and causes milk fever in the cow. Blood Mg levels are not regulated, they reflect dietary Mg intake. It is influenced indirectly by calcium-regulating hormones such as calcitonin and parathyroid hormones. [42].

Toxic heavy metals/metalloids (THMs) such as arsenic, cadmium, lead, and mercury are relatively dense metals/metalloids that are not beneficial to humans or livestock and cause carcinogenicity, organ toxicity, and other negative health effects in small amounts [43]. In addition to industrial, domestic, agricultural, medical, and technological applications of THMs, exposure of food-producing animals (FPAs) and humans to toxic metals may result in massive environmental pollution [44]. Based on these results, female camel milk had the lowest levels of lead and cadmium compared to camel milk in Iran [45]. According to the Food and Agriculture Organization, World Health Organization (WHO), and Codex standard (CXS 193-2007), the determined limit for lead and cadmium is 20 µg/kg and 10 µg/kg, respectively. So, the results of this study showed that lead and cadmium levels were below the recommended levels.

To our knowledge, no study has been conducted on the relationship between the concentration of minerals in the serum and milk of female camels and their calves in dromedary camels except selenium [46]. The mineral correlation test was performed in this study to determine the main relationships between these minerals in serum and milk for female camel and their calves. Manganese and cobalt were found to be strongly correlated with other minerals investigated. Therefore, the correlation analysis and results focused on Mn and Co in the serum and milk of female camels and their calves with other minerals. Manganese is a component of the bone matrix; it is also part of glycoproteins and enzymes [47]. Manganese is required for lipid metabolism and the synthesis of cholesterol, which is required for the synthesis of estrogen, progesterone, and testosterone. Thus, animals with a manganese deficiency have lower reproductive performance [47]. The most common alterations due to magnesium deficiency are loss of appetite, excessive rumen fermentation, decreased milk production and tetany [42]. Cobalt is an essential trace element required for the formation of vitamin B12 (cobalamin) in humans and animals [48]. Cobalt deficiency is only seen in ruminants that consume low levels of copper in their diets, which could lead to a deficiency of this vital vitamin, and cause metabolic disorders, inappetence, weight loss, and reproductive problems [48]. This study found a significant positive correlation between Mn and Co with Cu, Fe, Se, and Zn in calves' blood serum, female camel serum, and milk during the winter in the two groups. Previous studies have shown that Mn reaches the fetus through the placenta, which is related to the amount of Mn in the basal diet [49]. Fortunately, blood Mn concentration is not a good indicator of nutritional status for this trace element. The increase in Mn levels may be due to glucose levels [50]. Essawi and Gouda [51] reported in dromedary camels that newborn weight is not significantly associated with female camel serum levels of the various minerals.

5. Conclusions

The nutrient requirement of female camels during lactation in the winter season is higher than to be covered by diet. TMR supplementation improves nutritional milk value, especially protein milk percentage, inorganic matter percentages, and mineral status of female camels' milk compared to female camels fed only alfalfa hay. Moreover, with TMR supplementation, blood serum concentrations of Co, Mn, and Zn of female camels and their calves were increased. Cobalt and Mn were found to have a high correlation with most other studied minerals between female camels, calves and milk. It is highly recommended that female camels must be supplemented with TMR during lactation in the winter season to improve their productivity and mineral status.

Author Contributions: Conceptualization, M.M.A. (Mutassim M. Abdelrahman) and B.F.; methodology, M.M.A. (Mutassim M. Abdelrahman) and R.S.A.; software, M.M.A. (Mutassim M. Abdelrahman) and A.M.M.; validation, M.M.A. (Mutassim M. Abdelrahman), A.M.M. and M.M.A. (Mohsen M. Alobre); formal analysis, R.S.A.; investigation, A.M.M. and M.M.A. (Mohsen M. Alobre); resources, A.M.M. and M.M.A. (Mohsen M. Alobre); data curation, M.M.A. (Mutassim M. Abdelrahman) and B.F.; writing—original draft preparation, A.S.A.; writing—review and editing, M.M.A. (Mutassim M. Abdelrahman) and A.M.M.; visualization, I.A.A.; supervision, M.M.A. (Mutassim M. Abdelrahman) and R.S.A.; project administration, M.M.A. (Mutassim M. Abdelrahman); funding acquisition, M.M.A. (Mutassim M. Abdelrahman) and I.A.A. All authors have read and agreed to the published version of the manuscript.

Funding: This research was funded by the National Plan of Science, Technology and Innovation (MAARIFAH), King Abdulaziz City for Science and Technology, Kingdom of Saud Arabia, Award number 13 AGR 1208-02.

Institutional Review Board Statement: The study was conducted according to the guidelines of the Declaration of Helsinki and approved by the Institutional Review Board (or Ethics Committee) of King Saud University (KSU-SE-22-21 and 24 March 2022).

Informed Consent Statement: Informed consent was obtained from all subjects involved in the study.

Data Availability Statement: The data presented in this study are available on request from the Corresponding author.

Acknowledgments: This project was funded by The National Plan of Science, Technology and Innovation (MAARIFAH), King Abdulaziz City for Science and Technology, Kingdom of Saud Arabia, Award number 13 AGR 1208-02.

Conflicts of Interest: The authors declare no conflict of interest.

References

1. Abdallah, H.; Faye, B. Typology of camel farming system in Saudi Arabia. *Emir. J. Food Agric.* **2013**, *25*, 250–260. [CrossRef]
2. Faye, B. Interaction Between Camel Farming and Environment. In *Handbook of Research on Health and Environmental Benefits of Camel Products*; IGI Global: Hershey, PA, USA, 2020; pp. 363–378. [CrossRef]
3. Faye, B. *The Camel, New Challenges for a Sustainable Development*; Springer: Berlin/Heidelberg, Germany, 2016; Volume 48, pp. 689–692. [CrossRef]
4. Faye, B. Camel farming sustainability: The challenges of the camel farming system in the XXIth century. *J. Sustain. Dev.* **2013**, *6*, 74–82. [CrossRef]
5. El-Zubeir, I.E.; Ehsan, M.N. Studies on some camel management practices and constraints in pre-urban areas of Khartoum State, Sudan. *Int. J. Dairy Sci.* **2010**, *5*, 276–284. [CrossRef]
6. Lacetera, N.; Bernabucci, U.; Ronchi, B.; Nardone, A. Effects of selenium and vitamin E administration during a late stage of pregnancy on colostrum and milk production in dairy cows, and on passive immunity and growth of their offspring. *Am. J. Vete Res.* **1996**, *57*, 1776–1780. [CrossRef]
7. Ronchi, B.; Lacetera, N.; Bernabucci, U.; Nardone, A.; Verini Supplizi, A. Distinct and common effects of heat stress and restricted feeding on metabolic status of Holstein heifers. *Zootec. Nutr. Anim.* **1999**, *25*, 11–20.
8. Moore, C.; Kay, J.; Collier, R.; VanBaale, M.; Baumgard, L. Effect of supplemental conjugated linoleic acids on heat-stressed Brown Swiss and Holstein cows. *J. Dairy Sci.* **2005**, *88*, 1732–1740. [CrossRef]
9. Dwyer, C.; Conington, J.; Corbiere, F.; Holmøy, I.; Muri, K.; Nowak, R.; Rooke, J.; Vipond, J.; Gautier, J.-M. Invited review: Improving neonatal survival in small ruminants: Science into practice. *Animal* **2016**, *10*, 449–459. [CrossRef]
10. Kamber, M.; Baume, N.; Saugy, M.; Rivier, L. Nutritional supplements as a source for positive doping cases? *Int. J. Sport Nutr. Exerc. Metab.* **2001**, *11*, 258–263. [CrossRef]
11. Faye, B.; Bengoumi, M. *Camel Clinical Biochemistry and Hematology*; Springer: Cham, Switzerland, 2018. [CrossRef]
12. Deen, D. Metabolic syndrome: Time for action. *Am. Fam. Phys.* **2004**, *69*, 2875–2882.
13. El-Bahrawy, K.; El-Hassanein, E. Seasonal variations of some blood and seminal plasma biochemical parameters of male Dromedary camels. *Am. Eurasian J. Agric. Environ. Sci.* **2011**, *10*, 354–360.
14. Ali, A.; Tharwat, M.; Al-Sobayil, F. Hormonal, biochemical, and hematological profiles in female camels (*Camelus dromedarius*) affected with reproductive disorders. *Anim. Reprod. Sci.* **2010**, *118*, 372–376. [CrossRef]

15. Underwood, E.J. The incidence of trace element deficiency diseases. *Philos. Trans. R. Soc. London B Biol. Sci.* **1981**, *294*, 3–8. [CrossRef]

16. McDowell, L.R.; Arthington, J.D. Minerals for grazing ruminants in tropical regions. In *Minerals for Grazing Ruminants in Tropical Regions*; Institute of Food and Agricultural Sciences, University of Florida: Gainesville, FL, USA, 2005; p. 86.

17. Bicknell, D. Trace mineral and reproduction. *Zimb. Herd Book* **1995**, *21*, 19.

18. Faye, B.; Bengoumi, M. Trace-elements status in camels. *Biol. Trace Elem. Res.* **1994**, *41*, 1–11. [CrossRef]

19. Faye, B.; Saint-Martin, G.; Cherrier, R.; Ruffa, A. The influence of high dietary protein, energy and mineral intake on deficient young camel (*Camelus dromedarius*)–II. Changes in mineral status. *Comp. Biochem. Physiol. Comp. Physiol.* **1992**, *102*, 417–424. [CrossRef]

20. Zong-Ping, L.; Zhuo, M.; You-Jia, Z. Studies on the relationship between sway disease of Bactrian camels and copper status in Gansu province. *Vet. Res. Commun.* **1994**, *18*, 251–260. [CrossRef]

21. Seboussi, R.; Faye, B.; Alhadrami, G.; Askar, M.; Bengoumi, M.; Elkhouly, A. Chronic selenosis in camels. *J. Camel Pract. Res.* **2009**, *16*, 25–38.

22. Diacono, E.; Faye, B.; Bengoumi, M.; Kessabi, M. Hydrotelluric and industrial fluorosis survey in the dromedary camel in the south of Morocco. In *Impact of Pollution on Animal Products*; Springer: Berlin/Heidelberg, Germany, 2008; pp. 85–90. [CrossRef]

23. Osman, T.; Al-Busadah, K. Normal concentrations of twenty serum biochemical parameters of she-camels, cows and ewes in Saudi Arabia. *Pak. J. Biol. Sci.* **2003**, *6*, 1253–1256. [CrossRef]

24. Abdelrahman, M.M.; Aljumaah, R.S.; Ayadi, M. Selenium and iodine status of two camel breeds (*Camelus dromedaries*) raised under semi intensive system in Saudi Arabia. *Ital. J. Anim. Sci.* **2013**, *12*, e14. [CrossRef]

25. Alhidary, I.A.; Abdelrahman, M.M.; Harron, R.M. Effects of a long-acting trace mineral rumen bolus supplement on growth performance, metabolic profiles, and trace mineral status of growing camels. *Trop. Anim. Health Prod.* **2016**, *48*, 763–768. [CrossRef]

26. Helrich, K. *Official methods of analysis of the Association of Official Analytical Chemists*; Association of Official Analytical Chemists: Arlington, VA, USA, 1990; Volume 2.

27. Zhao, D.-B.; Bai, Y.-H.; Niu, Y.-W. Composition and characteristics of Chinese Bactrian camel milk. *Small Rumi. Res.* **2015**, *127*, 58–67. [CrossRef]

28. Faraz, A.; Waheed, A.; Tauqir, N.; Mirza, R.; Ishaq, H.; Nabeel, M. Characteristics and composition of camel (*Camelus dromedarius*) milk: The white gold of desert. *Adv. Anim. Vet. Sci.* **2020**, *8*, 766–770. [CrossRef]

29. Nagy, P.; Fábri, Z.N.; Varga, L.; Reiczigel, J.; Juhász, J. Effect of genetic and nongenetic factors on chemical composition of individual milk samples from dromedary camels (*Camelus dromedarius*) under intensive management. *J. Dairy Sci.* **2017**, *100*, 8680–8693. [CrossRef] [PubMed]

30. Ayadi, M.; Aljumaah, R.S.; Musaad, A.; Samara, E.M.; Abelrahman, M.; Alshaikh, M.; Saleh, S.K.; Faye, B. Relationship between udder morphology traits, alveolar and cisternal milk compartments and machine milking performances of dairy camels (*Camelus dromedarius*). *Span J. Agri. Res.* **2013**, *3*, 790–797. [CrossRef]

31. Mehaia, M.A.; Hablas, M.A.; Abdel-Rahman, K.M.; El-Mougy, S.A. Milk composition of majaheim, wadah and hamra camels in Saudi Arabia. *Food Chem.* **1995**, *52*, 115–122. [CrossRef]

32. Ismaili, M.A.; Saidi, B.; Zahar, M.; Hamama, A.; Ezzaier, R. Composition and microbial quality of raw camel milk produced in Morocco. *J. Saudi Soc. Agric.* **2019**, *18*, 17–21. [CrossRef]

33. Haddadin, M.S.; Gammoh, S.I.; Robinson, R.K. Seasonal variations in the chemical composition of camel milk in Jordan. *J. Dairy Res.* **2008**, *75*, 8–12. [CrossRef]

34. Zhang, H.; Yao, J.; Zhao, D.; Liu, H.; Li, J.; Guo, M. Changes in chemical composition of Alxa Bactrian camel milk during lactation. *J. Dairy Sci.* **2005**, *88*, 3402–3410. [CrossRef]

35. Al-Otaibi, M.; El-Demerdash, H. Nutritive value and characterization properties of fermented camel milk fortified with some date palm products chemical, bacteriological and sensory properties. *Int. J. Nutr. Food Sci.* **2013**, *2*, 174–180. [CrossRef]

36. Dell'Orto, V.; Cattaneo, D.; Beretta, E.; Baldi, A. Effects of trace element supplementation on milk yield and composition in camels. *Int. Dairy J.* **2000**, *10*, 873–879. [CrossRef]

37. Agrawal, R.; Kochar, D.; Sahani, M.; Tuteja, F.; Ghorui, S. Hypoglycemic activity of camel milk in streptozotocin induced diabetic rats. *Int. J. Diab. Dev. Ctries.* **2004**, *24*, 47–49.

38. Al-Wabel, N. Mineral contents of milk of cattle, camels, goats and sheep in the central region of Saudi Arabia. *Asian J. Biochem.* **2008**, *3*, 373–375. [CrossRef]

39. Raziq, A.; de Verdier, K.; Younas, M.; Khan, S.; Iqbal, A.; Khan, M.S. Milk composition in the Kohi camel of mountainous Balochistan, Pakistan. *J. Camel. Sci.* **2011**, *4*, 49–62.

40. Soliman, G.Z. Comparison of chemical and mineral content of milk from human, cow, buffalo, camel and goat in Egypt. *Egypt. J. Hosp. Med.* **2005**, *21*, 116–130. [CrossRef]

41. Khaskheli, M.; Arain, M.; Chaudhry, S.; Soomro, A.; Qureshi, T. Physico-chemical quality of camel milk. *J. Agri. Soc. Sci.* **2005**, *2*, 164–166.

42. Goff, J.P. Macromineral physiology and application to the feeding of the dairy cow for prevention of milk fever and other periparturient mineral disorders. *Anim. Feed Sci. Technol.* **2006**, *126*, 237–257. [CrossRef]

43. Jaishankar, M.; Tseten, T.; Anbalagan, N.; Mathew, B.B.; Beeregowda, K.N. Toxicity, mechanism and health effects of some heavy metals. *Interdiscip. Toxicol.* **2014**, *7*, 60–72. [CrossRef]

44. Tchounwou, P.B.; Yedjou, C.G.; Patlolla, A.K.; Sutton, D.J. Heavy metal toxicity and the environment. *Exp. Suppl.* **2012**, *101*, 133–164.
45. Mostafidi, M.; Moslehishad, M.; Piravivanak, Z.; Pouretedal, Z. Evaluation of mineral content and heavy metals of dromedary camel milk in Iran. *Food Sci. Technol.* **2016**, *36*, 717–723. [CrossRef]
46. Seboussi, R.; Faye, B.; Alhadrami, G.; Askar, M.; Ibrahim, W.; Hassan, K.; Mahjoub, B. Effect of different selenium supplementation levels on selenium status in camel. *Biol. Trace Elem. Res.* **2008**, *123*, 124–138. [CrossRef]
47. Miyasaka, S.A. Nutrición Animal. Editorial Trillas Sa De Cv. México. 2015. Available online: https://books.google.co.ve/books?id=1vkrPwAACAAJ&num=20 (accessed on 17 September 2022).
48. Tokarnia, C.H.; Peixoto, P.V.; Barbosa, J.D.; Brito, M.F.; Döbereiner, J. Deficiência de cobre. *Deficiências Minerais Em Anim. Prod.* **2010**, *191*, 88–102.
49. Gamble, C.; Hansard, S.; Moss, B.; Davis, D.; Lidvall, E. Manganese utilization and placental transfer in the gravid gilt. *J. Anim. Sci.* **1971**, *32*, 84–87. [CrossRef] [PubMed]
50. Hidiroglou, M.; Knipfel, J. Maternal-fetal relationships of copper, manganese, and sulfur in ruminants. A review. *J. Dairy Sci.* **1981**, *64*, 1637–1647. [CrossRef]
51. Essawi, W.; Gouda, H. Inter-Relationship between some Trace Elements during Pregnancy and Newborn Birth Weight in Dromedary Camels. *Zagazig Vet. J.* **2020**, *48*, 319–327. [CrossRef]

Article

Moving Buffalo Farming beyond Traditional Areas: Performances of Animals, and Quality of Mozzarella and Forages

Francesco Serrapica [1], Felicia Masucci [1,*], Giuseppe De Rosa [1], Ada Braghieri [2,*], Fiorella Sarubbi [3], Francesca Garofalo [4], Fernando Grasso [1] and Antonio Di Francia [1]

[1]   Department of Agricultural Sciences, University of Naples Federico II, 80055 Portici, Italy
[2]   School of Agricultural, Forestry, Food and Environmental Sciences, University of Basilicata, 85100 Potenza, Italy
[3]   Institute for the Animal Production System in the Mediterranean Environment, National Research Council, 80055 Portici, Italy
[4]   Istituto Zooprofilattico Sperimentale del Mezzogiorno, 80055 Portici, Italy
*   Correspondence: masucci@unina.it (F.M.); ada.braghieri@unibas.it (A.B.); Tel.: +39-081-253-9307 (F.M.); +39-0971-205-101 (A.B.)

Abstract: An observational case study was designed to highlight issues associated with a possible expansion of dairy buffalo (Bubalus bubalis) farming outside the traditional coastal plains of southern Italy. Twenty pregnant buffaloes were transferred to a hilly inland farm. After calving, production and reproduction data were collected monthly throughout lactation. From 4 to 6 months of lactation, buffaloes were enrolled in a feeding trial to evaluate the effects of locally grown forages (maize silage vs. hay) on milk production and in vivo digestibility. Sensory properties of mozzarella cheese produced at a local dairy were also evaluated. No obvious effects of diet were found. Compared to the data recorded in the previous lactation completed in the farm of origin, milk yield was reduced by 37.2%, and milk protein by 6.1%, whereas milk fat improved (+4.5%). A lower pregnancy rate (−13.3%), increased days open (+122%), and a prolonged intercalving period (+26.9%) were also observed. Lactation length was shorter than the standard value of 270 d. The results showed that peculiar reproductive characteristics, lower environmental temperatures, and the specificity of the mozzarella production process are the main problems to be addressed in an expansion of buffalo farming outside traditional areas.

Keywords: dairy buffaloes; farming environment; reproductive and productive performances; feeding trial; mozzarella cheese; sensory properties

Citation: Serrapica, F.; Masucci, F.; De Rosa, G.; Braghieri, A.; Sarubbi, F.; Garofalo, F.; Grasso, F.; Di Francia, A. Moving Buffalo Farming beyond Traditional Areas: Performances of Animals, and Quality of Mozzarella and Forages. *Agriculture* **2022**, *12*, 1219. https://doi.org/10.3390/agriculture12081219

Academic Editor: Daniel Simeanu

Received: 15 July 2022
Accepted: 11 August 2022
Published: 13 August 2022

Publisher's Note: MDPI stays neutral with regard to jurisdictional claims in published maps and institutional affiliations.

1. Introduction

Italian buffalo (*Bubalus bubalis*, Mediterranean type) farming is a traditional enterprise almost exclusively devoted to producing mozzarella cheese [1]. Currently, roughly 75% of the European 420,000 dairy buffalo heads are raised in specific coastal plains of Campania, a region of southern Italy [2–4]. The growing demand for buffalo mozzarella in the last few decades has driven a rapid intensification of the sector in terms of increased use of off-farm inputs and implementation of genetic improvement techniques [5,6]. The shift from extensive to intensive modes of production has increased the stocking rate and then the environmental impact of buffalo farming in terms of worsening air and water quality, as well as animal welfare and health, even facilitating occasional animal disease outbreaks [7–10]. Moving buffalo farming from the traditional coastal plains to inland hilly zones could mitigate the issues related to buffalo farming along with revitalizing the economy and contrasting the depopulation of more marginal zones [11–13]. Nevertheless, this expansion should consider the different environmental and infrastructural conditions of hilly and inland areas compared to lowland farms. Buffaloes originate and are mainly distributed in tropic and sub-tropic environments, so that they present anatomic and physiological mechanisms, for instance the sparse hairs and the numerous melanin pigments on epidermis, able to counter hot and humid climate conditions rather than the cooler conditions of internal areas [14–16]. In addition, some specific features of buffalo reproduction, such as seasonality and estrus behavior, can reduce the reproductive efficiency and then profitability of farms [17,18].

This study aimed to assess the possibility of expanding buffalo farming beyond the traditional coastal areas and to highlight related issues, and an observational case study was designed by moving a group of lactating buffaloes in an internal hilly area. We also assessed the yield performance of the buffaloes fed with locally produced forages (hays vs. maize silage), based on different scenarios for irrigation water availability and farm size. The objectives of this research can be summarized as follows: (i) assess the adaptability of buffaloes to the conditions of internal hilly areas by surveying the productive and reproductive performance throughout the whole lactation; (ii) compare the yield performance of buffaloes fed diets based on hay or maize silage; (iii) evaluate the sensory characteristics of locally produced mozzarella.

2. Materials and Methods

2.1. Study Site

The study site was a dairy cattle farm (55 ha Utilized Agriculture Area, 120 cows) sited in an internal hilly area of Campania Region (41°17′ N, 15°02′ E, 540 m a.s.l.). The area is bounded by the pre-Apennine mountains, with a Mediterranean sub-continental climate (Köppen zone Csb) characterized by dry and warm summers. The rainfall pattern is very irregular and strongly influenced by the interaction between the wet air masses and orography [19]. The average annual rainfall ranges from 800 to 1100 mm, with the maximum monthly precipitation during November and December, and minimum values during July and August. The temperatures average 4 °C in winter and 20 °C in summer [20,21].

2.2. Productive and Reproductive Performances

In early summer 20 pluriparous (age 6.0 ± 1.9 years, parity 3.2 ± 1.8), pregnant (on average 7.7 months of gestation) dry buffaloes were moved from farm of origin (41°05′ N, 14°06′ E, 13 m a.s.l) and introduced in the experimental farm, where they were kept together in an open barn equipped with external paddock, water bowls and manger until calving that took place in the farm. The observation period started about two months after the relocation, in July, at the time of delivery, and covered the whole lactation period (i.e., until dry off time). The animals were fed the same maize silage-based total mixed ration (M-TMR) formulated for a milk yield of 8.0 kg/d which was modified at the 6th month of lactation for meeting the lower needs [22,23]. After a voluntary waiting period of 60 d from parturition, on the basis of visual observation of the estrous signals, the buffaloes were artificially inseminated by intrauterine deposition of frozen-thawed semen from two proven bulls. Pregnancy diagnosis was assessed by rectal palpations at 30-day intervals, and days open (d from calving to pregnancy diagnosis), calving interval, and conception rate (percentages of cows pregnant) were calculated. Every month, milk from each cow was measured and sampled to assess for fat, protein, lactose (Milkoscan 605, Foss, Hillerød, Denmark), and somatic cells count (SCC; Fossomatic 90, Foss Electric, Hillerød, Denmark). Mozzarella cheese yield was estimated based on the milk chemical composition by using the equation of Altiero et al. [24].

Climatic data (daily air temperature, relative humidity, and rainfall) from the experimental farm and the farm of origin were collected.

2.3. Feeding Trial

At the 4th month of lactation, the cows were weighted, scored for body condition score (BCS) [25], and randomly split into two groups homogenous for milk yield and quality.

One group continued to be fed the M-TMR whereas the other was fed an isonitrogenous and isoenergetic ration in which the maize silage was substitute by ryegrass hay (H-TMR) (Table 1). The trial lasted 8 weeks, after 10 d of adaptation. The groups were housed in two adjacent free stall barns and handled in similar way in terms of feeding and management.

Table 1. Ingredients (kg/d as-fed) and chemical composition (% of dry matter, DM, if not otherwise stated) of the total mixed rations fed to the study buffaloes.

	M-TMR	H-TMR
Ingredients		
Maize silage	15.4	-
Alfalfa hay	3.1	4.0
Ryegrass hay	2.0	5.0
Concentrates	6.3	7.0
Hydrogenated fat [1]	-	0.3
Chemical composition		
Dry matter (kg/d)	14.7	14.4
Ash	8.2	9.5
Crude protein	15.2	15.5
Ether extract	3.1	4.1
Starch	25.4	18.0
Neutral detergent fiber	42.7	45.5
Acid detergent fiber	27.8	31.5
Acid detergent lignin	6.3	7.4
NE_L (UFL/kg DM)	0.85	0.86

[1] Supplement containing 4% hydrogenated palm oil (Hidrofat© Nutrición Internacional, S.L., Madrid, Spain). M-TMR, maize silage based total mixed ration; H-TMR, hay-based total mixed ration; NE_L, net energy of lactation; UFL, Unité Fourragère Lait: 1 UFL = 7.11 MJ/kg of NE_L; DM, dry matter.

Dry matter intake (DMI) was measured every week on a group basis along with milk and rations sampling. The TMRs were analyzed according to AOAC [26] for dry matter (DM), ash, fat (Soxhlet apparatus), and crude protein (CP, Kjeldahl method). Neutral detergent fiber including residual ash (aNDF), acid detergent fiber (ADF), and acid detergent lignin (ADL) were determined according to Van Soest et al. [27]. Starch content was assessed by a Polax-2l polarimeter (Atago Co., Ltd., Tokyo, Japan) in a 200 mm long observation tube [28]. Net Energy for Lactation (NEL) was calculated based on chemical composition [29]. At the 7th week of the experimental period, the total tract in vivo digestibility was evaluated by using acid insoluble ash (AIA) as intrinsic marker [30] according to the procedure described by Vander Pol [31] and Masucci et al. [32]. Briefly, for three consecutive days, fecal grab samples were collected from the rectum of each animal at 0900 h–1300 h and 1700 h. Samples of the rations were also collected and DMI determined. The samples were dried to determine partial DM, composited per animal (fecal samples), and analyzed for CP, NDF, ADF, and AIA content.

2.4. Mozzarella Cheese Production

Because traditional mozzarella cheese is an artisanal product subject to variation in relation to cheesemaking conditions, the consistency of the sensory properties of mozzarella on different days of production was tested. It was not possible to produce mozzarella separately for the M-TMR and H-TMR groups, as no cheesemaking equipment was available to process small amounts of milk in separate tanks at the same time. Therefore, at the end of the feeding trial (sixth month of lactation) the buffaloes were reunited and fed the same maize silage-based diet. After two weeks of adaptation, the mozzarella production process was carried out on two different days (D1 and D2), one week apart, by the same staff from a local dairy. Milk was collected at evening and morning milking, transported to the dairy, and manufactured for mozzarella production according to the traditional procedure [33].

2.5. Chemical and Sensory Analyses

Before the mozzarella manufacturing process, milk was sampled to determine chemical composition and titratable acidity [26]. Mozzarella samples from the D1 and D2 batches of production were separately analyzed for macro-component and quantitative descriptive sensory analyses (QDA) after 24 and 48 h from manufacturing. Overnight, mozzarella samples were kept at 10 °C and allowed to equilibrate at room temperature (22–23 °C) for at least 4 h before analysis. Chemical composition was determined on 3 samples/batch, each consisting of 4 blended pieces of mozzarella. Moisture was quantified by oven drying while fat and protein were determined according to the Gerber and Kjeldahl methods, respectively [26].

To perform a quantitative descriptive analysis (QDA) [34], 15 subjects were recruited among regular eaters of buffalo mozzarella cheese (defined as consuming this cheese at least once a week). In total, 10 panelists (6 females and 4 males, mean age 29 year) were selected by assessing their capacity to identify the 4 basic tastes (sourness, sweetness, bitterness, and saltiness), according to the ISO recommendations [35]. During a preliminary phase, the panelists were asked to taste buffalo mozzarella samples and, on the basis of available literature [34,36], they developed and agreed on a 19-attribute consensus list (Table 2) concerning appearance (5), odor/flavor (5), taste (4), and texture (5).

Table 2. List of attributes and reference samples used by the 10-member trained panel for Buffalo mozzarella cheese sensory profiling [1].

Attribute	Reference Sample		Definition
	Lower Anchor	**Upper Anchor**	
Appearance			
Color	7.5Y 9/2 (ivory)	N 9.5 (white)	Overall intensity of white color (from ivory to white)
Brightness	Ripened scamorza cheese	Buffalo mozzarella cheese	Shininess or glossiness of surface
Stringy appearance	Ripened scamorza cheese	Nodino cheese	Typical filamentous or fibrillar structure of milk casein after hot water stretching
Whey loss	Ripened scamorza cheese	Buffalo mozzarella cheese	Amount of whey visible on the surface of the cheese sample after cutting
Skin thickness	Mozzarella cheese	Burrata cheese	Skin thickness of the buffalo mozzarella cheese evaluable after cutting
Odor/Flavor			
Overall Odor	Ricotta cheese	Buffalo mozzarella cheese	Level of overall odor
Overall Flavor	Ricotta cheese	Buffalo mozzarella cheese	Level of overall flavor
Milk	Mixture of water (80%) and milk (20%)	Whole milk	Odor/flavor arising from milk at room temperature
Cream	Mixture of water (80%) and cream (20%)	Whole cream	Odor/flavor arising from cream at room temperature
Yogurt	Whole milk	Plain whole yogurt	
Taste			
Saltiness	1.5 mL stock solution 100 mL^{-1}	3 mL stock solution 100 mL^{-1}	Fundamental taste associated with sodium chloride
Sweetness	8 mL stock solution 100 mL^{-1}	20 mL stock solution 100 mL^{-1}	Fundamental taste associated with sucrose
Bitterness	4 mL stock solution 100 mL^{-1}	8 mL stock solution 100 mL^{-1}	Fundamental taste associated with quinine
Sourness	8 mL stock solution 100 mL^{-1}	16 mL stock solution 100 mL^{-1}	Fundamental taste associated with citric acid
Texture			
Tenderness	Ripened provolone cheese	20 g round size mozzarella cheese	Minimum force required to chew cheese sample: the lower the force the higher the tenderness
Screechiness	Asiago cheese	Nodino cheese	Mouth sensation describing the typical filamentous or fibrillary texture of milk casein after hot water stretching
Elasticity	Parmesan cheese seasoned 36 mo	Emmental cheese	Degree to which the original shape of a product is restored after compression between the teeth
Moisture	Ripened scamorza cheese	Ricotta cheese	Moisture released by the product in the mouth during early mastication
Oiliness	Ricotta cheese	80 g butter mixed with 40 g ricotta cheese	Amount of oily/fatty feeling in the mouth during chewing

[1] Color definitions as in Munsell Book of Color (X Rite color. Europe GmbH). Stock solutions for sweetness, sourness, saltiness, and bitterness were, respectively: 50 g sucrose 250 mL^{-1} solution; 2.5 g citric acid 250 mL^{-1} solution; 25 g sodium chloride 250 mL^{-1} solution; 51.25 g quinine hydrochloride 250 mL^{-1} solution.

Subsequently, assessors were trained with a reference frame, using specific products to each identified attribute (Table 2). During the panel training, at least two points of the scale were anchored to the reference material. In particular, they were repeatedly exposed to the reference samples (3 times) while overtly showing the corresponding intensity levels. Subsequently, panelists re-assessed the two levels of intensity of each attribute in blind conditions.

Then, a QDA was conducted in sensory booths [35,37], to assess the sensory profile of the products. The tests started at about 10.00 h and panelists evaluated two of 50 g buffalo mozzarella cheese at a serving temperature of 13 °C. To mask appearance differences in the sample, assessors evaluated the first cheese under red light for odor/flavor, taste, and texture attributes, and the second cheese under white fluorescent lighting, for appearance attributes. The panelists evaluated 3 replications of each product, at 24 h and 48 h from mozzarella manufacturing. The interval between consecutive samples was roughly 10 min and panelists drank a sip of water after each sample.

As QDA is based on the use of a 100 mm unstructured intensity linear scale, with anchor points at each end (0 = absent and 100 = very intense), panelists were trained to identify the intensity ranges for low, medium, and high intensity [36].

2.6. Statistical Analysis

A descriptive statistic was used for productive (milk yield and quality over the whole lactation) and reproductive performances. Data from the feeding trial were analyzed using SAS statistical package (version 6.09, SAS Institute, Cary, NC, USA). Milk yield and quality of the TMR-M and TMR-H groups underwent analysis of variance (ANOVA) for repeated measures (mix proc) with diet (TMR-M and TMR-H) as a non-repeated factor and sampling time and diet × sampling time as repeated factors. The cow variance was considered as random and as the error term to test the main effect of the diet. Dry matter intake, and animals' BCS were analyzed by ANOVA (GLM procedure) to determine the fixed effects of the dietary treatment (TMR-M and TMR-H). Chemical composition of mozzarella cheese was analyzed by one-way ANOVA (GLM procedure) to determine the fixed effects of time of conservation (24 and 48 h). Sensory profile data were subjected to ANOVA with assessor (10), replication (3), storage time (24 and 48 h), day of production (1 and 2), and the interaction as factors. Data are reported as least square means (LSM) and standard errors of means (SEM). Statistical significance was declared at $p < 0.05$ and tendencies discussed at $p < 0.10$.

3. Results

3.1. Climatic Data

Data retrieved from the climatical stations are summarized in Table 3.

Table 3. Seasonal climatic data (mean ± standard deviation) of the experimental farm and origin farm during the experimental period.

Item	Summer [1]	Fall [2]	Winter [3]	Spring [4]
Study site				
AT [5], °C	20.1 ± 1.9	12.4 ± 4.9	2.9 ± 1.4	8.1 ± 3.2
RH [6], %	74.1 ± 3.0	83.7 ± 7.1	86.5 ± 4.6	82.0 ± 6.6
R [7], mm	26.2 ± 3.7	70.5 ± 30.7	59.0 ± 23.3	45.6 ± 26.3
Farm of origin				
AT [5], °C	26.1 ± 1.6	20.4 ± 3.4	10.0 ± 3.1	13.5 ± 2.3
RH [6], %	58.8 ± 3.1	72.7 ± 4.9	72.7 ± 2.7	66.8 ± 6.6
R [7], mm	7.4 ± 5.4	32.4 ± 2.9	34.4 ± 15.7	32.9 ± 15.7

[1] Summer = June, July, and August; [2] Fall = September, October, and November; [3] Winter = December, January, and February; [4] Spring = March, April, and May. [5] AT, daily air temperature; [6] AR, daily relative humidity; [7] R, monthly rainfall.

The study area was steadily colder (−7 °C per day), wetter (+14% of daily relative humidity), and rainier (+23.5 mm of rain per month) than the coastal area from which the animals were moved. The largest seasonal variations were in fall, when the daily temperatures and total rainfall differed by about 8 °C and 114 mm, respectively.

3.2. Maize-Based vs. Hay-Based Rations

The main differences between the chemical and nutritive characteristics of TMRs were for the contents in starch (higher in M-TMR), and NDF and ADF (higher of H-TMR), while the energy concentrations were quite close between the two diets (Table 1). The effects of dietary treatment on DMI and milk production are shown in Table 4. The M-TMR group showed a slightly but significantly ($p < 0.05$) higher DMI, whereas milk yield, milk fat, and protein were rather similar between the dietary groups along with SCC and theorical mozzarella cheese yield.

Table 4. Dry matter intake, body condition score, milk yield, and quality (LSM) of buffaloes fed total mixed ration based on maize silage (M-TMR) and hay (H-TMR).

Item	Diets		SEM	p Value		
	M-TMR	**H-TMR**		**D**	**T**	**D × T**
DMI, kg/d	14.70	14.10	0.19	0.0357	-	-
BCS	7.07	7.06	0.26	0.9579	-	-
Milk yield, kg/d	6.09	6.69	0.74	0.5639	<0.0001	0.3500
Fat, %	9.05	8.76	0.21	0.3406	<0.0001	0.9347
Protein, %	4.61	4.60	0.05	0.9093	<0.0001	0.9959
Lactose, %	4.65	4.58	0.07	0.4555	<0.0001	0.5292
SCC, n. cells/mL	742.150	903.830	194.56	0.5603	0.2777	0.9475

M-TMR, maize silage based total mixed ration; H-TMR, hay-based total mixed ration; SEM, standard error of mean; D, diet; T, time; DMI, dry matter intake; BCS, body condition score.

The apparent total tract digestibility as influenced by the dietary treatments is in Table 5. No significant differences were detected between the groups, albeit NDF and ADF digestibility tended to be better in the H-TMR group ($p < 0.0527$ and $p < 0.0661$, respectively).

Table 5. In vivo digestibility coefficients (LSM) of buffaloes fed total mixed ration based on maize silage (M-TMR) and hay (H-TMR).

Item	Diets		SEM	p Value
	M-TMR	**H-TMR**		
Dry matter	76.26	75.65	0.91	0.6244
Organic matter	77.97	77.6	0.91	0.7661
Crude protein	76.25	77.47	0.89	0.323
Neutral detergent fiber	67.11	70.56	1.23	0.0527
Acid detergent fiber	63.26	66.89	1.38	0.0661

M-TMR, maize silage based total mixed ration; H-TMR, hay-based total mixed ration; SEM, standard error of mean.

3.3. Productive and Reproductive Performances

Since the lack of statistically significant differences between the two feeding groups, milk yield and quality, over the whole lactation were calculated regardless the diet (Table 6).

Compared to the data recorded in the previous lactation completed in the farm of origin, milk yield was reduced by 37.2% and milk protein by 6.1%, whereas milk fat improved (+4.5%) because of the lower milk quantity. In addition, milk yield was well below the average values reported for buffalo cows bred in Campania [38–41]. Lactation length was shorter than the standard value of 270 d (Table 6). Compared to the data available for buffaloes farmed in similar conditions and seasons [42–45], and the data recorded in the previous lactation, a lower pregnancy rate (−13.3%), increased days open (+122%), and a prolonged intercalving period (+26.9%) were observed (Table 6).

Table 6. Productive and reproductive performances (mean ± standard deviation) of lactating buffaloes.

Item	Present Study	Previous Lactation	Δ [1] (%)
Milk yield, kg/lactation	1546 ± 622	2463 ± 632	−37.2
Milk fat, %	8.64 ± 0.83	8.27 ± 0.28	+4.5
Milk protein, %	4.43 ± 0.32	4.72 ± 0.07	−6.1
Days open, day	189 ± 112	85± 32	+122.3
Lactation length, days	256 ± 80	314 ± 27	−18.5
Calving interval, day	499 ± 112	393 ± 32	+26.9
Pregnancy rate [2], %	65	75	−13.3

[1] Δ, calculated as percentage difference between the means recording during the study and in previous lactation completed in a plain farm. [2] Calculates on a group basis as the number of cows confirmed pregnant divided by the total number of cows × 100.

3.4. Mozzarella Cheese Quality

Milk chemical composition (fat 8.61 vs. 8.76%, protein 4.79 vs. 4.81%, lactose 4.88 vs. 5.01%, respectively, for D1 and D2), titratable acidity (7.15 vs. 7.13 °SH/100 mL), as well as mozzarella yield (23.7 vs. 22.5%) were rather similar across two cheesemaking days.

By contrast, the chemical composition of the two batches of mozzarella varied over the two days of production. In fact, the D1 mozzarella had significantly higher moisture ($p < 0.01$), protein ($p < 0.05$), and fat ($p < 0.01$) contents than D2 (Table 7).

Table 7. Chemical composition and sensory profile of mozzarella cheese (LSM) affected by cheesemaking day and conservation time.

	Cheesemaking Day		Storage Time (h)		SEM	p Value		
	D1	D2	24	48		Day	ST	Day × ST
Chemical composition								
Moisture	54.17	51.01	52.53	52.65	0.41	0.0006	0.8398	0.5998
Fat	28.97	25.32	27.02	27.27	0.51	0.001	0.7385	0.9111
Protein	14.77	14.35	14.55	14.57	0.09	0.0136	0.9029	0.7153
Appearance								
Color	34.8	26.2	34.0	27.0	2.99	0.0457	0.1035	0.0791
Brightness	46.6	39.4	46.1	39.8	2.83	0.0753	0.1204	0.7259
Whey loss	34.1	19.1	28.5	24.7	2.61	0.0001	0.3088	0.4597
Stringy appearance	35.0	25.5	31.0	29.4	2.45	0.008	0.6513	0.2985
Skin thickness	33.0	23.0	26.9	29.2	2.40	0.0044	0.1115	0.2659
Odor/Flavor								
Overall odor	46.5	46.2	45.0	45.0	2.06	0.9132	0.3677	0.6966
Yogurt odor	34.5	35.3	34.7	35.1	2.31	0.8092	0.9189	0.4872
Overall flavor	49.8	41.0	46.3	44.4	2.22	0.0063	0.5428	0.7856
Milk flavor	43.5	37.1	41.0	39.6	2.02	0.0285	0.6228	0.1128
Cream flavor	39.6	31.0	35.7	34.8	2.12	0.0056	0.7712	0.745
Yogurt flavor	25.5	22.6	2.1	2.1	2.09	0.3457	0.3181	0.1499
Taste								
Saltiness	39.0	27.3	30.5	35.8	1.99	<0.0001	0.0687	0.1147
Sweetness	17.8	25.5	21.7	21.6	1.98	0.0073	0.9512	0.1251
Bitterness	9.3	8.5	9.8	8.0	1.26	0.6619	0.3218	0.8313
Sourness	35.9	17.8	29.1	24.6	2.15	<0.0001	0.1477	0.139
Texture								
Tenderness	43.1	21.0	33.8	30.2	2.29	<0.0001	0.2698	0.1959
Screechiness	30.4	46.3	41.0	35.8	2.86	0.0002	0.2068	0.1813
Elasticity	37.7	44.0	38.1	43.6	2.58	0.1072	0.1617	0.1775
Moisture	54.0	21.1	37.3	37.8	2.58	<0.0001	0.8814	0.1777
Oiliness	39.0	29.4	32.3	36.1	2.39	0.0062	0.2741	0.8426

D1, first day of mozzarella cheese production; D2, second day of mozzarella cheese production; SEM, standard error of mean; D, cheesemaking day; T, storage time.

As for sensory analyses, no significant product x replication or product x assessor interactions were detected in preliminary ANOVA, suggesting the efficacy of the training program and of the reference frame developed in this study, both allowing to reach high reliability of the panel (i.e., products were not evaluated differently in different replications or by different assessors).

Apart from brightness (Appearance), overall odor, yoghurt odor and yoghurt flavor (Odor/Flavor), bitterness (Taste), and elasticity (Texture), the sensory attributes of mozzarella cheese produced in the two days significantly differed. Mozzarella D1 showed higher intensities perceived of the attributes related to appearance (i.e., color $p < 0.05$, whey releasing $p < 0.001$, stringy appearance, skin thickness $p < 0.01$), and flavor (milk flavor $p < 0.05$, overall flavor, cream flavor $p < 0.01$). Moreover, D1 was less sweet ($p < 0.01$), but sourer ($p < 0.001$) and saltier ($p < 0.01$). Among the texture attributes, oiliness, tenderness ($p < 0.001$), and moisture ($p < 0.01$) rated higher in D1, and accordingly, lower intensity was perceived for screechiness ($p < 0.01$) (Table 7).

Storage time did not affect either chemical composition or sensorial parameters. The typical sensory defect of mozzarella due to storage, such as outer skin adherence (22.52 ± 1.82 vs. 26.07 ± 1.75, $p = 0.1926$, respectively, for 24 and 48 h of storage time), was not significantly increased.

4. Discussion

No great effects were observed by using a hay-based diet instead of maize, as is usually the case on traditional lowland buffalo farms. The lower NDF content of the maize-based diet may have led to the higher DMI observed for M-TMR group since it is the main determinant of the rumen fill and then of the ingestion capacity in ruminants [46]. However, the higher DMI did not influence milk yield and quality and with them mozzarella cheese yield because of the similar milk protein and fat contents. Since the level of fibrous fractions in rations can greatly influence digestibility [47–49], a better digestibility of the less fibrous M-TMR diet would have been expected. However, the higher starch levels of this diet have likely lowered rumen pH and so impaired the activity of cellulolytic microorganisms [50]. Anyway, the tendency to a lower fiber digestibility observed for the M-TMR diet had no obvious effects either on the milk yield or on the animals' BCS. Overall, the hay-based diet has yielded similar results to that based on maize silage, but its tendentially higher cost [51] makes it unprofitable if maize silage is available.

No complications were observed at calving time and the neonatal deaths (3 calves) were within the usual range for buffalo farming [32,52], indicating that cows' handling and calf care practices were appropriate. An undesirable result, however, is the rather low milk production largely determined by the short duration of lactation, which does not seem to be due to feeding errors, since correct BCS and DMI and good milk production were observed during the feeding trial. Nevertheless, yield losses induced by environmental change and weather distress cannot be excluded. In this regard, after a sudden drop ($-3\ ^\circ$C) in the minimum temperature, buffaloes at the early and intermediate stages of lactation reduced milk production up to 20% for several days after the event, thus indicating that low temperatures can have a cumulative effect [53].

The poor reproductive performances do not find an easy explanation. As buffalo is a short-day breeder and the cows calve in late summer, an early postpartum resumption of ovarian activity and a better reproduction efficiency would have been expected [54]. The shortness of the sexual receptivity period combined with the difficulties in estrus detection of buffaloes may have led to improper artificial insemination timing [54,55]. Indeed, the buffalo estrus phase averages 20 h with high incidences of short (<12 h) and medium (13–24 h) durations [56,57]. In addition, the visual observations for estrus detection are less efficient in buffaloes since, compared to cattle, the typical signs and behaviors of estrus as frequent urination, temporary teat engorgement, vulvar edema, and vocalizations, restlessness, tail raising are extremely weak and poorly represented, and many cows show estrus signs in the late-night and early morning hours [58]. The poor reproduction efficiency may have been amplified by weather-induced stress. According to Zicarelli [51,59], buffalo cows are prone to change their reproductive pattern when exposed to sudden climatic variation, becoming acyclic when cold wind and heavy rain associated with thermal drops occur. Overall, failure of instrumental insemination and cooler environmental temperatures are likely to account for the poor production and reproductive performance, while the BCS at the end of the observation period and the feeding trial data do not indicate errors in diet or feeding. In any case, these problems could be overcome by adopting insemination protocols for buffaloes and sheltering the animals from cool environments.

Both chemical and sensory characteristics of mozzarella cheese varied widely over the two days of production and appear to be closely related to each other. The higher moisture content of D1 mozzarella might explain the higher intensities perceived for whey releasing and moisture while the higher rate for cream and milk flavor, and tenderness may be ascribed to higher fat content of these samples, as assumed by Stevens and Shah [60]. The sensory quality of a fresh cheese depends on several factors linked to both the characteristics of the raw milk and the cheese-making technology [61]. The differences both in the chemical composition and in the sensory profile of the mozzarella samples seem to indicate a non-uniformity of the D1 and D2 cheesemaking processes. Factors as natural milk treatment, whey starter culture, curd pH and temperature, and management of the stretching phase can greatly influence the composition of the mozzarella, as they impact on the amount of moisture and fat trapped in the typical fibrous texture of mozzarella, as well as attributes such as structure, the stringy appearance, and skin thickness [62,63]. In addition, skin thickness seems to be also related the preservative liquid composition, and, in particular, to higher citric acid concentration [64]. These results indicate the need for adequate training of local cheesemakers on the mozzarella cheese-making process, as any perceived reduction in sensory characteristics typical of traditional dairy products may not be accepted by consumers [65].

As for the storage time effect, mozzarella is a high perishability cheese due to the high moisture content and water activity, with mass transfer between the cheese matrix and its serum phase. Probably, the storage times considered were rather short to affect many sensory attributes. In fact, other authors [66] found significant changes in volatile profile of traditional mozzarella cheese and lactose free mozzarella after 13 days and 8 days of storage, respectively; these modifications resulted from amino acid and fatty acid metabolism occurred in the samples and caused a sensory decay of positive descriptors associated with fresh cheese products and higher intensities of negative attributes, such as bitter taste, associated with release of bitter tasting peptides due to the proteolytic activity of spoilage microorganisms [67]. Moreover, a decrease in sensory hardness was significantly observed by Alinovi et al. [68] after 7 d of refrigerated storage and they related it to casein hydrolysis.

5. Conclusions

During the first year of farming in a cooler inland hilly area, multiparous buffaloes showed a reduction in milk production by more than one-third and an increase in calving interval by one-fourth compared with the previous lactation completed on the farm of origin in a coastal lowland area.

Instrumental insemination failure and facing lower environmental temperatures are probably the origin of these results, while the BCS at the end of the observation period and the data from the feeding trial do not indicate dietary or feeding mistakes. Overall, the distinctive reproduction characteristics of buffaloes, the cooler environmental temperatures and the specificity of the mozzarella production process are the main problems to be faced in an extension of buffalo farming beyond the traditional areas.

Author Contributions: Conceptualization, F.M., F.S. (Francesco Serrapica), and A.D.F.; methodology, F.M. and A.B.; data curation, F.M., F.S. (Francesco Serrapica), A.B., and G.D.R.; formal analysis, F.S. (Francesco Serrapica), F.S. (Fiorella Sarubbi), F.G. (Francesca Garofalo), and F.G. (Fernando Grasso); investigation, F.M., A.B., and F.S. (Francesco Serrapica); resources, F.M. and A.D.F.; writing—original draft preparation, F.M. and F.S. (Francesco Serrapica); writing—review and editing, F.M., G.D.R., F.G. (Fernando Grasso), and A.B.; supervision, A.D.F.; project administration, A.D.F.; funding acquisition, A.D.F. All authors have read and agreed to the published version of the manuscript.

Funding: This research was supported by the project "Messa a punto di metodologie Interdisciplinari per la valorizzazione del Territorio e della Qualità e tracciabilità geografica dei loro prodotti agricoli (animali e vegetali)—MITEQ".

Institutional Review Board Statement: This study was conducted according to the guidelines of the Declaration of Helsinki and approved by the Ethical Animal Care and Use Committee of Federico II University of Naples (protocol code: PG/2021/0075836 of 23 July 2021).

Informed Consent Statement: Informed consent was obtained from the farm owner involved in the study.

Data Availability Statement: The datasets of the present study are available from the corresponding author on reasonable request.

Acknowledgments: Gratitude is due to Fabio Napolitano who is continuing to inspire our work. The authors thank Amelia Riviezzi for her expert technical assistance and Maria Luisa Varricchio for valuable assistance during cheesemaking and sample collection.

Conflicts of Interest: The authors declare no conflict of interest.

References

1. Uzun, P.; Serrapica, F.; Masucci, F.; Barone, C.M.A.; Yildiz, H.; Grasso, F.; Di Francia, A. Diversity of Traditional Caciocavallo Cheeses Produced in Italy. *Int. J. Dairy Technol.* **2020**, *73*, 234–243. [CrossRef]
2. National Database of Istituto Zooprofilattico Sperimentale Dell'Abruzzo e Del Molise Giuseppe Caporale. Available online: https://www.Vetinfo.It/ (accessed on 2 June 2022). (In Italian).
3. FAOSTAT. Database. Available online: http://www.Fao.Org/Faostat/En/#home (accessed on 3 June 2022).
4. Vargas-Ramella, M.; Pateiro, M.; Maggiolino, A.; Faccia, M.; Franco, D.; De Palo, P.; Lorenzo, J.M. Buffalo milk as a source of probiotic functional products. *Microorganisms* **2021**, *9*, 2303. [CrossRef] [PubMed]
5. Cesarani, A.; Biffani, S.; Garcia, A.; Lourenco, D.; Bertolini, G.; Neglia, G.; Misztal, I.; Macciotta, N.P.P. Genomic Investigation of Milk Production in Italian Buffalo. *Ital. J. Anim. Sci.* **2021**, *20*, 539–547. [CrossRef]
6. Serrapica, F.; Masucci, F.; Romano, R.; Napolitano, F.; Sabia, E.; Aiello, A.; Di Francia, A. Effects of Chickpea in Substitution of Soybean Meal on Milk Production, Blood Profile and Reproductive Response of Primiparous Buffaloes in Early Lactation. *Animals* **2020**, *10*, 515. [CrossRef]
7. Martucciello, A.; Galletti, G.; Pesce, A.; Russo, M.; Sannino, E.; Arrigoni, N.; Ricchi, M.; Tamba, M.; Brunetti, R.; Ottaiano, M.; et al. Short Communication: Seroprevalence of Paratuberculosis in Italian Water Buffaloes (Bubalus Bubalis) in the Region of Campania. *J. Dairy Sci.* **2021**, *104*, 6194–6199. [CrossRef]
8. Esposito, C.; Fiorito, F.; Miletti, G.; Serra, F.; Balestrieri, A.; Cioffi, B.; Cerracchio, C.; Galiero, G.; De Carlo, E.; Amoroso, M.G.; et al. Involvement of Herpesviruses in Cases of Abortion among Water Buffaloes in Southern Italy. *Vet. Res. Commun.* **2022**, *1*, 1–11. [CrossRef]
9. Napolitano, F.; Bragaglio, A.; Sabia, E.; Serrapica, F.; Braghieri, A.; De Rosa, G. The Human- Animal Relationship in Dairy Animals. *J. Dairy Res.* **2020**, *87*, 47–52. [CrossRef]
10. Napolitano, F.; Serrapica, F.; Braghieri, A.; Masucci, F.; Sabia, E.; De Rosa, G. Human-Animal Interactions in Dairy Buffalo Farms. *Animals* **2019**, *9*, 246. [CrossRef]
11. Troiano, C.; Buglione, M.; Petrelli, S.; Belardinelli, S.; De Natale, A.; Svenning, J.-C.; Fulgione, D. Traditional Free-Ranging Livestock Farming as a Management Strategy for Biological and Cultural Landscape Diversity: A Case from the Southern Apennines. *Land* **2021**, *10*, 957. [CrossRef]
12. Serrapica, F.; Masucci, F.; Romano, R.; Santini, A.; Manzo, N.; Seidavi, A.; Omri, B.; Salem, A.Z.M.; Di Francia, A. Peas May Be a Candidate Crop for Integrating Silvoarable Systems and Dairy Buffalo Farming in Southern Italy. *Agrofor. Syst.* **2020**, *94*, 1345–1352. [CrossRef]
13. Serrapica, F.; Masucci, F.; De Rosa, G.; Calabrò, S.; Lambiase, C.; Di Francia, A. Chickpea Can Be a Valuable Local Produced Protein Feed for Organically Reared, Native Bulls. *Animals* **2021**, *11*, 2353. [CrossRef] [PubMed]
14. Matera, R.; Cotticelli, A.; Gómez Carpio, M.; Biffani, S.; Iannacone, F.; Salzano, A.; Neglia, G. Relationship among Production Traits, Somatic Cell Score and Temperature–Humidity Index in the Italian Mediterranean Buffalo. *Ital. J. Anim. Sci.* **2022**, *21*, 551–561. [CrossRef]
15. De Rosa, G.; Grasso, F.; Braghieri, A.; Bilancione, A.; Di Francia, A.; Napolitano, F. Behavior and Milk Production of Buffalo Cows as Affected by Housing System. *J. Dairy Sci.* **2009**, *92*, 907–912. [CrossRef] [PubMed]
16. Mota-Rojas, D.; Braghieri, A.; Álvarez-Macías, A.; Serrapica, F.; Ramírez-Bribiesca, E.; Cruz-Monterrosa, R.; Masucci, F.; Mora-Medina, P.; Napolitano, F. The Use of Draught Animals in Rural Labour. *Animals* **2021**, *11*, 2683. [CrossRef] [PubMed]

17. Nava-Trujillo, H.; Valeris-Chacin, R.; Morgado-Osorio, A.; Zambrano-Salas, S.; Tovar-Breto, L.; Quintero-Moreno, A. Reproductive Performance of Water Buffalo Cows: A Review of Affecting Factors. *J. Buffalo Sci.* **2020**, *9*, 133–151. [CrossRef]

18. Yáñez-Pizaña, A.; Tarazona-Morales, A.; Roldan-Santiago, P.; Ballesteros-Rodea, G.; Pineda-Reyes, R.; Orozco-Gregorio, H. Physiological and Behavioral Changes of Water Buffalo in Hot and Cold Systems: Review. *J. Buffalo Sci.* **2020**, *9*, 110–120. [CrossRef]

19. Pelosi, A.; Furcolo, P. An Amplification Model for the Regional Estimation of Extreme Rainfall within Orographic Areas in Campania Region (Italy). *Water* **2015**, *7*, 6877–6891. [CrossRef]

20. Chirico, G.B.; Pelosi, A.; De Michele, C.; Bolognesi, S.F.; D'Urso, G. Forecasting Potential Evapotranspiration by Combining Numerical Weather Predictions and Visible and Near-Infrared Satellite Images: An Application in Southern Italy. *J. Agric. Sci.* **2018**, *156*, 702–710. [CrossRef]

21. Uzun, P.; Masucci, F.; Serrapica, F.; Varricchio, M.L.; Pacelli, C.; Claps, S.; Di Francia, A. Use of Mycorrhizal Inoculum under Low Fertilizer Application: Effects on Forage Yield, Milk Production, and Energetic and Economic Efficiency. *J. Agric. Sci.* **2018**, *156*, 127–135. [CrossRef]

22. Bartocci, S.; Terramoccia, S.; Tripaldi, C. The Utilisation of a High Level Energy/Protein Diet for Lactating Mediterranean Buffaloes: Intake Capacity and Effects on Quanti–Qualitative Milk Parameters. *Livest. Sci.* **2006**, *99*, 211–219. [CrossRef]

23. Campanile, G.; De Filippo, C.; Di Palo, R.; Taccone, W.; Zicarelli, L. Influence of Dietary Protein on Urea Levels in Blood and Milk of Buffalo Cows. *Livest. Prod. Sci.* **1998**, *55*, 135–143. [CrossRef]

24. Altiero, V.; Moio, L.; Addeo, F. Previsione Della Resa in Mozzarella Sulla Base Del Contenuto in Grasso e Proteine Del Latte Di Bufala. *Sci. Tec. Latt.-Casearia* **1989**, *40*, 425–433.

25. Wagner, J.J.; Lusby, K.S.; Oltjen, J.W.; Rakestraw, J.; Wettemann, R.P.; Walters, L.E. Carcass Composition in Mature Hereford Cows: Estimation and Effect on Daily Metabolizable Energy Requirement During Winter. *J. Anim. Sci.* **1988**, *66*, 603–612. [CrossRef] [PubMed]

26. AOAC (Association of Official Analytical Chemists). *Official Methods of Analysis*, 17th ed.; AOAC International: Gaithersburg, MD, USA, 2002.

27. Van Soest, P.J.; Robertson, J.B.; Lewis, B.A. Methods for Dietary Fiber, Neutral Detergent Fiber, and Nonstarch Polysaccharides in Relation to Animal Nutrition. *J. Dairy Sci.* **1991**, *74*, 3583–3597. [CrossRef]

28. ISO (International Organization for Standardization). *Animal Feeding Stuffs. Determination of Starch Content–Polarimetric Method*; ISO: Geneva, Switzerland, 2000.

29. Sauvant, D.; Noziere, P. La quantification des principaux phénomènes digestifs chez les ruminants: Les relations utilisées pour rénover les systèmes d'unités d'alimentation énergétique et protéique. *INRA Prod. Anim.* **2013**, *26*, 327–346. [CrossRef]

30. Foley, A.E.; Hristov, A.N.; Melgar, A.; Ropp, J.K.; Etter, R.P.; Zaman, S.; Hunt, C.W.; Huber, K.; Price, W.J. Effect of Barley and Its Amylopectin Content on Ruminal Fermentation and Nitrogen Utilization in Lactating Dairy Cows. *J. Dairy Sci.* **2006**, *89*, 4321–4335. [CrossRef]

31. Vander Pol, M.; Hristov, A.N.; Zaman, S.; Delano, N. Peas Can Replace Soybean Meal and Corn Grain in Dairy Cow Diets. *J. Dairy Sci.* **2008**, *91*, 698–703. [CrossRef]

32. Masucci, F.; De Rosa, G.; Grasso, F.; Napolitano, F.; Esposito, G.; Di Francia, A. Performance and Immune Response of Buffalo Calves Supplemented with Probiotic. *Livest. Sci.* **2011**, *137*, 24–30. [CrossRef]

33. Sacchi, R.; Marrazzo, A.; Masucci, F.; Di Francia, A.; Serrapica, F.; Genovese, A. Effects of Inclusion of Fresh Forage in the Diet for Lactating Buffaloes on Volatile Organic Compounds of Milk and Mozzarella Cheese. *Molecules* **2020**, *25*, 1332. [CrossRef]

34. Uzun, P.; Masucci, F.; Serrapica, F.; Napolitano, F.; Braghieri, A.; Romano, R.; Manzo, N.; Esposito, G.; Di Francia, A. The Inclusion of Fresh Forage in the Lactating Buffalo Diet Affects Fatty Acid and Sensory Profile of Mozzarella Cheese. *J. Dairy Sci.* **2018**, *101*, 6752–6761. [CrossRef]

35. ISO (International Organization for Standardization). *Sensory Analysis–General Guidelines for the Selection, Training and Moni-Toring of Selected Assessors and Expert Sensory Assessor*; ISO: Geneva, Switzerland, 2012.

36. Braghieri, A.; Zotta, T.; Morone, G.; Piazzolla, N.; Majlesi, M.; Napolitano, F. Starter Cultures and Preservation Liquids Modulate Consumer Liking and Shelf Life of Mozzarella Cheese. *Int. Dairy J.* **2018**, *85*, 254–262. [CrossRef]

37. ISO (International Organization for Standardization). *Sensory Analysis—General Guidance for the Design of Test Rooms*; ISO: Geneva, Switzerland, 1998.

38. Cappelli, G.; Di Vuolo, G.; Gerini, O.; Noschese, R.; Bufano, F.; Capacchione, R.; Rosini, S.; Limone, A.; De Carlo, E. Italian Tracing System for Water Buffalo Milk and Processed Milk Products. *Animals* **2021**, *11*, 1737. [CrossRef] [PubMed]

39. Associazione Nazionale Allevatori Specie Bufalina (ANASB). Dati Relativi a Soci, Consistenze, Valutazioni Morfologiche, Aziende Visitate e Depositi DNA Effettuati Dalle Aziende Iscritte al Libro Genealogico, dal 2013 al 2020. Available online: https://www.Anasb.It/Statistiche/ (accessed on 30 May 2022). (In Italian).

40. Zicarelli, L. Current Trends in Buffalo Milk Production. *J. Buffalo Sci.* **2020**, *9*, 121–132. [CrossRef]

41. Parlato, E.; Zicarelli, L. Effect of Calving Interval on Milk Yield in Italian Buffalo Population. *J. Buffalo Sci.* **2016**, *5*, 18–22. [CrossRef]

42. Vecchio, D.; Neglia, G.; Rendina, M.; Marchiello, M.; Balestrieri, A.; Di Palo, R. Dietary Influence on Primiparous and Pluriparous Buffalo Fertility. *Ital. J. Anim. Sci.* **2007**, *6*, 512–514. [CrossRef]

43. Zicarelli, L. Reproductive Seasonality in Buffalo. *Reprod. Seas. Buffalo* **1997**, *4*, 29–52.

44. Catillo, G.; Moioli, B.; Napolitano, F. Estimation of Genetic Parameters of Some Productive and Reproductive Traits in Italian Buffalo. Genetic Evaluation with BLUP-Animal Model. *Asian-Australas. J. Anim. Sci.* **2001**, *14*, 747–753. [CrossRef]
45. Neglia, G.; Pacelli, C.; Zicarelli, G.; Sabia, E.; Tripaldi, C.; Terzano, G.M. Fertility Parameters in Lactating Mediterranean Buffaloes Fed Two Different Diets. *Ital. J. Anim. Sci.* **2007**, *6*, 607–610. [CrossRef]
46. Oba, M.; Allen, M.S. Evaluation of the Importance of the Digestibility of Neutral Detergent Fiber from Forage: Effects on Dry Matter Intake and Milk Yield of Dairy Cows. *J. Dairy Sci.* **1999**, *82*, 589–596. [CrossRef]
47. Kleefisch, M.-T.; Zebeli, Q.; Humer, E.; Kröger, I.; Ertl, P.; Klevenhusen, F. Effects of the Replacement of Concentrate and Fibre-Rich Hay by High-Quality Hay on Chewing, Rumination and Nutrient Digestibility in Non-Lactating Holstein Cows. *Arch. Anim. Nutr.* **2017**, *71*, 21–36. [CrossRef]
48. Souza, J.G.; Ribeiro, C.V.D.M.; Harvatine, K.J. Meta-Analysis of Rumination Behavior and Its Relationship with Milk and Milk Fat Production, Rumen pH, and Total-Tract Digestibility in Lactating Dairy Cows. *J. Dairy Sci.* **2022**, *105*, 188–200. [CrossRef] [PubMed]
49. Cavallini, D.; Mammi, L.M.E.; Biagi, G.; Fusaro, I.; Giammarco, M.; Formigoni, A.; Palmonari, A. Effects of 00-Rapeseed Meal Inclusion in Parmigiano Reggiano Hay-Based Ration on Dairy Cows' Production, Reticular pH and Fibre Digestibility. *Ital. J. Anim. Sci.* **2021**, *20*, 295–303. [CrossRef]
50. Chen, H.; Wang, C.; Huasai, S.; Chen, A. Effects of Dietary Forage to Concentrate Ratio on Nutrient Digestibility, Ruminal Fermentation and Rumen Bacterial Composition in Angus Cows. *Sci. Rep.* **2021**, *11*, 17023. [CrossRef]
51. Borreani, G.; Coppa, M.; Revello-Chion, A.; Comino, L.; Giaccone, D.; Ferlay, A.; Tabacco, E. Effect of different feeding strategies in intensive dairy farming system on milk fatty acid profile, and implications on feeding costs in Italy. *J. Dairy Sci.* **2013**, *96*, 6840–6855. [CrossRef] [PubMed]
52. Capozza, P.; Martella, V.; Lanave, G.; Catella, C.; Diakoudi, G.; Beikpour, F.; Camero, M.; Di Martino, B.; Fusco, G.; Balestrieri, A.; et al. An Outbreak of Neonatal Enteritis in Buffalo Calves Associated with Astrovirus. *J. Vet. Sci.* **2021**, *22*, e84. [CrossRef] [PubMed]
53. Upadhyay, R.C.; Singh, S.V.; Kumar, A.; Gupta, S.K. Ashutosh Impact of Climate Change on Milk Production of Murrah Buffaloes. *Ital. J. Anim. Sci.* **2007**, *6*, 1329–1332. [CrossRef]
54. Presicce, G. Reproduction in the Water Buffalo. *Reprod. Domest. Anim.* **2007**, *42*, 24–32. [CrossRef] [PubMed]
55. Neglia, G.; de Nicola, D.; Esposito, L.; Salzano, A.; D'Occhio, M.J.; Fatone, G. Reproductive Management in Buffalo by Artificial Insemination. *Theriogenology* **2020**, *150*, 166–172. [CrossRef] [PubMed]
56. Barile, V.L. Technologies Related with the Artificial Insemination in Buffalo. *J. Buffalo Sci.* **2012**, *1*, 139–146. [CrossRef]
57. Terzano, G.M.; Barile, V.L.; Borghese, A. Overview on Reproductive Endocrine Aspects in Buffalo. *J. Buffalo Sci.* **2012**, *1*, 126–138. [CrossRef]
58. Bertoni, A.; Napolitano, F.; Mota-Rojas, D.; Sabia, E.; Álvarez-Macías, A.; Mora-Medina, P.; Morales-Canela, A.; Berdugo-Gutiérrez, J.; Legarreta, I.G. Similarities and Differences between River Buffaloes and Cattle: Health, Physiological, Behavioral and Productivity Aspects. *J. Buffalo Sci.* **2020**, *9*, 92–109. [CrossRef]
59. Zicarelli, L. Influence of Environmental Temperature on Milk Production in the Italian Mediterranean Buffalo. *J. Buffalo Sci.* **2021**, *10*, 41–49. [CrossRef]
60. Stevens, A.; Shah, N.P. Textural and Melting Properties of Mozzarella Cheese Made with Fat Replacers. *Milchwissenschaft* **2002**, *57*, 387–390.
61. Martin, B.; Verdier-Metz, I.; Buchin, S.; Hurtaud, C.; Coulon, J.-B. How Do the Nature of Forages and Pasture Diversity Influence the Sensory Quality of Dairy Livestock Products? *Anim. Sci.* **2005**, *81*, 205–212. [CrossRef]
62. Kindstedt, P.; Carić, M.; Milanović, S. Pasta-Filata Cheeses. In *Cheese: Chemistry, Physics and Microbiology. Volume 2: Major Cheese Groups*; Fox, P.F., McSweeney, P.L.H., Cogan, T.M., Guinee, T.P., Eds.; Elsevier Academic Press: London, UK, 2004; pp. 251–277.
63. Feng, R.; Barjon, S.; van den Berg, F.W.J.; Lillevang, S.K.; Ahrné, L. Effect of Residence Time in the Cooker-Stretcher on Mozzarella Cheese Composition, Structure and Functionality. *J. Food Eng.* **2021**, *309*, 110690. [CrossRef]
64. Mizuno, R.; Abe, T.; Koishihara, H.; Okawa, T. The Effect of Preservative Liquid Composition on Physicochemical Properties of Mozzarella Cheese. *Food Sci. Technol. Res.* **2016**, *22*, 261–266. [CrossRef]
65. Vecchio, R.; Lombardi, A.; Cembalo, L.; Caracciolo, F.; Cicia, G.; Masucci, F.; Di Francia, A. Consumers' Willingness to Pay and Drivers of Motivation to Consume Omega-3 Enriched Mozzarella Cheese. *Br. Food J.* **2016**, *118*, 2404–2419. [CrossRef]
66. Cincotta, F.; Condurso, C.; Tripodi, G.; Merlino, M.; Prestia, O.; Stanton, C.; Verzera, A. Comparison of Lactose Free and Traditional Mozzarella Cheese during Shelf-Life by Aroma Compounds and Sensory Analysis. *LWT* **2021**, *140*, 110845. [CrossRef]
67. Nielsen, S.D.; Jansson, T.; Le, T.T.; Jensen, S.; Eggers, N.; Rauh, V.; Sundekilde, U.K.; Sørensen, J.; Andersen, H.J.; Bertram, H.C.; et al. Correlation between Sensory Properties and Peptides Derived from Hydrolysed-Lactose UHT Milk during Storage. *Int. Dairy J.* **2017**, *68*, 23–31. [CrossRef]
68. Alinovi, M.; Wiking, L.; Corredig, M.; Mucchetti, G. Effect of Frozen and Refrigerated Storage on Proteolysis and Physicochemical Properties of High-Moisture Citric Mozzarella Cheese. *J. Dairy Sci.* **2020**, *103*, 7775–7790. [CrossRef]

 *agriculture*

Article

Qualitative and Nutritional Evaluation of Paddlefish (*Polyodon spathula*) Meat Production

Daniel Simeanu [1], Răzvan-Mihail Radu-Rusu [2,*], Olimpia Smaranda Mintas [3,*] and Cristina Simeanu [2]

[1] Department of Control, Expertise and Services, Faculty of Food and Animal Sciences, "Ion Ionescu de la Brad" University of Life Sciences, 8 Mihail Sadoveanu Alley, 700489 Iasi, Romania

[2] Department of Animal Resources and Technology, Faculty of Food and Animal Sciences, "Ion Ionescu de la Brad" University of Life Sciences, 8 Mihail Sadoveanu Alley, 700489 Iasi, Romania

[3] Department of Livestock and Agrotourism, Faculty of Environmental Protection, University of Oradea, 1 University Street, 410087 Oradea, Romania

* Correspondence: radurazvan@uaiasi.ro (R.-M.R.-R.); olimpia.mintas@uoradea.ro (O.S.M.)

Abstract: *Polyodon spathula* is a valuable species of sturgeon native to North America that has acclimatized very well in Europe. Detailed knowledge of the quantitative and qualitative productive performance of paddlefish meat is of interest. Through this article, we aimed to highlight the chemical composition, cholesterol, and collagen content of fillets issued from paddlefish aged two and three summers and to highlight, as well as the nutritional value, the profile of fatty acids and amino acids, the sanogenic indices and the biological value of proteins for the epaxial and hypaxial muscle groups. The chemical analysis of the fillets by age indicated slightly higher values in summer three, compared to summer two: +5.32% dry matter, +0.89% protein, +41.21% fat, therefore +10.94% gross energy and for collagen by 2.94%; instead, for water, minerals and the W/P ratio the values were lower by 1.52%, 10.08%, and 2.29%. The nutritional assessment revealed that paddlefish has a meat with high PUFA content (approx. 22% of total fatty acids) and good values of sanogenic indices (Polyunsaturation Index = 7.01–8.77; Atherogenic Index = 0.57; Thrombogenic Index = 0.38–0.39; Hypocholesterolemic Fatty Acids = 33.01–41.34; Hypocholesteromic/Hypercolesteromic Fatty Acids ratio = 1.9). Also, the proteins of these fish are of good quality for young and adult consumers (EAA index = 156.11; Biological Value = 158.46; Nutritional Index (%) = 28.30) and good enough for children (Essential Amino Acids Index = 96.41; Biological Value = 93.39; Nutritional Index (%) = 17.45).

Keywords: paddlefish; meat quality; chemical composition; fatty acids; biological value

Citation: Simeanu, D.; Radu-Rusu, R.-M.; Mintas, O.S.; Simeanu, C. Qualitative and Nutritional Evaluation of Paddlefish (*Polyodon spathula*) Meat Production. *Agriculture* **2022**, *12*, 1965. https://doi.org/10.3390/agriculture12111965

Academic Editor: Javier Álvarez-Rodríguez

Received: 19 October 2022
Accepted: 18 November 2022
Published: 21 November 2022

Publisher's Note: MDPI stays neutral with regard to jurisdictional claims in published maps and institutional affiliations.

1. Introduction

Only one species of sturgeon from the *Polyodontidae* family is found in the waters of the North American continent, namely the *Polyodon spathula* or the commonly named paddlefish, spoon-billed cat, or spoonbill (Walbaum, 1972) [1]. These fish are distributed in the Mississippi River basin, from the Great Lakes area down to Florida. They can reach approx. 1.5–2 m body length and a body mass of about 50–70 kg upon maturity. Morphologically, *Polyodon spathula* is close to other sturgeons, with the exception of the head area where there is an elongated, dorsally-ventrally compressed rostrum that has the shape of a paddle. The length of this rostrum can reach 1/3 of the total body size in adult specimens. In this species, tactile and electrical receptors were found, deployed as a network on the operculum and the rostrum [2–6].

The sturgeon species *Polyodon spathula* usually does not have scales. However, small scales have been identified in the area of the lateral line, at the base of the dorsal fin and at the base of the pectoral fins [7,8]. The skin is smooth and its color varies from light gray to black on the dorsal area and on the flanks, and grayish-white on the abdomen. The paddle fish has poorly developed vision; the eyes are small and placed at the base of the rostrum [9,10]. The skeleton presents bone pieces only in the head area (jaws), while the rest is cartilaginous. This fish has a well-developed filtering apparatus, being a planktonophagous species. The mouth is large, located in a ventral position, and the digestive tube has a formation specific to the *Acipenseriform* family, namely the spiral valve [11–13]. Paddlefish consume planktonic organisms—especially zooplankton, aquatic insects and larvae [14–17]. They reach sexual maturity at the age of 7–8 years in males and 10–13 years in females [18,19]. The eggs are small, with a diameter of 2.5–3 mm, and females lay about 100,000 eggs per reproduction season [20–22].

Reproduction takes place in spring, when the water temperature reaches 11–14 °C. Fish migrate even hundreds of kilometers in the upper part of the river or on its affluents [23–26]. Young sturgeons grow very fast in the first year of life, even during the winter period [11,27], after which the development becomes slower. The growth rate is around 5 cm body length/year throughout the first five years of life. Growth of body mass increases after this age, becoming double or even triples at 10 years old [28–31].

The freshwater sturgeon species, *Polyodon spathula*, has been important and acclimatized in Eastern Europe since 1992. At the moment, in Romania, thanks to the studies carried out by researchers from the Research and Development Station for Fisheries in Nucet and from the Research and Development for Aquaculture and Aquatic Ecology, Ciurea, Iași, it was facilitated the development of polyculture technologies, along with other species, such as the common carp and Asian cyprinids and also techniques for reproduction, pre-development and growth of young fish in protected areas, throughout the first year of life [32,33]. The success achieved in the research facilities also encouraged farmers to turn to this species of sturgeon for rearing for meat production. Thus, the qualitative meat production of these sturgeons, in general, but especially those obtained outside the usual breeding area—North America, must be studied in particular. *Polyodon spathula* meat has a taste and texture similar to that of other sturgeons, firm, white in color and boneless [34–37]. The carcasses of paddlefish have a "bullet" shape (no head, visceral mass and no fins) and represent 57% of the fish body. If the processing is carried out in the form of a fillet, the dressed yield decreases to 27% after removal of the skin and red meat [38,39]. Sturgeons' meat has been known and accepted since the beginning of the colonization of the North American continent in the late 1800s, but the meat of *Polyodon spathula* was less accepted. Currently, this meat is marketed under the title of "boneless catfish", to associate this meat with a popular product in the southern United States of America. As a result, the paddlefish meat still remains unknown to many consumers, and the market is very limited [40]. Meat from *Polyodon spathula* contains small amount of lipids, 1 to 4.5%, depending on the body mass and age of the individuals [38]. Concerning total protein content, the flank muscles (fillet) contain between 18 and 20% protein, placing this fish in the class of fish with a high protein content [41–43]. The meat has a stable shelf life since it can be kept refrigerated for up to seven days or even up to seven months under freezing conditions [38].

There are some issues when approaching the assessment of paddlefish meat production, both quantitatively and as quality, due to the fact that this species is reared mostly in polyculture in Romania and in small flocks, there are no studies on consumers' acceptability level and preferences. Also, the technological aspects at the fishery farm level are more difficult to manage, in comparison with other sturgeon species (water surface covering with nests to protect the paddlefish against its predators—waterfowl, difficulties to provide supplemental feeding to other species in polyculture), and farmers prefer to not choose paddlefish when populate the ponds, despite its natural good potential for good feed conversion weight gain using mostly natural feed sources (mostly zooplankton).

Within the presented context, our findings contribute to the most detailed knowledge of the *Polyodon spathula* sturgeon meat and to the completion of the scientific literature with original data on the most detailed on the chemical composition of the fillet, of the main muscles groups, and on the biological value of the proteins issued from paddlefish aged second and third summers raised in a fish farm in Romania.

2. Materials and Methods

2.1. Biological Material

A total of 500 sturgeons, *Polyodon spathula* species (300 individuals aged 2 Summers and 200 individuals aged 3 Summers) were used as main study populations. Fish were farmed to be sold on market in Hudesti Fishery, Botosani County, located in the North-East of Romania, at these geo coordinates: lat. N 48.1212, long. E 26.5519. The sturgeons were reared throughout the 2010–2012 timespan in a pool with a surface of 30 hectares, in polyculture with certain indigenous and Asian cyprinid species: *Cyprinus carpio* (indigenos carp, 39% of the fish in the pond), *Hypophtalmichthys molitrix* (Asian silver carp, 30% of the fish in the pond), *Ctenopharingodon idella* (Asian grass carp, 18% of the fish in the pond), *Aristichthys nobilis* (Asian bighead carp, 12% of the fish in the pond).

No supplemental feeding was designed and provided for paddlefish, knowing the species fed mostly through filtration and its natural feed source consists in zooplankton (and mostly in cladocerans). Concentrated feed was provided for other fish species in the pond and secondarily, it served to better develop the zooplankton eaten by paddlefish. The diets supplementarily fed to cyprinids consisted in a mixture of soybean meal, sunflower meal, corn, oat, and wheat flours and the proportions were optimized to achieve different nutritional levels (Table 1) in accordance with season and pond richness in natural resources.

Table 1. Nutritional features of the supplemental feed provided in fishery, throughout May-October, 2010–2012.

Month	Metabolizable Energy kcal/kg	Crude Protein %	Lysine %	Methionine + Cystine %	Crude Fat %	Crude Fiber %	Ca %	P %
May	2645	18.06	0.69	0.74	1.56	10.83	0.14	0.72
June	2617	20.53	0.76	0.83	1.33	11.88	0.16	0.82
July	2603	21.74	0.83	0.86	1.22	12.39	0.17	0.87
August	2600	22.00	0.84	0.89	1.20	12.50	0.18	0.88
September	2831	22.94	0.87	0.90	1.03	9.49	1.17	0.82
October	2893	26.34	1.15	0.99	1.16	9.92	0.19	0.84

In order to better depict the environmental conditions provided to policultural rearing of the paddlefish, Table 2 displays the water quality traits (average values) throughout the warm seasons of 2010–2012.

Table 2. Water quality traits in the paddlefish rearing pond.

Water Trait	May	June	July	August	September
Colour	Green brown	Green	Green	Yellow green	Yellow green
Transparency (cm)	24.58	19.67	18.34	16.21	17.76
Temperature (°C)	18.86	20.73	23.84	25.72	24.65
Oxygen (mg/L)	7.35	6.65	4.47	5.19	5.88
pH value (pHU)	7.39	7.23	7.17	6.51	6.75
Calcium (mg/L)	52.54	60.78	56.76	46.94	40.75
Magnesium (mg/L)	25.42	28.31	19.78	18.10	18.34
Chlorides (mg/L)	17.98	12.45	11.78	13.12	15.32
Nitrates (mg/L)	0.32	0.24	0.14	0.16	0.16
Nitrites (mg/L)	0.32	0.27	0.26	0.18	0.16
Phosphates (mg/L)	0.02	0.02	0.03	0.03	0.03
Organic matters (mg/L)	68.7	71.6	79.8	98.5	59.6

Out of the 500 sturgeons in the pond, upon harvesting for selling them on the market, 10% were weighted and 20 fishes per each age (summer two and summer three) were chosen for meat sampling after refrigeration and analytical laboratory investigations. No animals were used for applying experimental factors and to measure their effect throughout certain specific reasoning criteria. The specimens used for meat sampling were raised in common with the other marketable fishes and the samples were taken after the exitus was installed in refrigeration marketing condition.

From the specimens chosen for analysis, samples were collected from in the form of fillets (50% of fish) and from the groups of lateral muscles (50% of fish). Each lateral muscle is divided along its length into two masses, one dorsal (epaxial) and one ventral (hypaxial), by a connective wall (transverse myoseptum) fixed to the integument and the axial skeleton and located slightly below the lateral line. The muscles located above the lateral line are called epaxial muscles and are divided into dorsal epaxial muscles and costal epaxial muscles, while those located below the lateral line are called hypaxial muscles and are divided into costal hypaxial muscles and abdominal hypaxial muscles [1] (Figure 1).

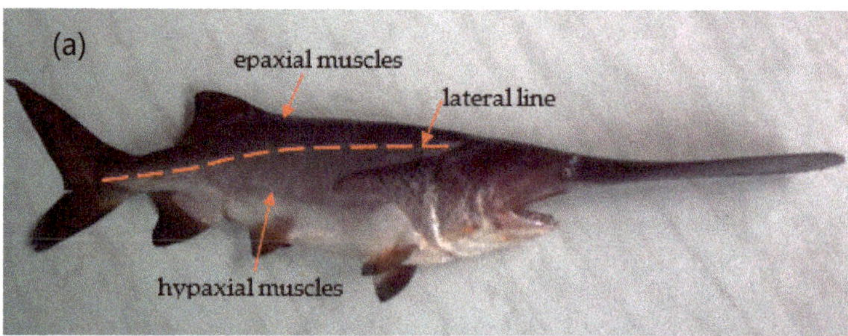

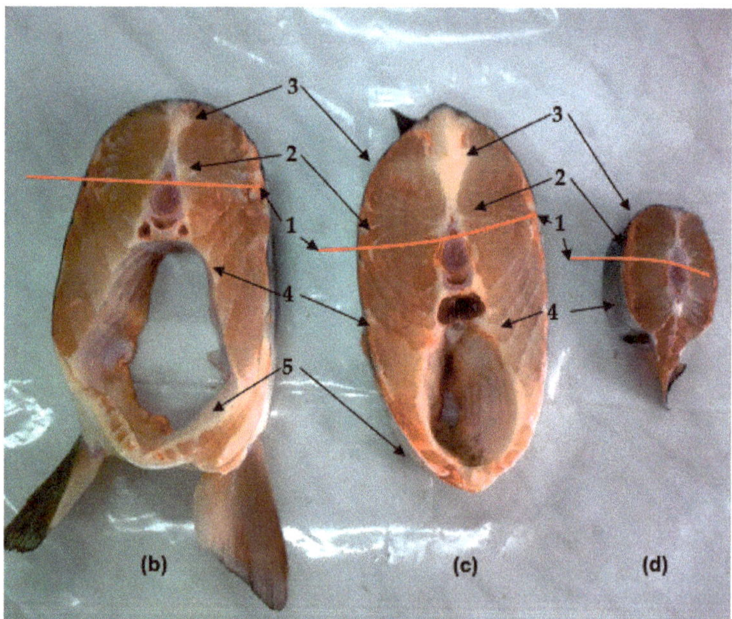

Figure 1. Side muscles in *Polyodon spathula*: (**a**) topographic placement; (**b**,**c**)—cross sections through the abdominal area; (**d**)—cross section through the caudally area; 1—side line; 2—costal epaxial muscles; 3—dorsal epaxial muscles; 4—costal hypaxial muscles; 5—abdominal hypaxial muscles.

2.2. Assessment of Proximate Chemical Composition and Energetic Value

Water was assessed by oven dehydration method (drying of sample at +105 °C temperature) following the SR ISO 1442/1997 analytical protocol.

Total minerals were measured gravimetrically by calcinations method in electric furnace, using a working temperature of +550 °C (SR ISO 936:1998 standard).

Protein was assessed via Kjeldhal method, carried on in accordance with the 981:10; AOAC Official methods of analysis/1990, compatible with SR ISO 937:2007, adapted to Velp Scientifica analytical devices (DK6 digester; UDK7 distiller) (manufacturer VELP Scientifica SRL, Usmate Velate, Italy) [43–46].

Water/protein ratio (W/P) was calculated by reporting the water analytical content of each sample to the crude protein content of each sample.

Total lipids were measured on the Velp Scientifica–SER 148 extractor (manufacturer VELP Scientifica SRL, Usmate Velate, Italy), using the Soxhlet method (AOAC Officinal methods of analysis/1990 [43–46]), compatible with the SR ISO 1443:2008 protocols [43,47].

Nitrogen free extract (remains of carbohydrates not analyzed individually) was calculated algebraically, by difference between the total solids (dry matter) content, total minerals content, total proteins content, and total lipids content.

Meat gross energy (GE) content was derived through mathematical computation, using the raw calories content of one grams of each organic matter component when burnt into a calorimeter, in accordance with the relation (1) [42,43]:

$$GE\ (kcal/100\ g) = g\ TP \times 5.70\ kcal + g\ TL \times 9.50\ kcal + g\ NFE \times 4.2\ kcal \tag{1}$$

where: GE = gross energy, TP = total proteins, TL = total lipids, NFE = nitrogen free extract.

Omega Bruins Food-Check Near InfraRed (NIR) spectrophotometer (manufacturer Bruins Instruments GmbH, Puchheim, Germany) was used to run rapid scanning transmission tests for collagen content assessment [42,43].

Each sample was analyzed in 30 repetitions, for proximate composition, collagen content and 30 repetitions have been carried out for W/P ratio and GE calculations.

2.3. Fatty Acids Profile Analysis and Lipids Nutritional Quality

Fatty acid methyl esters (FAME) in paddlefish meat were extracted and quantified by gas chromatography and mass spectrometry detection, using the Perkin Elmer Chromatographic system coupled with mass spectrometer detector (GC-MS) (Clarus 680 gas-chromatograph and Clarus SQ8T quadrupole mass spectrometer (manufacturer, Perkin Elmer Inc., Boston, MA, USA, for both devices). An elite-wax with stationary polar phase Polyethylene glycol (PEG) chromatographic column was used (total length 30 m, 1.0 μm film thickness and internal diameter of 0.25 mm, injecting port temp. = 220 °C, injected sample = 1.0 μL, He carrier flew at 1.5 mL/min rate, while used splitting ratio was 40:1). Temperature gradient was set at 100 °C, throughout 2 min, standstill, and stationary 1 min at 250 °C. Mass Spectrometer was characterized by the following operational values: 150 °C the temperature of the transfer line; 150 °C the temperature of the source; 1500 multiplier; solvent delay of 0–1.5 min for the solvent [48–51].

The method uses the chromatographic isolation of the fatty acid melange in samples, after esterifying them with $CHNaO_2$ (sodium formate or methanoate), on a capillary column, and fragmenting the molecules by ionization. Identification and measuring of FAME quantities are carried on using target and qualifying ions. The FAME in *Polyodon spathula* meat issued from prior lipid saponification phase, followed by boron trifluoride (15% vol.) catalyze on the esterification reaction. Fatty acids values in the samples were achieved by running the comparison between FAME retention time with a homologated standard (FAME Supelco 37 Mix). Each fatty acid was quantified as g FAME/100 g of total identified FAME [48–51].

Seven analytical repetitions were carried out per sample to quantify the fatty acids and cholesterol content in paddlefish meat.

To outline the lipid profile, the fatty acids were grouped accordingly: saturated fatty acids–SFA, as sum (Σ SFA = C10:0 + C12:0 + C14:0 + C15:0 + C16:0 + C18:0 + C20:0 + C22:0); mono unsaturated fatty acids–MUFA, as sum (Σ MUFA = C16:1 + C18:1 *cis*−9 + C20:1 n−9 + C22:1 n−9); poly unsaturated fatty acids, as sum ((Σ PUFA = C18:2 n−6 + C20:2 n−6 + C18:3 n−3 + C20:3 n−3 + C20:5 n−3 + C18:3 n−6), total unsaturated fatty acids, as sum of MUFA and PUFA [49,50,52].

Quantities of Omega-3 and Omega-6 PUFA series were expressed as ration (n−3/n−6). Paddlefish meat polyunsaturation index (PI) was calculated by using the relation (2), published by Timmons [49,52,53]:

$$PI = C18{:}2\ n{-}6 + (C18{:}3\ n{-}3 \times 2). \tag{2}$$

The atherogenic index (AI) and thrombogenic index (TI) of fats were calculated, using the data issued from FAME GC analysis of the paddlefish meat, applying the relations (3) and (4) published by Ulbricht and Southgate [48,52,54,55]:

$$AI = (C12{:}0 + C16{:}0 + 4 \times C14{:}0)/[\Sigma\ MUFA + \Sigma\ (n{-}6) + \Sigma\ (n{-}3)], \tag{3}$$

$$TI = (C14{:}0 + C16{:}0 + C18{:}0)/[0.5 \times \Sigma\ MUFA + 0.5 \times \Sigma\ (n{-}6) + 3 \times \Sigma\ (n{-}3) + \Sigma(n{-}3)/\Sigma(n{-}6)]. \tag{4}$$

The relation (5) proposed by Fernandez et al. [55] was used to calculate the ratio between fatty acids with hypocholesterolemic (h) and hypercholesterolemic (H) effect.

$$h/H\ (hypocholesterolemic/Hypercholesterolemic) = (C18{:}1 + PUFA)/(C12{:}0 + C14{:}0 + C16{:}0). \tag{5}$$

2.4. Amino Acid Analysis and Protein Quality Assessment

Amino acids in paddlefish meat were identified and quantifies using the Model 118/119 CL, CR Mikrotechna AAA 881 automatic analyzer (manufacturer Mikrotechna, Prague, Czech Republic). Samples were hydrolyzed in HCl 6 M at 110 °C, continuously throughout 24 h, using nitrogen atmosphere. Cysteine and methionine were assessed in form of cysteic acid and methionine sulfone, following sample oxidation with performic acid. Tryptophan was assessed after NaOH hydrolysis at 110 °C, throughout 22 h, in accordance with the protocol specified by Official Methods of Analysis of the Association of Analytical Chemists [44,46]. Amino acid quantification was expressed in the g/16 g N system, equivalated to g/100 g of total protein [43,56–58].

Seven analytical repetitions were carried out per sample to quantify the Amino Acids.

Quality of the proteins was assessed by methods using parameters such as: inner content in essential amino acids (AA), fraction of amino acids of exogenous origin (EAA), protein chemical score (CS) (relation 6) and the essential amino acid index (EAAI) (relation 7) [43,48,58,59].

$$CS = \frac{content\ in\ amino\ acid\ A\ in\ the\ studied\ protein}{content\ in\ amino\ acid\ A\ the\ standard\ protein} \times 100 \tag{6}$$

Usually, egg protein is used as standard or etalon protein, in accordance with certain reference international scientific bodies: Food and Agriculture Organization of the United Nation (FAO)/World Health Organization (WHO)/United Nations University (UNU). Nutritional value of proteins was computed, in this research, according to the standard requirements:

- Standard 1—children consumers: Tryptophan = 1.7; Threonine = 4.3; Isoleucine = 4.6; Leucine = 9.3; Lysine = 6.6; Methionine + Cysteine = 4.2; Phenylalanine + Tyrosine = 7.2; Valine = 5.5; EAA = 37.9 g/16 g Nitrogen) [56–58,60]
- Standard 2—youth consumers: Tryptophan = 1.0; Threonine = 4.0; Isoleucine = 4.0; Leucine = 7.0; Lysine = 5.5; Methionine + Cysteine = 3.5; Phenylalanine + Tyrosine = 6; Valine = 5; EAA = 36 g/16 g Nitrogen) [56–58,61]
- Standard 3—adult consumers: Tryptophan = 0.6; Threonine = 2.6; Isoleucine = 3; Leucine = 4.4; Lysine = 3.1; Methionine + Cysteine = 2.7; Phenylalanine + Tyrosine = 3.3; Valine = 2.3; EAA = 22 g/16 g Nitrogen) [56–58,61]

The EAAI (essential amino acid index) was calculated using relation (7) and the values the chemical indexes for the eight essential amino acids, in accordance to Oser's methodology [43,48,58,62]:

$$EAAI = \sqrt[n]{CS1 \times CS2 \times CS3 \times \ldots \times CSn} \tag{7}$$

Oser's [48,58,63] mathematical relation (8) was applied to compute the proteins Biological Value (BV):

$$BV = 1.09\,(EAAI) - 11.7. \tag{8}$$

The mathematical relation (9) proposed by Crisan and Sands, 1978 [43,48,58,64] was used to compute the nutritional index (NI) of paddlefish meat:

$$NI\,(\%) = \frac{EAAI \times \%\ \text{protein}}{100}. \tag{9}$$

2.5. Data Analysis

The analytical data were input in a Microsoft Excel 2016 database then processed to calculate main statistic descriptors (mean, variance, standard deviation, standard error of mean, coefficient of variation), and analysis of variance—one-way ANOVA, followed by post hoc Tukey testing—using the GraphPad Prism 9.4.0 (673) software (manufacturer, GraphPad Software, San Diego, CA, USA).

Statistics outputs are based on 30 analytical repetitions in proximate content, W/P ratio, collagen and gross energy and on 7 analytical repetitions for fatty acids, cholesterol and amino acids contents.

3. Results

3.1. Chemical Proximate Composition

Results of the proximate chemical analysis of *Polyodon spathula* sturgeon fillets from the 2nd and 3rd summer are displayed in Table 3, along with data on the gross energy value, collagen content and water/protein ratio.

Table 3. Chemical composition and certain nutritional features of *Polyodon spathula* fillet, aged two and three summers.

Age	Statistics	Water	Dry Matter	Protein	Total Fat	Minerals	Energy	Collagen	W/P Ratio
S2	Mean ± SD	77.83 ± 0.27	22.17 ± 0.27	17.94 ± 0.18	2.82 ± 0.14	1.41 ± 0.30	102.05 ± 0.81	3.93 ± 0.06	4.34 ± 0.05
	CV	0.35	1.21	0.99	5.02	21.13	0.79	1.61	0.09
S3	Mean ± SD	76.65 ± 0.25	23.35 ± 0.25	18.10 ± 0.13	3.98 ± 0.12	1.27 ± 0.34	113.22 ± 1.32	4.04 ± 0.08	4.24 ± 0.03
	CV	0.33	1.07	0.74	3.12	26.82	1.17	1.86	0.79
Comparisons	±% S3 vs. S2	−1.52%	+5.32%	+0.89%	+41.21%	−10.08%	+10.94%	+2.94%	−2.29%
	p values	2.53×10^{-10}	2.53×10^{-10}	0.9928	2.53×10^{-10}	0.9979	2.53×10^{-10}	0.9998	0.9998

S2—2nd summer. S3—3rd summer SD—standard deviation. CV—coefficient of variation. $n = 30$.

Same parameters were analysed separately per muscular groups (epaxial dorsal, epaxial costal, hypaxial costal and hypaxial abdominal) on samples taken from paddlefish aged two summers (Table 4).

Chemical analysis results from the same muscle groups issued from *Polyodon spathula* sturgeons aged three summers are presented in Table 5.

Achieved data was compared between the two investigated ages, both as relative differences (± %) and as analysis of variance (Table 6).

3.2. Fatty Acids Profiling and the Sanogenic Indices of the Polyodon Spathula Sturgeons Meat

Profile of fatty acids and the sanogenic indices assessment are presented in Table 7. The analysis was carried on for each age and on epaxial and hypaxial muscles groups.

Table 4. Chemical composition and certain nutritional features of *Polyodon spathula* meat, aged two summers.

Muscles	Descriptive Statistics	Water (g/100 g)	Dry Matter (g/100 g)	Protein (g/100 g)	Total Fat (g/100 g)	Minerals (g/100 g)	Energy (Kcal/100 g)	Collagen (g/100 g)	W/P Ratio
ED	Mean ± SD	77.17 ± 0.46	22.83 ± 0.46	18.17 ± 0.30	3.29 ± 0.22	1.37 ± 0.41	107.28 ± 1.56	3.92 ± 0.12	4.25 ± 0.09
	CV	0.59	2.00	1.65	6.55	30.20	1.46	3.99	2.07
EC	Mean ± SD	78.36 ± 0.62	21.64 ± 0.62	18.06 ± 0.29	2.19 ± 0.16	1.39 ± 0.62	96.88 ± 1.44	3.79 ± 0.12	4.34 ± 0.08
	CV	0.80	2.16	1.63	7.10	45.02	1.48	3.18	1.86
HC	Mean ± SD	77.96 ± 0.37	22.04 ± 0.37	17.59 ± 0.27	2.98 ± 0.20	1.47 ± 0.35	101.99 ± 2.04	4.08 ± 0.05	4.43 ± 0.08
	CV	0.48	1.70	1.56	6.62	33.46	2.00	1.33	1.79
HA	Mean ± SD	44.05 ± 0.42	55.95 ± 0.42	16.77 ± 0.31	38.15 ± 0.42	1.03 ± 0.68	415.71 ± 4.26	4.26 ± 0.08	2.63 ± 0.05
	CV	0.95	0.75	1.84	1.11	65.58	1.02	1.81	1.94
p values (ANOVA)	ED vs. EC	4.10×10^{-14}	4.10×10^{-14}	3.63×10^{-11}	2.08×10^{-15}	0.9989	2.41×10^{-15}	3.74×10^{-6}	6.80×10^{-5}
	ED vs. HC	2.21×10^{-8}	2.21×10^{-8}	2.15×10^{-15}	0.0001	0.8843	9.08×10^{-12}	9.36×10^{-9}	4.45×10^{-14}
	ED vs. HA	2.15×10^{-15}	2.15×10^{-15}	4.92×10^{-8}	2.16×10^{-15}	0.0968	2.29×10^{-15}	2.27×10^{-15}	2.43×10^{-15}
	EC vs. HC	0.0079	0.0079	2.15×10^{-15}	1.74×10^{-14}	0.9360	3.90×10^{-11}	9.05×10^{-15}	2.37×10^{-5}
	EC vs. HA	2.31×10^{-15}	2.31×10^{-15}	2.37×10^{-15}	2.51×10^{-15}	0.0690	2.38×10^{-15}	2.46×10^{-15}	2.05×10^{-15}
	EC vs. HA	2.19×10^{-15}	2.19×10^{-15}	0.9935	2.29×10^{-15}	0.0142	2.19×10^{-15}	6.96×10^{-11}	2.61×10^{-15}

ED—Epaxial dorsal. EC—Epaxial costal. HC—hypaxial costal. HA—Hypaxial abdominal. SD—standard deviation. CV—coefficient of variation. W/P—Water/Protein ratio. $n = 30$.

Table 5. Chemical composition and certain nutritional features of *Polyodon spathula* meat, aged three summers.

Muscles	Descriptive Statistics	Water (g/100 g)	Dry Matter (g/100 g)	Protein (g/100 g)	Total Fat (g/100 g)	Minerals (g/100 g)	Energy (Kcal/100 g)	Collagen (g/100 g)	W/P Ratio
ED	Mean ± SD	76.62 ± 0.50	23.38 ± 0.50	18.80 ± 0.34	3.37 ± 0.22	1.22 ± 0.59	110.64 ± 2.37	4.03 ± 0.09	4.08 ± 0.07
	CV	0.66	2.16	1.79	6.48	48.69	2.14	2.29	1.81
EC	Mean ± SD	76.47 ± 0.48	23.53 ± 0.48	18.19 ± 0.28	4.09 ± 0.20	1.25 ± 0.59	114.56 ± 2.00	3.96 ± 0.14	4.21 ± 0.08
	CV	0.63	2.05	1.54	4.78	46.95	1.74	3.61	1.82
HC	Mean ± SD	76.85 ± 0.43	23.15 ± 0.43	17.32 ± 0.24	4.49 ± 0.22	1.34 ± 0.59	114.44 ± 2.29	4.14 ± 0.07	4.44 ± 0.06
	CV	0.86	1.86	1.36	5.01	44.26	2.00	1.58	1.42
HA	Mean ± SD	43.92 ± 0.37	56.09 ± 0.37	16.77 ± 0.31	38.22 ± 0.69	1.61 ± 0.70	414.16 ± 6.31	4.32 ± 0.05	2.70 ± 0.04
	CV	0.85	0.66	1.84	1.82	43.55	1.52	1.13	1.48
p values (ANOVA)	ED vs. EC	2.09×10^{-15}	2.09×10^{-15}	1.97×10^{-13}	5.98×10^{-10}	0.9968	0.0004	0.0172	3.92×10^{-11}
	ED vs. HC	2.84×10^{-14}	2.84×10^{-14}	2.61×10^{-15}	2.36×10^{-14}	0.8653	0.0006	3.85×10^{-5}	2.33×10^{-15}
	ED vs. HA	2.28×10^{-15}	2.28×10^{-15}	2.42×10^{-15}	2.03×10^{-15}	0.0721	2.08×10^{-15}	7.12×10^{-15}	2.11×10^{-15}
	EC vs. HC	0.0083	0.0083	4.16×10^{-15}	0.0007	0.9410	0.9992	2.67×10^{-11}	2.41×10^{-15}
	EC vs. HA	2.11×10^{-15}	2.11×10^{-15}	2.63×10^{-15}	2.24×10^{-15}	0.1156	2.17×10^{-15}	2.17×10^{-15}	2.58×10^{-15}
	EC vs. HA	2.25×10^{-15}	2.25×10^{-15}	2.17×10^{-15}	5.98×10^{-10}	0.3375	2.34×10^{-15}	2.76×10^{-10}	2.36×10^{-15}

ED—Epaxial dorsal. EC—Epaxial costal. HC—hypaxial costal. HA—Hypaxial abdominal. SD—standard deviation. CV—coefficient of variation. W/P—Water/Protein ratio. $n = 30$.

Table 6. Analysis of variance between ages (two summers vs. three summers) on the chemical composition and certain nutritional features of *Polyodon spathula* meat.

Muscles	Percent Differences	Water	Dry Matter	Protein	Total Fat	Minerals	Energy	Collagen	W/P Ratio
ED	±% S3 vs. S2	−0.71%	+2.40%	+3.43%	−2.34%	−11.12%	+3.13%	+2.86%	−4.01%
	p values	0.2701	0.2701	0.1017	0.9999	0.9999	2.53×10^{-10}	0.9999	0.9999
EC	±% S3 vs. S2	−2.41%	+8.72%	+0.70%	+86.76%	−10.09%	+18.25%	+4.49%	−3.09%
	p values	1.0167	1.0167	0.9999	2.53×10^{-10}	0.9999	2.53×10^{-10}	0.9999	0.9999
HC	±% S3 vs. S2	−1.41%	+5.00%	−1.55%	+50.67%	−9.10%	+12.21%	+1.57%	+0.13%
	p values	1.0329	1.0329	0.9968	2.86×10^{-10}	0.9999	2.53×10^{-10}	0.9998	1.0064
HA	±% S3 vs. S2	−0.32%	+0.25%	−3.04%	+0.18%	+56.27%	−0.37%	+1.57%	+2.80%
	p values	1.0428	1.0428	0.9998	1.2613	0.9991	0.1007	0.9999	0.9999

ED—Epaxial dorsal. EC—Epaxial costal. HC—hypaxial costal. HA—Hypaxial abdominal. S2—2nd summer. S3—3rd summer.

Table 7. Profile of fatty acids (g/100 g total FAME), cholesterol content (mg/100 g) and sanogenic indices of *Polyodon spathula* sturgeons meat, aged two and three summers.

Fatty Acid		Descriptive Statistics	Summer 2		Summer 3		*p* Values
			Epaxial Muscles	Hypaxial Muscles	Epaxial Muscles	Hypaxial Muscles	
Myristic	C14:0	Mean ± SD CV	2.02 ± 0.06 2.81	2.16 ± 0.07 3.44	2.44 ± 0.07 2.76	2.53 ± 0.06 2.55	ES3 vs. ES2, $p = 1.81 \times 10^{-14}$ HS3 vs. HS2, $p = 1.62 \times 10^{-13}$
Pentadecanoic	C15:0	Mean ± SD CV	0.40 ± 0.01 3.14	0.43 ± 0.01 2.86	0.49 ± 0.01 2.94	0.50 ± 0.02 4.27	ES3 vs. ES2, $p = 1.57 \times 10^{-9}$ HS3 vs. HS2, $p = 1.74 \times 10^{-7}$
Palmitic	C16:0	Mean ± SD CV	15.31 ± 0.21 1.35	16.38 ± 0.57 3.51	18.53 ± 0.59 3.18	19.18 ± 0.61 3.19	ES3 vs. ES2, $p = 1.83 \times 10^{-14}$ HS3 vs. HS2, $p = 1.79 \times 10^{-14}$
Palmitoleic	C16:1	Mean ± SD CV	8.10 ± 0.26 3.27	8.67 ± 0.25 2.91	9.80 ± 0.36 3.69	10.14 ± 0.31 3.10	ES3 vs. ES2, $p = 1.80 \times 10^{-14}$ HS3 vs. HS2, $p = 1.82 \times 10^{-14}$
Margaric	C17:0	Mean ± SD CV	0.67 ± 0.03 4.36	0.71 ± 0.03 4.08	0.80 ± 0.03 3.14	0.83 ± 0.02 2.42	ES3 vs. ES2, $p = 7.56 \times 10^{-13}$ HS3 vs. HS2, $p = 4.17 \times 10^{-12}$
Stearic	C18:0	Mean ± SD CV	1.45 ± 0.04 2.84	1.55 ± 0.05 3.52	1.76 ± 0.05 2.72	1.82 ± 0.06 3.52	ES3 vs. ES2, $p = 1.76 \times 10^{-14}$ HS3 vs. HS2, $p = 1.80 \times 10^{-14}$
Oleic	C18:1 *cis*–9	Mean ± SD CV	15.18 ± 0.55 3.61	16.24 ± 0.36 2.19	18.36 ± 0.82 4.44	19.01 ± 0.80 4.21	ES3 vs. ES2, $p = 1.81 \times 10^{-14}$ HS3 vs. HS2, $p = 1.79 \times 10^{-14}$
Asclepic	C18:1 *cis*–11	Mean ± SD CV	4.30 ± 0.12 2.72	4.60 ± 0.20 4.27	5.21 ± 0.15 2.90	5.39 ± 0.16 2.91	ES3 vs. ES2, $p = 1.80 \times 10^{-14}$ HS3 vs. HS2, $p = 1.82 \times 10^{-14}$
Linoleic	C18:2, n–6	Mean ± SD CV	1.71 ± 0.06 3.39	1.83 ± 0.05 2.62	2.07 ± 0.06 2.74	2.14 ± 0.06 2.82	ES3 vs. ES2, $p = 1.78 \times 10^{-14}$ HS3 vs. HS2, $p = 1.82 \times 10^{-14}$
α-linolenic	C18:3, n–3	Mean ± SD CV	2.65 ± 0.11 4.08	2.83 ± 0.09 3.13	3.20 ± 0.12 3.80	3.32 ± 0.11 3.44	ES3 vs. ES2, $p = 1.81 \times 10^{-14}$ HS3 vs. HS2, $p = 1.84 \times 10^{-14}$
γ-linolenic	C18:3, n–6	Mean ± SD CV	0.42 ± 0.01 3.54	0.45 ± 0.02 4.01	0.51 ± 0.02 3.24	0.53 ± 0.02 3.19	ES3 vs. ES2, $p = 1.57 \times 10^{-9}$ HS3 vs. HS2, $p = 1.52 \times 10^{-8}$
Arahic	C20:0	Mean ± SD CV	0.09 ± 0.01 5.99	0.09 ± 0.01 5.82	0.10 ± 0.01 5.28	0.11 ± 0.01 4.92	ES3 vs. ES2, $p = 0.6656$ HS3 vs. HS2, $p = 0.1282$
Gadoleic	C20:1	Mean ± SD CV	0.47 ± 0.01 2.76	0.50 ± 0.01 2.66	0.57 ± 0.03 4.80	0.59 ± 0.02 4.16	ES3 vs. ES2, $p = 1.90 \times 10^{-10}$ HS3 vs. HS2, $p = 1.57 \times 10^{-8}$
Eicosadienoic	C20:2	Mean ± SD CV	0.29 ± 0.01 3.24	0.31 ± 0.01 2.37	0.35 ± 0.01 2.85	0.36 ± 0.01 3.22	ES3 vs. ES2, $p = 2.45 \times 10^{-6}$ HS3 vs. HS2, $p = 3.74 \times 10^{-5}$
Dihomo-γ-linolenic	C20:3, n–6	Mean ± SD CV	0.12 ± 0.01 4.61	0.13 ± 0.01 4.15	0.15 ± 0.01 4.22	0.15 ± 0.01 4.61	ES3 vs. ES2, $p = 0.0108$ HS3 vs. HS2, $p = 0.1282$
Eicosatrienoic	C20:3, n–3	Mean ± SD CV	0.34 ± 0.01 3.18	0.36 ± 0.01 3.32	0.41 ± 0.01 2.70	0.42 ± 0.01 2.72	ES3 vs. ES2, $p = 1.74 \times 10^{-7}$ HS3 vs. HS2, $p = 2.37 \times 10^{-6}$
Arachidonic	C20:4, n–6	Mean ± SD CV	0.94 ± 0.02 2.35	1.01 ± 0.03 2.88	1.14 ± 0.04 3.47	1.18 ± 0.04 3.57	ES3 vs. ES2, $p = 1.81 \times 10^{-14}$ HS3 vs. HS2, $p = 1.95 \times 10^{-14}$
Eicosapentaenoic	C20:5, n–3	Mean ± SD CV	3.01 ± 0.11 3.58	3.22 ± 0.10 3.16	3.64 ± 0.13 3.54	3.77 ± 0.12 3.11	ES3 vs. ES2, $p = 1.80 \times 10^{-14}$ HS3 vs. HS2, $p = 1.77 \times 10^{-14}$
Clupanodonic	C22:5, n–3	Mean ± SD CV	1.75 ± 0.06 3.66	1.87 ± 0.08 4.05	2.11 ± 0.07 3.12	2.19 ± 0.07 3.18	ES3 vs. ES2, $p = 1.81 \times 10^{-14}$ HS3 vs. HS2, $p = 1.79 \times 10^{-14}$
Docosahexaenoic	C22:6, n–3	Mean ± SD CV	2.31 ± 0.05 2.18	2.47 ± 0.09 3.67	2.80 ± 0.11 3.85	2.90 ± 0.11 3.76	ES3 vs. ES2, $p = 1.82 \times 10^{-14}$ HS3 vs. HS2, $p = 1.77 \times 10^{-14}$
Cholesterol		Mean ± SD CV	48.86 ± 3.07 6.28	52.27 ± 1.42 2.71	59.10 ± 1.85 3.13	61.18 ± 2.18 3.56	ES3 vs. ES2, $p = 1.82 \times 10^{-14}$ HS3 vs. HS2, $p = 1.79 \times 10^{-14}$
Σ SFA			19.94	21.33	24.12	24.97	
Σ MUFA			28.05	30.01	33.94	35.13	
Σ PUFA			13.53	14.48	16.37	16.95	
n–3			10.05	10.76	12.16	12.59	
n–6			3.19	3.42	3.86	4.00	
n–3/n–6			3.15	3.15	3.15	3.15	
n–6/n–3			0.32	0.32	0.32	0.32	

Table 7. *Cont.*

Fatty Acid	Descriptive Statistics	Summer 2		Summer 3		p Values
		Epaxial Muscles	Hypaxial Muscles	Epaxial Muscles	Hypaxial Muscles	
PUFA/SFA		0.68	0.68	0.68	0.68	
USFA/SFA		2.09	2.09	2.09	2.09	
PI		7.01	7.50	8.48	8.77	
AI		0.57	0.57	0.57	0.57	
TI		0.38	0.39	0.39	0.39	
HFA		17.33	18.55	20.97	21.71	
hFA		33.01	35.32	39.94	41.34	
h/H		1.90	1.90	1.90	1.90	

SD—standard deviation. CV—coefficient of variation. ES2—epaxial muscles, summer 2. ES3—epaxial muscles, summer 3. HS2—hypaxial muscles, summer 2. HS3—hypaxial muscles, summer 3. PI: polyunsaturated index. TI: thrombogenic index. AI: Atherogenic Index. HFA: Hypercholesterolemic Fatty Acids (C12:0 + C14:0 + C16:0). hFA: hypocholesterolemic Fatty Acids (C18:1 + polyunsaturated FA). h/H: hypocholesterolemic/hypercholesterolemic FA. $n = 7$.

3.3. Amino Acids Profiling and the Nutritional Assessment of Polyodon Spathula Sturgeons Meat

The amino acids content of *Polyodon spathula* sturgeons fillets, aged 2 and 3 summers is displayed in Table 8.

Table 8. Amino acids content (g/100 g) in the fillets sampled from *Polyodon spathula* sturgeons aged two and three summers.

Amino Acids	Summer 2			Summer 3			p Values
	Mean	±SD	CV	Mean	±SD	CV	
Tryptophan	0.202	0.005	2.65	0.204	0.006	2.95	>0.9999
Threonine	0.790	0.017	2.18	0.799	0.028	3.50	0.9999
Isoleucine	0.830	0.035	4.21	0.839	0.035	4.13	0.9999
Leucine	1.463	0.049	3.37	1.480	0.052	3.53	0.9924
Lysine	1.654	0.051	3.09	1.673	0.053	3.14	0.9628
Methionine	0.533	0.023	4.25	0.539	0.019	3.60	>0.9999
Cysteine	0.193	0.008	4.19	0.195	0.007	3.48	>0.9999
Phenylalanine	0.703	0.026	3.67	0.711	0.024	3.41	0.9999
Tyrosine	0.608	0.020	3.28	0.615	0.026	4.16	0.9999
Valine	0.928	0.027	2.94	0.939	0.035	3.68	0.9998
Arginine	1.077	0.036	3.38	1.090	0.047	4.32	>0.9999
Histidine	0.530	0.011	2.13	0.536	0.022	4.05	0.9997
Alanine	1.088	0.046	4.21	1.101	0.036	3.24	0.8220
Aspartic acid	1.843	0.053	2.86	1.865	0.051	2.73	0.1275
Glutamic acid	2.688	0.095	3.55	2.719	0.091	3.34	0.9999
Glycine	0.864	0.035	4.01	0.874	0.038	4.34	0.9999
Proline	0.637	0.024	3.69	0.644	0.022	3.47	0.9999
Serina	0.735	0.023	3.18	0.744	0.023	3.05	>0.9999
$\sum$ amino acids	17.365			17.568			
$\sum$ essential amino acids	7.102			7.185			
$\sum$ semi-essential amino acids	0.801			0.810			
$\sum$ non-essential amino acids	9.463			9.573			

In order to nutritionally evaluate the investigated fish proteins, the content of amino acids was calculated as g amino acids/100 g protein, so that the obtained data can be compared with the FAO/WHO standards for different categories of consumers (Table 9).

Table 9. Amino acids content (g/100 g protein) of the etalon protein and of the fillets sampled from *Polyodon spathula* sturgeons, aged two and three summers.

Amino Acids	FAO/WHO Etalon Protein			S2	S3
	Standard 1 Children	Standard 2 Youth	Standard 3 Adults		
Tryptophan	1.7	1.0	0.6	1.1	1.1
Threonine	4.3	4.0	2.6	4.4	4.4
Isoleucine	4.6	4.0	3.0	4.6	4.6
Leucine	9.3	7.0	4.4	8.1	8.1
Lysine	6.6	5.5	3.1	9.2	9.2
Methionine + Cystine	4.2	3.5	2.7	4.0	4.0
Phenylalanine + Tyrosine	7.2	6.0	3.3	7.3	7.3
Valine	5.5	5.0	2.3	5.2	5.2
EAA (g/16 g N)	43.4	36	22	43.9	43.9

S2—2nd summer. S3—3rd summer. EAA (g/16 g N)—exogenous amino acids.

Nutritional evaluation of the proteins in the meat of *Polyodon spathula* sturgeons, aged two and three summers, is presented in Table 10.

Table 10. Nutritional assessment of the proteins in the *Polyodon spathula* sturgeons meat, aged two and three summers.

Amino Acids	S2—Chemical Indices			S3—Chemical Indices		
	Standard 1	Standard 2	Standard 3	Standard 1	Standard 2	Standard 3
Tryptophan	66.00	112.20	187.00	65.88	112.00	186.67
Threonine	102.07	109.73	168.81	102.05	109.70	168.77
Isoleucine	100.24	115.28	153.70	100.15	115.18	153.57
Leucine	87.40	116.11	184.73	87.39	116.10	184.70
Lysine	139.23	167.07	296.42	139.20	167.04	296.35
Methionine + Cystine	96.02	115.23	149.37	95.98	115.17	149.30
Phenylalanine + Tyrosine	101.15	121.38	220.70	101.14	121.37	220.67
Valine	93.75	103.12	224.17	93.76	103.14	224.22
EAAI (%)	96.43	118.81	193.47	96.40	118.76	193.39
BV	93.41	117.80	199.18	93.38	117.75	199.10
NI (%)	17.36	21.39	34.82	17.55	21.63	35.22

S2—2nd summer. S3—3rd summer. EAAI (%)—essential amino acids index. BV—biological value. NI (%)—nutritional index.

4. Discussion

4.1. Chemical Proximate Composition

The most valuable part of a fish, the musculature, reaches up to 40–50% of its weight and includes: mostly the trunk musculature (the lateral muscles-the fillet, the red muscles and the muscles of the odd fins); head musculature (mandible and gill muscles); musculature of the girdles and of the even fins [1].

Water content of the fillet muscles in *Polyodon spathula*, varied between 75.65% in S3 and 77.83% in S2 (Table 3); water having a variable content depending on the age of the fish (higher at younger ages); the assessed values felt within the limits specified by literature [38,39]. From a statistical point of view, high significant differences were revealed between the two age categories ($p < 0.001$) (Table 3).

The proportion of proteins in the fillet of the studied fish ranged between 17.94% and 18.01%, data which are similar to those in other studies [38,65]. The assessed protein content places this species in the group of protein fish (15–20% protein) [66,67]. Even in the case of protein content, no significant statistical differences were found between the two ages ($p > 0.05$) (Table 6).

In general, in fish, the lipid content varies within very wide limits (0.1–28%) [43,66], fish being classified in: fatty fish, with more than 8% fat; fish with medium fattening status, between 4 and 8%; lean fish, with less than 4% fat [42,67]. In the conditions of our study, when no supplemental feeding was directly provided to paddlefish, the fillets had a lipid content between 2.82 and 3.98%, values that place the analyzed sturgeons in the class of fish with low lipid content. In comparison with other original findings on paddlefish aged one summer [42,67], lipid level in summer two and three was higher by 15–62%, following the natural trend of lipid accumulation as body reserves in parallel with fish ageing. In this situation, the obtained values obtained felt within the limits specified in the literature [38,65,68,69]. There are authors stating that meat of cultured sturgeons should have a lipid content between 5 and 10% [70]. High significant differences ($p < 0.001$) were found for the fillet total fat between the two analyzed ages (Table 6).

In comparable studies [71], data related to proximate composition of other sturgeon species revealed meat moisture content of 75.5% in *Acipenser baerii* (Siberian sturgeon) and of 77.7% in *Acipenser transmontanus* (White sturgeon), while in our findings, the fillet had 76.6–77.8 water content. Related to ash level, paddlefish samples from our findings ranged between 1.2–1.4%, while in the Siberian sturgeon it reached 1.3% and in White sturgeon 1.1%. Paddlefish meat analyzed in our study was lighter in terms of lipids accumulation (2.8–3.9%), in comparison with the Siberian sturgeon (5.6%) and richer than the White sturgeon (2.6%). Total protein content was pretty similar (17.9–18.1%), in comparison with 17.6% (*A. baerii*) and 18.6% (*A. transmontanus*).

Among the most present and most mechanically and enzymatically degrading resistant proteins in the connective tissues, collagen participates in maintaining the structural integrity of the tissues. From a chemical point of view, collagen is an incomplete protein, with low biological value [42,43]. The percentage of collagen in the fillets sampled from the studied fish was 3.93% in the third summer and 4.04% in the fourth summer. The collagen content is 3–10% for most fish species, and in the case of homoeothermic animals, it can reach up to 17% of the total protein [42,43,66,67]. So, in the analyzed sturgeons, the proportion in collagen fell within the quoted values. Statistically, no significant differences were found between the two ages ($p > 0.05$) (Table 6).

Water-protein ratio (W/P) is a criterion for evaluating the quality of fish meat. Accordingly, fish are ranked into five categories: 1st—fish with high nutritional value (W/P = 2.5–3.5); 2nd—fish with good nutritional value (W/P = 3.5–4.2); 3rd—fish with mediocre nutritional value (W/P = 4.2–4.7); 4th—fish with low nutritional value (W/P = 4.7–5.2) and 5th—fish in a state of advanced starvation (W/P above 5.2) [42,43,67]. Studied *Polyodon spathula* sturgeons felt into the category of fish with mediocre food value (3rd). In previous studies [43], W/P ratio has been found to be 4.33 (mediocre) in paddlefish aged one summer and 3.79 (good nutritional value) in paddlefish aged four summers, suggesting that meat becomes more qualitative starting with summer two.

Gross energy value of the fillet from paddlefish aged two summers was 102.05 kcal/100 g and of those aged three summers was 113.22 kcal/100 g (+10.94% energy, compared to previous year) ($p < 0.001$), suggesting an increase of nutritional value consequently to ageing but also with body development, a fact that is mainly due to the increase of lipids amount in the meat. The meat issued from three summers aged paddlefish was richer in calories than the meat produced by other (105 kcal/100 g) [69–71].

Assessment of the chemical composition of epaxial muscles group (dorsal ED and costal EC) and of the hypaxial muscle group (costal HC and abdominal HA) was carried on to identify the occurrence of any differences between the groups. In the paddlefish aged two summers, water content was quite low HA muscles (44.05%) compared to the ED, EC and HC muscles, which had 56.21–57.1% lower values. In all comparisons run between the four muscles, only the costal ones (EC and HC) did not differ significantly ($p > 0.05$) for the water and total solids content, while the other differences were highly significant ($p < 0.001$) (Table 4). The HA muscles had a much higher energy value (more than 400 kcal by 100 g, 3.87–4.29 folds compared to other muscle groups) ($p < 0.001$) due to the much higher presence of lipids in the abdominal muscles compared to the other muscle groups ($p < 0.001$). These parameters also influence the W/P ratio which was lower for HA muscles (2.63), by 59.37–61.88% compared to ED, EC and HC muscles ($p < 0.001$) (Table 4).

In the samples issued from *Polyodon spathula* sturgeons aged three summers, there were high significant differences ($p < 0.001$) found for water and total solids content between muscles ED, EC, HC, and HA (the first three muscle groups comprise 57.15–57.43% more water than the HA muscles). On the contrary, the HA muscles were 8.51–11.34% richer than the other analyzed muscles, that led to a gross energy content also higher in HA ($p < 0.001$) and to a much lower W/P ratio (by 60.8–66.2%), compared to the other three other studied muscle groups ($p < 0.001$) (Table 5).

When the influence of fish age on the proximate composition of the meat was analyzed (Table 6), only the total fat content presented significant differences ($p < 0.001$), particularly for the EC and HC muscles, while the gross energy content varied significantly ($p < 0.001$) in relation with age (S3 vs. S2) in ED, EC, and HC muscles. Therefore, fish ageing induces excessive lipids accumulation not only on filled but also on other muscle groups, hence the higher gross energy meat content in summer three, compared to summer two (especially in those muscles having less caloric values in summer two, due to lower lipid content, in comparison with HA muscles, which accumulated fat earlier before summer three).

4.2. Fatty Acids Profiling and the Sanogenic Indices of the Polyodon Spathula Sturgeons Meat

According to the data in Table 7, among the 20 identified fatty acids, Palmitic acid had the highest occurrence (15.31–19.18 g/100 g total fatty acids), closely followed by the Oleic acid (15.18–19.01 g/100 g total fatty acids), a fact also highlighted by other authors for paddlefish meat [39,41] or in the case of other sturgeons [71–74].

Out of the total of fatty acids identified in the analyzed samples, the highest proportion was taken by the MUFA (approx. 45.6%), followed by the SFA (approx. 32.4%), then by PUFA (approx. 22%), which indicates the presence of good quality fat in the meat of *Polyodon spathula* sturgeons aged two and three summers.

The high content of MUFA and PUFA, known to have a beneficial effect on human health (especially due to their protective role against cardiovascular diseases, as well as the values for the cholesterol content [39,41]) make the meat of the paddlefish an important source of "good fats". The degree of assimilation of fish fats, in human consumers compared to other fats, is very high, a fact explained, first of all, by the particular presence of linoleic, linolenic, arachidonic, eicosapentaenoic, docosapentaenoic, and docosahexaenoic acids. Nutritionally, the increased n–3/n–6 PUFA ratio occurring in *Polyodon spathula* sturgeon meat may also have a protective effect against breast cancer [75].

In other sturgeon species, total SFA reached 19.2 g/total FA (*Acipenser baerii*)—close to our findings for two summers paddlefish and 24.5 g/total FA (*Acipenser transmontanus*), comparable to the samples in our study issued from summer three individuals [71].

As expected, the analysis of meat proximate composition revealed increasing accumulation of lipids from the 2nd to the 3rd summer. Consequently, MUFA increased from 28.05 to 33.94% of FAME (epaxial muscles) and from 30.01 to 35.13% of FAME (hypaxial muscles). Total MUFA varied between 45.5 g/total FA in *A. baerii* and 31.3 g/total FA in *A. transmontanus*, both species richer than the samples from *P. spathula*. Siberian sturgeon meat and Asian sturgeon meat were two to three folds richer in PUFA (35.3 g/total FA–44.2 g/total FA), compared to PUFA levels found in paddlefish investigated meat [71].

The high values of the polyunsaturation index (PI) of the II and III summer paddlefish meat (7.01–8.77) indicate the high level of PUFA, an aspect considered important for human health due to the implications in regulating the level of blood cholesterol [76,77].

From the perspective of human health, the thrombogenic index (TI) and atherogenic index (AI) highlight the predisposition to the incidence of cardiovascular diseases and express the relationship between saturated (pro-thrombo/atherogenic) and unsaturated (anti-thrombo/atherogenic) lipids [52,78]. The AI value calculated for sturgeon *Polyodon spathula* aged 2 or 3 summers (AI = 0.57) is similar to those of carp (0.57) [79] and with approx. 14% lower of that found in trout (0.65) [80]. The highest AI values were reported by Kucukgulmez et al. [80] in two saltwater fish species (AI = 1.22). The TI values calculated for the studied sturgeons were very low compared to other freshwater fish species (carp TI = 0.63, trout TI = 0.49) [79], which reveals a very low tendency for blood clotting in those consuming such meat.

Ployodon spathula sturgeon meat is characterized by the fairly high presence of fatty acids with hypocholesterolemic effects (hFA) (33.01–41.34), an aspect that also results from the high h/H ratio (hypocholesterolemic/hypercholesterolemic FA) (1, 9). The value of the h/H index suggests the presence of sufficiently valuable lipids, with the potential to lower consumers' blood plasma cholesterol [55].

4.3. Amino Acids Profiling and the Nutritional Assessment of Polyodon Spathula Sturgeons Meat

Data in Table 8 reveal slightly higher values of amino acids content in three summers aged sturgeons, in close correlation with the dynamics of meat water and protein content. The most abundant amino acids were Glutamic acid (2.688% in S2 and 2.719% in S3), Aspartic acid (1.843% in S2 and 1.865% in S3), Lysine (1.654% in S2 and 1.673% in S3), followed by Leucine, Alanine, Arginine, Valine, Glycine, Isoleucine, Threonine, and by Tryptophan (0.202% in S2 and 0.204% in S3). The sum of amino acids was quite similar in summers 2 and 3 compared to previous original research results where levels of 17.44 g/100 g total amino acids were reported [43]. The amino acids on the first three positions had the same sequence in the case of the data reported by other authors [41,81,82]. The presence of Glutamic acid gives the meat of sturgeon *Polyodon spathula* a special taste. Also, the special role of this amino acid in brain metabolism has been highlighted since it participates in the synthesis of several physiological substances [83–87].

The analysis by category of amino acids (essential, semi-essential and non-essential) shows revealed that essential amino acids represent approx. 40.9%, the semi-essentials represent 4.61%, while the non-essentials ones represent 54.49%.

In order to estimate the biological value of the proteins from the meat of *Polyodon spathula* sturgeons, aged 2 and 3 summers by chemical methods, the content of essential and semi-essential amino acids (g/100 g protein) was assessed, then reported to the values specified within FAO and WHO standards for three categories of consumers (Table 10). These comparisons revealed high values for all amino acids and for both ages of fish in youth and adult consumers; for children the requirements for Tryptophan, Valine and Methionine + Cystine were not covered.

By calculating the Oser index or EAA index [43,48,63], it was possible to show that these fish contain high quality protein since the proportion of essential and semi-essential amino acids passed above 100% of etalon protein (118.76–193.47%) in the case of young people and adults. On the contrary, for children, the values were slightly lower, compared to etalon protein (96.40% in S3 meat and 96.43% in S2 meat), suggesting an insufficient degree of coverage of the essential amino acid requirements for this category of consumers. This fact was also highlighted by the calculation of BV and NI (%).

The obtained results are consistent with other reported data on the amino acid content and nutritional value of the meat of *Polyodon spathula* sturgeons [41,88–90], as well as of other sturgeons [91].

5. Conclusions

Our study brings scientific novelty through a detailed approach of paddlefish meat nutritional quality, of detailed muscle groups, supported with particular analysis of protein biological values in comparison with international standards for different age groups of consumers, as well as of sanogenic indices derived from the lipidic profile of the meat.

Proximate chemical composition analysis of the meat from *Polyodon spathula* sturgeons, aged two and three summers highlighted that between the two ages there are no significant differences for the chemical composition of the fillets, whereas significant differences occurred between muscle groups. In the case of both ages, there were significant differences between HA muscles and the other three analyzed groups (HC, ED, and EC), mainly due to the accumulation of fat throughout one year or growth. This fact has also significantly impacted the gross energy content of the analyzed meat.

Nutritional evaluation of the paddlefish meat indicated that the fats are of good quality with a significant presence of PUFA, and good values for sanogenic indices were found as well. Also, the protein quality is good for youth and adults and good enough for children.

According to our findings, the optimal age to capitalize paddlefish meat is the 3rd summer, when the best meat yields, proximate composition and nutritional value were met. The study mainly focused on these nutritional aspects and less on the technological factors related to quantitative productions. It would be interesting to overcome such shortcomings in future studies, taking into account all factors and building-up, eventually, a mathematical model to quantify the degrees of influence on quantitative and qualitative paddlefish meat production.

As a follow-up research project, it would be interesting to run textural instrumental analysis on paddlefish meat, accompanied by an analysis of the influence of cooking methods on technological and nutritional quality, in parallel with sensory evaluation and public tasting campaigns to support the degree of awareness and acceptability in consumers for paddlefish (knowing that consumers do not prefer this kind of meat despite its high nutritional features).

Author Contributions: Conceptualization, C.S. and D.S.; methodology, D.S., O.S.M. and R.-M.R.-R.; software, D.S. and R.-M.R.-R.; validation, C.S. and R.-M.R.-R.; formal analysis, O.S.M. and D.S.; investigation, C.S., O.S.M. and D.S.; data curation, C.S. and R.-M.R.-R.; writing—original draft preparation, D.S. and C.S.; writing—review and editing, D.S. and R.-M.R.-R.; visualization, D.S. and R.-M.R.-R.; supervision, D.S. All authors have read and agreed to the published version of the manuscript.

Funding: This research received no external funding.

Institutional Review Board Statement: Not applicable. No animals were used for applying experimental factors on them and to measure their effect throughout certain specific reasoning criteria. The fish individuals used for meat sampling were raised in common with the other marketable fishes in the farm and the samples were taken after the exitus was installed in refrigeration marketing condition.

Informed Consent Statement: Not applicable.

Data Availability Statement: Data supporting reported results available, upon request, at the authors.

Conflicts of Interest: The authors declare no conflict of interest.

References

1. Shelton, W.L. Biology. In *Paddlefish Aquaculture*; Mims, S.D., Shelton, W.L., Eds.; John Wiley & Sons, Inc.: Hoboken, NJ, USA, 2015; pp. 11–75.
2. Jennings, C.A.; Zigler, S.J. Biology and Life History of Paddlefish in North America: An Update. In *Paddlefish Management, Propagation, and Conservation in the 21st Century: Building from 20 Years of Research and Management*; Book Series: American Fisheries Society Symposium; Paukert, C.P., Scholten, G.D., Eds.; American Fisheries Society: Bethesda, MD, USA, 2009; Volume 66, pp. 1–22.
3. Jaric, I.; Riepe, C.; Gessner, J. Sturgeon and paddlefish life history and management: Experts' knowledge and beliefs. *J. Appl. Ichthyol.* **2018**, *34*, 244–257. [CrossRef]
4. Schwinghamer, C.W.; Phelps, Q.E.; Tripp, S.; Yasger, T.; Storts, N. The Role of Propagation in Paddlefish Restoration and Conservation: A Case Study of Missouri. In Proceedings of the American Fisheries Society Symposium 2019, Haines City, FL, USA, 3–5 April 2019; Volume 88, pp. 211–228.
5. Riveros, G.A.; Acosta, F.J.; Patel, R.R.; Hodo, W. Computational mechanics of the paddlefish rostrum. *Eng. Comput.* **2020**, *37*, 1317–1340. [CrossRef]
6. Thurmer, C.R.; Patel, R.R.; Riveros, G.A.; Alexander, Q.G.; Ray, J.D.; Netchaev, A.; Brown, R.D.; Leathers, E.G.; Klein, J.D.; Hoover, J.J. Instrumenting *Polyodon spathula* (Paddlefish) Rostra in Flowing Water with Strain Gages and Accelerometers. *Biosens.-Basel* **2020**, *10*, 37. [CrossRef] [PubMed]
7. Wagner, G. Notes on Polyodon. *Science* **1994**, *19*, 554–555. [CrossRef]

8.   Murray, A.M.; Brinkman, D.B.; DeMar, D.G.; Wilson, G.P. Paddlefish and sturgeon (Chondrostei: *Acipenseriformes: Polyodontidae* and *Acipenseridae*) from lower Paleocene deposits of Montana, USA. *J. Vertebr. Paleontol.* **2020**, *40*, e1775091. [CrossRef]

9.   Russel, T.R. Biology and life history of the paddlefish—A review. In *The Paddlefish: Status, Management and Propagation*; Dillard, J.G., Graham, L.K., Russell, T.R., Eds.; Special Publication 7; American Fisheries Society, North Central Division: Bethesda, MD, USA, 1986; pp. 2–20.

10.   Jaric, I.; Bronzi, P.; Cvijanovic, G.; Lenhardt, M.; Smederevac-Lalic, M.; Gessner, J. Paddlefish (*Polyodon spathula*) in Europe: An aquaculture species and a potential invader. *J. Appl. Ichthyol.* **2019**, *35*, 267–274. [CrossRef]

11.   Mims, S.D.; Shelton, W.L. Paddlefish. In Proceedings of the Symposium on Aquaculture in the 21st Century, Phoenix, Arizona, 22 August 2001.

12.   Shelton, W.L.; Mims, S.D.; Semmens, K.J.; Cuevas-Uribe, R. Artificial Propagation of Paddlefish: An Overview of Developments. In Proceedings of the Symposium on Paddlefish Management, Oklahoma City, OK, USA, 4 February 2017.

13.   Yang, G.; Tao, Z.Y.; Xiao, J.; Tu, G.H.; Kumar, V.; Wen, C.G. Characterization of the gastrointestinal microbiota in paddlefish (*Polyodon spathula*). *Aquac. Rep.* **2020**, *17*, 100402. [CrossRef]

14.   Rosen, R.A.; Hales, D.C. Feeding of paddlefish—*Polyodon spathula*. *Copeia* **1981**, *2*, 441–455. [CrossRef]

15.   Zhu, Y.J.; Li, X.M.; Yang, D.G. Food preference of paddlefish, *Polyodon spathula (Walbaum*, 1792), in polyculture with bighead carp *Aristichthys nobilis* (Richardson, 1845) in non-fed ponds. In Proceedings of the 7th International Symposium on Sturgeons— Sturgeons, Science and Society at the Cross-Roads—Meeting the Challenges of the 21st Century, Nanaimo, BC, Canada, 21–25 July 2013.

16.   Fang, C.; Ma, M.Y.; Ji, H.; Ren, T.J.; Mims, S.D. Alterations of digestive enzyme activities, intestinal morphology and microbiota in juvenile paddlefish, *Polyodon spathula*, fed dietary probiotics. *Fish Physiol. Biochem.* **2015**, *41*, 91–105. [CrossRef]

17.   Brooks, H.; Haines, G.E.; Lin, M.C.; Sanderson, S.L. Physical modeling of vortical cross-step flow in the American paddlefish, *Polyodon spathula. PLoS ONE* **2018**, *13*, e0193874. [CrossRef]

18.   Graham, L.K. Contemporary status of the North American paddlefish, *Polyodon spathula*. In Proceedings of the International Conference on Sturgeon Biodiversity and Conservation, New York, NY, USA, 28–30 July 1994.

19.   Scarnecchia, D.L.; Schooley, J.D. Trophies, Technology, and Timescape in Fisheries Management, as Exemplified through Oklahoma's World Record Paddlefish *Polyodon spathula. Fisheries* **2022**, *47*, 381–394. [CrossRef]

20.   Dzyuba, V.; Shelton, W.L.; Kholodnyy, V.; Boryshpolets, S.; Cosson, J.; Dzyuba, B. Fish sperm biology in relation to urogenital system structure. *Theriogenology* **2019**, *132*, 153–163. [CrossRef] [PubMed]

21.   Schooley, J.D.; Geik, A.; Scarnecchia, D.L. First observations of intersex development in paddlefish *Polyodon spathula. J. Fish Biol.* **2020**, *97*, 919–925. [CrossRef] [PubMed]

22.   Scarnecchia, D.L.; Schooley, J.D.; Slominski, A.; Backes, K.M.; Brown, B.; Gordon, B.D.; Lim, Y. Patterns and Scaling of Reproductive Output in Paddlefish *Polyodon spathula* and Its Conservation Implications. *Trans. Am. Fish. Soc.* **2022**, *151*, 347–355. [CrossRef]

23.   Tripp, S.J.; Phelps, Q.E.; Hupfeld, R.N.; Herzog, D.P.; Ostendorf, D.E.; Moore, T.L.; Brooks, R.C.; Garvey, J.E. Sturgeon and Paddlefish Migration: Evidence to Support the Need for Interjurisdictional Management. *Fisheries* **2019**, *44*, 183–193. [CrossRef]

24.   Schwinghamer, C.W.; Tripp, S.; Phelps, Q.E. Using Ultrasonic Telemetry to Evaluate Paddlefish Spawning Behavior in Harry S. Truman Reservoir, Missouri. *N. Am. J. Fish. Manag.* **2019**, *39*, 231–239. [CrossRef]

25.   Elliott, C.M.; DeLonay, A.J.; Chojnacki, K.A.; Jacobson, R.B. Characterization of Pallid Sturgeon (*Scaphirhynchus albus*) Spawning Habitat in the Lower Missouri River. *J. Appl. Ichthyol.* **2020**, *36*, 25–38. [CrossRef]

26.   Hershey, H.; DeVries, D.R.; Wright, R.A.; McKee, D.; Smith, D.L. Evaluating Fish Passage and Tailrace Space Use at a Low-Use Low-Head Lock and Dam. *Trans. Am. Fish. Soc.* **2022**, *151*, 50–71. [CrossRef]

27.   Patterson, J.T.; Mims, S.D.; Wright, R.A. Effects of body mass and water temperature on routine metabolism of American paddlefish *Polyodon spathula. J. Fish Biol.* **2013**, *82*, 1269–1280. [CrossRef]

28.   Crance, J.H. Habitat Suitability Index Curves for Paddlefish, Developed by the Delphi Technique. *N. Am. J. Fish. Manag.* **1987**, *7*, 123–130. [CrossRef]

29.   Miller, M.J. The Ecology and Functional Morphology of Feeding of North America Sturgeon and Paddlefish. In *Sturgeons and Paddlefish of North America*; LeBreton, G.T.O., Beamish, F.W.H., McKinley, S.R., Eds.; Kluwer Academic: Dordrecht, The Netherlands, 2004; pp. 87–102.

30.   Hupfeld, R.N.; Phelps, Q.E.; Tripp, S.J.; Herzog, D.P. Quantitative Evaluation of Paddlefish Sport Fisheries in Missouri's Large Reservoirs: Implications for the Management of Trophy Sport Fisheries. *N. Am. J. Fish. Manag.* **2018**, *38*, 295–307. [CrossRef]

31.   Elnakeeb, M.A.; Vasilyeva, L.M.; Sudakova, N.V.; Anokhina, A.Z.; Gewida, A.G.A.; Amer, M.S.; Naiel, M.A.E. Paddlefish, *Polyodon spathula*: Historical, current status and future aquaculture prospects in Russia. *Int. Aquat. Res.* **2021**, *13*, 89–107. [CrossRef]

32.   Costache, M. Research on Sperm Quality of North American Sturgeon Species *Polyodon Spathula*. In Proceedings of the 16th International Multidisciplinary Scientific Geoconference (SGEM 2016), Albena, Bulgaria, 30 June–6 July 2016.

33.   Simeanu, C.; Avarvarei, B.V.; Simeanu, D. *Polyodon spathula*—A specie of interest for pisciculture in Eastern European countries, Study on Meat Quality and Environmental Impact in Romania. *Res. Asp. Biol. Sci.* **2022**, *5*, 140–154. [CrossRef]

34.   Mims, S.D.; Shelton, W.L. Monosex culture of paddlefish and shovelnose sturgeon. In Proceedings of the Symposium on Harvest, Trade and Conservation of North American Paddlefish and Sturgeon, Chattanooga, TN, USA, 7–8 May 1998; Williamson, D.F., Ed.; TRAFIC North America/World Wildlife Fund: Washington, DC, USA, 1998; pp. 67–76.

35. Wang, B.W.; Wang, C.Z.; Mims, S.D.; Xiong, Y.L.L. Characterization of the proteases involved in hydrolyzing paddlefish (*Polyodon spathula*) myosin. *J. Food Biochem.* **2000**, *24*, 503–515. [CrossRef]

36. Mims, S.D.; Onders, R.J.; Shelton, W.L. Propagation and Culture of Paddlefish. In Proceedings of the Symposium on Paddlefish Conservation and Management, Omaha, NE, USA, 5 December 2006.

37. Onders, R.J.; Mims, S.D. Paddlefish Production for Meat and Caviar. In *Paddlefish Aquaculture*; Mims, S.D., Shelton, W.L., Eds.; John Wiley & Sons, Inc.: New Jersey, NJ, USA, 2015; pp. 129–151.

38. Lou, X.; Wang, C.; Xiong, Y.L.; Wang, B.; Liu, G.; Mims, S.D. Physicochemical Stability of Paddlefish (*Polyodon spathula*) Meat Under Refrigerated and Frozen Storage. *J. Aquat. Food Prod. Technol.* **2000**, *9*, 27–39. [CrossRef]

39. Herring, J.L.; Mims, S.D. Paddlefish Food Products. In *Paddlefish Aquaculture*; Mims, S.D., Shelton, W.L., Eds.; John Wiley & Sons, Inc.: New Jersey, NJ, USA, 2015; pp. 179–208.

40. Wang, C.; Mims, S.D.; Xiong, Y.L. Consumer acceptability of paddlefish, a potential aquaculture species. *Meat Focus Int.* **1995**, *4*, 8–9.

41. Shi, P.S.; Wang, Q.; Zhu, Y.T.; Gu, Q.H.; Xiong, B.X. Comparative study on muscle nutritional composition of juvenile bighead carp (*Aristichthys nobilis*) and paddlefish (*Polyodon spathula*) fed live feed. *Turk. J. Zool.* **2013**, *37*, 321–328. [CrossRef]

42. Simeanu, D.; Creanga, S.; Simeanu, C. Research on the meat quality produced by *Polyodon Spathula* sturgeons species related to human nutritional requirements. *Res. J. Biotechnol.* **2015**, *10*, 36–43.

43. Simeanu, C.; Simeanu, D.; Popa, A.; Usturoi, A.; Bodescu, D.; Dolis, M.G. Research Regarding Content in Amino-acids and Biological Value of Proteins from *Polyodon spathula* Sturgeon Meat. *Rev. Chim.* **2017**, *68*, 1063–1069. [CrossRef]

44. AOAC. *Official Methods of Analysis of the Association of Official Analytical Chemists*, 15th ed.; AOAC: Washington, DC, USA, 1990.

45. Raţu, R.N.; Ciobanu, M.M.; Radu-Rusu, R.M.; Usturoi, M.G.; Ivancia, M.; Doliş, M.G. Study on the chemical composition and nitrogen fraction of milk from different animal species. *USAMV Bucharest-Sci. Papers. Ser. D Anim. Sci.* **2021**, *LXIV*, 374–379.

46. Dolis, M.G.; Diniţă, G.; Pânzaru, C. Contributions to study of mulberry leaf use by Bombyx mori larvae. *Sci. Pap. Ser. D Anim. Sci.* **2022**, *LXV*, 358–370.

47. Papuc, C.; Crivineanu, M.; Nicorescu, V.; Papuc, C.; Predescu, C. Increase of the Stability to Oxidation of Lipids and Proteins in Carp Muscle (*Cyprinus carpio*) Subject to Storage by Freezing by Polyphenols Extracted from Sea Buckthorn Fruits (*Hippophae rhamnoides*). *Rev. Chim.* **2012**, *63*, 1198–1203.

48. Mierliţă, D.; Simeanu, D.; Pop, I.M.; Criste, F.; Pop, C.; Simeanu, C.; Lup, F. Chemical Composition and Nutritional Evaluation of the Lupine Seeds (*Lupinus albus L.*) from Low-Alkaloid Varieties. *Rev. Chim.* **2018**, *69*, 453–458. [CrossRef]

49. Mierliţă, D.; Pop, I.M.; Lup, F.; Simeanu, D.; Vicas, S.I.; Simeanu, C. The Fatty Acids Composition and Health Lipid Indices in the Sheep Raw Milk Under a Pasture-Based Dairy System. *Rev. Chim.* **2018**, *69*, 160–165. [CrossRef]

50. Criste, F.L.; Mierliţă, D.; Simeanu, D.; Boişteanu, P.C.; Pop, I.M.; Georgescu, G.; Nacu, G. Study of Fatty Acids Profile and Oxidative Stability of Egg Yolk from Hens Fed a Diet Containing White Lupine Seeds Meal. *Rev. Chim.* **2018**, *69*, 2454–2460. [CrossRef]

51. Ciobanu, M.M.; Boişteanu, P.C.; Simeanu, D.; Postolache, A.N.; Lazăr, R.; Vîntu, C.R. Study on the profile of fatty acids of broiler chicken raised and slaughtered in industrial system. *Rev. Chim.* **2019**, *70*, 4089–4094.

52. Struţi, D.I.; Mierliţă, D.; Simeanu, D.; Pop, I.M.; Socol, C.T.; Papuc, T.; Macri, A.M. The effect of dehulling lupine seeds (*Lupinus albus L.*) from low-alkaloid varieties on the chemical composition and fatty acids content. *Rev. Chim.* **2020**, *71*, 59–70. [CrossRef]

53. Timmons, J.S.; Weiss, W.P.; Palmquist, D.L.; Harper, W.J. Relationships among dietary roasted soybeans, milk components, and spontaneous oxidized flavor of milk. *J. Dairy Sci.* **2001**, *84*, 2440–2449. [CrossRef]

54. Ulbricht, T.L.V.; Southgate, D.A.T. Coronary heart disease: Seven dietary factors. *Lancet* **1991**, *338*, 985–992. [CrossRef]

55. Fernandez, M.; Ordóñez, J.A.; Cambero, I.; Santos, C.; Pin, C.; De La Hoz, L. Fatty acid compositions of selected varieties of Spanish dry ham related to their nutritional implications. *Food. Chem.* **2007**, *9*, 107–112.

56. FAO; WHO. *Protein Quality Evaluation*; Report of Joint FAO/WHO Export Consultant, FAO Food Nutrition Paper 51; FAO: Rome, Italy, 1991; pp. 19–21.

57. FAO. *Protein and Amino Acid Requirements in Human Nutrition*; Report of Joint WHO/FAO/UNU Expert Consultation, WHO Technical Report Series 935; FAO: Rome, Italy, 2007; p. 15.

58. Simeanu, D.; Nistor, A.C.; Avarvarei, B.V.; Boişteanu, P.C. Chemical Composition and Nutritional Evaluation of Pasteurized Egg Melange. *Rev. Chim.* **2019**, *70*, 1390–1395. [CrossRef]

59. Borad, S.G.; Kumar, A.; Singh, A.K. Effect of processing on nutritive values of milk protein. *Crit. Rev. Food Sci. Nutr.* **2017**, *57*, 3690–3702. [CrossRef] [PubMed]

60. Kassis, N.M.; Beamer, S.K.; Matak, K.E.; Tou, J.C.; Jaczynski, J. Nutritional composition of novel nutraceutical egg products developed with omega-3-rich oils. *LWT-Food Sci. Technol.* **2010**, *43*, 1204. [CrossRef]

61. Kotlarz, A.; Sujak, A.; Strobel, W.; Grzesiak, W. Chemical composition and nutritive value of protein of pea seeds-effect of harvesting year and variety. *Veg. Crops Res. Bull.* **2011**, *75*, 57–69. [CrossRef]

62. Sujak, A.; Kotlarz, A.; Strobel, W. Compositional and nutritional evaluation of several lupin seeds. *Food Chem.* **2006**, *98*, 711–719. [CrossRef]

63. Oser, B.L. An integrated essential amino acid index for predicting the biological value of proteins. In *Protein and Amino Acid Nutrition*; Albanese, A.A., Ed.; Academic Press: New York, NY, USA, 1959; pp. 281–296.

64. Crisan, E.V.; Sands, A. Edible mushrooms: Nutritional value. In *The Biology and Cultivation of Edible Mushrooms*; Academic Press: New York, NY, USA, 1978; pp. 137–165.
65. Onders, R.J.; Mims, S.D.; Wilhelm, B.A.; Robinson, J.D. Growth, survival and fillet composition of paddlefish, *Polyodon spathula* (Walbaum) fed commercial trout or catfish feeds. *Aquac. Res.* **2005**, *36*, 1602–1610. [CrossRef]
66. Măgdici, E.; Pagu, I.B.; Nistor, C.E.; Cioran, C.; Avarvarei, B.V.; Hoha, G.V.; Păsărin, B. Qualitative evaluation of European catfish meat preserved by cold smoking. In Proceedings of the European Biotechnology Congress, Bucharest, Romania, 7–9 May 2015.
67. Nistor, C.E.; Băcilă, V.; Vărzaru Maros, I.; Hoha, G.V.; Păsărin, B. Identification of aminoacid profile and proximate composition of three trout breeds reared in the north eastern region of Romania. *Sci. Pap.-Ser. D-Anim. Sci.* **2019**, *62*, 233–240.
68. Decker, E.A.; Crum, A.D.; Mims, S.D.; Tidwell, J.H. Processing Yields and Composition of Paddlefish (*Polyodon spathula*) a Potential Aquaculture Species. *J. Agric. Food Chem.* **1991**, *39*, 686–688. [CrossRef]
69. Badiani, A.; Stipa, S.; Nanni, N.; Gatta, P.P.; Manfredini, M. Physical indices, processing yields, compositional parameters and fatty acid profile of three species of cultured sturgeon (genus *Acipenser*). *J. Sci. Food Agric.* **1997**, *74*, 257–264. [CrossRef]
70. Vilkova, D.; Chène, C.; Kondratenko, E.; Karoui, R. A comprehensive review on the assessment of the quality and authenticity of the sturgeon species by different analytical techniques. *Food Control* **2022**, *133*, 108648. [CrossRef]
71. Lopez, A.; Vasconi, M.; Bellagamba, F.; Mentasti, T.; Moretti, V.M. Sturgeon Meat and Caviar Quality from Different Cultured Species. *Fishes* **2020**, *5*, 9. [CrossRef]
72. Nikolova, L.; Georgiev, G.; Bonev, S. Slaughter yield and morpho-physiological characteristics of sturgeon hybrid (F-1 *Acipenser baerii* x *Acipenser gueldenstaedtii*) reared in net cages. *J. Environ. Prot. Ecol.* **2021**, *22*, 254–262.
73. Paltenea, E.; Talpeş, M.; Ionescu, A.; Zara, M.; Vasile, A.; Mocanu, E. Quality assessment of fresh and refrigerated culture sturgeon meat. *Sci. Pap. Anim. Sci. Biotechnol.* **2007**, *40*, 433–443.
74. Pelic, M.; Knezevic, S.V.; Balos, M.Z.; Popov, N.; Novakov, N.; Cirkovic, M.; Pelic, D.L. Fatty acid composition of Acipenseridae—Sturgeon fish. In Proceedings of the 60th International Meat Industry Conference (MEATCON), Kopaonik, Serbia, 22–25 September 2019.
75. Yang, B.; Ren, X.L.; Fu, Y.Q.; Gao, J.L.; Li, D. Ratio of n-3/n-6 PUFAs and risk of breast cancer: A meta-analysis of 274135 adult females from 11 independent prospective studies. *BMC Cancer* **2014**, *14*, 105. [CrossRef]
76. Jenkinson, A.; Franklin, M.; Wahle, K.; Duthie, G. Dietary intakes of polyunsaturated fatty acids and indices of oxidative stress in human volunteers. *Eur. J. Clin. Nutr.* **1999**, *53*, 523–528. [CrossRef]
77. Wang, S.P.; Chen, Y.H.; Li, H. Association between the levels of polyunsaturated fatty acids and blood lipids in healthy individuals. *Exp. Ther. Med.* **2012**, *4*, 1107–1111. [CrossRef]
78. Ghaeni, M.; Ghahfarokhi, K.N. Fatty acids profile, atherogenic (IA) and thrombogenic (IT) health lipid indices in *Leiognathusbindus* and *Upeneussulphureus*. *J. Mar. Sci. Res. Dev.* **2013**, *3*, 1–3. [CrossRef]
79. Busova, M.; Kourimska, L.; Tucek, M. Fatty acids profile, atherogenic and thrombogenic indices in freshwater fish common carp (*Cyprinus carpio*) and rainbow trout (*Oncorhynchus mykiss*) from market chain. *Cent. Eur. J. Public Health* **2020**, *28*, 313–319. [CrossRef]
80. Küçükgülmez, A.; Yanar, Y.; Çelik, M.; Ersor, B. Fatty Acids Profile, Atherogenic, Thrombogenic, and Polyene Lipid Indices in Golden Grey Mullet (*Liza aurata*) and Gold Band Goatfish (*Upeneus moluccensis*) from Mediterranean Sea. *J. Aquat. Food Prod. Technol.* **2018**, *27*, 912–918. [CrossRef]
81. Iwasaki, M.; Harada, R. Proximate and amino acid composition of the roe and muscle of selected marine species. *J. Food Sci.* **1985**, *50*, 1585–1587. [CrossRef]
82. Song, C.; Zhuang, P.; Zhang, L.Z.; Liu, J.; Luo, G. Comparison of nutritive components in muscles between wild and farmed juveniles of Chinese sturgeon *Acipenser sinensis*. *Acta Zool. Sin.* **2007**, *53*, 502–510.
83. Zhang, C.Y.; Li, L.; Li, C.P. *Biochemistry*, 2nd ed.; People Sanitation Press: Beijing, China, 1988; pp. 305–561.
84. Jones, C.; Palmer, A.; Griffiths, R.D. Randomized clinical outcome study of critically ill patients given glutamine- supplemented enteral nutrition. *Nutrition* **1999**, *15*, 108–115. [CrossRef]
85. Kim, J.D.; Lall, S.P. Amino acid composition of whole body tissue of Atlantic halibut (*Hippoglossus hippoglossus*), yellowtail flounder (*Pleuronectes ferruginea*), and Japanese flounder (*Paralichthys olivaceus*). *Aquaculture* **2000**, *187*, 367–373. [CrossRef]
86. Ruiz-Capillas, C.; Moral, A. Free amino acids in muscle of Norway lobster (*Nephrops norvegicus L.*) in controlled and modified atmospheres during chilled storage. *Food Chem.* **2004**, *86*, 85–91. [CrossRef]
87. Wang, T.T.; Liu, H.F.; Sun, Y.X.; Zhang, Y.H.; Li, D.W. Analysis of composition of muscle in *Acipenser schrencki* and several kinds of fish muscle. *Heilongjiang Fish* **2006**, *4*, 9–10.
88. Chen, J.; Liang, Y.Q.; Huang, D.M.; Hu, X.J.; Yang, H.Y.; Yu, F.H.; Fan, Y.L.; Zhu, B.K. Studies on the body composition of different growth development of *Polyodon spathula*. *J. Hydro Ecol.* **2008**, *1*, 65–68.
89. Shen, S.; Zhou, J.C.; Zhao, S.M.; Xiong, S.B. The nutritional composition and evaluation of muscle of *Polyodom spathula*. *Acta Nutr. Sin.* **2009**, *31*, 295–297.
90. Zhuang, P.; Song, C.; Zhang, L.Z. Evaluation of nutritive quality and nutrient components in the muscle of *Glossogobius giuris*. *J. Fish. China* **2010**, *34*, 559–564. [CrossRef]
91. Yin, H.B.; Sun, Z.W.; Sun, D.J.; Qiu, L.Q. Comparison of nutritive compositions in muscles among six farmed sturgeon species. *J. Dalian Fish. Univ.* **2004**, *19*, 92–96.

Article

Quantitative and Qualitative Assessment of European Catfish (*Silurus glanis*) Flesh

Cristina Simeanu [1], Emanuel Măgdici [1], Benone Păsărin [1], Bogdan-Vlad Avarvarei [1,*] and Daniel Simeanu [2]

[1] Department of Animal Resources and Technology, Faculty of Food and Animal Sciences, "Ion Ionescu de la Brad" University of Life Sciences, 8 Mihail Sadoveanu Alley, 700489 Iasi, Romania

[2] Department of Control, Expertise and Services, Faculty of Food and Animal Sciences, "Ion Ionescu de la Brad" University of Life Sciences, 8 Mihail Sadoveanu Alley, 700489 Iasi, Romania

* Correspondence: bvavarvarei@uaiasi.ro

Abstract: Quantitative and qualitative flesh production in the *Silurus glanis* species was comparatively studied between two fish groups: one from aquaculture (AG) and the other from a natural environment, the Prut River (RG). Morphometry was carried out on the fish, and then biometric and conformational indices were calculated. Better values were found in the aquaculture catfish. The Fulton coefficient was 0.82 in the Prut River fish and 0.91% in the farmed ones. The fleshy index reached 19.58% in the AG fish and 20.79% in the RG fish, suggesting better productive capabilities in the AG fish. Postslaughter, the flesh yield and its quality were assessed at different moments throughout the refrigeration period (0–15 days), and chemical compound loss occurred. In the AG samples, the water content decreased by 8.87%, proteins by 27.66%, and lipids by 29.58%. For the RG samples, the loss reached 8.59% in water, 25.16% in proteins, and 29%in lipids. By studying the fatty acids profile and sanogenic indices, good levels of PUFA (31–35%) were found, and the atherogenic index reached 0.35–0.41 while the thrombogenic index ranged between 0.22 and 0.27. Consequently, it can be stated that fish origin and especially the refrigeration period influence the flesh proximate composition and nutritional value of European catfish.

Keywords: European catfish; somatometry; corporal indice; flesh yield; nutritional quality

Citation: Simeanu, C.; Măgdici, E.; Păsărin, B.; Avarvarei, B.-V.; Simeanu, D. Quantitative and Qualitative Assessment of European Catfish (*Silurus glanis*) Flesh. *Agriculture* **2022**, *12*, 2144. https://doi.org/10.3390/agriculture12122144

Academic Editor: Tizhong Shan

Received: 28 October 2022
Accepted: 9 December 2022
Published: 13 December 2022

Publisher's Note: MDPI stays neutral with regard to jurisdictional claims in published maps and institutional affiliations.

1. Introduction

The *Silurus glanis* L., 1758 species is the main European representative of the *Siluriformes* order. They are predatory fishes with quite aggressive behaviour, hunting during the night or even in daytime when waters are murky, relying mainly on nonvisual sensitive organs [1,2]. They easily adapt to environmental conditions and are mostly found in freshwaters from Central and Eastern Europe and Southwest Asia but occasionally can be found in the salty waters of the Black Sea and Baltic Sea [3,4]. They are a large and very aggressive fish and have reached certain lakes and rivers in Western Europe where they developed quite explosively and sometimes detrimentally to the local fish species [5,6]. European catfish farming has been an increasing trend throughout the last years due to three main factors: it has a tasteful flesh, particularly appreciated by consumers, especially in Eastern Europe and Asia [3,7,8]; it is a useful species in fisheries practising polyculture, providing a good health state of biocenosis [9,10]; and last but not least, it is very appreciated in sport fishing in countries like Spain, the Netherlands, France, and Italy [11–13] due to its impressive dimensions and the struggle during the drill. Catfish rearing in different aquaculture systems is closely correlated to the natural environment. According to the FAO statistics on aquaculture and fisheries, the catfish spreading and rearing area in Europe covers 19 countries (Belarus, Bulgaria, Croatia, Czech Republic, France, Germany, Greece, Hungary, Lithuania, Moldova Rep., North Macedonia, Poland, Romania, Russian Federation, Serbia, Slovakia, Slovenia, Switzerland, and Ukraine), and the reported production was of 7554 tons in 2019 [14]. The second continent in terms of yield of *Silurus glanis* is Asia, which is mostly spread across seven countries (Azerbaijan, Iran, Kazakhstan, Tajikistan, Turkey, Turkmenistan, and Uzbekistan), and the reported production in 2019 was 3851 tons [14].

In Romania, catfish farming is less developed in comparison with cypriniculture and salmoniculture. From 2010–2019, catfish flesh yield in Romania reached 1826 tons (165 tons in 2019), ranking Romania in tenth place worldwide. A very important role in achieving such production belongs to the Danube River and Danube Delta, from where, according to the NAFA (National Agency for Fishery and Aquaculture), the largest quantity is harvested [14]. Catfish have morphological, anatomical, and physiological adaptations that allow them to occupy other habitats different from flowing waters, such as natural and artificial lakes, favouring their farming in monoculture or polyculture systems along with carp or other cyprinids. In this way, fish farms represent the second provenance source of European catfish flesh in Romania.

European catfish is one of the species leaving strong impressions in consumers' sensory memory, and consumers have benefitted from an increasing trend among farmers throughout the last years. The flesh is highly appreciated by consumers due to its average content of lipids and high content of proteins but also due to its remarkable sensory traits, especially developed after some special cooking processes [15,16].

In the current paper, we aimed to:

- Obtain quantitative data by gravimetry and morphometry run for certain anatomical parts of European catfish originating from aquaculture and a natural environment;
- Calculate biometric indices to assess fish productive potential, maintenance state, and adaptability to provided environment conditions;
- Track the dynamics of flesh chemical composition, fatty acids profile, and sanogenic indices under the influence of the lastingness of refrigeration period (up to 15 days).

The practical utility of the study is given by the knowledge and understanding of the qualitative and quantitative characteristics of *Silurus glanis* flesh issued from different rearing environments and preserved by refrigeration, so we could indicate which meat is of better quality in terms of origin and shelf life.

2. Materials and Methods

2.1. Biological Studied Material

Three hundred individuals of *Silurus glanis* of various sizes and weights issued from two different rearing environments represented by "Acvares" fish farm situated in Iași County, Romania (aquaculture, geo coordinates (47°19′23.4″ N–27°31′08.8″ E)) and by Prut River on sector Bivolari–Gorban, Iași County, Romania (geo coordinates: 47°32′01″ N 27°26′29″ E–46°53′43″ N 28°05′07″ E) (capture) were studied. Biological material from aquaculture was gathered using a trawl and was stored alive in a submersed storage cage in a waiting pool.

Fish from Prut River were captured by using sport fishing equipment composed of a rod and reel. The sector of Prut River where fish were captured is the eastern boundary of Iasi County, Romania.

The groups were coded in relation to fish origin:

- ➤ AG—aquaculture group, individuals/samples from farmed fish;
- ➤ RG—river group, individuals/samples from Prut River (capture).

Body mass of the fish ranged between 1300 g and 2200 g. This weight range for studied specimens is specified by the literature (better development rhythm combined with economics of production) [17–22] and the market demand (buyers' preference on average fish weight).

2.2. Physical–Chemical Parameters of Water

Water temperature (°C) had close values for both rearing systems. In the studied period (March-October), in March, water reached a temperature of 10.2 °C in system AG and 9.2 °C in system RG. The highest thermal values were recorded in July (25.5 °C) for fish farms, respectively, in August (24.6 °C) for Prut River. At the beginning of the studied period (March–August), a slight difference in temperature was observed, with higher values in farm starting in September; these were slightly higher in Prut River. Water pH varied between 7.2 and 7.9 on farms while in Prut River water, it ranged between 7.1 and 7.5, considered normal for good development of studied species [23]. Quantity of dissolved oxygen ranged between 4.09 mg/L and 8.85 mg/L in farm water while in natural environment, dissolved oxygen varied from 8.06 mg/L to 10.12 mg/L, values considered normal for the regular development of catfish. The other physical–chemical parameters of the water analysed in the current study were chlorides (Cl⁻, 60.21–105.91 mg/L in farm and 74.26–101.34 mg/L in Prut River); nitrites, NO_2, 0.02–0.15 mg N/L in farm and 0.08–0.21 mg/L in Prut River; nitrates, NO_3, with values from 0 up to 2.51 mg N/L in farm and between 1.12–2.14 mg N/L in Prut River; ammonium (NH_4^+) with values from 0.03 till 0.14 mg N/L in farm and between 0.14–0.35 mg N/L in Prut River; and phosphates (PO_4^3) with values from 0 till 0.12 mg P/L in fish farm and between 0.10–0.21 mg P/L in Prut River. The water from both environments was suitable for a normal development of fish, placed in 2nd and 3rd quality categories specific for fish farming systems [23].

Rearing of European catfish on farm was realised in 2 ponds of 30 hectares each in polyculture with carp (*Ciprinus carpio*) (82%) and silver carp (*Hypophthalmichtys molitrix*) and bighead carp (*Aristichtys nobilis*) (12%) aged one year older than the European catfish with body masses of around 500 g. Rate of European catfish in both ponds was 6% (750 individuals). The catfish had masses of around 250 g at brooding moment and between 1.3 kg and 2.2 kg at the end of growth.

2.3. Catfish Feeding

Feeding of catfishes from aquaculture was realised with mixed feed, providing between 120–180 kg/day (monthly variable) in 6 portions of 20–30 kg. Mixed feed was purchased from the specialised market with the following characteristics: granule dimension–6 mm, dry matter 89%, crude protein 54%, crude fat 20%, crude ash 9%, crude fibre 1%, P 1.1%, vitamin A 15000 IU/kg, vitamin D3 1800 IU/kg, vitamin E 105 mg/kg, and vitamin C 280 mg/kg. Energetic value was 20.6 MJ/kg digestible energy.

Catfishes from Prut River benefited from feed sources that naturally occur in the river, consisting of earthworms, snails, insects, tadpoles, frogs, and fish, such as carp (*Ciprinus carpio*), Gibel carp (*Carassius auratus gibelio*), common bream (*Abramis brama*), common bleak (*Alburnus alburnus*), chub (*Leuciscus cephalus*), common rudd (*Scardinius erythrophthalmus*), and common dace (*Leuciscus leuciscus*).

2.4. Morphometric and Gravimetric Assessments

Morphometric assessments were based on the literature methodology [24–32], and the measurements were run using an ihtyometer. Other measuring instruments used in determination of metric characters were graded ruler, tape measure, square, callipers, and tape line.

Growth performances and the main corporal indices were assessed via morphometry based on the anatomical keypoints highlighted in Figure 1 [25,26].

Gravimetric assessments were run using a PGW 6002 I precision scale and a Partner AS220/C/2 analytical scale.

Dressed yield after slaughter was calculated using Equation (1) [20,21]:

$$DY\ (\%) = (\text{carcass mass or anatomical analysed portion} \times 100)/\text{initial mass of live fish} \quad (1)$$

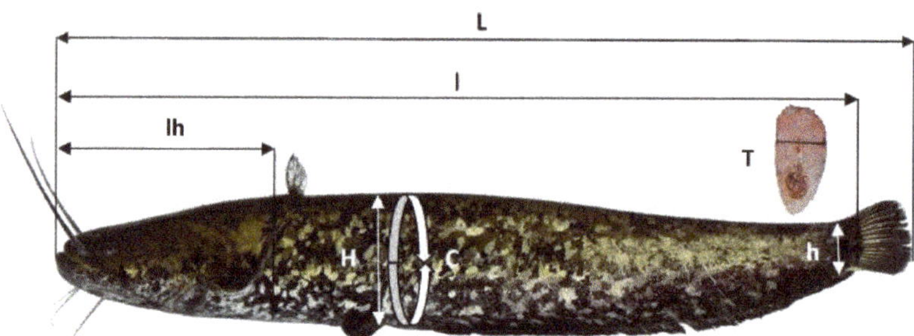

Figure 1. Body morphometry in European catfish (original photo): L—total length of fish; l—standard length of body; lh—length of head; H—maximal height of body; h—minimal height of body; C—maximal circumference of body; T—maximal thickness of body.

2.5. Body Indices and Coefficients

For reaching the aims of the current research, certain corporal indices were calculated:

Profile index (height) points out the corporal format of studied individuals facilitating their framing into a certain profile [20,21] in accordance with Equation (2):

$$PI = l/H, \tag{2}$$

where PI = profile index; l = body standard length (cm); and H = body maximal height (cm).

Fulton coefficient (maintenance index) indicates a direct proportionality ratio between its values and maintenance state of studied individuals and was calculated using Equation (3) [33]:

$$FC = (m * 100)/l^3, \tag{3}$$

where FC (Fulton coefficient) (maintenance index) (%); m = corporal mass (g); and l = fish standard length (cm).

Quality index (Kiselev) is based on Kiselev relation offering data regarding quality of fishery material. This index was calculated in accordance with Equation (4):

$$QI = l/C, \tag{4}$$

where QI = quality index; l = body standard length (cm); and C = body maximum circumference (cm).

Thickness index represents the existing ratio between body maximal height and its maximal thickness in accordance with Equation (5):

$$TI = (T/H) * 100, \tag{5}$$

where TI = thickness index (%); T = body maximal thickness (cm); and H = body maximal height (cm).

Fleshy index was calculated in accordance with Nistor et al. (2012) [34,35] and expresses the percentage rate of head from body standard length using Equation (6):

$$FI = (lh/l) * 100, \tag{6}$$

where FI = fleshy index (%); lh = head length; and l = body standard length (cm).

The equations used in body indices and coefficients computation were provided by the literature [36–38].

2.6. Sampling

Consecutive to morphometry, fish was preserved via refrigeration in sealed plastic recipients and then trenched and filleted to obtain skinless fillets. Samples had a mass of

100 g and were individually refrigerated at temperatures between 2 °C and 4 °C and an air relative moisture of 80–85%.

2.7. Assessment of Flesh Chemical Composition

Flesh water content was assessed via oven drying method, so the analysed sample was subjected to drying at 105 °C temperature as specified by Romanian Standard SR ISO 1442/1997.

Assessment of proteins was realised in accordance with AOAC official methods of analysis/1990 [39–42], compatible with Romanian Standard SR ISO 937:2007, by using a Velp Scientifica device (DK6 digestion unit and UDK7 distillation unit) following the Kjeldhal method.

Soxhlet method was applied to measure fat content on a Velp Scientific–SER 148 device following the manufacturer specifications as well as the AOAC official methods of analysis/1990 [39–42], compatible with Romanian Standard SR ISO 1443:2008 [39,43].

To determine the crude ash (total minerals) content, the sample was subjected to calcination method in an electric muffle furnace at a working temperature of +550 °C, in accordance with the Romanian Standard SR ISO 936:1998.

2.8. Analysis of Fatty Acids Profile and Nutritional Quality of Lipids

The extraction and quantification of FAME (fatty acid methyl esters) from European catfish flesh was realised by detection through gas chromatography and mass spectrometry on a Perkin Elmer chromatographic device connected with a mass spectrometer detector (GC-MS) [44–47].

Fatty acids were quantified as g FAME/100 g of total identified FAME [44–47]. For lipid profile, fatty acids were grouped as follows:

Saturated fatty acids (SFA) as Σ SFA = C10:0 + C12:0 + C14:0 + C15:0 + C16:0 + C18:0 + C20:0 + C22:0;

Monounsaturated fatty acids (MUFA) as Σ MUFA = C16:1 + C18:1 *cis*–9 + C20:1 n−9 + C22:1 n−9;

Polyunsaturated fatty acids (PUFA) as Σ PUFA = C18:2 n−6 + C20:2 n−6 + C18:3 n−3 + C20:3 n−3 + C20:5 n−3 + C18:3 n−6,

Total unsaturated fatty acids as sum of MUFA and PUFA [45,46,48].

The quantities of Ω-3 and Ω-6 PUFA series were expressed as a rate (n−3/n−6).

Polyunsaturation index (PI) of European catfish flesh was calculated in accordance with Equation (7), established by Timmons [45,48,49]:

$$PI = C18:2\ n-6 + (C18:3\ n-3 \times 2). \tag{7}$$

AI (atherogenic index) and TI (thrombogenic index) of fats were run on the basis of FAME GC analysis for European catfish flesh in accordance with Equations (8) and (9), established by *Ulbricht and Southgate* [44,48,50,51]:

$$AI = (C12:0 + C16:0 + 4 \times C14:0)/[\Sigma\ MUFA + \Sigma\ (n\text{-}6) + \Sigma\ (n\text{-}3)], \tag{8}$$

$$TI = (C14:0 + C16:0 + C18:0) / [0.5 \times \Sigma\ MUFA + 0.5 \times \Sigma\ (n\text{-}6) + 3 \times \Sigma\ (n\text{-}3) + \Sigma(n\text{-}3) / \Sigma(n\text{-}6)]. \tag{9}$$

Equation (10), published by Fernandez et al. [51], was utilised in calculation of rate between fatty acids with hypocholesterolemic (h) and hypercholesterolemic (H) effects.

$$h/H\ (hypocholesterolemic\ /\ Hypercholesterolemic) = (C18:1 + PUFA)/(C12:0 + C14:0 + C16:0) \tag{10}$$

2.9. Data Analysis

The main experimental data (50 repetitions per biometric traits, body indices, and yields of cut parts/group and 6 analytical repetitions for analytical chemistry investigations/group) were statistically processed to obtain the arithmetic mean and standard error of mean. Statistical significance of differences between samples was investigated via Fisher testing [52] included within the Data Analysis ToolPack—ANOVA single factor, Microsoft Excel 2019 software.

3. Results

3.1. Morphometry and Body Indices

The measurements of certain anatomical portions in the case of European catfish offer only strict quantitative information. Calculations of certain rates between dimensions also provide qualitative information regarding a productive potential [53,54].

The main biometric investigations are presented in Table 1. An average live weight of 1840.71 g was measured in the AG group, which is 3.12% higher than the RG one. Moreover, a 7.89% higher maximum body height (10.80 cm) was found in the AG fish compared to the RG fish. The average total length of the AG fish was 63.45 cm, 8.07% lower than in the RG fish. The values for the AG fish were lower by 2.16% for standard length, 7.77% for head length, 2.76% for maximum circumference, and 1.97% for maximum body thickness compared to the RG fish. The data in Table 1 highlight a better body development of aquaculture fish vs. river environment fish.

Table 1. Main biometric assessments in European catfish (*Silurus glanis*).

Biometric Traits	RG (n = 50)			AG (n = 50)			*p*-Value
	Mean ± SEM	Min.	Max.	Mean ± SEM	Min.	Max.	
Body mass (g)	1784.91 ± 37.43	1252.6	2192.5	1840.71 ± 30.25	1386.0	2152.4	0.0793
Total length (cm)	68.51 ± 0.61	58.53	75.12	63.45 ± 0.42	58.9	65.7	0.0041
Standard length (cm)	60.04 ± 0.40	54.3	63.7	58.74 ± 0.46	51.4	63.6	0.1059
Head length (cm)	12.48 ± 0.12	10.62	13.39	11.50 ± 0.11	9.6	12.4	0.0028
Body maximum height (cm)	10.01 ± 0.11	8.4	11.3	10.80 ± 0.10	9.5	11.8	0.0075
Body maximum circumference (cm)	32.18 ± 0.37	26.0	37.3	31.29 ± 0.21	28.6	32.9	0.0964
Body maximum thickness (cm)	7.08 ± 0.09	5.9	8.0	6.94 ± 0.06	6.1	7.8	0.1868

SEM: Standard error of mean. Analysis of variance: check *p* values per row.

The main body indices in the studied fish populations are comparatively presented in relation to their origin (RG—river; AG—aquaculture) in Table 2.

Table 2. Main body indices in *Silurus glanis* related to fish origin.

Calculated Index	RG (n = 50)			AG (n = 50)			*p*-Value
	Mean ± SEM	Min.	Max.	Mean ± SEM	Min.	Max.	
Profile index	6.00 ± 0.04	5.47	6.58	5.44 ± 0.03	5.00	5.91	0.0002
Fulton coefficient	0.82 ± 0.01	0.74	1.03	0.91 ± 0.01	0.78	1.10	0.0063
Quality index	1.87 ± 0.02	1.57	2.03	1.88 ± 0.01	1.77	2.01	0.0947
Thickness index	70.77 ± 1.04	59.02	84.28	64.29 ± 0.61	57.65	70.98	0.0029
Fleshy index	20.79 ± 0.19	18.18	23.19	19.58 ± 0.14	17.77	21.53	0.0082

SEM: Standard error of mean. Analysis of variance: check *p* values per row.

Calculation of the profile index for the AG fish revealed an average of 5.44, which is 9.33% lower than the RG fish, underlining a better development of dorsal muscle mass in the aquaculture fish. The Fulton coefficient presented 10.97% better values in the AG fish; moreover, the quality index had slightly higher values by 0.53% in the AG vs. RG group, suggesting as well a better development of the aquaculture catfish than the ones captured from the natural environment. The thickness index reached 70.77 in the RG fish, which is 10.07% higher than the AG fish, suggesting that river catfish are more robust and better adapted to the natural environment.

3.2. Quantitative Flesh Production

The processing yield for portions with high economical value in catfish is presented in Table 3.

Table 3. Processing yield for portions with high economical value at catfish.

Cut Part	RG (n = 50)			AG (n = 50)			*p*-Value
	Mean ± SEM	Min.	Max.	Mean ± SEM	Min.	Max.	
Live mass (g)	1784.91 ± 37.43	1252.59	2192.54	1840.71 ± 30.25	1386.04	2152.35	0.0793
Carcass mass (g)	1598.39 ± 21.97	1156.50	1928.69	1659.40 ± 27.93	1260.44	1940.07	0.3781
Carcass yield (%)	89.55 ± 0.71	86.89	91.68	90.15 ± 1.13	86.81	92.89	0.1028
Torso mass (g)	1159.48 ± 17.99	805.13	1385.39	1124.31 ± 25.31	835.72	1324.75	0.0083
Torso yield (%)	64.96 ± 0.81	61.70	67.24	61.08 ± 0.81	58.79	63.07	0.3973
Fillet mass (g)	825.17 ± 13.44	610.74	948.33	830.35 ± 14.99	621.89	960.41	0.1629
Fillet yield (%)	46.23 ± 0.50	44.12	48.80	45.11 ± 0.55	43.10	47.12	0.0793

SEM: Standard error of mean. Analysis of variance: check *p* values per row.

The carcass yield revealed 0.67% better values in the AG fish compared to the RG fish while the catfish from the river had better efficacy for torso and fillet yields by 6.35% and 2.48%, respectively.

3.3. Qualitative Flesh Production

A comparative evaluation of losses and water content from European catfish refrigerated flesh in different storage periods is presented in Table 4.

Table 4. Comparative evaluation of losses and water content in European catfish refrigerated fillets throughout different storage periods.

Storage Interval (Days)	n	Group	Losses (%)	Water (%)
			Mean ± SEM	Mean ± SEM
0	6	AG	100 ± 0.00	77.80 ± 1.00
	6	RG	100 ± 0.00	78.19 ± 2.02
p-value			-	0.3039
3	6	AG	97.61 ± 0.98	76.60 ± 1.23
	6	RG	98.12 ± 1.67	77.26 ± 1.40
p-value			0.1872	0.1762
6	6	AG	93.56 ± 2.85	74.26 ± 2.64
	6	RG	94.58 ± 1.34	74.88 ± 1.12
p-value			0.3633	0.1906

Table 4. *Cont.*

Storage Interval (Days)	n	Group	Losses (%) Mean ± SEM	Water (%) Mean ± SEM
9	6	AG	90.48 ± 2.72	72.74 ± 2.25
	6	RG	89.87 ± 3.25	72.63 ± 3.21
p-value			0.1964	0.3692
12	6	AG	88.79 ± 1.95	72.18 ± 2.07
	6	RG	89.05 ± 2.72	72.18 ± 2.61
p-value			0.1408	>0.9999
15	6	AG	87.12 ± 2.52	70.90 ± 2.17
	6	RG	87.87 ± 2.03	71.47 ± 2.42
p-value			0.2386	2.2861

SEM: Standard error of mean. Analysis of variance: check *p* values per column, for each storage period.

Certain amounts of water and nutrients were lost from the flesh throughout the experimental period. Water loss reached 8.87% in the AG samples and 8.59% in the RG ones.

The evaluation of dry matter constituents from the European catfish in relation to the lastingness of the storage period is presented in Table 5.

Table 5. Evaluation of dry matter constituents from European catfish flesh in different storage periods.

Storage Period (Days)	n	Group	Ash (%) Mean ± SEM	Proteins (%) Mean ± SEM	Lipids (%) Mean ± SEM
0	6	AG	1.07 ± 0.00	17.75 ± 1.08	3.38 ± 0.22
	6	RG	1.11 ± 0.04	18.08 ± 1.27	2.62 ± 0.16
p-value			0.0782	0.7320	0.0373
3	6	AG	1.05 ± 0.02	16.65 ± 0.82	3.31 ± 0.18
	6	RG	1.08 ± 0.04	17.01 ± 0.89	2.77 ± 0.13
p-value			0.0836	0.6592	0.0459
6	6	AG	1.05 ± 0.05	15.21 ± 0.75	3.04 ± 0.16
	6	RG	1.07 ± 0.04	16.22 ± 0.92	2.42 ± 0.12
p-value			0.0919	0.3182	0.0428
9	6	AG	1.03 ± 0.08	13.85 ± 0.72	2.87 ± 0.23
	6	RG	1.04 ± 0.05	14.05 ± 0.76	2.15 ± 0.14
p-value			0.1305	0.7855	0.0285
12	6	AG	1.02 ± 0.07	13.04 ± 0.71	2.56 ± 0.24
	6	RG	1.03 ± 0.09	13.85 ± 0.92	1.98 ± 0.12
p-value			0.1287	0.4348	0.0319
15	6	AG	1.00 ± 0.10	12.84 ± 0.56	2.38 ± 0.13
	6	RG	1.01 ± 0.07	13.53 ± 1.02	1.86 ± 0.10
p-value			0.1149	0.4817	0.0402

SEM: Standard error of mean. Analysis of variance: check *p* values per column, for each storage period.

The protein levels decreased by 27.66% in the AG samples and 25.16% in the RG samples while the loss of lipids was 29.58% in AG flesh and 29% in RG flesh. Total minerals (crude ash) decreased by 6.54% in AG samples and 9% in RG flesh by the end of the 15 days of storage.

3.4. Fatty Acids Profile and Sanogenic Indices

The profile of fatty acids from the European catfish flesh and sanogenic indices evaluation is presented in Table 6. Analysis was carried out on fillets obtained from the catfish with the age of two summers issued from the farm (AG) or the wild environment (Prut River) (RG).

Table 6. Fatty acids profile (g/100 g total FAME) and sanogenic indices of European catfish fillets.

Fatty Acids	AG (n = 6)	RG (n = 6)	*p*-Value
	Mean ± SEM	Mean ± SEM	
C 14:0	3.08 ± 0.078	1.54 ± 0.054	4.95×10^{-9}
C 14:1	ND	0.32 ± 0.012	-
C 16:0	11.54 ± 0.113	15.16 ± 0.399	1.46×10^{-5}
C 16:1	3.66 ± 0.069	8.39 ± 0.730	1.20×10^{-8}
C 18:0	2.92 ± 0.080	3.69 ± 0.196	5.20×10^{-5}
C 18:1 n−9	19.98 ± 0.397	24.83 ± 1.492	9.53×10^{-5}
C 18:2 n−6	8.25 ± 0.202	4.79 ± 0.397	3.75×10^{-8}
C 18:3 n−3	1.70 ± 0.021	3.83 ± 0.218	1.41×10^{-9}
C 20:0	0.18 ± 0.005	0.19 ± 0.017	0.1478
C 20:1 n−9	5.75 ± 0.195	1.82 ± 0.100	1.23×10^{-10}
C 20:2 n−6	0.44 ± 0.006	0.49 ± 0.010	0.0109
C 20:4 n−6	0.25 ± 0.008	2.12 ± 0.042	6.56×10^{-12}
C 20:3 n−3	5.51 ± 0.198	0.27 ± 0.007	2.93×10^{-13}
C 20:5 n−3	3.57 ± 0.083	3.09 ± 0.125	0.0001
C 22:0	0.49 ± 0.004	0.80 ± 0.010	9.14×10^{-9}
C 22:5 n−6	0.17 ± 0.001	ND	-
C 22:5 n−3	2.09 ± 0.014	1.91 ± 0.059	0.0293
C 22:6 n−3	7.17 ± 0.187	8.61 ± 1.173	0.0003
Σ SFA	18.21	21.37	
Σ MUFA	29.38	35.36	
Σ PUFA	25.58	25.11	
n−3	20.03	17.71	
n−6	9.12	7.39	
n−3/n−6	2.20	2.40	
n−6/n−3	0.46	0.42	
PUFA/SFA	1.40	1.17	
USFA/SFA	3.02	2.83	
PI	11.65	12.45	
AI	0.41	0.35	
TI	0.22	0.27	
HFA	14.62	16.70	
hFA	45.56	49.94	
h/H	3.12	2.99	

SEM: Standard error of mean. Analysis of variance: check *p* values per row. PI: polyunsaturated index, TI: thrombogenic index, AI: atherogenic index, HFA: hypercholesterolemic fatty acids (C12:0 + C14:0 + C16:0), hFA: hypocholesterolemic fatty acids (C18:1 + polyunsaturated FA), h/H: hypocholesterolemic/hypercholesterolemic FA, ND: not detectable.

Out of the five PUFAs found in the studied European catfish, the most occurring in the AG samples were linoleic acid (8.25 g/100 g total FAME), eicosatrienoic acid (5.51 g/100 g total FAME), and eicopentaenoic acid (3.57 g/100 g total FAME). In RG fish, the PUFA with the highest proportion was linoleic acid as well (4.79 g/100 g total FAME) followed by alfa linolenic acid (3.83 g/100 g total FAME) and eicosapentaenoic acid as well (3.09 g/100 g total FAME).

The sum of the fatty acids was 17.35% higher in the RG samples than the AG samples for SFA and 20.35% higher for MUFA while lower for PUFA by 1.84% in river-originating fish compared to aquacultured ones.

Significant amounts of polyunsaturated fatty acids were measured in European catfish flesh with levels above 25 g/100 g total FAME, suggesting a high quality of inner lipids, a fact also highlighted by the good values of the sanogenic indices in both fish groups: PI = 11.65, TI = 0.22, and AI = 0.41 in AG samples and PI = 12.45, TI = 0.27, and AI = 0.35 in RG flesh.

4. Discussions

4.1. Morphometry and Body Indices

Dimensional and gravimetric investigations were carried out on 50 individuals/group that were aged two summers with body mass values close to the means of the groups.

No statistically significant differences occurred for body mass between groups. Differences can be attributed to environmental conditions because the vastness of the Prut hydrographic basin, permanent water stream, and lower feed quantity versus the environment provided by the farm influence, as was demonstrated, a growth rhythm. The obtained values fell within the literature limits [55,56].

Total length (L) values were statistically different ($p < 0.01$) between groups.

Standard length (l) represents a very important morphometric assessment because it highlights the anatomical portion that presents a direct interest for consumers [57]. Between groups, it presented close values of 58.74 ± 0.46 cm (AG) and 60.04 ± 0.40 cm (RG) ($p > 0.05$).

Head length (lh) is an important biometric assessment in fish because this cut does not have a high demand in consumption. So, in artificial selection, individuals with a lower HL related to total body length are preferred [58]. For the AG group, the HLs reached 11.50 ± 0.11 cm (18.12% from L) while in the RG group, they reached 12.48 ± 0.12 cm (18.21% from L), with significant differences between groups ($p < 0.01$) (Table 1).

Maximal body heights (H) presented significant statistical differences between groups ($p < 0.01$) (9.5 to 11.8 cm in AG, 8.4 to 11.3 cm in RG).

Maximal body circumference (C) is run on the anatomical portion in which the body has the highest thickness and respective height [53]. No significant differences occurred between groups. In the AG group, C reached 31.29 cm and was 2.76% less than in the RG group.

Maximal body thickness (T) was maximal in the RG group at 7.08 ± 0.09 cm, respectively, 1.97% higher than the one in the AG group.

Profile index (PI) highlights the body format and represents a rate between the body standard length (cm) and its maximal height (cm). Low values for this index suggest a convex aspect of the dorsal line, which, in practice, indicates the muscle mass is well represented at the dorsal level [59]. Highly significant differences between means occurred between groups ($p < 0.001$), with a lower value in aquaculture fish, proving the influence of artificial selection of the European catfish population at the farm level.

The Fulton coefficient (FC) indicated a very good maintenance state, acquired by the remarkable adaptability of the studied individuals to environmental conditions as well as through an optimal nutrient uptake from the water supply [33,60]. This index ranged between 0.91 ± 0.01 (AG) and 0.82 ± 0.01 (RG) ($p < 0.05$).

The quality index (QI) suggested a rich muscular mass. Considering that European catfish have quite a serpent-shaped body, the values of those indexes as well as the other indices must not be compared with the values obtained for other fish species [61]. The QI reached 1.88 ± 0.01 in the AG group and 1.87 ± 0.02 in the RG group ($p > 0.05$), proving the quality of biological material in both groups.

The thickness index (TI) expresses the muscle width from the backbone region in relation to the body's maximal height, proving information regarding the fattening state or even about the fish's body format [21]. The means for this index were 64.29 ± 0.61 (AG) and 70.77 ± 1.04 (RG). Associating those values with the ones of the profile index, we can conclude that differences regarding body format exist between these two populations because at similar values of body mass and standard body length, the total maximal height for fishes from the Prut River was significantly reduced without a negative impact on the thickness of back musculature. This aspect can be interpreted as an adaptation to environmental conditions because a dorso-ventral body flattening opposes less resistance to a constant water flow.

The fleshy index (FI) represents the rate between head length and body standard length. This index has great importance because in production, the goal is to raise individuals with an optimal body format with economically relevant anatomical regions in greater proportion. The lower the FI, the higher the torso proportion from the fish's standard length [62,63]. The FI in the AG group reached 19.58 ± 0.14, and the RG group was 5.82% higher, indicating a higher rate of head participation in relation to fish standard length in river fish. These index values can also be considered as adaptations of individuals from the Prut River to environmental conditions, considering that feeding in the wild is not as facile as at the farm; therefore, a larger oral cavity can facilitate prey catching.

The achieved profile index values suggest a higher back and much more voluminous dorsal musculature in farmed catfish while the thickness index shows that catfish from the natural environment have a thicker body as an adaptation to environmental conditions. Higher values of the Fulton coefficient for farmed fish reveal their better development, and the lower values of the quality index (Kiselev) depict a richer muscular mass in farm vs. river-originating fish. The lower values of the fleshy index for the farm fish indicate better fleshiness in comparison with catfish from the natural environment. Considering mind–body indices and coefficients for both categories of fish, it can be stated that the aquaculture significantly and positively impacted body development in comparison with the natural environment.

4.2. Quantitative Flesh Production

Individuals captured from the Prut River had yields of 89.55% for carcasses, 64.96% for torsos, and 6.23% for fillets while the farmed fish achieved a 90.15% carcass yield, 61.08% torso yield, and 45.11% fillet yield. The obtained means were significantly different only for torso yields ($p < 0.05$). No major differences regarding the slaughtering yield occurred for the catfish originating from both environments. The obtained data fell within the limits in the literature for this species [8,19,64].

In a similar study, Jankovka et al. (2006) [19] analysed catfishes with different origins and mentioned close values between the yields of fish from a natural environment (carcass 90.75%; torso 60.08%; and fillets 42.79%) and those that were farmed (carcass 90.76%; torso 60.86%; and fillets 45.11%).

4.3. Qualitative Flesh Production

Due to its proximate composition, fish flesh is placed among the products with high biological value, mainly due to the protein quality which contains almost all the essential amino acids. The high quality of the fish flesh is supported by the reduced quantity of the connective tissue in muscles. Hussain et al. (2011) [65] assigned a digestibility coefficient for fish flesh of around 97% based specifically on the reduced quantity of the connective tissue.

Lipids from fish flesh have high amounts of unsaturated fatty acids, which are beneficial to consumers' health but provide a negative influence on the stability of muscle tissue, favouring its rapid alteration mostly through oxidation [66,67].

In skinless fillets, the mean value for relative moisture was 77.80% in AG samples and 78.19% in RG samples.

During storage periods, the mean values of water between the groups did not exceed a percentage difference higher than 0.66% (RG vs. AG). No statistically significant differences occurred between groups for the water content in the flesh; therefore, the rearing environment did not have an influential role in defining flesh proximate composition. A much more important role in the chemical composition of the flesh is attributed to the applied preservation method and storage period [60]. In a 15-day period, European catfish fillets refrigerated in chilled air flow recorded mass losses of 12.88% and 12.13% for the AG group and RG group, respectively, versus the initial rate. In the AG fish, the losses meant a decrease of 6.90 percentage points of moisture versus the first evaluation. In the RG samples, the water content decreased by 6.72 percentage points throughout the same storage period.

The literature provides much information on the water content of European catfish flesh [68,69]. In our study, flash moisture ranged between 77.68 ± 0.45% and 79.45 ± 0.37% in relation to feeding type. Close values for the same trait were reported by other authors [15,16,55].

Flesh protein quantity in the RG group had a decreasing trend vs. the AG group due, most probably, to the samples' more intense hydrolysis. This flesh denaturation favours the development of alternative bacteria. So, AG samples indicated a protein mean value of 17.75% compared to 18.08% in the RG group. At the end of the 15 storage days, the protein level reached 12.84% in the AG group and 13.53% in the RG group. During the six evaluation sessions for this constituent, no statistical significance occurred between groups. In a study on the chemical composition of flesh gathered from aquaculture featuring catfishes with live weights close to the aquaculture catfishes analysed in the current study (1813.51 g), Honzlova et al. (2021) [16] mentioned comparable values for protein content (16.35–18.12%).

Generally, in most fish species, there is a correlation between lipids content and the conditions provided in different rearing systems; this situation is also valid for European catfish [20,70,71]. In the present study, total lipids values greater by 22.5% occurred in aquaculture catfishes vs. the river-originating ones, thus highlighting the influence of feeding with mixed fodders on the flesh fattiness. Concretely, the recorded limits of lipids varied during the 15 storage days between 3.38% and 2.38% for AG fishes and 2.62–1.86% for RG catfishes. There were also observed fluctuations due to the activity of alteration microorganisms as well as oxidative processes that appeared on the sample surface [64,69,72]. Moreover, for the case of lipids, the literature indicates close values to those obtained in the current study [15,16]. Linhartova et al. (2018) [73] indicated catfish flesh lipids contents of 4.13% in intensive rearing system samples and 2.97% in the semi-intensive system.

Statistically, during the whole storage period, the differences with statistical significance ($p < 0.01$) can be accounted for by the influence of the rearing environments. So, due to water action as well as due to lower food quantity, related to water volume unit, fish from the wild environment are prone to exhibit higher effort in comparison with the ones from the farm where, due to a very high density of fish–prey, the latter have a much more sedentary lifestyle. Regarding the dynamics of these three traits, a descending trend was noticed, indirectly proportional to the prolongation of the storage period.

Ash quantity in AG samples decreased by 0.07 percentage points compared to the initial quantity while RG samples decreased by 0.10 percentage points.

4.4. Fatty Acids Profile and Sanogenic Indices

In accordance with the data presented in Table 6, out of the 18 identified fatty acids, oleic acid was most present (19.98–24.83 g/100 g total fatty acids) followed by palmitic acid (11.54–15.16 g/100 g total fatty acids), a fact also outlined by other authors [73].

Statistically, significant differences occurred between the aquaculture and captured fish for most of the analysed fatty acids ($p < 0.01$) except for arachidonic and eicosadienoic acid.

In lipids constitution, the highest proportion was occupied by MUFA (29.38–35.36%) followed by PUFA (25.11–25.58%) and SFA (18.21–21.37%), indicating a high quality of fats in the analysed European catfish fillets. Higher MUFA and SFA proportions were obtained in the wild samples versus the farmed fish which presented higher values for PUFA.

Close values were reported by Linharthova et al. (2018) [73] for catfishes intensively and semi-intensively reared: MUFA between 37.36 to 41.61%, PUFA 28.86 to 34.61%, and SFA 22.25 to 24.23%. High contents of MUFA and PUFA, known for their beneficial effects on human health, especially as protective against cardiovascular diseases [74,75], make the European catfish flesh an important source of "good fats". The assimilation degree of fish fat is better in human consumers versus other dietary fats due to the higher presence of linoleic, linolenic, arachidonic, eicosapentaenoic, docosapentaenoic, and docosahexaenoic acids. The high rate of PUFA $n-3/n-6$ in fillets obtained from *Silurus glanis* reared in an intensive rearing system and extensive system could also have a protective effect against breast cancer [76].

The high level of the polyunsaturated index (PI) in fillets from aquaculture and captured European catfish (11.65–12.45) outlines a high PUFA level, relevant for human health due to implications for the adjustment of cholesterol levels in blood [77,78]. Bearing in mind the fact that the PI for the RG samples was 6.9% higher, this places the fishes from the natural environment on a better rank compared to the ones from aquaculture.

From a human health perspective, the thrombogenic index (TI) and atherogenic index (AI) highlight the predisposition for cardiovascular disease occurrence in consumers and express the relation between saturated (prothrombo/atherogene) and unsaturated lipids (antithromobo/atherogene) [48,79]. The AI value calculated for *Silurus glanis* samples in AG (AI = 0.41) was 39.02% lower compared to carp (0.57) and 58.53% less than in trout (0.65). In samples from captured *Silurus glanis*, the AI value of 0.35 was also lower than in carp (by 62.85%) and in trout (by 85.71%) [80]. The highest values for the AI were reported by Kucukgulmez et al. (2018) [81] for two fish species from salty waters (AI = 1.22). The TI calculated for the studied catfish was lower in comparison with other freshwater fish species (carp TI = 0.63, trout TI = 0.49, and paddlefish TI = 0.39) [80,82], suggesting a possible lower tendency in consumers for blood clots formation. Comparatively analysing the data obtained in the current study, we observe the fact that fish from the RG had an 14.6% better AI in comparison with AG fish. Regarding the thrombogenic index, we observed that aquaculture fishes are superior to the ones from the natural environment by around 22.7%.

European catfish fillets were characterised by a quite high occurrence of fatty acids with hypocholesterolemic effects (hFA) (45.56–49.94), aspects resulting from the high rate of the h/H FA (2.99–3.12). The h/H index suggests the presence of enough valuable lipids with the potential to decrease consumers' blood serum cholesterol [51]. It was observed that catfish from the natural environment could have a better hypocolesterolemic effect (hFA higher at RG with 9.61% and h/H lower with 4.16%) than fish from aquaculture.

5. Conclusions

A morphometric assessment revealed that catfish from a natural environment had close values to the ones from the farm with a slight superiority of aquaculture specimens.

In terms of chemical composition, fish origin has particularly influenced the lipids content, found in higher amounts in farm fishes, due to environmental conditions as well as to the facile feed availability.

Flesh storage in a refrigerated state for 15 days leads to chemical modifications by losing tissue water, especially throughout the first storage phases, followed by a decrease of nutrients (proteins, fats) due to processes associated with the exudation and degradation of the flesh.

The fatty acids profile and sanogenic indices suggested the better quality of catfish versus other freshwater commonly consumed fish species due to the significant proportion of PUFA and better sanogenic indices. A better lipids profile and sanogenic indices values occurred in wild catfish versus farmed ones.

As a research follow-up, it would be suitable to have an analysis of the pollution degree of catfish environments and the possible transfer of such pollutants through the food chain of water–fish–human consumer.

Author Contributions: Conceptualization, C.S. and E.M.; methodology, B.P.; software, E.M. and B.-V.A.; validation, B.P. and D.S.; formal analysis, C.S. and E.M.; investigation, C.S., E.M., B.P. and B.-V.A.; data curation, E.M. and D.S.; writing—original draft preparation, C.S. and E.M.; writing—review and editing, C.S., B.-V.A. and D.S.; supervision, C.S. All authors have read and agreed to the published version of the manuscript.

Funding: This research received no external funding.

Institutional Review Board Statement: Not applicable. No animals were used for applying experimental factors on them or to measure their effect through reasoning criteria. The fish individuals used for morphometry and flesh sampling were issued from a fishery or public river and were part of a group marketed afterwards.

Informed Consent Statement: Not applicable.

Conflicts of Interest: The authors declare no conflict of interest.

References

1. Antognazza, C.M.; Costantini, T.; Campagnolo, M.; Zaccara, S. One Year Monitoring of Ecological Interaction of *Silurus glanis* in a Novel Invaded Oligotrophic Deep Lake (Lake Maggiore). *Water* **2022**, *14*, 105. [CrossRef]
2. Saez-Gomez, P.; Prenda, J. Freshwater Fish Biodiversity in a Large Mediterranean Basin (Guadalquivir River, S Spain): Patterns, Threats, Status and Conservation. *Diversity* **2022**, *14*, 831. [CrossRef]
3. Bergström, K.; Nordahl, O.; Söderling, P.; Koch-Schmidt, P.; Borger, T.; Tibblin, P.; Larsson, P. Exceptional longevity in northern peripheral populations of Wels catfish (*Silurus glanis*). *Sci. Rep.* **2022**, *12*, 8070. [CrossRef]
4. Severov, Y.A. Size–Age Structure, Growth Rate, and Fishery of European Catfish *Silurus glanis* in the Lower Kama Reservoir. *J. Ichthyol.* **2020**, *60*, 118–121. [CrossRef]
5. Kuzishchin, K.V.; Gruzdeva, M.A.; Pavlov, D.S. Traits of Biology of European Wels Catfish *Silurus glanis* from the Volga–Ahtuba Water System, the Lower Volga. *J. Ichthyol.* **2018**, *58*, 833–844. [CrossRef]
6. Tarkan, A.S.; Vilizzi, L.; Top, N.; Ekmekci, F.G.; Stebbing, P.D.; Copp, G.H. Identification of potentially invasive freshwater fishes, including translocated species, in Turkey using the Aquatic Species Invasiveness Screening Kit (AS-ISK). *Int. Rev. Hydrobiol.* **2017**, *102*, 47–56. [CrossRef]
7. Panicz, R.; Drozd, R.; Drozd, A.; Nędzarek, A. Species and sex-specific variation in the antioxidant status of tench, *Tinca tinca*; wels catfish, *Silurus glanis*; and sterlet, *Acipenser ruthenus* (Actinopterygii) reared in cage culture. *Acta Ichthyol. Et Piscat.* **2017**, *47*, 213–223. [CrossRef]
8. Lyach, R. Harvest Rates of Rheophilic Fish Vimba vimba, Chondrostoma nasus, and Barbus barbus Have a Strong Relationship with Restocking Rates and Harvest Rates of Their Predator Silurus glanis in Lowland Mesotrophic Rivers in Central Europe. *Sustainability* **2021**, *13*, 11379. [CrossRef]
9. Vejrik, L.; Vejrikovaa, I.; Kocvara, L.; Blabolil, P.; Peterka, J.; Sajdlova, Z.; Juza, T.; Smejkal, M.; Kolarik, T.; Barton, D.; et al. The pros and cons of the invasive freshwater apex predator, European catfish *Silurus glanis*, and powerful angling technique for its population control. *J. Environ. Manag.* **2019**, *241*, 374–382. [CrossRef]
10. Linhart, O.; Cheng, Y.; Rodina, M.; Gela, D.; Tučková, V.; Shelton, W.L.; Tinkir, M.; Memiş, D.; Xin, M.M. AQUA_2020_1080: Sperm management of European catfish (*Silurus glanis* L.) for effective reproduction and genetic conservation. *Aquaculture* **2020**, *529*, 735620. [CrossRef]
11. Lyach, R.; Remr, J. Changes in recreational catfish Silurus glanis harvest rates between years 1986-2017 in Central Europe. *J. Appl. Ichthyol.* **2019**, *35*, 1094–1104. [CrossRef]
12. Rees, E.A.; Edmonds-Brown, V.R.; Alam, M.F.; Wright, R.M.; Britton, J.R.; Davies, G.D.; Cowx, I.G. Socio-economic drivers of specialist anglers targeting the non-native European catfish (*Silurus glanis*) in the UK. *PLoS ONE* **2017**, *12*, e0178805. [CrossRef]
13. Riha, M.; Rabaneda-Bueno, R.; Jari, I.; Souza, A.T.; Vejrik, L.; Drastik, V.; Blabolil, P.; Holubova, M.; Juza, T.; Gjelland, K.O.; et al. Seasonal habitat use of three predatory fishes in a freshwater ecosystem. *Hydrobiologia* **2022**, *849*, 3351–3371. [CrossRef]
14. FAO. FAO Yearbook. Fishery and Aquaculture Statistics 2019/FAO Annuaire. Statistiques des Pêches et de L'aquaculture 2019/FAO Anuario. Estadísticas de Pesca y Acuicultura 2019. Rome/Roma. 2021. Available online: https://www.fao.org/fishery/static/Yearbook/YB2019_USBcard/navigation/index_intro_e.htm (accessed on 1 September 2022).

15. Hamzacebi, S.; Serezli, R. Comparison of growth performance of European catfish (*Silurus glanis* L.) rearing in freshwater and 5 parts per thousand salinity in recirculating system. *Su Urunleri Derg.* **2019**, *36*, 373–378. [CrossRef]
16. Honzlova, A.; Curdova, H.; Schebestova, L.; Stara, A.; Priborsky, J.; Koubova, A.; Svobodova, Z.; Velisek, J. A nitrogen factor for European pike-perch (*Sander lucioperca*), northern pike (*Esox lucius*), and sheatfish (*Silurus glanis*) fillets. *Acta Ichthyol. Et Piscat.* **2021**, *51*, 119–129. [CrossRef]
17. Zielasko, M.; Greiling, A.M.; Lubke, K.; Otto-Luebker, H.; Patzkewitsch, D.; Erhard, M.; Wedekind, H. Field study on reducing stress of catfish in a recirculation aquaculture system: An innovative tank design for autonomous movement from holding unit to stunning unit. *Bull. Eur. Assoc. Fish Pathol.* **2019**, *33*, 14–23.
18. Jankowska, B.; Zakes, Z.; Zmijewski, T.; Ulikowski, D.; Kowalska, A. Slaughter value and flesh characteristics of European catfish (*Silurus glanis*) fed natural and formulated feed under different rearing conditions. *Eur. Food Res. Technol.* **2006**, *224*, 453–459. [CrossRef]
19. Adamek, Z.; Grecu, I.; Metaxa, I.; Sabarich, L.; Blancheton, J.P. Processing traits of European catfish (*Silurus glanis Linnaeus*, 1758) from outdoor flow-through and indoor recycling aquaculture units. *J. Appl. Ichthyol.* **2015**, *31*, 38–44. [CrossRef]
20. Măgdici, E.; Nistor, C.E.; Pagu, I.B.; Hoha, G.V.; Avarvarei, B.V.; Păsărin, B. Research regarding the influence of slaughtering age on quantitative meat production at European catfish (*Silurus glanis*). *J. Biotehnol.* **2014**, *185*, S76. [CrossRef]
21. Bektas, Y.; Aksu, I.; Kalayci, G.; Turan, D. Low Genetic Diversity in Turkish Populations of Wels Catfish *Silurus glanis* L., 1758 (Siluridae, Pisces) Revealed by Mitochondrial Control Region Sequences. *Turk. J. Fish. Aquat. Sci.* **2020**, *20*, 767–776. [CrossRef]
22. Krasteva, V. Influence of initial stocking density on growth performance and survival of European catfish (*Silurus glanis* L.) larvae under controlled conditions. *Bulg. J. Agric. Sci.* **2020**, *26*, 248–253.
23. ORDER, no. 1146 from 10th of December 2002 Regarding Water Quality, Issued by Romanian Ministry of Water and Environment Protection, Published in Official Gazzette no. 197 from 27th of March 2003. Available online: https://legislatie.just.ro/Public/DetaliiDocumentAfis/42642 (accessed on 21 October 2022).
24. Lyach, R. Fisheries Management of the European Catfish *Silurus glanis* is Strongly Correlated to the Management of Non-Native Fish Species (Common Carp *Cyprinus carpio*, Rainbow Trout *Oncorhynchus mykiss*, and Grass Carp *Ctenopharyngodon idella*). *Sustainability* **2022**, *14*, 6001. [CrossRef]
25. Yazici, R.; Yazicioglu, O. Some morphometric relationships of Wels catfish (*Silurus glanis* L., 1758) inhabiting Siddikli dam (Kırşehir, Turkey). *J. Anatol. Environ. Anim. Sci.* **2020**, *5*, 199–204. [CrossRef]
26. Trif, N.; Bordeianu, M.; Codrea, V.A. A Maeotian (Late Miocene) freshwater fish-fauna from Romania. *Palaeoworld* **2022**, *31*, 140–152. [CrossRef]
27. Haubrock, P.J.; Azzini, M.; Balzani, P.; Inghilesi, A.F.; Tricarico, E. When alien catfish meet-Resource overlap between the North American *Ictalurus punctatus* and immature European *Silurus glanis* in the Arno River (Italy). *Ecol. Freshw. Fish* **2020**, *29*, 4–17. [CrossRef]
28. Yazici, R.; Yilmaz, M.; Yazicioglu, O. Precision of Age Estimates Obtained from Five Calcified Structure for Wels Catfish, *Silurus glanis*. *J. Ichthyol.* **2021**, *61*, 452–459. [CrossRef]
29. Cruz-Castan, R.; Meiners-Mandujano, C.; Curiel-Ramirez, S. Duration and intensity spawning, body size dependence: The case of little tunny Euthynnus alletteratus caught in the southwest Gulf of Mexico. *Rev. Biol. Mar. Y Oceanogr.* **2019**, *54*, 214–220. [CrossRef]
30. Lomnicky, G.A.; Hughes, R.M.; Peck, D.V.; Ringold, P.L. Correspondence between a recreational fishery index and ecological condition for USA streams and rivers. *Fish. Res.* **2021**, *233*, 105749. [CrossRef]
31. Solomon, S.O.; Okomoda, V.T.; Ogbenyikwu, A.I. Intraspecific morphological variation between cultured and wild *Clarias gariepinus* (Burchell) (Clariidae.; Siluriformes). *Arch. Pol. Fish.* **2015**, *23*, 53–61. [CrossRef]
32. Gonzalez-Martinez, A.; Lopez, M.; Molero, H.M.; Rodriguez, J.; Gonzalez, M.; Barba, C.; Garcia, A. Morphometric and Meristic Characterization of Native Chame Fish (*Dormitator latifrons*) in Ecuador Using Multivariate Analysis. *Animals* **2020**, *10*, 1805. [CrossRef]
33. Mocanu, E.E.; Savin, V.; Popa, M.D.; Dima, F.M. The Effect of Probiotics on Growth Performance, Haematological and Biochemical Profiles in Siberian Sturgeon (*Acipenser baerii* Brandt, 1869). *Fishes* **2022**, *7*, 239. [CrossRef]
34. Nistor, C.E.; Pagu, I.B.; Fotea, L.; Radu, C.; Păsărin, B. Study of some morphological characteristics *Onchorhynchus mykiss* breed farmed in salmonid exploitations from Moldova. *Lucr. Ştiinţifice-Ser. Zooteh.* **2012**, *57*, 245–249.
35. Nistor, C.E.; Pagu, I.B.; Simeanu, C.; Păsărin, B. Analysis of some morphological characteristics of *Salvelinus fontinalis* trout breed farmed in salmonid exploitations from Neamţ and Suceava Counties. *Lucr. Ştiinţifice-Ser. Zooteh.* **2012**, *58*, 225–229.
36. Simeanu, C.; Păsărin, B.; Simeanu, D. The study of some morphological characteristics of the sturgeon species of *Polyodon spathula* in different development stages. *Lucr. Ştiinţifice-Ser. Zooteh.* **2010**, *54*, 244–247.
37. Mocanu, E.E.; Dima, F.M.; Savin, V.; Popa, M.D. Maintenance Status, Haematological Profile, and Blood Biochemistry of Common Carp (*Cyprinus Carpio*) Fed on Diets with Added Grape Marc. *Rev. Romana Med. Vet.* **2022**, *32*, 63–69.
38. Önsoy, B.; Serhan, A.; Filiz, H.; Bilge, G. Determination of the best length measurement of fish. *North-West. J. Zool. Oradea Rom.* **2011**, *7*, 178–180.
39. Simeanu, C.; Simeanu, D.; Popa, A.; Usturoi, A.; Bodescu, D.; Doliş, M.G. Research regarding content in amino-acids and biological value of proteins from *Polyodon spathula* sturgeon meat. *Rev. Chim.* **2017**, *68*, 1063–1069. [CrossRef]
40. AOAC. *Official Methods of Analysis of the Association of Official Analytical Chemists*, 15th ed.; AOAC: Washington, DC, USA, 1990.

41. Rațu, R.N.; Ciobanu, M.M.; Radu-Rusu, R.M.; Usturoi, M.G.; Ivancia, M.; Doliş, M.G. Study on the chemical composition and nitrogen fraction of milk from different animal species. *USAMV Buchar. -Sci. Papers Ser. D Anim. Sci.* **2021**, *64*, 374–379.
42. Doliș, M.G.; Diniță, G.; Pânzaru, C. Contributions to study of mulberry leaf use by *Bombyx mori* larvae. *Sci. Papers. Ser. D Anim. Sci.* **2022**, *65*, 358–370.
43. Bland, J.M.; Grimm, C.C.; Bechtel, P.J.; Deb, U.; Dey, M.M. Proximate Composition and Nutritional Attributes of Ready-to-Cook Catfish Products. *Foods* **2021**, *10*, 2716. [CrossRef]
44. Mierliță, D.; Simeanu, D.; Pop, I.M.; Criste, F.; Pop, C.; Simeanu, C.; Lup, F. Chemical composition and nutritional evaluation of the lupine seeds (*Lupinus albus* L.) from low-alkaloid varieties. *Rev. Chim.* **2018**, *69*, 453–458. [CrossRef]
45. Mierliță, D.; Pop, I.M.; Lup, F.; Simeanu, D.; Vicas, S.I.; Simeanu, C. The fatty acids composition and health lipid indices in the sheep raw milk under a pasture-based dairy system. *Rev. Chim.* **2018**, *69*, 160–165. [CrossRef]
46. Criste, F.L.; Mierliță, D.; Simeanu, D.; Boișteanu, P.C.; Pop, I.M.; Georgescu, G.; Nacu, G. Study of fatty acids profile and oxidative stability of egg yolk from hens fed a diet containing white lupine seeds meal. *Rev. Chim.* **2018**, *69*, 2454–2460. [CrossRef]
47. Ciobanu, M.M.; Boișteanu, P.C.; Simeanu, D.; Postolache, A.N.; Lazăr, R.; Vîntu, C.R. Study on the profile of fatty acids of broiler chicken raised and slaughtered in industrial system. *Rev. Chim.* **2019**, *70*, 4089–4094. [CrossRef]
48. Struți, D.I.; Mierliță, D.; Simeanu, D.; Pop, I.M.; Socol, C.T.; Papuc, T.; Macri, A.M. The effect of dehulling lupine seeds (*Lupinus albus L.*) from low-alkaloid varieties on the chemical composition and fatty acids content. *Rev. Chim.* **2020**, *71*, 59–70. [CrossRef]
49. Timmons, J.S.; Weiss, W.P.; Palmquist, D.L.; Harper, W.J. Relationships among dietary roasted soybeans, milk components, and spontaneous oxidized flavor of milk. *J. Dairy Sci.* **2001**, *84*, 2440–2449. [CrossRef]
50. Ulbricht, T.L.V.; Southgate, D.A.T. Coronary heart disease: Seven dietary factors. *Lancet* **1991**, *338*, 985–992. [CrossRef]
51. Fernandez, M.; Ordóñez, J.A.; Cambero, I.; Santos, C.; Pin, C.; De La Hoz, L. Fatty acid compositions of selected varieties of Spanish dry ham related to their nutritional implications. *Food Chem.* **2007**, *9*, 107–112. [CrossRef]
52. Petrie, A.; Watson, P. *Statistics for Veterinary and Animal Science*, 3rd ed.; Wiley-Blackwell: Hoboken, NJ, USA, 2013.
53. Jawad, L.; Al-Janabi, M. Morphometric characteristics of catfish *Silurus triostegus* (Haekel, 1843) from the Tigris and Shatt al-Arab rivers, Iraq. *Croat. J. Fish.* **2016**, *74*, 179–185. [CrossRef]
54. Dutta, R.; Ahmed, A.M.; Pokhrel, H.; Sarmah, R.; Nath, D.; Mudoi, L.P.; Baruah, D.; Bhagabati, S.K.; Songtheng, P. First report of an endangered silurid catfish, *Pterocryptis barakensis* (Siluridae) from Brahmaputra drainage, North Eastern Himalayan region of India. *J. Appl. Ichthyol.* **2020**, *36*, 528–532. [CrossRef]
55. Kucukgul, A.; Yungul, M.; Kahraman, Z.; Dorucu, M. The Effect of Water Parameters and Dissolved Minerals on the Hematological Parameters of during Breeding Period Catfish (*Silurus glanis*). *Braz. Arch. Biol. Technol.* **2019**, *62*, e19180400. [CrossRef]
56. Sandor, Z.J.; Revesz, N.; Lefler, K.K.; Colovic, R.; Banjac, V.; Kumar, S. Potential of corn distiller's dried grains with solubles (DDGS) in the diet of European catfish (*Silurus glanis*). *Aquac. Rep.* **2021**, *20*, 100653. [CrossRef]
57. Kumar, S.; Zs, J.S.; Nagy, Z.; Fazekas, G.; Havasi, M.; Sinha, A.K.; De Boeck, G.; Gal, D. Potential of processed animal protein versus soybean meal to replace fish meal in practical diets for European catfish (*Silurus glanis*): Growth response and liver gene expression. *Aquac. Nutr.* **2017**, *23*, 1179–1189. [CrossRef]
58. Zibiene, G.; Zibas, A. Impact of commercial probiotics on growth parameters of European catfish (*Silurus glanis*) and water quality in recirculating aquaculture systems. *Aquac. Int.* **2019**, *27*, 1751–1766. [CrossRef]
59. Gumus, E.; Yilayaz, A.; Kanyilmaz, M.; Gumus, B.; Balaban, M. Evaluation of body weight and color of cultured European catfish (*Silurus glanis*) and African catfish (*Clarias gariepinus*) using image analysis. *Aquac. Eng.* **2021**, *93*, 102147. [CrossRef]
60. Nikolic, D.; Skoric, S.; Raskovic, B.; Lenhardt, M.; Krpo-Cetkovic, J. Impact of reservoir properties on elemental accumulation and histopathology of European perch (*Perca fluviatilis*). *Chemosphere* **2020**, *244*, 125503. [CrossRef]
61. Balzani, P.; Kouba, A.; Tricarico, E.; Kourantidou, M.; Haubrock, P.J. Metal accumulation in relation to size and body condition in an all-alien species community. *Environ. Sci. Pollut. Res.* **2022**, *29*, 25848–25857. [CrossRef]
62. Păsărin, B.; Hoha, G.; Costăchescu, E.; Avarvarei, B.V. Research regarding morphology, structure and physic-chemical features of the muscles provided from farming trout (*Onchorhyncus mykiss* and *Salvelinus fontinalis*). *Curr. Opin. Biotechnol.* **2011**, *22*, S98–S99. [CrossRef]
63. Sac, G.; Gaygusuz, O.; Gaygusuz Cigdem, G.; Ozulug, M. Length-weight relationship of seven freshwater fish species from Turkish Thrace. *J. Appl. Ichthyol.* **2019**, *35*, 808–811. [CrossRef]
64. Wasenitz, B.; Karl, H.; Palm, H.W. Composition and quality attributes of fillets from different catfish species on the German market. *J. Food Safe. Food Qua. -Arc. Lebensm.* **2018**, *69*, 57–65. [CrossRef]
65. Hussain, S.M.; Afzal, M.; Salim, M.; Javid, A.; Khichi, T.A.A.; Hussain, M.; Raza, S. Apparent digestibility of fish meal, blood meal and meat meal for *Labeo rohita* fingerlings. *J. Anim. Plant Sci.* **2011**, *21*, 807–911.
66. Olayemi, F.F.; Adedayo, M.R.; Bamishaiye, E.I.; Awagu, E.F. Proximate composition of catfish (*Clarias gariepinus*) smoked in Nigerian Stored Products Research Institute (NSPRI): Developed kiln. *Int. J. Fish. Aquac.* **2011**, *3*, 96–98.
67. Yan, X.B.; Pan, S.M.; Li, Z.H.; Dong, X.H.; Tan, B.P.; Long, S.S.; Li, T.; Suo, X.X.; Yang, Y.Z. Amelioration of Flesh Quality in Hybrid Grouper (female *Epinephelus fuscoguttatus* x male *E. lanceolatu*) Fed with Oxidized Fish Oil Diet by Supplying *Lactobacillus pentosus*. *Front. Mar. Sci.* **2022**, *9*, 926106. [CrossRef]
68. Gandotra, R.; Sharma, S.; Koul, M.; Gupta, S. Effect of chilling and freezing on fish muscle. *IOSR J. Pharm. Biol. Sci. (IOSRJPBS)* **2012**, *2*, 5–9. [CrossRef]

69. Jankowska, B.; Zakes, Z.; Zmijewski, T.; Ulikowski, D.; Kowalska, A. Fatty acids profile in dorsal and ventral sections of fillets from European catfish (*Silurus glanis* L.) fed various feeds. *Arch. Pol. Fish.* **2005**, *13*, 17–29.
70. Zhou, P.C.; Chu, Y.M.; Lv, Y.; Xie, J. Quality of frozen mackerel during storage as processed by different freezing methods. *Int. J. Food Prop.* **2022**, *25*, 593–607. [CrossRef]
71. Kok, F.; Goksoy, E.; Gonulalan, Z. The microbiological, chemical and sensory features of vacuumed-packed Wels catfish (*Silurus glanis* L.) pastrami stored under ambient conditions (20 °C). *J. Anim. Vet. Adv.* **2009**, *8*, 817–824.
72. Benjakul, S.; Bauer, F. Biochemical and Physicochemical Changes in Catfish (*Silurus glanis* Linne) Muscle as Influenced by Different Freeze-thaw Cycles. *Food Chem.* **2001**, *72*, 207–217. [CrossRef]
73. Linhartova, Z.; Krejsa, J.; Zajic, T.; Masilko, J.; Sampels, S.; Mraz, J. Proximate and fatty acid composition of 13 important freshwater fish species in Central Europe. *Aquac. Int.* **2018**, *26*, 695–711. [CrossRef]
74. Bazarsadueva, S.; Radnaeva, L.; Shiretorova, V.; Dylenova, E. The comparison of fatty acid composition and lipid quality indices of roach, perch, and pike of lake Gusinoe (Western Transbaikalia). *Int. J. Environ. Res. Public Health* **2021**, *18*, 9032. [CrossRef]
75. Saliu, F.; Leoni, B.; Della Pergola, R. Lipid classes and fatty acids composition of the roe of wild *Silurus glanis* from subalpine freshwater. *Food Chem.* **2017**, *232*, 163–168. [CrossRef] [PubMed]
76. Dydjow-Bendek, D.; Zagozdzon, P. Total Dietary Fats, Fatty Acids, and Omega-3/Omega-6 Ratio as Risk Factors of Breast Cancer in the Polish Population—A Case-Control Study. *In Vivo* **2020**, *34*, 423–431. [CrossRef] [PubMed]
77. Gladyshev, M.I.; Sushchik, N.N.; Glushchenko, L.A.; Zadelenov, V.A.; Rudchenko, A.E.; Dgebuadze, Y.Y. Fatty acid composition of fish species with different feeding habits from an Arctic Lake. *Dokl. Biochem. Biophys.* **2017**, *474*, 220–223. [CrossRef] [PubMed]
78. Luczynska, J.; Paszczyk, B.; Nowosad, J.; Jan Luczynski, M. Mercury, Fatty Acids Content and Lipid Quality Indexes in Muscles of Freshwater and Marine Fish on the Polish Market. Risk Assessment of Fish Consumption. *Int. J. Environ. Res. Public Health* **2017**, *14*, 1120. [CrossRef]
79. Ghaeni, M.; Ghahfarokhi, K.N. Fatty acids profile, atherogenic (IA) and thrombogenic (IT) health lipid indices in *Leiognathusbindus* and *Upeneussulphureus*. *J. Mar. Sci. Res. Dev.* **2013**, *3*, 1–3. [CrossRef]
80. Busova, M.; Kourimska, L.; Tucek, M. Fatty acids profile, atherogenic and thrombogenic indices in freshwater fish common carp (*Cyprinus carpio*) and rainbow trout (*Oncorhynchus mykiss*) from market chain. *Cent. Eur. J. Public Health* **2020**, *28*, 313–319. [CrossRef] [PubMed]
81. Kucukgulmez, A.; Yanar, Y.; Çelik, M.; Ersor, B. Fatty acids profile, atherogenic, thrombogenic, and polyene lipid indices in Golden grey mullet (*Liza aurata*) and Gold band goatfish (*Upeneus moluccensis*) from Mediterranean Sea. *J. Aquat. Food Prod. Technol.* **2018**, *27*, 912–918. [CrossRef]
82. Simeanu, D.; Radu-Rusu, R.M.; Mintas, O.S.; Simeanu, C. Qualitative and Nutritional Evaluation of Paddlefish (*Polyodon spathula*) Meat Production. *Agriculture* **2022**, *12*, 1965. [CrossRef]

 *agriculture*

MDPI

Article

Meat Quality in Rabbit (*Oryctolagus cuniculus*) and Hare (*Lepus europaeus Pallas*)—A Nutritional and Technological Perspective

Gabriela Frunză [1], Otilia Cristina Murariu [1,*], Marius-Mihai Ciobanu [1], Răzvan-Mihail Radu-Rusu [2,*], Daniel Simeanu [3] and Paul-Corneliu Boișteanu [3]

[1] Department of Food Technologies, Faculty of Agriculture, "Ion Ionescu de la Brad" University of Life Sciences, 3 Mihail Sadoveanu Alley, 700489 Iasi, Romania
[2] Department of Animal Resources and Technology, Faculty of Food and Animal Sciences, "Ion Ionescu de la Brad" University of Life Sciences, 8 Mihail Sadoveanu Alley, 700489 Iasi, Romania
[3] Department of Control, Expertise and Services, Faculty of Food and Animal Sciences, "Ion Ionescu de la Brad" University of Life Sciences, 8 Mihail Sadoveanu Alley, 700489 Iasi, Romania
* Correspondence: otiliamurariu@uaiasi.ro (O.C.M.); radurazvan@uaiasi.ro (R.-M.R.-R.)

Abstract: This study aimed to nutritionally and technologically characterize the meat produced by rabbit (*Oryctolagus cuniculus*, Flemish Giant breed, 50 farmed individuals) and hare (*Lepus europaeus Pallas*, 50 hunted individuals). Muscles were sampled from several carcass regions: dorsal torso—*Longissimus dorsi* (LD), thigh—*Semimembranosus* (SM), and upper arm—*Triceps brachii* (TB). To better depict the meat's nutritional quality, the proximate composition and fatty acid profile were assessed, and then gross energy content and lipid sanogenic indices (Polyunsaturation—PI, atherogenic—AI, thrombogenic—TI, hypocholersyerolemic/hypercholesterolemic ratio—h/H, Nutritional Value Index—NVI) were calculated. pH values at 24 and 48 h post-slaughter, cooking loss (CL), and water-holding capacity (WHC) were the investigated technological quality traits. Gross energy was higher in rabbit TB samples, compared with hare, due to more accumulated lipids ($p < 0.001$). pH value was higher for TB muscles in both species; the WHC was higher for hare ($p < 0.001$), and CL was higher for rabbit ($p < 0.001$). The PI values were 6.72 in hare and 4.59 in rabbit, AI reached 0.78 in hare and 0.73 in rabbit, TI was calculated at 0.66 in hare and 0.39 in rabbit, and the h/H ratio reached 3.57 in hare and 1.97 in rabbit, while the NVI was 1.48 in hare and 1.34 in rabbit samples. Meat from both species is nutritionally valuable for human consumers, meeting nutritional values better than the meat of farmed or other wild species of fowl and mammals. Hare meat was found to be healthier than rabbit in terms of lower fat content, lighter energy, and better lipid health indices.

Keywords: meat; rabbit; hare; nutritional quality; lipid health indices; water-holding capacity; cooking loss

Citation: Frunză, G.; Murariu, O.C.; Ciobanu, M.-M.; Radu-Rusu, R.-M.; Simeanu, D.; Boișteanu, P.-C. Meat Quality in Rabbit (*Oryctolagus cuniculus*) and Hare (*Lepus europaeus Pallas*)—A Nutritional and Technological Perspective. *Agriculture* **2023**, *13*, 126. https://doi.org/10.3390/agriculture13010126

Academic Editor: Wataru Mizunoya

Received: 22 December 2022
Accepted: 28 December 2022
Published: 3 January 2023

1. Introduction

Accentuated growth of both world population and life expectancy could lead to a future "food crisis". Whilst the demand for animal protein increases, the conventional sources are insufficient, and areas for agriculture and fodder crop usage have become less available. The UN predicted a world population size of 9.6 billion by 2050, suggesting a necessity to increase food and feed production [1,2]. Rabbit breeding in developing countries has helped many people out of poverty. According to the FAO, the world supply of rabbit meat issued from Europe increased (>80%) between 1961–1985. The rabbit industry in Asia has also grown rapidly throughout the past 30 years [2]. Recent studies [3–7] report that rabbit meat yield and consumption have risen in countries such as China and Mexico, while in the European countries that were the usual consumers (Italy, Poland, France, and Spain) a significant reduction was observed [1–7]. In 2018, global rabbit meat output reached 1.39 million tonnes, out of which the European countries represented 19.43%, while Asian countries were predominant (72.71%) [2]. In addition, young people orient toward other types of meat and pre-cooked products, rather than consuming rabbit meat, which is more laborious to cook [4]. Rabbits are usually sold as pre-packed whole carcasses or cut up (hind legs and loin). Designing food products containing rabbit meat could be a response to increasing consumers' repeatable satisfaction when buying such products [4,5]. Rabbit meat is considered a functional food due to its high nutritional properties [8–12]. It is a source of low allergenic valuable proteins with high nutritional value (essential amino acids), and it has a sanogenic lipid profile, i.e., low levels of fat and cholesterol (dietetic meat) [9,13,14], due to its high content of unsaturated fatty acids (UFA, especially ω-3 and ω-6) and a good ratio of polyunsaturated fatty acids (n-6/n-3, PUFA) [15–17]. It is also a very good source of minerals (P, K, Ca, Se, and Co) and has the highest concentration of Fe (2.9 mg/100 g for hare to 4.9 mg/100 g hare and rabbit meat together) [8,9] vs. any other type of meat (2.6 mg/100 g beef, 1.9 mg/100 g lamb, 1.3 mg/100 g chicken, 0.9/100 g mg pork) [8,9]. It is also a great source of vitamins: B3, B6, B12 (the highest content of B12: 8.7–11.9 mg/100 g, threefold more than beef) [10], and E. Its low Na level makes it recommendable for children, pregnant women, people with cardiovascular diseases, and elderly people [13,18,19].

One of the most popular small game species is the brown hare (*Lepus europaeus* Pallas) [20–30], sometimes reared in farms for the restocking of hunting and protected areas across Europe [20,21]. The potential of adding hare meat into the human diet is high, due to its sensory characteristics [22], low fat content, high unsaturated fatty acids [21,22], valuable proteins, mineral content, and vitamins [23,24], while its energetic value is similar to other meats [20,23]. Hare meat is classified as red meat, mainly in terms of its high Fe content [20,27], but its availability is restricted by hunting seasons. The hare in general, and wild rabbit in particular, consume a wide variety of plants and grains that qualitatively and nutritionally differ by season, which may cause large variation in the meat composition [28]. Very few data are available on the characterization of hare meat [20,22–30]. Only four articles approached the chemical quality of hare meat (*Lepus europaeus*) from hunting (from our knowledge), from Austria [24,25], Croatia, and Slovakia [26,27]; another three studies describe the quality of hare meat collected from farmed brown hare in Italy [20,30] and Poland [31].

The lack of data on the characterization of hare meat in comparison with rabbit (*Oryctolagus cuniculus*) meat led us to carry out this study. Due to its spectacular size, the Flemish Giant breed is the most farmed one in Romania. Its traits include: massive head; long ears (17–21 cm) with a rounded tip and worn in a "V" shape; and variable fur colouring, mostly agouti but also black, dark grey (kangaroo), brown, or chinchilla. Adults can weight up to 7.5 kg, and exceptionally above 12 kg. The aim of this study was to nutritionally and technologically compare the meat of farmed rabbits (*Oryctolagus cuniculus*—Flemish Giant breed) with the meat of hunted hares (*Lepus europaeus* Pallas).

2. Materials and Methods

2.1. Meat

Meat issued from 50 rabbits (25 males and 25 females), slaughtered at 11 months old, with average carcass weights of 10.9 kg and from 50 hares (23 males and 27 females), aged around 9 months, that had been shot during the regular hunting season (November to January) in Iasi County, Romania. Their carcasses weighted around 3.6 kg. Meat was sampled right after slaughter, from 3 muscular groups: *Longissimus dorsi* (LD), *Semimembranosus* (SM), and *Triceps brachii* (TB); they were chosen due to the expected different physical-chemical properties, as well as to cover the main anatomical regions of carcasses (dorsal torso or episoma—LD, hind leg—SM, foreleg—TB). Muscles from one half of each individual carcass were sampled to run physical-chemical tests, while the ones from the other half were used for technological assessments. Samples were minced per muscle group and homogenised prior to analysis. Afterwards, quantities as required by each method were used to run 20 analytical repetitions per trait.

2.2. The Nutritional Assessment of Hare and Rabbit Meat

2.2.1. Chemical Properties and Energy Value of Hare and Rabbit Meat

For the proximate composition analysis, the muscle samples were preliminarily finely ground and homogenized using an electric shredder. The water, protein, and lipid contents were assessed on the Omega Bruins Food-Check Near InfraRed (NIR) spectrophotometer (Bruins Instruments GmbH, Puchheim, Germany); the crude ash content was assessed by furnace muffle calcination in a Nabertherm B180 device (Nabertherm GmbH, Lilienthal, Germany) (550 °C for 24 h after a preliminary carbonization on Bunsen burner flame) [32,33]. The nitrogen-free extract (NFE) was calculated by difference, using the Equation (1).

$$\text{NFE (g/100 g)} = 100 - \text{Water} - \text{Ash} - \text{Proteins} - \text{Lipids} \tag{1}$$

The gross energy value was calculated via the Atwater Equation (2), which uses the caloric value of each organic matter compound in the analysed matrix (total proteins, lipids, nitrogen-free extract—NFE) [34].

$$\text{GE (kcal/100 g meat)} = \text{g proteins} \times 4.27 \text{ kcal} + \text{g lipids } 9.02 \text{ kcal} + \text{g NFE} \times 3.87 \text{ kcal} \tag{2}$$

2.2.2. Fatty Acid Content

The assessment of fatty acids was performed on the FOSS 6500 NIR spectrophotometer (FOSS co., Hillerod, Denmark). The samples (harvested immediately after slaughter, stored at −80 °C, thawed at 2–4 °C for 24 h, then chopped with a food processor) were placed in sterile Petri dishes, weighed, then lyophilized at −110 °C for 24 h, using the CoolSafe ScanVac freeze dryer (LaboGene co., Lillerod, Denmark), weighed again, then vacuumed and stored in a freezer at −80 °C until analysis. The following saturated fatty acids (SFA) were assessed: C14:0 (myristic acid), C15:0 (pentadecanoic acid), C16:0 (palmitic acid), C17:0 (heptadecanoic acid), and C18:0 (stearic acid). Among the monounsaturated fatty acids (MUFA, ω7 and ω9) these were analysed: 16:1 n-7 (palmitoleic acid, n-7 fatty acid), C18:1 n-7 (vaccenic acid cis isomer of oleic acid), and C18:1 n-9 (oleic acid). A total of nine polyunsaturated fatty acids (PUFA, ω3 and ω6) were also assessed: C18:2 n-6 (linoleic), C18:3 n-3 (linolenic/ALA), C20:2 n-6 (eicosadienoic), C20:3 n-6 (eicosatrienoic), C20:4 n-6 (arachidonic/AA), C20:5 n-3 (eicosapentaenoic/EPA), C22:4 n-6 (docosatetraenoic), C22:5 n-3 (docosapentaenoic/DPA), and C22:6 n-3 (docosahexaenoic/DHA).

2.2.3. Health Lipid Indices Calculation

Rabbit and hare meat *health lipid quality* was assessed by calculating certain sanogenic indices provided by literature (Equations (3)–(9)):

- The amounts of SFA, MUFA, PUFA (issued from analytical findings, summed up);
- The desirable fatty acids [35]

$$\text{(DFA) DFA} = 18{:}0 + \text{MUFA} + \text{PUFA} \tag{3}$$

- The essential fatty acids [35]

$$\text{(EFA) EFA} = \text{C18:2 n-6} + \text{C18:3 n-3} + \text{C20:4 n-6} \tag{4}$$

- The Polyunsaturation Index (PI) [35,36]

$$\text{PI} = \text{C18:2 n-6} + (\text{C18:3 n-3} \times 2) \tag{5}$$

- The Atherogenic Index (AI) [37,38],

$$\text{AI} = [(4 \times \text{C14:0}) + \text{C16:0} + \text{C18:0}]/\text{MUFA} + \text{PUFA n-6} + \text{PUFA n-3} \tag{6}$$

- The Thrombogenic Index (TI) [37,39],

$$TI = (14{:}0 + 16{:}0 + 18{:}0)/[(0.5 \times MUFA) + (0.5 \times n\text{-}6\ PUFA) + (3 \times n\text{-}3\ PUFA) \\ + (n\text{-}3\ PUFA/n\text{-}6\ PUFA)] \quad (7)$$

- The ratio between the hypocholesterolemic and Hypercholesterolemic fatty acids (h/H) [38–40]:

$$h/H = (C18{:}1 + PUFA)/(C14{:}0 + C16{:}0) \quad (8)$$

- The Nutritive Value Index (NVI) [41,42],

$$NVI = (C\ 18{:}0 + C18{:}1)/C\ 16{:}0 \quad (9)$$

2.3. The Technological Assessment of Hare and Rabbit Meat

2.3.1. pH Value

The pH value of meat was measured at 24 and 48 h post-slaughter (on chilled samples, at 2–4 °C), using the digital pH meter HI99163 (Hanna Instruments Ltd., Leighton Buzzard, UK), with a penetration probe. Calibration of the pH meter was performed at 4.0 and 7.0 pH at ambient temperature.

2.3.2. Cooking Loss

Cooking loss (CL%) of meat was assessed gravimetrically. The samples were weighed with an analytical scale, then individually packed in thermo-resisting polyethylene bags, labelled, and subjected to heat treatment (80 °C for one hour in a water bath), then forcedly cooled in ice flakes for 30 min and rested at ambient temperature (21 °C) for another 30 min. Then, the samples were weighed again after removal, with paper filter, of the meat juice resulting from the heat treatment. Cooking loss was expressed as percentage.

2.3.3. Water-Holding Capacity

The water-holding capacity (WHC%) was carried out by a method of compression of the meat over filter paper between two plates [43]. Measurements were performed on the muscles stored at 2 °C, 24 h after slaughter. WHC was assessed on a sample of environ 3 g of meat, placed on previously desiccated and weighed filter-paper (7 cm diameter). The paper with the sample was placed between two glass plates and immediately loads of 2.25 kg were applied, for 5 min. After that, the damp paper filter was rapidly weighed after removal of the compressed meat. The percentage of WHC was calculated as a ratio per cent of weight of released water (damp filter paper weight–dry filter paper weight) to initial weight of meat.

2.4. Data Analysis

The results obtained were statistically processed through the main descriptors computation (Arithmetic mean, SD—standard deviation. V%—coefficient of variation) and analysis of variance, using the GraphPad Prism 9.4.1 software, running the unpaired two-tailed *t* test with Welch's correction, designed for one-to-one group comparisons assuming that the SDs are not equal (20 analytical results from each group of data).

3. Results

3.1. Proximate Composition and Gross Energy Content of Meat

Table 1 presents the proximate composition of hare and rabbit meat (g/100 g), while Table 2 displays the gross energy content (kcal/100 g).

Table 1. Proximate composition of hare and rabbit meat (g/100 g).

Proximate Compound	Muscles	Species	Mean	±SD	V%	*p* Value
Water (g/100 g)	SM	hare	75.15 [a]	±0.28	0.37	0.0213
		rabbit	74.85 [b]	±1.61	2.15	
	LD	hare	75.10	±0.41	0.55	0.1736
		rabbit	74.97	±1.24	1.65	
	TB	hare	74.76 [a]	±0.25	0.33	<0.001
		rabbit	74.18 [d]	±2.36	3.18	
Ash (g/100 g)	SM	hare	1.23	±0.01	0.81	0.4270
		rabbit	1.17	±0.01	0.85	
	LD	hare	1.24 [a]	±0.01	0.81	0.0238
		rabbit	1.21 [b]	±0.02	1.65	
	TB	hare	1.26 [a]	±0.02	1.59	0.0080
		rabbit	1.22 [c]	±0.01	0.82	
Proteins (g/100 g)	SM	hare	21.59	±0.09	0.42	0.7747
		rabbit	21.57	±0.41	1.90	
	LD	hare	21.53	±0.17	0.79	0.5918
		rabbit	21.62	±0.55	2.54	
	TB	hare	21.45	±0.08	0.37	0.4706
		rabbit	21.52	±0.13	0.60	
Lipids (g/100 g)	SM	hare	1.90 [a]	±0.13	6.84	0.0017
		rabbit	2.31 [c]	±0.21	9.09	
	LD	hare	1.64 [a]	±0.06	3.66	0.0019
		rabbit	1.93 [c]	±0.03	1.55	
	TB	hare	2.10 [a]	±0.16	7.62	0.0045
		rabbit	2.57 [c]	±0.12	4.67	
Nitrogen Free Extract (g/100 g)	SM	hare	0.13	±0.01	5.12	0.2617
		rabbit	0.10	±0.01	5.67	
	LD	hare	0.49 [a]	±0.02	3.81	0.0003
		rabbit	0.27 [d]	±0.01	4.61	
	TB	hare	0.43 [a]	±0.02	4.39	0.0019
		rabbit	0.51 [c]	±0.02	4.05	

LD—*Longissimus dorsi*; SM—*Semimembranosus*; TB—*Triceps brachii*; SD—standard deviation; V%– coefficient of variation; ANOVA: means signalled with non-identical superscripts, between species, within the same muscle group, differ significantly for: $p < 0.05$ [a vs. b]; $p < 0.01$ [a vs. c]; $p < 0.001$ [a vs. d].

Table 2. Gross energy content in hare and rabbit meat (kcal/100 g).

Muscles	Species	Mean	±SD	V%	*p* Value
SM	hare	109.83 [a]	±6.75	6.15	0.0009
	rabbit	113.33 [d]	±6.37	5.62	
LD	hare	108.62 [a]	±7.85	7.23	0.0014
	rabbit	110.77 [c]	±9.40	8.49	
TB	hare	112.20 [a]	±5.09	4.54	<0.001
	rabbit	117.05 [d]	±4.42	3.78	

LD—*Longissimus dorsi*; SM—*Semimembranosus*; TB—*Triceps brachii*; SD—standard deviation; V%– coefficient of variation; ANOVA: means signalled with non-identical superscripts, between species, within the same muscle group, differ significantly for: $p < 0.01$ [a vs. c]; $p < 0.001$ [a vs. d].

The highest water content was found in hare Semimebranosus (SM) (75.15 g/100 g), followed by Longissimus dorsi (LD) (75.10 g/100 g), while the lowest one occurred in rabbit Triceps brachii (TB) (71.185 g/100 g), varying inversely proportional with lipid content.

The highest protein content was found in rabbit LD samples (21.62 g/100 g meat), and close values, between 21.45 and 21.59 g proteins/100 g, occurred in other samples. No statistically significant differences occurred between species for the protein content, and the analytical homogeneity was high (V% varied between 0.37 and 2.54%).

Little difference was identified for ash content in hare samples, for all muscular groups (1.23–1.26 g/100 g); however, a higher than total mineral level was analysed in rabbit meat (1.17–1.22 g/100 g).

Rabbit meat was 17.7–22.4% higher in lipids (TB, 2.57 g/100 g; SM, 2.31 g/100 g; LD, 1.93 g/100 g) than hare (TB, 2.10 g/100 g; SM, 1.90 g/100 g; LD, 1.64 g/100 g) ($p < 0.01$).

Consequently, higher gross energy was found in rabbit meat (110.71 Kcal/100 g in LD to 117.05 Kcal/100 g in TB) compared to hare meat (108.62 Kcal/100 g in LD to 112.20 Kcal/100 g in TB) ($p < 0.01$ for LD and $p < 0.001$ for SM and TB) (Table 2).

3.2. Fatty Acid Content

Table 3 presents the fatty acid content in Semimembranosus, Longissimus dorsi, and Triceps brachii muscles sampled from hares and rabbits. The most frequently occurring fatty acid in SM was linoleic acid/C18:2 n-6 in hare (509.01 mg/100 g meat), followed by oleic acid/C18:1 n-9 in rabbit (484.12 mg/100 g meat). In rabbit, palmitic acid/C16:0 also occurred in a significant quantity (450.06 mg/100 g meat) alongside essential linoleic fatty acid/C18:2 n-6 (342.86 mg/100 g meat). Generally, in hare, the MUFA and PUFA were in higher quantities vs. rabbit, except for oleic acid (285.45 mg/100 g hare meat vs. 484.12 mg/100 g rabbit meat) and palmitoleic acid (1.05 mg/100 g hare meat vs. 80.13 mg/100 g meat).

Table 3. Fatty acid content (mg/100 g) in hare and rabbit meat.

	Fatty Acids	Species	*Semimembranosus* mm.				*Longissimus dorsi* mm.				*Triceps brachii* mm.			
			Mean	±SD	V%	*p* Value	Mean	±SD	V%	*p* Value	Mean	±SD	V%	*p* Value
SFA	C14:0	hare	5.55 [a]	±0.95	17.12	<0.001	1.91 [a]	±0.35	18.32	<0.001	0.72 [a]	±0.13	18.06	<0.001
		rabbit	45.03 [d]	±8.65	19.21		31.67 [d]	±3.20	10.10		66.02 [d]	±1.66	2.51	
	C15:0	hare	7.81 [a]	±1.09	13.96	0.0016	9.01 [a]	±1.52	16.87	0.0009	9.12 [a]	±1.26	13.82	0.0064
		rabbit	9.02 [c]	±2.37	26.27		6.06 [d]	±0.65	10.73		13.94 [c]	±3.23	23.17	
	C16:0	hare	302.47 [a]	±5.60	1.85	<0.001	329.07	±3.42	1.04	0.7453	297.04 [a]	±4.25	1.43	<0.001
		rabbit	450.06 [d]	±7.38	1.64		344.87	±4.31	1.25		687.94 [d]	±9.01	1.31	
	C17:0	hare	17.5 [a]	±2.70	15.43	0.0500	18.98 [a]	±1.14	6.01	<0.001	21.36 [a]	±1.59	7.44	0.0087
		rabbit	10.97 [b]	±2.42	22.06		6.93 [d]	±0.56	8.08		18.31 [c]	±4.24	23.16	
	C18:0	hare	101.92	±9.57	9.39	0.2637	112.16 [a]	±3.20	2.85	0.0038	114.94 [a]	±1.99	1.73	0.0028
		rabbit	119.88	±8.56	7.14		87.22 [c]	±5.92	6.79		177.07 [c]	±4.39	2.48	
MUFA	C16:1 n-7	hare	1.05 [a]	±0.04	3.81	<0.001	2.45 [a]	±0.16	6.53	<0.001	6.12 [a]	±0.31	5.07	<0.001
		rabbit	80.13 [d]	±3.10	3.87		51.01 [d]	±4.11	8.06		123.66 [d]	±3.90	3.15	
	C18:1 n-7	hare	24.78	±0.66	2.66	0.2415	27.01 [a]	±1.10	4.07	0.0009	27.94 [a]	±0.36	1.29	0.0025
		rabbit	28.11	±1.38	4.91		19.22 [d]	±1.63	8.48		45.88 [c]	±1.11	2.42	
	C18:1 n-9	hare	285.45 [a]	±4.02	1.41	0.0005	328.4	±4.56	1.39	0.6478	347.54 [a]	±6.88	1.98	0.0003
		rabbit	484.12 [d]	±5.03	1.04		309.17	±3.83	1.24		741.22 [d]	±9.78	1.32	
PUFA	C18:2 n-6	hare	509.01 [a]	±6.06	1.19	<0.001	559.55 [a]	±7.55	1.35	<0.001	639.05	±8.18	1.28	0.2204
		rabbit	342.86 [d]	±8.37	2.44		233.47 [d]	±4.18	1.79		577.12	±6.29	1.09	
	C18:3 n-3	hare	44.78 [a]	±4.01	8.95	0.0064	50.37 [a]	±9.46	18.78	<0.001	58.59	±1.07	1.83	0.9533
		rabbit	32.88 [c]	±2.43	7.39		20.11 [d]	±0.33	1.64		59.01	±1.91	3.24	
	C20:2 n-6	hare	8.95 [a]	±0.09	1.01	<0.001	10.34 [a]	±0.16	1.55	<0.001	10.41 [a]	±0.67	6.44	<0.001
		rabbit	3.98 [d]	±0.68	17.09		2.91 [d]	±0.18	6.19		8.29 [d]	±0.25	3.02	
	C20:3 n-6	hare	0.54 [a]	±0.01	1.85	<0.001	0.47 [a]	±0.04	8.51	<0.001	1.72 [a]	±0.39	22.67	<0.001
		rabbit	3.89 [d]	±0.27	6.94		4.16 [d]	±0.12	2.88		3.99 [d]	±0.50	12.53	
	C20:4 n-6	hare	60.91 [a]	±2.26	3.71	<0.001	64.21 [a]	±2.47	3.85	<0.001	56.24 [a]	±0.80	1.42	0.0206
		rabbit	53.01 [d]	±4.16	7.85		54.32 [d]	±1.18	2.17		51.78 [b]	±4.72	9.12	
	C20:5 n-3	hare	2.82 [a]	±0.39	13.83	<0.001	2.75 [a]	±0.04	1.45	<0.001	2.99 [a]	±0.69	23.08	<0.001
		rabbit	11.05 [d]	±1.17	10.59		10.04 [d]	±0.31	3.09		9.97 [d]	±1.86	18.66	
	C22:4 n-6	hare	16.77 [a]	±0.39	2.33	<0.001	16.9 [a]	±0.22	1.30	<0.001	16.5 [a]	±0.26	1.58	<0.001
		rabbit	15.07 [d]	±0.57	3.78		15.09 [d]	±0.14	0.93		15.44 [d]	±0.60	3.89	
	C22:5 n-3	hare	20.13 [a]	±1.21	6.01	<0.001	22.54 [a]	±1.51	6.70	<0.001	13.45 [a]	±1.66	12.34	<0.001
		rabbit	8.06 [d]	±1.13	14.02		9.26 [d]	±0.28	3.02		7.14 [d]	±1.06	14.85	
	C22:6 n-3	hare	38.53 [a]	±1.93	5.01	<0.001	42.84 [a]	±0.82	1.91	<0.001	27.38 [a]	±3.01	10.99	0.0004
		rabbit	22.89 [d]	±2.68	11.71		24.44 [d]	±0.68	2.78		21.89 [d]	±2.82	12.88	

V%—coefficient of variation; ANOVA: means signalled with non-identical superscripts, between species, within the same muscle group, differ significantly for: $p < 0.05$ [a vs. b]; $p < 0.01$ [a vs. c]; $p < 0.001$ [a vs. d].

In LD muscles, the highest content of fatty acids occurred in hare, and linoleic acid (559.55 mg/100 g meat) was more than double compared to rabbit (233.47 mg/100 g meat); for hare in general, the LD PUFA existed in higher amounts.

In TB samples, oleic acid was the most commonly occurring in rabbit (741.22 mg/100 g meat vs. 347.54 mg/100 g in hare) followed by linolenic acid in hare (639.05 mg/100 g meat vs. 577.12 mg/100 g meat in rabbit) and by C16:0 in rabbit (687.94 mg/100 g meat vs 297.04 mg/100 g meat for hare).

In general, SFAs were more present in rabbit than in hare, especially for C16:0 (687.94 vs. 297.04 mg/100 g in TB, 450.06 vs. 302.47 mg/100 g in SM, and 344.87 vs. 329.07 mg/100 g in LD). PUFA occurred at higher rates in hare; linoleic acid reached 639.05 mg/100 g in TB, 559.55 mg/100 g in LD, and 509.01 mg/100 g in SM. For MUFA, the highest values were found for oleic acid in rabbit (741.22 vs. 347.54 mg/100 g in TB; 484.12 vs. 285.45 mg/100 g in SM). In LD muscles, the oleic acid content was higher in hare, although it was close to the amounts in rabbit (328.40 vs. 309.17 mg/100 g). Palmitoleic acid (16:1 n-7) was higher in rabbit (51.01 mg/100 g).

3.3. Health Lipid Indices for Hare and Rabbit Meat

The total fatty acids and health lipid indices for hare and rabbit meat are presented in Table 4. The EFA value in hare was 680.90 mg/100 g meat, and in rabbit it was 474.85 mg/100 g meat. In hare, the %EFA for SM was 42.45%, in LD it was 42.16%, and in TB it was 45.66. In rabbit, the %EFA was 24.91% for SM, in LD it was 25.03%, and in TB it was 26.17%. Overall, the meat was 43.42% EFA in hare and 25.37% EFA in rabbit.

Table 4. Total fatty acids and health lipid indices for hare and rabbit meat (mg/100 g meat).

Health Lipid Indices	SM	LD	TB	Average/3 Carcass Areas
Σ SFA hare	435.25	471.13	443.18	449.85
Σ SFA rabbit	634.96	476.75	963.28	691.66
Σ MUFA hare	310.28	357.86	381.60	349.91
Σ MUFA rabbit	592.36	379.4	910.76	627.51
Total PUFA hare	702.44	769.97	826.33	766.25
Total PUFA rabbit	493.69	373.80	754.63	540.71
Σ PUFA n-6 hare	596.18	651.47	723.92	657.19
Σ PUFA n-6 rabbit	418.81	309.95	656.62	461.79
Σ PUFA n-3 hare	106.26	118.50	102.41	109.06
Σ PUFA n-3 rabbit	74.88	63.85	98.01	78.91
EFA hare	614.70	674.13	753.88	680.90
EFA rabbit	428.75	307.90	687.91	474.85
DFA hare	1114.64	1239.99	1322.87	1225.83
DFA rabbit	1205.93	840.42	1842.46	1296.27
Σ Total fatty acids hare	1447.97	1598.96	1651.11	1566.01
Σ Total fatty acids rabbit	1721.01	1229.95	2628.67	1859.88
% EFA hare	42.45	42.16	45.66	43.42
% EFA rabbit	24.91	25.03	26.17	25.37
% DFA hare	76.98	77.55	80.12	78.22
% DFA rabbit	70.08	68.33	70.09	69.50
Σ n6/Σn3 hare	5.61	5.50	7.07	6.06
Σ n6/Σn3 rabbit	5.59	4.85	6.70	5.71
Σ PUFA/ΣSFA hare	1.61	1.63	1.86	1.70
Σ PUFA/ΣSFA rabbit	0.778	0.784	0.783	0.78
PI hare	5.99	6.60	7.56	6.72
PI rabbit	4.09	2.74	6.95	4.59
AI hare	0.73	0.78	0.83	0.78
AI rabbit	0.68	0.50	1.02	0.73
TI hare	0.62	0.68	0.67	0.66

Table 4. *Cont.*

Health Lipid Indices	SM	LD	TB	Average/3 Carcass Areas
TI rabbit	0.37	0.32	0.49	0.39
h/H hare	3.29	3.40	4.04	3.58
h/H rabbit	2.03	1.86	2.04	1.98
NVI hare	1.36	1.42	1.65	1.48
NVI rabbit	1.41	1.21	1.40	1.34

LD—*Longissimus dorsi*; SM—*Semimembranosus*; TB—*Triceps brachii*; SFA = Saturated fatty acids; MUFA = monounsaturated fatty acids; PUFA = polyunsaturated fatty acids; EFA = essential fatty acids; %EFA = EFA $\times$ 100/Σ Total fatty acids; DFA = desirable fatty acids; %DFA = DFA $\times$ 100/Σ total fatty acids; PI = Polyunsaturation Index; AI = Atherogenic Index; TI = Thrombogenic Index; h/H = Ratio between the hypocholesterolemic and hypercholesterolemic fatty acids; NVI = Nutritive Value Index.

The desirable fatty acids (DFA%) ranged from 68.33% in rabbit LD muscles (the lowest value) to 80.12% in hare TB muscles (the highest value). Higher values were observed for all hare samples.

The Polyunsaturation Index was 5.99 in hare SM muscles, 6.60 in LD, and 7.56 in TB (Table 4). In rabbit, PI reached 4.09 in SM, 2.74 in LD, and 6.95 in TB.

The values from the Atherogenic Index in hare varied from 0.73 in SM to 0.83 in TB; in rabbit, the variability was higher, within the AI limits of 0.68 to 1.02.

The Thrombogenic Index in hare meat was relatively similar for the three muscle groups (0.62 for SM, 0.68 for LD, and 0.67 for TB) and much higher compared to rabbit samples (0.37 in SM, 0.49 in TB, and 0.32 in LD).

Higher values were observed in hare meat for the h/H index (3.29 in SM, 3.40 in LD, and 4.04 in TB) compared to rabbit (2.03 in SM, 1.86 in LD, and 2.04 in TB).

The Nutritional Value Index in hare samples varied between 1.36 in SM and 1.65 in TB, while in rabbit it ranged from 1.21 in LD to 1.41 in SM.

Overall, for all three muscle groups, each species' meat can be characterized thus: the PI was 6.72 in hare and 4.59 in rabbits; the AI reached 0.78 in hare and 0.73 in rabbit; the TI was calculated at 0.66 in hare and at 0.39 in rabbit; and the h/H index was 3.57 in hare and 1.97 in rabbit; while the NVI in hare was 1.48, and in rabbit it reached 1.34, indicating a better sanogenic value for hare and rabbit than the meat issued from other species of mammals.

3.4. The Technological Assessment of Hare and Rabbit Meat

3.4.1. pH Value

The pH24 and pH48 measured values of hare and rabbit meat are presented in Table 5.

Table 5. The pH24 and pH48 for hare and rabbit meat.

Muscles	Period	Species	Mean	$\pm$SD	V%	*p* Values
Longissimus dorsi	24 h	hare	5.631 [a]	$\pm$0.10	1.78	0.0008
		rabbit	5.724 [d]	$\pm$0.12	2.10	
	48 h	hare	5.665 [a]	$\pm$0.11	1.94	0.0003
		rabbit	5.762 [d]	$\pm$0.14	2.43	
Semimembranosus	24 h	hare	5.685 [a]	$\pm$0.08	1.41	0.0066
		rabbit	5.796 [c]	$\pm$0.12	2.07	
	48 h	hare	5.769 [a]	$\pm$0.11	1.91	0.0039
		rabbit	5.818 [c]	$\pm$0.13	2.23	
Triceps brachii	24 h	hare	6.033	$\pm$0.08	1.33	0.2529
		rabbit	6.002	$\pm$0.15	2.50	
	48 h	hare	6.138 [a]	$\pm$0.08	1.30	0.0034
		rabbit	6.087 [c]	$\pm$0.08	1.31	

SD—standard deviation; V%—coefficient of variation; ANOVA: means signalled with non-identical superscripts, between species, within the same muscle group, differ significantly for: $p < 0.01$ [a vs. c]; $p < 0.001$ [a vs. d].

3.4.2. Cooking Loss

The cooking loss (%) values of hare and rabbit meat samples are presented in Table 6.

Table 6. Cooking loss (%) for rabbit and hare meat.

Muscles	Species	Mean	±SD	V%	*p* Values
Longissimus dorsi	rabbit	31.41 [a]	±2.75	8.76	<0.001
	hare	27.52 [d]	±1.31	4.76	
Semimembranosus	rabbit	36.2 [a]	±2.25	6.22	<0.001
	hare	31.04 [d]	±1.39	4.48	
Triceps brachii	rabbit	30.23 [a]	±2.23	7.38	<0.001
	hare	28.64 [d]	±1.24	4.33	

SD—standard deviation; V%—coefficient of variation; ANOVA: means signalled with non-identical superscripts, between species, within the same muscle group, differ significantly for: $p < 0.001$ [a vs. d].

3.4.3. Water-Holding Capacity (%)

The water-holding capacity (%) values of hare and rabbit meat are presented in Table 7.

Table 7. Water-holding capacity (%) of rabbit and hare meat.

Muscles	Species	Mean	±SD	V%	*p* Values
Longissimus dorsi	rabbit	7.79 [a]	±0.30	3.85	<0.001
	hare	14.58 [d]	±0.60	4.12	
Semimembranosus	rabbit	12.17 [a]	±0.80	6.57	<0.001
	hare	18.23 [d]	±1.43	7.84	
Triceps brachii	rabbit	8.77 [a]	±0.58	6.61	<0.001
	hare	15.73 [d]	±1.26	8.01	

SD—standard deviation; V%—coefficient of variation; ANOVA: means signalled with non-identical superscripts, between species, within the same muscle group, differ significantly for: $p < 0.001$ [a vs. d].

The lowest values of WHC% were found in rabbit LD (7.79%), followed by TB (8.77%) and SM (12.17%). For hare, the lowest WHC% was found in LD (14.58%), followed by TB (15.73%) and SM samples (18.23%).

4. Discussion

4.1. The Nutritional Assessment of Hare and Rabbit Meat

4.1.1. Chemical Properties and Energy Value of Hare and Rabbit Meat

Significant differences between species were found for water content in SM samples (+0.4% in hare, $p < 0.05$) and in TB, as well (+0.8% in hare, $p < 0.01$) (Table 1). The water content for hare meat was close to that measured by other authors [26] (75.34%) and was in line with wider findings [30] (73.3–75.5 in hind leg and LD muscles); meat water found in our study was higher than reported values in other articles [20] (73.3%), [31] (73% in foreleg, 74% in hind leg, and 73.4% in LD muscles) and [27] (72.83%).

Differences between species for ash were significant in LD muscles ($p < 0.05$) and TB muscles ($p < 0.01$) (Table 1). The measured ash in hare meat was lower than that reported in certain articles [31] but aligned with other findings, 1.16% ash [23]. Vizzarri et al. [20] found 1.06% ash in LD muscles, and Trocino et al. [30] found 1.28–1.42% ash in subadult and adult hares. The lipid content in hare meat was close to that reported by other authors [20,23,27,30,31] and slightly lower than that in other studies [31].

Slightly lower protein levels occurred in our study, in comparison with levels assessed by other authors in Croatia, Italy, and Poland [26,30,31]. The quantity of lipids was slightly higher in our study, while other studies found higher lipid content than in ours [31]. In the meat of hare that had been shot, other authors described the proximate composition as follows: water 75.32%, protein 23.08%, fat 1.09%, and ash 1.16% [26].

The interspecific differences related to calorie content were significant for LD samples ($p < 0.05$) and for SM and TB muscles ($p < 0.01$) where high lipid content was measured (Table 2).

In farmed hare meat, a study found levels of 73.3% water, 22% protein, 2.1% lipids, and 1.35% ash in the hind legs and, respectively, 75.5% water, 23% protein, 1.0% lipids, and 1.44% ash in LD [30].

Protein levels in rabbit varies in accordance with the carcass part, ranging between 18.6 g/100 g in the forelegs and 22.4 g/100 g in LD muscles [41–44].

The assessed fat content was relatively close to the results found in the literature [42–50]. Pla M. et al., 2008, [45] found an average of 1.2 g/100 g lipids in LD muscle and 3.03 g/100 g in the hind legs. Similar values of ash content were found in average sized rabbit breeds in Italy [30].

The water values were placed within the limits obtained by other authors, from 69.7 g/100 g in foreleg and hind leg muscles to 75.3 g/100 g in LD muscles [42,46–50].

Rabbit and hare meat provides moderate energy value, due to its high protein and low fat content.

4.1.2. Fatty Acid Content

Regarding the fatty acid content (Table 3), the differences between species were significant for SM muscles ($p < 0.05$) but not for C18:0 (SFA) and C18:1 n-7 (MUFA), where differences of lower amplitude were observed ($p > 0.05$).

In LD muscles, the differences between species were statistically significant, with the exception of C18:1 n-9 (MUFA) fatty acids.

In TB muscles, the differences between species were statistically significant, but not for C18:2 n-6 and C18:3 n-3 (PUFA) fatty acids.

4.1.3. Health Lipid Indices

The fatty acid profile and health lipid indices, according to our findings, are presented in Table 4. For humans, EFAs are very important because they cannot be synthesized in the body and humans require EFA uptake through food. Previous studies reported that EFAs and DFAs have an important role in biological activity. For example, the proportion of DFAs in the breasts and thighs of three breeds of chickens and broilers [51] ranged from 65.15% to 69.83% and from 70.23% to 72.25%, respectively. Other authors [52] suggested that the DFAs could be useful in describing the potential health effects of different types of lipids [35].

In the present study, DFAs were 78.22% in hare and 69.50% in rabbit.

The PUFA:SFA ratio for hare meat reached 1.61 in SM muscles, 1.63 in LD muscles, and 1.86 in TB muscles, with an average of 1.70. Higher values were revealed in our research than those reported in other studies [31] (1.17 in LD, 1.20 in the hind leg, and 1.40 in the foreleg) and was close to values reported in other articles [30] (0.83 for hind leg of youth and 1.70 for adult). The high PUFA/SFA ratio in hare and rabbit meat is favourable for human health. The α-linolenic acid (C18:3 9 c12 c15 c) and PUFAs C20:5 EPA, C22:5 DPA, and C22:6 DHA received the most attention due to their importance for human health [53]; they are effective in reducing triacylglycerol in blood and preventing cardiovascular diseases. EPA and DHA also reduce inflammation and play a role in decreasing the incidence of childhood allergic diseases. In addition, EPA and DHA have biological activities that might influence tumoral cell proliferation and viability; DHA can promote tumours cell apoptosis, possibly by inducing oxidative stress [54,55]. The European Food Safety Authority recommends an intake of 250 mg of EPA plus DHA per day as sufficient for preventing morbidity in humans with no pathology installed [56].

The n-6/n-3 ratio represents an important lipid quality index and can help to prevent or treat many diseases. Eating habits have dramatically changed throughout the years, and Western diets have become poor in n-3 PUFAs and rich in n-6 PUFAs, resulting in an unhealthy n-6: n-3 ratio of 17 to 20:1, which seems much higher than the recommended ratio of 2.5 to 8:1 [57–59]. A ratio of 10:1 was recommended for n-6/n-3 to avoid the risk of cardiovascular disease, obesity, and chronic diseases [57].

The values of the TI and h/H nutritional indexes were similar to those computed for the muscle of the loin (*Longissimus lumborum*) from rabbits fed a pelleted diet supplemented with fresh alfalfa and thyme [60,61]. There was a higher n-3 family PUFA content in rabbit meat derived from a grass-based diet [61–63].

The atherogenic index (AI) and thrombogenic index (TI) are important tools indicating a potential for stimulating platelets' aggregation and estimating the likelihood of food to favour the onset of coronary heart disease. They also provide an indication about the nutritional quality of lipids, where low values suggest healthier food with better nutritional quality of fatty acids and, subsequently, a greater potential for preventing coronary diseases [41].

Fatty acid content of food has become increasingly important, and fatty acids are involved in health issues in humans. The nutritional value of fats is proven by the content and structure of fatty acids, as well as by the ratio between them [12,13,64,65]. The content and profile of the fatty acids in meat is dependent on species, breed, gender, anatomical part, muscle group, and rearing system conditions (especially animal feeding).

Palmitic acid (C16:0) and stearic acid (C18:0) were the major saturated fatty acids from the samples analysed; these were also observed by other authors in dry-cured pork and rabbit-meat burgers [4,5,66,67].

In rabbit meat, palmitoleic acid was clearly higher than in the hare meat in all three muscular groups, probably because adipocytes in hares are scarce, knowing the liveweight of hares was smaller by more than 50% (3.6 kg in hare compared with 10.9 in rabbit). Palmitoleic acid (produced and released by adipocytes) has been shown to enhance whole-body glucose disposal, to attenuate hepatic steatosis (protecting pancreatic beta-cells from palmitic acid-induced death), to improve the circulating lipid profile (in rodents and humans), and to act as regulator of physiological cardiac hypertrophy [68,69].

Oleic acid (18:1 n-9 MUFA), the most commonly occurring fatty acid in the samples, can be nutritionally beneficial: moderate and constant intake of oleic acid can reduce hypercholesterolemia [70] and insulin levels in the body, in addition to preventing the occurrence of pancreatic cancer [71] and assisting in the control of obesity [72]. The meat samples from rabbits presented a considerable amount of palmitoleic acid C16:1 n-7 (especially in TB muscles, 123.66 mg/100 g) that provides nutritional benefits along with a lower occurrence of atherosclerosis [73], type II diabetes [74], and obesity [16,75].

Concerns about the quality of rabbit meat have been observable in the literature for more than 25 years, coming in anticipation of consumers' wishes, who, unfortunately, do not eat a lot of rabbit meat anymore [76,77].

The relationship between the dietary fatty acid (FA) profile and rabbit meat FA profile has been widely and thoroughly evaluated [42,78,79] throughout the past 40 years. For non-ruminant animals, the FA profile of meat partially reflects their diet composition, and many studies aimed to incorporate the n-3 PUFA in rabbit diets, to improve quality of tissular lipids. Rabbit meat is a good source of unsaturated fatty acids (UFA) and linoleic acid (18:2 n-6; LA); the manipulation of diet is very effective in increasing n-3 PUFA, thus obtaining functional meat [10].

There are authors [80] claiming that cardiovascular disease (CVD) accounts for 45% of deaths in Europe and for 32% of deaths worldwide [81,82], whilst atherosclerosis seems to be the most important cause of cardiovascular mortality in developed countries [83]. A diet rich in saturated fatty acids (SFAs) and cholesterol along with a low intake of fibre and PUFAs is associated with atherosclerosis; therefore, a diet rich in PUFAs and low in SFAs [84] is recommended. The impact of fatty acids on atherosclerosis still remains controversial. Recent studies indicated no correlation between the consumption of SFAs and the overall mortality and also showed that some diets containing SFAs, such as dairy products, may be associated with a reduction in CVD risk [85,86].

The positive influence of PUFAs on CVD risk mitigation appear to be obvious; concerning the impact of the n-6/n-3 ratio, nutritionists recommend consuming large amounts of n-3 and give less importance to n-6; however, certain studies involving humans stated the important roles of both n-3 and n-6 fatty acids, with no correlation between the n-6/n-3 ratio, obesity, and CVD risk [57,80,86].

Due to the high PUFA level in hare meat, the AI was lower but close to rabbit meat (on average, AI was 0.78 vs 0.73), which should be attractive to consumers because of the role of PUFAs in decreasing CVD; the TI tended to be higher in hare meat (on average 0.66) than in rabbit meat (0.39 for all muscle groups). The low values of both the AI and TI may be features specific to the species.

Usually, game meat comes from hunting, but hare farmed for restocking purposes can be used for meat production due to their slaughter results (high proportion of meat in hind legs and loins) and high nutritional value of meat with especially high PUFA and EFA proportions [20,30]. This fact would offer more commercial opportunities, in addition to restocking, to hare farmers [30]. Under controlled conditions, it might be an alternative method for producing a high-quality meat that could protect consumers' health.

Indices related to human health calculated for rabbit meat are in line with other studies [77].

4.2. The Technological Assessment of Hare and Rabbit Meat

4.2.1. pH Value

The pH values were higher in TB muscles (Table 5) for both species (above 6.00 pH units). The pH of meat is mainly influenced by the metabolic specialisation of muscle (glycolytic and oxidative fibres). TB muscles have a higher oxidative metabolism, lower glycolytic potential, and higher pH value than LD and SM muscles. No significant differences occurred between species in TB muscles for pH24, and they became significant for pH48 ($p < 0.01$). In LD muscles, the species differed significantly ($p < 0.001$), as did SM muscles ($p < 0.001$) for both assessments (pH24 and pH48). The results are in line with other studies for rabbit meat issued from medium-sized breeds [87,88].

4.2.2. Cooking Loss

The highest values for cooking loss (CL%) occurred in rabbit SM muscles (Table 6). In hare, they were generally low compared to those measured in the rabbit sample, probably due to the much larger size of the muscle fibres in the latter, compared to smaller thickness of the hare muscle fibres [89,90]. Significant differences ($p < 0.001$) occurred between species for all three muscular groups, but they were in line with other studies for hare [30] and rabbit [41].

4.2.3. Water-Holding Capacity

WHC% was higher in hare than in rabbit, for all muscles groups studied ($p < 0.001$) (Table 7). In SM muscle, a higher average value of WHC (12.17% for rabbit vs 18.23% for hare) is probably due to the smaller diameter of myocites and also to higher water content. The values obtained in our study are lower than those reported by other authors for rabbit [87] in *L. lumborum* muscles (14.5–15.4%) and in *L. dorsi* muscles (16.25% and 32%) [91,92], with differences generated by breeds (medium size/hybrid vs rabbit breed) and age at slaughter (42 days, 64 days-old, and 240 days vs. 330 days-old).

Further research is needed to elucidate the role of feeding, habitat, and locomotion on meat quality of hare and rabbit. The differences found in this study could be influenced by age (rabbits were examined at older age—11 months and hares at younger age—9 months). The rabbits had a limited range of motion, and the hares moved across larger areas. Thus, in other studies [93], the percentage of fat deposits was highest in caged rabbits (compared to a bigger proportion of hind parts in rabbits reared in large pens). Water-holding capacity and lipid content were not affected by the housing systems.

For hare, the effects of diet and gender did not produce important differences in productive performances and on the quantitative and qualitative parameters of their meat. Thus, the hare reared in cages had a better performance, with fleshier carcasses and higher accumulations of fat. Rearing of hare in pens favours the production of meat with better dietary characteristics (advisable for diets oriented for the prevention of CVD) [94].

Due to its low contents of fat and cholesterol as well as to the high proportion of PUFA, rabbit meat is considered a "healthy" meat [95–97]. However, its consumption is sometimes rejected because its cooking is considered time-consuming and requires culinary skills. In order to promote the consumption of rabbit meat, the processing companies have been trying to develop ready-to-cook and ready-to-eat products. A possible way to improve rabbit meat utilization for convenience foods preparation could be realised by freezing it when the market price is lower (during summer) and using it as raw matter in preparing further processed products (minced meat, hamburgers, sausages, charcuterie) when the price becomes higher again [41,98].

The meat of goat, cattle, buffaloes, fowl, and sheep is insufficient to satisfy the growing demand for animal protein, and it has become necessary to explore alternatives animal protein sources to reduce this deficit. The rabbit, a small animal (that does not produce CO_2 in high quantities), can be considered an alternative sustainable source of protein, comparable to chicken from a nutritional point of view, and its nutritional and dietary qualities should be promoted through public campaigns for benefits regarding consumer health. Hare meat is a full-value meat, easily digestible with typical aroma for the given species, and it has finer muscle fibres than the meat of other mammals. Venison ranks among the richest proteinaceous meat, along with fish, and thanks to its relatively low fat content, rabbit has higher nutritional and dietetic qualities than other farmed species [20].

5. Conclusions

Gross energy content was higher in all muscles in rabbit versus hare meat, especially in *Triceps brachii* muscles from rabbit compared to hare, due to more lipid accumulation.

On the technological meat quality, pH value was higher in *Triceps brachii* muscles from both species, while water-holding capacity was better in hare samples, and cooking loss was better in rabbit samples. As follow-up, the connective matrix of skeletal muscles should be investigated, regarding its composition and proportion in meat, to explain the differences between species, in which now favours hare muscles.

For all muscle groups, the mean value of AI was 0.78 in hare and 0.73 in rabbit; the TI had a mean of 0.66 in hare and of 0.39 in rabbit; and the h/H index was 3.57 in hare and 1.97 in rabbit. The NVI reached 1.48 in hare and 1.34 in rabbit. Overall, hare meat was found as more dietetic (lower fat, lower gross energy, better sanogenic lipidic profile indices) than rabbit meat, whose fatty acid profile is influenced by farm feeding.

From a nutritional point of view, the consumption of both hare and rabbit meat can positively contribute to improvements in human health, because of its favourable health indices and high nutritional value.

Author Contributions: Conceptualization, G.F. and P.-C.B.; methodology, G.F. and P.-C.B.; software, M.-M.C. and R.-M.R.-R.; validation, O.C.M., D.S. and R.-M.R.-R.; formal analysis, G.F., O.C.M. and M.-M.C.; investigation, G.F.; data curation, P.-C.B.; writing—original draft preparation, G.F. and R.-M.R.-R.; writing—review and editing, G.F., O.C.M. and R.-M.R.-R.; visualization, D.S. and P.-C.B.; supervision, G.F. All authors have read and agreed to the published version of the manuscript.

Funding: This research received no external funding.

Institutional Review Board Statement: No animals were used directly to apply experimental factors on them. The study focused on the analysis of meat sampled from carcasses issued from farmed rabbit designed for the meat market and from hunted hares.

Informed Consent Statement: Not applicable.

Data Availability Statement: Data supporting reported results available, upon request, from the authors.

Conflicts of Interest: The authors declare no conflict of interest.

References

1. Müller, J.; Evans, C.L.R.; Roberts, P.R. Entomophagy and Power. *J. Insects Food Feed* **2016**, *2*, 121–136. [CrossRef]
2. Wu, L. Rabbit meat trade of major countries: Regional pattern and driving forces. *World Rabbit Sci.* **2022**, *30*, 69–82. [CrossRef]
3. FAOSTAT. The Statistics Division of the FAO. 2020. Available online: http://www.fao.org/faostat/en/#data (accessed on 10 October 2022).
4. Mancini, S.; Mattioli, S.; Nuvoloni, R.; Pedonese, F.; Dal Bosco, A.; Paci, G. Effects of Garlic Powder and Salt on Meat Quality and Microbial Loads of Rabbit Burgers. *Foods* **2020**, *9*, 1022. [CrossRef] [PubMed]
5. Mancini, S.; Mattioli, S.; Nuvoloni, R.; Pedonese, F.; Dal Bosco, A.; Paci, G. Effects of garlic powder and salt additions on fatty acids profile, oxidative status, antioxidant potential and sensory properties of raw and cooked rabbit meat burgers. *Meat Sci.* **2020**, *169*, 108226. [CrossRef] [PubMed]
6. Petracci, M.; Soglia, F.; Leroy, F. Rabbit meat in need of a hat-trick: From tradition to innovation (and back). *Meat Sci.* **2018**, *146*, 93–100. [CrossRef]
7. Krunt, O.; Zita, L.; Kraus, A.; Bures, D.; Needham, T.; Volek, Z. The effect of housing system on rabbit growth performance, carcass traits, and meat quality characteristics of different muscles. *Meat Sci.* **2022**, *193*, 108953. [CrossRef]
8. U.S. Department of Agriculture. Food Data Central Database. Available online: https://fdc.nal.usda.gov/ (accessed on 5 April 2022).
9. Długaszek, M.; Kopczyński, K. Elemental Composition of Muscle Tissue of Wild Animals from Central Region of Poland. *Int. J. Environ. Res.* **2013**, *7*, 973–978.
10. Dalle Zotte, A.; Szendro, Z. The role of rabbit meat as functional food. *Meat Sci.* **2011**, *88*, 319–331. [CrossRef]
11. Pogány Simonová, M.; Chrastinová, L'.; Lauková, A. Enterocin 7420 and Sage in Rabbit Diet and Their Effect on Meat Mineral Content and Physico-Chemical Properties. *Microorganisms* **2022**, *10*, 1094. [CrossRef]
12. Tufarelli, V.; Tateo, A.; Schiavitto, M.; Mazzei, D.; Calzaretti, G.; Laudadio, V. Evaluating productive performance, meat quality and oxidation products of Italian White breed rabbits under free-range and cage rearing system. *Anim. Biosci.* **2022**, *35*, 884–889. [CrossRef]
13. Laudadio, V.; Tufarelli, V. Pea (*Pisum sativum* L.) seeds as an alternative dietary protein source for broilers: Influence on fatty acid composition, lipid and protein oxidation of dark and white meats. *J. Am. Oil Chem. Soc.* **2011**, *88*, 967–973. [CrossRef]
14. Luo, G.; Zhu, T.; Ren, Z. METTL3 Regulated the Meat Quality of Rex Rabbits by Controlling PCK2 Expression via a YTHDF2–N6-Methyladenosine Axis. *Foods* **2022**, *11*, 1549. [CrossRef] [PubMed]
15. Alves dos Santos, J.J.; Fonseca Pascoal, L.A.; Brandão Grisi, C.V.; da Costa Santos, V.; de Santana Neto, D.C.; Filho, J.J.; Ferreira Herminio, M.P.; Fabricio Dantas, A. Soybean oil and selenium yeast levels in the diet of rabbits on performance, fatty acid profile, enzyme activity and oxidative stability of meat. *Livest. Sci.* **2022**, *263*, 105021. [CrossRef]
16. Pedro, D.; Saldana, E.; Lorenzo, J.M.; Pateiro, M.; Dominguez, R.; Dos Santos, A.B.; Campagnol, C.B.P. Low-sodium dry-cured rabbit leg: A novel meat product with healthier properties. *Meat Sci.* **2021**, *173*, 108372. [CrossRef]
17. Szendrő, K.; Szabó-Szentgróti, E.; Szigeti, O. Consumers' Attitude to Consumption of Rabbit Meat in Eight Countries Depending on the Production Method and Its Purchase Form. *Foods* **2020**, *9*, 654. [CrossRef]
18. Trombetti, F.; Minardi, P.; Mordenti, A.L.; Badiani, A.; Ventrella, V.; Albonetti, S. The Evaluation of the Effects of Dietary Vitamin E or Selenium on Lipid Oxidation in Rabbit Hamburgers: Comparing TBARS and Hexanal SPME-GC Analyses. *Foods* **2022**, *11*, 1911. [CrossRef]
19. Minardi, P.; Mordenti, A.L.; Badiani, A.; Pirini, M.; Trombetti, F.; Albonetti, S. Effect of the dietary antioxidants supplementation on rabbit performances, meat quality and oxidative stability of muscles. *World Rabbit Sci.* **2020**, *28*, 145–159. [CrossRef]
20. Vizzarri, F.; Nardoia, M.; Palazzo, M. Effect of dietary *Lippia citriodora* extract on productive performance and meat quality parameters in hares (*Lepus europaeus Pall.*). *Arch. Anim. Breed.* **2014**, *57*, 20. [CrossRef]
21. Rigo, N.; Trocino, A.; Poppi, L.; Giacomelli, M.; Grilli, G.; Piccirillo, A. Performance and mortality of farmed hares. *Animal* **2015**, *9*, 1025–1031. [CrossRef]
22. Rødbotten, M.; Kubberød, E.; Lea, P.; Ueland, Ø. A sensory map of the meat universe. Sensory profile of meat from 15 species. *Meat Sci.* **2004**, *68*, 137–144. [CrossRef]
23. Konjević, D. Hare brown (*Lepus europaeus Pallas*) and potential in diet of people today. *Prof. Work* **2007**, *9*, 288–291.
24. Valencak, T.G.; Arnold, W.; Tataruch, F.; Ruf, T. High content of polyunsaturated fatty acids in muscle phospholipids of a fast runner, the European brown hare (*Lepus europaeus*). *J. Comp. Physiol. B* **2003**, *173*, 695–702. [CrossRef] [PubMed]
25. Valencak, T.G.; Gamsjäger, L.; Ohrnberger, S.; Culbert, N.J.; Ruf, T. Healthy n-6/n-3 fatty acid composition from five European game meat species remains after cooking. *BMC Res. Notes* **2015**, *8*, 273. [CrossRef] [PubMed]
26. Skrivanko, M.; Hadžiosmanović, M.; Cvrtila, Z.; Zdolec, N.; Filipović, I.; Kozačinski, L.; Florijančić, T.; Bošković, I. The hygiene and quality of hare meat (*Lepus europaeus Pallas*) from Eastern Croatia. *Arch. Lebensm.* **2008**, *59*, 180–184. [CrossRef]

27. Mertin, D.; Slamečka, J.; Ondruška, I.; Zaujec, K.; Jurčík, R.; Gašparík, J. Comparison of meat quality between european brown hare and domestic rabbit. *Slovak J. Anim. Sci.* **2012**, *45*, 89–95.

28. Strmiskova, G.; Strmiska, F. Contents of mineral substances in venison. *Food/Nahrung* **1992**, *36*, 307–308. [CrossRef]

29. Papadomichelakis, G.; Zoidis, E.; Pappas, A.C.; Hadjigeorgiou, I. Seasonal variations in the fatty acid composition of Greek wild rabbit meat. *Meat Sci.* **2017**, *134*, 158–162. [CrossRef]

30. Trocino, A.; Birolo, M.; Dabbou, S.; Gratta, F.; Rigo, N.; Xiccato, G. Effect of age and gender on carcass traits and meat quality of farmed brown hares. *Animal* **2018**, *9*, 1–8. [CrossRef]

31. Króliczewska, B.; Miśta, D.; Korzeniowska, M.; Pecka-Kiełb, E.; Zachwieja, A. Comparative evaluation of the quality and fatty acid profile of meat from brown hares and domestic rabbits offered the same diet. *Meat Sci.* **2018**, *145*, 292–299. [CrossRef]

32. AOAC International. *Official Methods of Analyses*, 17th ed.; Association of Official Analytical Chemists: Washington, DC, USA, 2000.

33. AOAC International. *Official Methods of Analyses*, 18th ed.; Association of Official Analytical Chemists: Gaithersburg, MD, USA, 2005.

34. FAO. Food Energy–Methods of Analysis and Conversion Factors. Food and Agriculture Organization of the United Nations, Rome, 2003. Report of a Technical Workshop. Available online: http://www.fao.org/uploads/media/FAO_2003_Food_Energy_02.pdf (accessed on 5 May 2022).

35. Chen, Y.; Qiao, Y.; Xiao, Y.; Chen, H.; Zhao, L.; Huang, M.; Zhou, G. Differences in physicochemical and nutritional properties of breast and thigh meat from crossbred chickens, commercial broilers and spent hens. *Asian-Australas. J. Anim. Sci.* **2016**, *29*, 855–864. [CrossRef]

36. Timmons, J.S.; Weiss, W.P.; Palmquist, D.L.; Harper, W.J. Relationships among dietary roasted soybeans, milk components, and spontaneous oxidized flavor of milk. *J. Dairy Sci.* **2001**, *84*, 2440–2449. [CrossRef] [PubMed]

37. Ulbricht, T.L.V.; Southgate, D.A.T. Coronary heart disease: Seven dietary factors. *Lancet* **1991**, *338*, 985–992. [CrossRef] [PubMed]

38. Fernandez, M.; Ordóñez, J.A.; Cambero, I.; Santos, C.; Pin, C.; De La Hoz, L. Fatty acid compositions of selected varieties of Spanish dry ham related to their nutritional implications. *Food. Chem.* **2007**, *9*, 107–112. [CrossRef]

39. Struți, D.I.; Mierliță, D.; Simeanu, D.; Pop, I.M.; Socol, C.T.; Papuc, T.; Macri, A.M. The effect of dehulling lupine seeds (*Lupinus albus* L.) from low-alkaloid varieties on the chemical composition and fatty acids content. *J. Chem.* **2020**, *71*, 59–70. [CrossRef]

40. Mierliță, D.; Pop, I.M.; Lup, F.; Simeanu, D.; Vicas, S.I.; Simeanu, C. The Fatty Acids Composition and Health Lipid Indices in the Sheep Raw Milk under a Pasture-Based Dairy System. *J. Chem.* **2018**, *69*, 160–165. [CrossRef]

41. Abdel-Naeem, H.H.; Sallam, K.I.; Zaki, H.M. Effect of different cooking methods of rabbit meat on topographical changes, physicochemical characteristics, fatty acids profile, microbial quality, and sensory attributes. *Meat Sci.* **2021**, *181*, 108612. [CrossRef]

42. Cullere, M.; Dalle Zotte, A. Rabbit meat production and consumption: State of knowledge and future perspectives. *Meat Sci.* **2018**, *143*, 137–146. [CrossRef]

43. Pla, M.; Apolinar, R. The filter-paper press as method for measuring water holding capacity of rabbit meat. *J. World Rabbit Sci. Assoc.* **2000**, *8*, 659–662.

44. Blasco, A.; Nagy, I.; Hernández, P. Genetics of growth, carcass and meat quality in rabbits. *Meat Sci.* **2018**, *145*, 178–185. [CrossRef]

45. Birolo, M.; Xiccato, G.; Bordignon, F.; Dabbou, S.; Zuffellato, A.; Trocino, A. Growth Performance, Digestive Efficiency, and Meat Quality of Two Commercial Crossbred Rabbits Fed Diets Differing in Energy and Protein Levels. *Animals* **2022**, *12*, 2427. [CrossRef]

46. Pla, M. A comparison of the carcass traits and meat quality of conventionally and organically produced rabbits. *Livest. Sci.* **2008**, *115*, 1–12. [CrossRef]

47. Zomeño, C.; Blasco, A.; Hernández, P. Divergent selection for intramuscular fat content in rabbits. II. Correlated responses on carcass and meat quality traits. *J. Anim. Sci.* **2013**, *91*, 4532–4539. [CrossRef] [PubMed]

48. Martínez-Álvaro, M.; Hernández, P.; Agha, S.; Blasco, A. Correlated responses to selection for intramuscular fat in several muscles in rabbits. *Meat Sci.* **2018**, *139*, 187–191. [CrossRef] [PubMed]

49. Hernandez, P.; Cesari, V.; Blasco, A. Effect of genetic rabbit lines on lipid content, lipolytic activities and fatty acid composition of hind leg meat and perirenal fat. *Meat Sci.* **2008**, *78*, 485–491. [CrossRef] [PubMed]

50. Martínez-Álvaro, M.; Hernández, P.; Blasco, A. Divergent selection on intramuscular fat in rabbits: Responses to selection and genetic parameters. *J. Anim. Sci.* **2016**, *94*, 4993–5003. [CrossRef] [PubMed]

51. Sosa-Madrid, B.S.; Varona, L.; Blasco, A.; Hernández, P.; Casto-Rebollo, C.; Ibáñez-Escriche, N. The effect of divergent selection for intramuscular fat on the domestic rabbit genome. *Animal* **2020**, *14*, 2225–2235. [CrossRef]

52. Rikimaru, K.; Takahashi, H. Evaluation of the meat from Hinai-jidori chickens and broilers: Analysis of general biochemical components, free amino acids, inosine 5′-monophosphate, and fatty acids. *J. Appl. Poult. Res.* **2010**, *19*, 327–333. [CrossRef]

53. Banskalieva, V.; Sahlu, T.; Goetsch, A. Fatty acid composition of goat muscles and fat depots: A review. *Small Rumin. Res.* **2000**, *37*, 255–268. [CrossRef]

54. Chikwanha, O.C.; Vahmani, P.; Muchenje, V.; Gugan, M.E.R.; Mapiye, C. Nutritional enhancement of sheep meat fatty acid profile for human health and wellbeing. *Food Res. Int.* **2018**, *104*, 25–38. [CrossRef]

55. Calder, P.C. Functional roles of fatty acids and their effects on human health. *J. Parenter. Enter. Nutr.* **2015**, *39* (Suppl. S1), 18S–32S. [CrossRef]

56. Miles, E.A.; Calder, P.C. Omega-6 and omega-3 polyunsaturated fatty acids and allergics diseases in infancy and childhood. *Curr. Pharm. Des.* **2014**, *20*, 946–953. [CrossRef] [PubMed]
57. European Food Safety Authority. Scientific opinion on dietary reference values for fats, including saturated fatty acids, polyunsaturated fatty acids, monounsaturated fatty acids, trans fatty acids, and cholesterol. *Eur. Food Saf. Auth. J.* **2010**, *8*, 1461.
58. Simopoulos, A.P. An Increase in the Omega-6/Omega-3 Fatty Acid Ratio Increases the Risk for Obesity. *Nutrients* **2016**, *8*, 128. [CrossRef] [PubMed]
59. World Health Organization/Food and Agriculture Organization (WHO/FAO). Expert Report: Diet, Nutrition and Prevention of Chronic Diseases. Report of a Joint WHO/FAO Expert Consultation. WHO Technical Report Series 916. 2003. Available online: http://apps.who.int/iris/bitstream/handle/10665/42665/WHO_TRS_916.pdf;jsessionid=376354A9A8120A5273B0C2 103D6CCB0B?sequence=1 (accessed on 10 October 2022).
60. World Health Organization/Food and Agriculture Organization (WHO/FAO). *Interim Summary of Conclusions and Dietary Recommendations on Total Fat and Fatty Acids from the Joint FAO/WHO Expert Consultation on Fats and Fatty Acids in Human Nutrition;* WHO: Geneva, Switzerland, 2008; pp. 10–14. Available online: https://www.foodpolitics.com/wp-content/uploads/FFA_ summary_rec_conclusion.pdf (accessed on 10 October 2022).
61. Capra, G.; Martínez, R.; Fradiletti, F.; Cozzano, S.; Repiso, L.; Márquez, R.; Ibáñez, F. Meat quality of rabbits reared with two different feeding strategies: With or without fresh alfalfa ad libitum. *World Rabbit Sci.* **2013**, *21*, 23–32. [CrossRef]
62. Dal Bosco, A.; Gerencsér, Z.; Szendrő, Z.; Mugnai, C.; Cullere, M.; Kovàcs, M.; Ruggeri, S.; Mattioli, S.; Castellini, C.; Dalle Zotte, A. Effect of dietary supplementation of Spirulina (*Arthrospira platensis*) and Thyme (*Thymus vulgaris*) on rabbit meat appearance, oxidative stability and fatty acid profile during retail display. *Meat Sci.* **2014**, *96*, 114–119. [CrossRef]
63. Dal Bosco, A.; Castellini, C.; Martino, M.; Mattioli, S.; Marconi, O.; Sileoni, V.; Ruggeri, S.; Tei, F.; Benincasa, P. The effect of dietary alfalfa and flax sprouts on rabbit meat antioxidant content, lipid oxidation and fatty acid composition. *Meat Sci.* **2015**, *106*, 31–37. [CrossRef]
64. Attia, Y.A.; Al-Harthi, M.A.; Korish, M.M.; Shiboob, M.M. Fatty acid and cholesterol profiles and hypocholesterolemic, atherogenic, and thrombogenic indices of table eggs in the retail market. *Lipids Health Dis.* **2015**, *14*, 136. [CrossRef]
65. Attia, Y.A.; Al-Harthi, M.A.; Korish, M.M.; Shiboob, M.M. Fatty acid and cholesterol profiles, hypocholesterolemic, atherogenic, and thrombogenic indices of broiler meat in the retail market. *Lipids Health Dis.* **2017**, *16*, 40. [CrossRef]
66. Carrapiso, A.I.; Tejeda, J.F.; Noguera, J.L.; Ibanez-Escriche, N.; Gonzalez, E. Effect of the genetic line and oleic acid-enriched mixed diets on the subcutaneous fatty acid composition and sensory characteristics of dry-cured shoulders from Iberian pig. *Meat Sci.* **2020**, *159*, 107933. [CrossRef]
67. Liu, S.; Wang, G.; Xiao, Z.; Pu, Y.; Ge, C.; Liao, G. H-NMR-based water-soluble low molecular weight compound characterization and free fatty acid composition of five kinds of Yunnan dry-cured hams. *LWT* **2019**, *108*, 174–182. [CrossRef]
68. Alonso-Vale, M.I.; Cruz, M.; Bolsoni-Lopes, A.; de Sá, R.; de Andrade, P. Palmitoleic Acid (C16:1n7) Treatment Enhances Fatty Acid Oxidation and Oxygen Consumption in White Adipocytes. *FASEB J.* **2015**, *29*, 884.25. [CrossRef]
69. Betz, I.R.; Qaiyumi, S.J.; Goeritzer, M.; Thiele, A.; Brix, S.; Beyhoff, N.; Grune, J.; Klopfleisch, R.; Greulich, F.; Uhlenhaut, N.H.; et al. Cardioprotective Effects of Palmitoleic Acid (C16:1n7) in a Mouse Model of Catecholamine-Induced Cardiac Damage Are Mediated by PPAR Activation. *Int. J. Mol. Sci.* **2021**, *22*, 12695. [CrossRef] [PubMed]
70. Smith, S.B.; Lunt, D.K.; Smith, D.R.; Walzem, R.L. Producing high-oleic acid beef and the impact of ground beef consumption on risk factors for cardiovascular disease: A review. *Meat Sci.* **2020**, *163*, 108076. [CrossRef] [PubMed]
71. Banim, P.J.R.; Luben, R.; Khaw, K.-T.; Hart, A.R. Dietary oleic acid is inversely associated with pancreatic cancer—Data from food diaries in a cohort study. *Pancreatology* **2018**, *18*, 655–660. [CrossRef] [PubMed]
72. Jagannathan, L.; Socks, E.; Balasubramanian, P.; Mcgowan, R.; Herdt, T.M.; Kianian, R.; Mohankumar, P.S. Oleic acid stimulates monoamine efflux through PPAR-α: Differential effects in diet-induced obesity. *Life Sci.* **2020**, *255*, 11786715. [CrossRef] [PubMed]
73. Magtanong, L.; Ko, P.J.; To, M.; Cao, J.Y.; Forcina, G.C.; Tarangelo, A.; Ward, C.C.; Cho, K.; Patti, G.J.; Nomura, D.K.; et al. Exogenous monounsaturated fatty acids promote a ferroptosis-resistant cell state. *Cell Chem. Biol.* **2019**, *26*, 420–432. [CrossRef]
74. Guillocheau, E.; Penhoat, C.; Drouin, G.; Godet, A.; Catheline, D.; Legrand, P.; Rioux, V. Current intakes of trans-palmitoleic (trans-C16:1 n-7) and trans-vaccenic (trans-C18:1 n-7) acids in France are exclusively ensured by ruminant milk and ruminant meat: A market basket investigation. *Food Chem. X* **2020**, *5*, 10008130. [CrossRef]
75. Frigolet, M.E.; Gutierrez-Aguilar, R. The role of the novel Lipokine Palmitoleic acid in health and disease. *Adv. Nutr.* **2017**, *8*, 173S–181S. [CrossRef]
76. Weber, J.; Bochi, V.C.; Ribeiro, C.P.; Victorio, A.M.; Emanuelli, T. Effect of different cooking methods on the oxidation, proximate and fatty acid composition of silver catfish (*Rhamdia quelen*) fillets. *Food Chem.* **2008**, *106*, 140–146. [CrossRef]
77. Guedes, C.M.; Almeida, M.; Closson, M.; Garcia-Santos, S.; Lorenzo, J.M.; Domínguez, R.; Ferreira, L.; Trindade, H.; Silva, S.; Pinheiro, V. Effect of Total Replacement of Soya Bean Meal by Whole Lupine Seeds and of Gender on the Meat Quality and Fatty Acids Profile of Growing Rabbits. *Foods* **2022**, *11*, 2411. [CrossRef]
78. Cullere, M.; Zotte, A.D.; Tasoniero, G.; Giaccone, V.; Szendrő, Z.; Szín, M.; Odermatt, M.; Gerencsér, Z.; Dal Bosco, A.; Matics, Z. Effect of diet and packaging system on the microbial status, pH, color and sensory traits of rabbit meat evaluated during chilled storage. *Meat Sci.* **2018**, *141*, 36–43. [CrossRef] [PubMed]
79. Crovato, S.; Pinto, A.; Di Martino, G.; Mascarello, G.; Rizzoli, V.; Marcolin, S.; Ravarotto, L. Purchasing Habits, Sustainability Perceptions, and Welfare Concerns of Italian Consumers Regarding Rabbit Meat. *Foods* **2022**, *11*, 1205. [CrossRef] [PubMed]

80. Bujok, J.; Miśta, D.; Wincewicz, E.; Króliczewska, B.; Dzimira, S.; Żuk, M. Atherosclerosis Development and Aortic Contractility in Hypercholesterolemic Rabbits Supplemented with Two Different Flaxseed Varieties. *Foods* **2021**, *10*, 534. [CrossRef] [PubMed]
81. European Heart Network. European Cardiovascular Disease Statistics. 2017. Available online: http://www.ehnheart.org/cvdstatistics.html (accessed on 10 October 2022).
82. Roth, G.A.; Huffman, M.D.; Moran, A.E.; Feigin, V.; Mensah, G.A.; Naghavi, M.; Murray, C.J. Global and regional patterns in cardiovascular mortality from 1990 to 2013. *Circulation* **2015**, *27*, 1667–1678. [CrossRef] [PubMed]
83. Barquera, S.; Pedroza-Tobías, A.; Medina, C.; Hernández-Barrera, L.; Bibbins-Domingo, K.; Lozano, R.; Moran, A.E. Global Overview of the Epidemiology of Atherosclerotic Cardiovascular Disease. *Arch. Med. Res.* **2015**, *46*, 328–338. [CrossRef]
84. Wang, D.D.; Hu, F.B. Dietary Fat and Risk of Cardiovascular Disease: Recent Controversies and Advances. *Annu. Rev. Nutr.* **2017**, *37*, 423–446. [CrossRef]
85. De Souza, R.J.; Mente, A.; Maroleanu, A.; Cozma, A.I.; Ha, V.; Kishibe, T.; Uleryk, E.; Budylowski, P.; Schunemann, H.; Beyene, J.; et al. Intake of saturated and trans unsaturated fatty acids and risk of all cause mortality, cardiovascular disease, and type 2 diabetes: Systematic review and meta-analysis of observational studies. *BMJ* **2015**, *351*, h3978. [CrossRef]
86. Visioli, F.; Poli, A. Fatty Acids and Cardiovascular Risk. Evidence, Lack of Evidence, and Diligence. *Nutrients* **2020**, *12*, 3782. [CrossRef]
87. Apata, E.S.; Eniolorunda, O.O.; Amao, K.E.; Okubanjo, A.O. Quality evaluation of rabbit meat as affected by different stunning methods. *Int. J. Agric. Sci.* **2012**, *2*, 054–058.
88. Blasco, A.; Piles, M. Muscular pH of the rabbit. *Ann. Zootech.* **1990**, *39*, 133–136. [CrossRef]
89. Tărnăuceanu, G.; Lazăr, R.; Boișteanu, P.C. Researches Regarding Histological Characterization of Muscles Harvested from Hares (*Lepus europaeus* Pallas). *Lucrări Ştiinţifice-Univ. Ştiinţe Agric. Med. Vet. Ser. Zooteh.* **2012**, *58*, 193–198.
90. Tărnăuceanu, G.; Pop, C.; Boișteanu, P.C. Histologic characterization of muscles collected from rabbits Flemish Giant breed. *Sci. Pap. Anim. Sci. Biotechnol.* **2015**, *48*, 307–311.
91. Lafuente, R.; López, M. Effect of electrical and mechanical stunning on bleeding, instrumental properties and sensory meat quality in rabbits. *Meat Sci.* **2014**, *98*, 247–254. [CrossRef] [PubMed]
92. Simonová Pogány, M.; Chrastinová, Ľ.; Mojto, J.; Lauková, A.; Szabóová, R.; Rafay, J. Quality of Rabbit Meat and Phyto-Additives. *Czech J. Food Sci.* **2010**, *28*, 161–167. [CrossRef]
93. Combes, S.; Postollec, G.; Cauquil, L.; Gidenne, T. Influence of cage or pen housing on carcass traits and meat quality of rabbit. *Animal* **2010**, *4*, 295–302. [CrossRef]
94. Vicenti, A.; Ragni, M.; di Summa, A.; Marsico, G.; Vonghia, G. Influence of Feeds and Rearing System on the Productive Performances and the Chemical and Fatty Acid Composition of Hare Meat. *Food Sci. Tech. Intern.* **2003**, *9*, 279–284. [CrossRef]
95. Hernandez, P.; Gondret, F. Rabbit meat quality and safety. In *Recent Advances in Rabbit Science*; ILVO: Melle, Belgium, 2006; pp. 267–290.
96. Liste, G.; Villaroel, M.; Chacon, G.; Sanudo, C.; Olleta, J.L.; Garcia-Belenguer, S. Effect of lairage duration on rabbit welfare and meat quality. *Meat Sci.* **2009**, *82*, 71–76. [CrossRef]
97. Tărnăuceanu, G.; Pop, C. Water holding capacity of meat from rabbits (*Belgian Giant breed*), Bulletin of University of Agricultural Sciences and Veterinary Medicine Cluj-Napoca. *Anim. Sci. Biotech.* **2016**, *73*, 111–112. [CrossRef]
98. Puolanne, E.; Halonen, M. Theoretical aspects of water-holding in meat. *Meat Sci.* **2010**, *86*, 151. [CrossRef]

 *agriculture*

Diets Composed of Tifton 85 Grass Hay (*Cynodon* sp.) and Concentrate on the Quantitative and Qualitative Traits of Carcass and Meat from Lambs

Yohana Corrêa [1], Edson Santos [1], Juliana Oliveira [1], Gleidson Carvalho [2], Luís Pinto [2], Danillo Pereira [1], Dallyson Assis [2], Gabriel Cruz [3], Natalia Panosso [1], Alexandre Perazzo [4], Guilherme Leite [1], Paulo Azevedo [1], Anny Lima [4], Daniele Ferreira [4], Fagton Negrão [5] and Anderson Zanine [4,*]

1   Department of Animal Science, Federal University of Paraíba, Areia 58397000, Brazil;
    yohana.correa@academico.ufpb.br (Y.C.); edson@cca.ufpb.br (E.S.); juliana@cca.ufpb.br (J.O.);
    danillomarte.zootec@gmail.com (D.P.); natalia.panosso@ufv.br (N.P.);
    guilherme.leite@academico.ufpb.br (G.L.); azevedo@cca.ufpb.br (P.A.)
2   Department of Animal Science, Federal University of Bahia, Salvador 40170110, Brazil;
    gleidsongiordano@ufba.br (G.C.); luisfbp@gmail.com (L.P.); dcoura2@hotmail.com (D.A.)
3   Department of Animal Science, Federal University of Viçosa, Viçosa 36570900, Brazil; gabriel.cruz@ufv.br
4   Department of Animal Science, Federal University of Maranhão, Chapadinha 65500000, Brazil;
    alexandreperazzo@ufpi.edu.br (A.P.); anny.graycy@ufma.br (A.L.); dany_dosanjos@yahoo.com.br (D.F.)
5   Department of Animal Science, Federal Institute of Education, Science and Technology Rondônia,
    Colorado do Oeste 76993000, Brazil; fagton.negrao@ifro.edu.br
*   Correspondence: anderson.zanine@ufma.br

Abstract: The high intake of fermentable carbohydrates may cause nutritional disorders and negatively affect animal performance. Thus, the research study aimed to determine the better roughage: concentrate ratio to improve the carcass traits and physicochemical quality of meat from feedlot-finished Santa Ines lambs. Diets were composed of Tifton 85 grass hay (*Cynodon* sp.) and concentrate (soybean meal, corn meal, urea, and mineral mixture) and consisted of five roughage: concentrate ratios of 88:12 (C12), 69:31 (C31), 50:50 (C50), 31:69 (C69), and 12:88 (C88). After 63 days the animals were slaughtered and carcass traits, the yield of commercial cuts, and physicochemical properties of meat were evaluated. The higher percentage of concentrate on roughage provided higher DM intake, better feed conversion, higher conformation, finishing, and carcass yield that resulted in heavier commercial cuts with higher fat content in the meat. The addition of 50% concentrate to the roughage improved the carcass traits, commercial cuts, and physicochemical parameters of the meat in a similar way to the diet with 88% concentrate, but with leaner meats, meeting the demands of the current consumer market.

Keywords: carcass yield; commercial cuts; low cost; neutral detergent fiber; non-fiber carbohydrate

Citation: Corrêa, Y.; Santos, E.; Oliveira, J.; Carvalho, G.; Pinto, L.; Pereira, D.; Assis, D.; Cruz, G.; Panosso, N.; Perazzo, A.; et al. Diets Composed of Tifton 85 Grass Hay (*Cynodon* sp.) and Concentrate on the Quantitative and Qualitative Traits of Carcass and Meat from Lambs. *Agriculture* **2022**, *12*, 752. https://doi.org/10.3390/agriculture12060752

Academic Editor: Daniel Simeanu

Received: 3 May 2022
Accepted: 23 May 2022
Published: 25 May 2022

Publisher's Note: MDPI stays neutral with regard to jurisdictional claims in published maps and institutional affiliations.

1. Introduction

Young lambs with high potential for body weight gain are usually finished between 60 and 150 days of age and need diets with high protein and energy content according to the NRC [1]. However, in Brazil, the ruminant's production usually uses forages as exclusive feed because it has a lower cost, but longer periods are required for finishing animals under these conditions, even when well-managed, it cannot fully meet the animal's nutritional demands, resulting in slower body development, animals with lighter carcass weight and consequently, low yield and decreased producer's revenue [2].

In contrast, feedlot systems for finishing use diets based on high concentrates ratio with a greater non-fiber carbohydrates (NFC) content and have short-term finishing periods resulting in rapid body development, reduced time to slaughter, heavier carcasses with better quality, and a higher percentage of marbling fat. Although, the high expense of feed provided to the animals do not always provide greater profitability to the producer [3,4].

However, the ruminal microbiota need availability of energy and protein in synchrony for growth and replication, resulting in better use of nutrients and SCFA production, which will provide a better performance, especially in parameters of carcass traits and meat quality. The intake of substantial amounts of readily fermentable carbohydrates can lead to changes in the ruminal fermentation pattern, resulting in increased production of short-chain fatty acids (SCFA), especially lactic acid that cause a drop in ruminal pH, directly affecting the ruminal microbiota [5] that may cause the animal to experience clinical or subclinical acidosis. SCFA production is closely related to the availability and synchronicity of nutrients which provide the proliferation and growth of the ruminal microbiota [6].

In view of the above for finishing lambs, it is believed that the addition of concentrate to the roughage will improve the synchronicity of rumen microorganisms, resulting in better performance, carcass traits, and meat physicochemical quality.

Thus, the present research aimed to determine the better roughage: concentrate ratio to improve the carcass traits and physicochemical properties of meat from feedlot-finished Santa Ines lambs.

2. Materials and Methods

The experiment was conducted at the Experimental Farm of Veterinary Medicine and Animal Science of the Federal University of Bahia, which is located at 174 km from the BR 101 highway, at 12°23′58″ S and 38°52′44″ W, Bahia State, Brazil. The research was approved by the ethics committee for the use of animals (CEUA), under protocol number 68/2018.

2.1. Animals, Diets, and General Procedures

Sixty intact Santa Ines sheep were used, with an initial average body weight (IBW) of 25.84 ± 0.53 kg, identified with numbered ear tags, dewormed, weighed, randomly assigned to their treatments, and housed in individual pens of 1.0 × 1.0 m, that were equipped with slatted wood floors, containing feed troughs and water troughs. The experiment lasted 78 days (15 for adaptation and 63 days for data collection). During the adaptation period, all animals were treated for internal and external parasites with ivermectin (Ivomec gold; Merial, Salvador, Bahia, Brazil) and vaccinated against clostridiosis using Polivalente (Sintoxan; Merial, Sao Paulo, Brazil). The animals received water ad libitum and the diets were supplied twice a day (8:00 and 15:00) i.e., 50% of the diet was allocated in the morning and the other half in the afternoon. Feed refusals were collected and weighed daily, and the amount of feed offered was adjusted to allow an approximate 20% refusal. Before the experiment, the feed components were chemically analyzed (Tables 1 and 2) separately with triplicate samples.

Table 1. Chemical composition of feed ingredients.

Items (g/kg DM)	Ingredient			
	Tifton Hay 85	**Soybean Meal**	**Corn Meal**	**Urea**
Dry matter, g/kg as fed	890	918	917	99.0
Organic matter	850	865	865	-
Crude protein	44.7	471	75.3	281
Ether extract	18.5	26.1	45.9	-
Neutral detergent fiber	762	144	134	-
Neutral detergent fiber$_{ap}$ [1]	682	101	118	-
Acid detergent fiber	346	77.4	36.0	-
Non-fiber carbohydrate	21.5	34.9	70.9	-

[1] Corrected to ash and protein.

Table 2. Ingredient proportions and chemical composition of experimental diets.

Items	Experimental Diets				
	C12	**C31**	**C50**	**C69**	**C88**
Ingredients (g/kg DM)					
Tifton hay	880	690	500	310	120
Soybean meal	80.0	80.0	76.0	74.0	72.0
Corn meal	10.0	200	394	586	778
Urea	15.0	15.0	15.0	15.0	15.0
Mineral mixture [1]	15.0	15.0	15.0	15.0	15.0
Chemical Composition (g/kg DM)					
Dry matter (g/kg as fed)	896	901	906	911	916
Ash	55.0	57.2	60.5	62.8	64.1
Organic matter	826	829	832	834	837
Crude protein	120	126	130	135	140
Ether extract	18.8	24.0	29.3	34.6	39.8
Neutral detergent fiber	683	564	445	325	206
Neutral detergent fiber$_{ap}$ [2]	609	502	395	288	181
Acid detergent fiber	311	252	193	134	75.1
Non-fiber carbohydrate	224	318	413	508	602
Metabolizable energy (MJ/kg)	8.75	9.50	9.88	10.2	12.5

[1] Guaranteed levels (for active elements): 120 g calcium, 87 g phosphorus, 147 g sodium, 18 g sulfur, 590 mg copper, 40 mg cobalt, 20 mg chrome, 1800 mg iron, 80 mg iodine, 1300 mg manganese, 15 mg selenium, 3800 mg zinc, 300 mg molybdenum, and maximum 870 mg fluoride. Solubility of phosphorus citric acid: 2–95%. [2] Corrected to ash and protein.

The experimental diets were formulated according to the requirements recommended by National Research Council [1] and consisted of five Tifton hay (*Cynodon* sp.) to concentrate ratios of 88:12 (C12), 69:31 (C31), 50:50 (C50), 31:69 (C69), and 12:88 (C88) (Table 2). The concentrates were composed of soybean meal, corn meal, urea, and mineral mixture. The Tifton-85 hay was chopped into short lengths (2–3 cm) and mixed with the concentrate before offering.

Samples ingredients, refusals, and feces were pre-dried in a forced-air ventilation oven at 55 °C for 72 h and ground in a Wiley knife mill with a sieve size of 1 mm for ingredients and refusal samples and 3 mm for feces samples.

2.2. Diet Composition Actually Consumed and Growth Performance

After a period 15 days for animals' adaptation to diets and stables, the animals were weighed to obtain the initial body weight (IBW).

The dry matter intake (DMI) was calculated by the difference between the dry matter (DM) present in the offered diet and the DM obtained from the animals' refusals. In the same way, metabolizable energy intake (ME intake) was calculated.

The diet composition actually consumed (DCC) by lambs was estimated based on the ratio of the intake of each nutrient by the equation:

$$DCC = (Nutrient/DMI) \times 100 \tag{1}$$

After the experimental period (63 days), the animals were weighed (body weight at slaughter—BWS) after 16 -h solid fasting. Total weight gain (TWG) was calculated as the difference between BWS and IBW. Average daily gain (ADG) for individual lambs was calculated using the sum of average daily divided by the number of days that the animals were confined. Feed conversion ratio (FCR) for each lamb was calculated as a ratio of daily DMI to ADG.

2.3. Slaughter Procedures

Animals arrived at the slaughterhouse and were fasted for solids for 16 h, with water *ad libitum*, placed in a duly covered and walled corral. The slaughter was carried out in the slaughterhouse and meat packing plant of the Cooperativa Agroindustrial de Pintadas (COOAP) in the municipality of Pintadas, state of Bahia, according to the standards of the State Inspection Service (SIE) of Bahia, according to the standards for humane slaughter, stunning animals by electronarcosis, followed by bleeding, by cutting the jugular veins and carotid arteries, then the animals were skinned and eviscerated. Head, limbs, and tail were removed, carcasses were identified, washed, and weighed to obtain the hot carcass weight (HCW).

After weighing, the carcasses were transferred to a cold room at $\pm 4\ ^\circ$C and stored under chilling for 24 h hung by the common Achilles tendon, with appropriate hooks, keeping 17 cm between the tarsometatarsal joints.

2.4. Carcass Traits and Commercial Cuts

During the chilling period, the pH of the carcass was recorded 24 h *postmortem*, using a portable pH meter coupled to a penetration electrode previously calibrated with buffer solutions of pH 4.00 and 7.00. Afterwards, carcasses were weighed to obtain the cold carcass weight (CCW) and the weight loss by cooling (WLC) was calculated using equation:

$$\text{WLC} = ((\text{HCW} - \text{CCW})/\text{HCW}) \times 100 \tag{2}$$

Then, the following carcass morphological measurements were evaluated: internal and external carcass length; leg length; chest and rump width; chest depth; thoracic and rump perimeter. All length and perimeter measurements were taken using a measuring tape, and width and depth measurements were taken with the aid of a handmade compass, whose registered opening was measured with a ruler. The carcass compactness index (CCI) was calculated by the equation:

$$\text{CCI} = \text{CCW}/\text{ICL} \tag{3}$$

Then, the subjective assessment of the conformation and the state of fatness (marbling, pelvic-renal fat, and fattening degree) was carried out on the carcass, following the methodology of Cezar and Sousa [7], considering the conformation index varying from 1 = poor (concave) to 5 = excellent (convex); pelvic-renal fat of 1 = little fat to 3 = much fat; fattening degree of 1 = very thin to 5 = very fat; and marbling of 1 = nonexistent to 5 = abundant.

After the subjective assessment of the carcass pelvic-renal fat, the kidneys and pelvic renal fat were removed, whose weights were recorded and subtracted from the hot and cold carcass weights. Then, the hot carcass yields (HCY) and cold carcass yields (CCY) were calculated by equations:

$$\text{HCY (\%)} = (\text{HCW}/\text{BWS}) \times 100 \tag{4}$$

$$\text{CCY (\%)} = (\text{CCW}/\text{BWS}) \times 100 \tag{5}$$

The cold carcasses were sawed into two symmetrical sides along the backbone and the left sides from the carcasses were cut into five commercial cuts (leg, loin, rib, shoulder, and neck) and weighed separately. The weight of reconstituted half carcass was obtained by the sum of five commercial cuts weights and used to calculate commercial cuts yield, as proposed by Cezar and Sousa [7].

2.5. Physicochemical Properties from the Proximate Composition of Meat

In the left half carcass, a cross section was made between the 12th and 13th ribs, and using a digital caliper, the cover fat thickness was measured on the *Longissimus lumborum* muscle.

The *Longissimus lumborum* samples were dissected with the aid of a scalpel to remove the subcutaneous fat and epimysium, individually packaged, identified, and stored at $-20\ ^\circ$C until the physicochemical analyses. Loins were thawed under chilling (8 $^\circ$C) the night before the beginning of the analyses.

Two samples that were 2.5 cm thick from *Longissimus lumborum* muscle were used to evaluate the cooking weight loss (CWL). The samples were weighed before and after cooking and recorded.

A stain-less-steel thermocouple (Gulterm 700; Gulton do Brazil, Brazil) was placed into the geometric center of each sample to check and record its internal temperature. The samples were cooked on a grill (George Foreman Jumbo Grill GBZ6BW, Rio de Janeiro, Rio de Janeiro, Brazil) until the internal temperature reached 71 °C. Upon reaching the temperature, the samples were taken, placed into a plastic bag, and cooled to 10 °C in an ice water bath. The cooking weight loss of each sample was obtained by the difference between the weights before and after cooking and expressed as a percentage.

These same samples were kept at 4 °C overnight for instrumental texture analysis conducted according to the method of the AMSA [8].

On the following day, the samples were brought to room temperature before Warner-Bratzler Shear Force (WBSF) analysis and using a cork borer at least three cores of 1.27 cm in diameter and 2.0 cm in length, parallel to the muscle fibers, were removed from each sample.

The WBSF was measured using a texture analyzer (Texture Analyzer TX-TX2; Mecmesin, NV, USA) fitted with a Warner $\pm$ Bratzler-type shear blade with a load of 25 kgf (kilogram-force) and a cutting speed of 20 cm/min [9] that sheared each core perpendicular to the fiber direction and was expressed in Newtons (N).

Meat color was evaluated with a Minolta CR300 colorimeter, operating in the CIE system (L*, a*, b*), where L* is lightness, a* is redness, and b* is yellowness. The colorimeter was calibrated with a white tile and the illuminant used was C and the observation was at 2°. Before analysis, samples were exposed to room temperature for 30 min for the formation of oxymyoglobin, the main pigment responsible for the bright red color of meat [10]. After this time, and as described by Miltenburg et al. [11], L*, a* and b* were measured at three different points on the muscle surface, and the average of the triplicates of each coordinate per animal sample was subsequently calculated.

The saturation index (Chroma) and Hue angle (h_{ab}) were determined using a* as (a) and b* as (b) data according to the equations determined by Hunt and King [12]:

$$Chroma = \sqrt{(a*)^2 + (b*)^2} \tag{6}$$

$$h_{ab} = tan^{-1}\left(\frac{b*}{a*}\right) \cdot \left(\frac{180}{x}\right) \tag{7}$$

Longissimus lumborum muscle samples for proximate composition analysis were ground and homogenized in a food processor. The moisture, ash, and protein contents were evaluated according to AOAC [13], in procedures 985.41, 920.153, and 928.08, respectively. Total lipids were extracted according to the methodology described by Folch et al. [14] by extraction with chloroform: methanol (2:1) solution, followed by evaporation of the solvent in an oven at 105 °C.

2.6. Experimental Design and Statistical Analysis

Animals were distributed in a completely randomized experimental design, with 5 treatments and 12 replications, totaling 60 animals. The following mathematical model was used:

$$Xij = \mu + Ti + \varepsilon ij \tag{8}$$

where Xij = observation of treatment i, in repetition j; μ = overall mean; Ti = effect of treatment i; εij = effect of uncontrolled factors in the plot.

Data obtained were tested by analysis of variance and the means compared using the Tukey test at 5% significance, according to the forage:concentrate ratio in the diet for lambs finished in feedlot, using the software Statistical Analysis System [15] (SAS, version 9.1). The subjective assessment of the carcass also included the Levene test to verify the variance homogeneity using the "HOVTEST" command.

3. Results

3.1. Diet Composition Actually Consumed and Growth Performance

Effective consumption of nutritional fractions and performance of Santa Ines lambs are presented in Table 3.

Table 3. Composition of the diet actually consumed by Santa Ines lambs who received different forage-to-concentrate ratios.

Item	Experimental Diets [1]					SEM [2]	*p*-Value [3]
	C12	C31	C50	C69	C88		
Initial body weight, kg	25.17	25.53	26.28	25.78	26.45	0.99	-
Dry matter intake, kg	1.07c	1.23bc	1.44ab	1.70a	1.51ab	0.08	<0.01
ME intake, MJ/day	9.34d	11.71cd	14.28bc	17.43ab	18.88a	0.81	<0.01
Composition of the diet actually consumed [4]							
Crude protein	14.16b	14.24a	14.15c	14.13d	14.10e	<0.01	<0.01
Ether extract	2.37e	2.69d	3.02c	3.34b	3.66a	<0.01	<0.01
Neutral detergent fiber	68.05a	56.05b	43.99c	32.02d	20.06e	0.01	<0.01
Non-fibrous carbohydrate	15.3e	26.91d	38.7c	50.41b	62.12a	0.01	<0.01
Body weight at slaughter, kg	29.68b	33.33b	38.88a	42.08a	39.91a	1.35	<0.01
Total weight gain, kg	4.51d	7.80c	12.60b	16.30a	13.46b	0.61	<0.01
Average daily gain, kg	0.07d	0.12c	0.20b	0.26a	0.21b	0.01	<0.01
Feed conversion ratio, kg/kg	15.06a	10.05b	7.27c	6.63c	7.27c	0.40	<0.01

[1] C12, C31, C50, C69, and C88 refer to diets with Tifton-85 hay to concentrate ratios of 88:12, 69:31, 50:50, 31:69, and 12:88, respectively. [2] Standard error of the mean. [3] Significance at $p < 0.05$. Averages followed by different letters on the line differ among themselves by Tukey's test ($p < 0.05$). [4] Value expressed as the percentage of each nutrient ingested in relation to DMI.

The dry matter intake differed between treatments ($p < 0.01$), with higher DMI in the treatment with 69% concentrate, but not statistically different from the treatments with 50% and 88% concentrate. The lowest DMI was observed for the treatment with 12% concentrate.

There was greater metabolizable energy (ME) intake ($p < 0.01$) observed for the C88 treatment, but it did not differ statistically from the treatment with 69% concentrate. The lowest ME intake was observed for the C12 treatment.

The effective consumption of CP ($p < 0.01$) was higher in the treatment with 31% concentrate and lower in the treatment with 88% concentrate. The effective consumption of EE ($p < 0.01$) and NFC ($p < 0.01$) showed the same behavior, with higher averages for the treatment with 88% and lower averages for the treatment with 12% concentrate. The effective consumption of NDF ($p < 0.01$) had the highest average for the treatment with 12% concentrate and the lowest average for the treatment with 88%.

The daily weight gain ($p < 0.01$) and consequently the total weight ($p < 0.01$) of the animals, presented greater means for the treatment with 69% concentrate and the lowest for the treatment with 12% concentrate in the diet.

The worst feed conversion ($p < 0.01$) was obtained for the treatment with 12% concentrate and there was no significant difference for treatments with 50%, 69% and 88% concentrate.

3.2. Carcass Traits and Commercial Cuts

The data of morphometric measurements and carcass traits of Santa Ines lambs are presented in Table 4.

Most morphometric measurements ($p < 0.05$) of the carcass had a significant difference between treatments except leg length ($p = 0.25$).

Regarding the development of the carcass, we can observe through the morphometric measures, that there was no significant difference ($p < 0.05$) between treatments with 50%, 69%, and 88% concentrate. The animals receiving 12% concentrate had lower carcass development. This behavior can also be evaluated using the CCI variable.

Table 4. Effect of dietary forage-to-concentrate ratios on morphometric measurements and carcass traits of lambs.

Item	Experimental Diets [1]					SEM [2]	*p*-Value [3]
	C12	C31	C50	C69	C88		
Morphometric measurements, cm							
Chest width	13.93b	14.92b	16.43a	17.29a	17.31a	0.33	<0.01
Rump width	20.42b	22.15ab	22.71ab	24.00a	24.02a	0.64	<0.01
Thoracic perimeter	64.5c	68.79b	73.25a	74.68a	74.68a	0.98	<0.01
External length	59.17b	61.75ab	64.08a	63.55a	64.18a	0.96	<0.01
Internal length	49.08b	51.5ab	53.42a	54.18a	53.54a	0.76	<0.01
Leg length	39.33	40.67	41.25	40.64	40.20	0.61	0.25
Chest depth	17.33b	18.33ab	19.04a	17.86ab	17.95ab	0.30	<0.01
Rump perimeter	49.08c	52.83bc	55.79ab	57.82a	58.14a	1.08	<0.01
Carcass compactness index [4], kg/cm	0.22c	0.27b	0.32a	0.34a	0.34a	0.01	<0.01
Hot carcass weight, kg	10.83c	13.98b	17.08a	18.30a	18.32a	0.67	<0.01
Cold carcass weight, kg	10.78c	13.93b	17.03a	18.27a	18.26a	0.67	<0.01
Hot carcass yield, %	36.47d	41.95c	43.91b	43.88bc	45.90a	0.47	<0.01
Cold carcass yield, %	36.33c	41.80b	43.78a	43.81a	45.76a	0.47	<0.01
Fat thickness, mm	0.84	1.04	1.14	1.21	1.13	0.09	0.07
Carcass conformation index [5]	2.77c	3.52b	4.02ab	4.18a	4.14ab	0.16	<0.01
Carcass fattening degree [6]	2.54c	3.63b	4.06ab	4.25ab	4.48a	0.17	<0.01
Pelvic-renal fat [7]	1.17b	1.75ab	2.25a	2.35a	2.36a	0.16	<0.01
Marbling [8]	1.00	1.25	1.25	1.36	1.64	0.13	0.06

[1] C12, C31, C50, C69, and C88 refer to diets with Tifton-85 hay to concentrate ratios of 88:12, 69:31, 50:50, 31:69, and 12:88, respectively. [2] Standard error of the mean. [3] Significance at *p* < 0.05. Averages followed by different letters on the line differ among themselves by Tukey's test (*p* < 0.05). [4] Carcass compactness index. [5] Carcass conformation index (1 = Poor (concave) to 5 = Excellent (convex)). [6] Fattening degree (1 = very thin to 5 = very fat). [7] Pelvic-renal fat (1= little fat to 3 = much fat). [8] Marbling (1= nonexistent to 5= abundant).

Body weight at slaughter did not present a significant difference (*p* < 0.05) between treatments with 50%, 69%, and 88% concentrate and animals receiving 12% concentrate in their diet presented lower weight (*p* < 0.01). The same behavior can be observed for hot carcass weight (*p* < 0.01) and cold carcass weight (*p* < 0.01).

Hot (*p* < 0.01) and cold (*p* < 0.01) carcass yields presented greater averages for treatment with 88% concentrate and lower yield for the carcass for the animals that received 12% concentrate in their diet.

The fat thickness presented an average of 1.07 cm and did not differ statistically (*p* = 0.07) between the treatments.

Within the subjective assessments of carcasses only marbling (*p* = 0.06) did not show a significant difference between treatments and obtained an average of 1.30.

The carcass conformation index (*p* < 0.01) showed the highest average for the treatment with 69% concentrate; however, it did not differ statistically from the treatments with 50% and 88% concentrate.

The highest average for carcass finishing degree (*p* < 0.01) was obtained by treatment with 88% concentrate, but it did not differ statistically from the treatments with 50% and 69% concentrate. The lowest carcass finishing degree was obtained by the C12 treatment.

When evaluating pelvic-renal fat (*p* < 0.01), it was observed that the lowest average obtained was for the treatment with 12% concentrate. The other treatments did not differ statistically.

The weight and yield of commercial cuts are presented in Table 5.

The reconstituted half carcass (*p* < 0.01) showed the same behavior as the CCW (Table 4), not differing between treatments C50, C69, and C88 with a lower average for treatment C12.

The neck, shoulder, loin, and leg cuts showed a lower average for treatment with 12% concentrate and did no differ between treatments C50, C69, and C88.

Table 5. Effect of dietary forage-to-concentrate ratios on weight and yield of commercial cuts from ambs.

Items	Experimental Diets [1]					SEM [2]	*p*-Value [3]
	C12	C31	C50	C69	C88		
Reconstituted half carcass, kg	5.06c	6.40b	7.90a	8.24a	8.32a	0.33	<0.01
Commercial cut weight, kg							
Neck	0.53c	0.65b	0.78a	0.77a	0.76a	0.03	<0.01
Shoulder	1.10c	1.37b	1.68a	1.77a	1.78a	0.06	<0.01
Rib	1.42b	1.87b	2.49a	2.62a	2.68a	0.12	<0.01
Loin	0.25c	0.32b	0.38a	0.41a	0.42a	0.01	<0.01
Leg	1.76c	2.2b	2.57a	2.68a	2.69a	0.09	<0.01
Commercial cut yield, %							
Neck	10.47	10.11	9.87	9.32	9.10	0.49	0.33
Shoulder	21.74	21.41	21.27	21.44	21.33	1.14	0.99
Rib	28.06	29.20	31.52	31.80	32.20	1.85	0.57
Loin	4.94	5.00	4.81	4.94	5.05	0.26	0.98
Leg	34.78	34.31	32.53	32.50	32.31	1.76	0.77

[1] C12, C31, C50, C69, and C88 refer to diets with Tifton-85 hay to concentrate ratios of 88:12, 69:31, 50:50, 31:69, and 12:88, respectively. [2] Standard error of the mean. [3] Significance at $p < 0.05$. Averages followed by different letters on the line differ among themselves by Tukey's test ($p < 0.05$).

The treatments with 12% and 31% concentrate did not differ statistically for the loin cut. Additionally, there was no significant difference between treatments with 50%, 69% and 88% concentrate.

No significant differences ($p > 0.05$) were observed between treatments for commercial cuts yields, with means of 9.16%, 20.05%, 28.57%, 4.63%, and 31.24% for yields of neck, shoulder, rib, loin, and leg, respectively.

3.3. Physicochemical Properties from the Proximate Composition of Meat

The physicochemical composition of the *Longissimus lumborum* muscle is presented in Table 6. Most of the variables evaluated showed no significant difference ($p > 0.05$).

Table 6. Effect of dietary forage-to-concentrate ratios on physicochemical composition of the *Longissimus lumborum* muscle from lambs.

Item	Experimental Diets [1]					SEM [2]	*p* Value [3]
	C12	C31	C50	C69	C88		
pH$_{0h}$	7.33a	7.28ab	7.23ab	7.04b	7.02b	0.07	<0.01
pH$_{24h}$	6.17	5.98	6.03	6.16	6.01	0.08	0.37
Color parameter							
L* (lightness)	38.97	37.54	38.30	37.14	38.19	0.47	0.08
a* (redness)	22.70	22.11	22.29	21.61	22.21	0.35	0.34
b* (yellowness)	6.19	5.58	5.83	5.18	5.63	0.38	0.53
Chroma (saturation)	23.53	22.80	23.04	22.22	22.91	0.416	0.35
Hue (h$_{ab}$)	15.25	14.16	14.66	13.48	14.22	0.761	0.65
Cooking weight loss, %	19.02	22.75	24.55	28.39	30.52	2.957	0.09
Shear force, N	20.99	21.39	16.28	24.53	20.21	2.060	0.14
Chemical composition, %							
Moisture	74.48a	73.51ab	72.95bc	72.83bc	71.95c	0.44	<0.01
Ash	1.07	1.12	1.12	1.09	1.14	0.02	0.39
Protein	21.93	22.55	22.97	22.51	22.92	0.45	0.37
Lipid	2.43d	2.41d	2.55c	2.63b	2.82a	0.01	<0.01

[1] C12, C31, C50, C69, and C88 refer to diets with Tifton-85 hay to concentrate ratios of 88:12, 69:31, 50:50, 31:69, and 12:88, respectively. [2] Standard error of the mean. [3] Significance at $p < 0.05$. Averages followed by different letters on the line differ among themselves by Tukey's test ($p < 0.05$).

The pH_{0h} showed a significant difference ($p < 0.01$) between treatments, with the highest average for the treatment with 12% concentrate. There was no significant difference between treatments for pH_{24h} ($p = 0.37$), presenting a mean value of 6.07.

For the color parameters, none of the variables showed a significant difference between treatments, presenting mean values of 38.03, 22.18, 5.68, 22.90, and 14.35 for lightness, redness, yellowness, Chroma, and Hue angle, respectively.

Cooking losses and shear force did not differ statistically between treatments and presented averages of 25.05 and 2.11, respectively.

The moisture showed a significant difference ($p < 0.01$) between treatments, with the highest average for the treatment with 12% concentrate and lower mean for C88 treatment.

Ash and protein did not differ statistically between treatments and presented averages of 1.11% and 21.86%, respectively.

The lipids content in the muscle showed the highest mean for the treatment with 88% concentrate and the lowest mean for the C12 treatment.

4. Discussion

4.1. Diet Composition Actually Consumed and Growth Performance

The lower DMI was observed for the animals receiving 12% concentrate, due to the high concentration of NDF that this diet presented (88%). This possibly limited DM intake by physical factors in the gastrointestinal tract, which reduced the passage rate of digesta and prevented the emptying of the gastrointestinal tract (GIT) that caused a distension of the rumen-reticulum, resulting in a decreased DMI [16–18] and inability to ingest more feed to nourish themselves.

The metabolizable energy (ME) intake by animals was greater than ME offered by the diets, in this way, it is confirmed that the reduction in DMI was not limited by the energy density of the ration and rather limited by physical factors in the gastrointestinal tract.

The diet effectively consumed allows us to observe the behavior of animal selectivity in relation to their offered, that is, we can observe through the DMI the amount of each nutrient that was consumed per day. In this way, it was observed that the research animals were selective because the diet actually consumed (Table 3) was not equal to that offered (Table 2).

According to the chemical composition of diets, the CP content ranged from 14.0% to 12.0%, EE of 3,98% to 1.88%, NFC of 60.2% to 22.4% and NDF of 18.1% to 60.9% for treatments with 88%, 68%, 50%, 31%, and 12% concentrate, respectively. When the composition of the diet effectively consumed was observed, the CP contents ranged from 14.10% to 14.24%, EE of 3.66% to 2.37%, NFC of 62.12% to 15.3%, and NDF of 20, 06% to 68.05%. This difference between the offered and actually consumed diet indicates that there was a preference of animals to the concentrate, since the concentrate has a higher percentage of CP, EE, and NFC and lower NDF in relation to forage.

Although these animals received diets with different roughage: concentrate ratio, they depend, in common on the amount of DM ingested and the levels of protein and energy satisfactory [19] for the development and efficiency of the rumen microbiota. These satisfactory levels result in higher synthesis of microbial protein and a better use of energy and protein by the animal so that they can achieve greater growth performance through development of muscle and adipose tissue, and consequently increase the BWS [20] resulting in a better feed conversion ratio. This can be observed for animals receiving 69% concentrate in diet.

According to Abbasi et al. [21] working with dietary energy levels of goat kids, and Cui et al. [22] working with dietary energy and protein levels of lambs, the low dietary energy level (9.1 vs. 10.7 MJ ME/kg and 10.9 vs. 8.6 MJ ME/kg, respectively) can reduce the average daily gain and worsen the feed conversion ratio. This was observed for the 12% concentrate treatment.

4.2. Carcass Traits and Commercial Cuts

Measurements and evaluations performed on the carcass allow us to understand the body development of animals, seeking to obtain a deposition maximum of muscle and intermediate fat in the carcasses. In view of the data obtained, the animals from the treatments with 50%, 69%, and 88% concentrate showed similar and better body development than the animals from the treatments C12 and C31. This is probably related to DMI for those animals that received diets above 50% roughage, preventing them from expressing their production potential, because with the high NDF content in these diets, the passage of digesta through the gastrointestinal tract was reduced, limiting DMI from animals by filling [17,20].

Within the measurements and evaluations carried out on the carcass, the carcass compactness index and carcass conformation index stand out as being parameters that estimate the amount of muscle deposited in the carcass, which is the edible part with the highest financial return for the producer. The animals from treatments C50, C69, and C88 did not differ statistically; this indicates that all animals from these treatments were similar in body performance with similar muscle composition and carcass conformation. This resulted in higher hot and cold carcass weight, and better cold carcass yield (also known as commercial yield). The same behavior was observed by Brand et al. [23] evaluating carcass characteristics and fat deposition of Merino, South African Mutton Merino, and Dorper lambs housed in a feedlot.

Lower weights and yields of carcass obtained for the animals that received 12% concentrate can be explained by the NDF and ADF content in the diet, which is higher than the other treatments. This fiber content provided the animals with a longer digesta retention time due to reduced passage rate, resulting in a greater volume/weight of the gastrointestinal tract. As the carcass yield calculation is obtained through the ratio between the carcass weight and the final body weight of the animal, the volume/weight of the GIT and its content are considered only in the final weight of the animal, which reduces the weight and consequently the carcass yield. With higher energy availability in feed, there is greater availability of nutrients for the development of muscle and adipose tissue, resulting in higher carcass weights, and consequently heavier commercial cuts [18].

Regarding the deposition of fat in the carcass, the fat thickness and marbling did not differ between treatments, indicating that this distribution occurred in a similar way, regardless of the roughage:concentrate ratio that the animal received. However, the subjective evaluation of carcass finishing indicated that animals from treatments C50, C69, and C88 had better distribution of fat in the carcass than animals that received 12% and 31% concentrate. This divergence may be related to the measured and/or visualized location, as the fat thickness is measured only between the 12th and 13th ribs, while the fat finish assessment examines the complete carcass.

Regarding the deposition of pelvic-renal fat, tropical animals have the physiological ability to deposit intra-abdominal fat, which occurs with maturity and acts as energy reserves that are mobilized during the period of feed shortage to reduce their energy deficit [18]. Renal pelvic fat has a high correlation with carcass fattening degree; that is, the greater the deposition of fat in this region, the greater the carcass fattening degree [7,24].

Carcass fattening degree is interesting because it protects carcasses from cold drying [25] and keeps meat with optimal concentrations of moisture, providing, together with the fat, a greater juiciness in the meat [26,27]. However, it is necessary to consider to what extent the amount of fat is interesting in the carcass fattening. Because it is highly correlated with the pelvic-renal fat, indicating feed energy waste, it fails to convert the energy absorbed from the food into muscle and/or marbling fat (edible part of the carcass) to convert it into a type of fat that is neither commercialized nor used in cooking. The animals that received 31%, 50%, 69% and 88% concentrate showed similar pelvic-renal fat deposition.

4.3. Physicochemical Properties from the Proximate Composition of Meat

As can be seen by pH_{0h}, diets with different roughage:concentrate ratios provided different amounts of muscle glycogen reserve during the animals' development. The diet with the lowest proportion of concentrate (C12) had a higher pH_{0h} compared to the other treatments. This was due to the high NDF and ADF content in the diet, which reduced the passage rate of digesta, distended the rumen-reticulum, and limited the intake of animals due to physical factors of the gastrointestinal tract. This probably prevented the animals from having their energy demand fully satiated, resulting in less muscle glycogen deposition.

The muscle pH in live animals is between 7.08 and 7.30, soon after slaughter the pH reduces to 7.0 and continues to reduce until reaching values between 5.4 and 5.8 for sheep meat [7,27]. This reduction in pH occurs due to the accumulation of lactic acid produced, via anaerobic mechanism, which is inversely proportional to the concentration of muscle glycogen present at the time of slaughter. The pH_{24h} did not differ between treatments and presented an average of 6.07, above the expected for sheep meat. The animals were probably subjected to some stress during transport and/or at the slaughterhouse, which made them use muscle glycogen for energy production via aerobic mechanism, resulting in lower lactic acid production and higher pH_{24h} [28].

In this research, the pH_{24h} obtained was higher than the isoelectric point of muscle proteins (5.2–5.3), which indicates that because it is above the neutral charge of proteins, these meats had an excessive negative charge in their fibers, which provided repulsion of the filaments, in this way, increased the space for the water molecules to bind, contributing to greater juiciness, less mechanical force when cutting, and smaller proportional losses of water during cooking. The mean values of CL were within the acceptable range of up to 35% [29].

The shear force did not differ between treatments, which was expected, since this variable is more affected by the age, physical activity, and sex of the animal [18]. The shear force is directly related to the connective tissue protein, that is, the size of the fiber bundles, the quantity and size of the fibers, the collagen solubility, if there is a presence of cross-links, among others. The mean values indicate that the meat obtained for most treatments was considered tender (<22.27 N), except the meat from animals that received a diet with 69% concentrate, which had an average value of 24.50 N, but was considered with medium tenderness (22.27 and 35.61 N) [7].

The pH_{24h} of the meat affects the load of the muscle proteins that alters the spacing between the fibers of the meat. This change in the structure affects how the light is reflected and absorbed and therefore affects the visual appearance of the meat, which is measured through the indices of brightness (L*), the intensity of yellow (b*), and intensity of red (a*). The different roughage:concentrate ratios did not affect the coordinates referring to color, which is important because it is the primary attribute considered at the time of purchase [30]. The averages obtained corroborate those in the literature for sheep meat, which are from 30.03 to 49.47 for L*, from 8.24 to 23.53 for a* and from 3.38 to 11.10 for b* [27]; also working with Santa Ines sheep, means similar to those of this study were found.

When evaluating the chemical composition of Santa Ines sheep meat, we observed that the different roughage:concentrate ratios did not provide a significant difference for the ash and protein contents and the averages obtained for the moisture and lipid contents were inversely proportional, as also described by Silva et al. [31]. The moisture and lipid contents in the meat are important for providing the consumer with the feeling of juiciness since the moisture in the meat is released at the time of mastication and the juiciness sensation persists due to the concentration of lipids in the meat, which stimulates salivation [26], ensuring the juiciness for longer.

The diet with 88% concentrate presented lower NDF content and higher NFC and ME content. This promoted greater use of the diet and energy supply to ruminants due to the production of ruminal propionic acid and a lower acetate:propionate ratio, which allows a higher concentration of circulating glucose, favoring insulin secretion and induction of lipogenesis, which resulted in greater storage of energy in the form of adipose tissue [32]. In this way, the meats of the animals that received this treatment probably provided greater succulence to the consumer compared to treatments C12 and C31. This is because, despite the meat from these last two treatments having a higher moisture content, the prolonged sensation of juiciness is obtained with the concentration of lipids in meat, which for these treatments had a lower proportion.

The Santa Ines breed is specialized in producing lean meats [24]; however, total lipid contents obtained in meats were adequate to guarantee the desirable organoleptic properties of meat [33].

5. Conclusions

Higher levels of concentrate favored carcass traits, commercial cuts, and physicochemical parameters of the meat. However, considering the demands of the producer (higher production with lower cost) and the demand of market consumer, the addition of 50% of concentrate to the roughage presented animal performance and carcass traits similar to the highest level of concentrate tested, also presenting as the most interesting option.

Author Contributions: Conceptualization, Y.C., E.S., J.O. and G.C. (Gleidson Carvalho); data curation, Y.C., N.P., A.P. and A.L.; formal analysis, A.P. and A.L.; funding acquisition, J.O.; investigation, Y.C., D.P., D.A., G.C. (Gabriel Cruz) and G.L.; methodology, E.S., J.O. and G.C. (Gleidson Carvalho); resources, J.O. and F.N.; software, L.P. supervision, E.S., G.C. (Gleidson Carvalho) and P.A.; writing—original draft, Y.C., J.O. and P.A.; writing—review and editing, J.O., A.P., A.L., D.F., F.N. and A.Z. All authors have read and agreed to the published version of the manuscript.

Funding: This research was funded by the Coordination for the Improvement of Higher Education Personnel (CAPES-Brazil—Finance Code 001) and National Council for Scientific and Technological Development (CNPq-Brazil) for the fellowship grant; and by the Maranhão State Research Foundation (FAPEMA-Brazil) and Federal Institute of Education, Science and Technology Rondônia (IFRO/EDITAL Nº 13/2021-REIT/PROPESP/IFRO-Support for Scientific and Literary Communication).

Institutional Review Board Statement: The study was conducted according to the guidelines of the Committee on the Ethics of Animal Experiments Guide of the Federal University of Bahia (protocol number 68/2018).

Informed Consent Statement: Not applicable.

Data Availability Statement: Not Applicable.

Conflicts of Interest: The authors declare no conflict of interest.

References

1. NRC (National Research Council). *Nutrient Requirements of Small Ruminants: Sheep, Goats, Cervids, and New World Camelids*; The National Academies Press: Washington, DC, USA, 2007.
2. Oliveira, K.A.; de Lima Macedo, G.; da Silva, S.P.; Araújo, C.M.; Varanis, L.F.M.; Sousa, L.F. Nutritional and metabolic parameters of sheep fed with extrused roughage in comparison with corn silage. *Semin. Cienc. Agrar.* **2018**, *39*, 1795–1804. [CrossRef]
3. Jacques, J.; Berthiaume, R.; Cinq-Mars, D. Growth performance and carcass characteristics of Dorset lambs fed different concentrates: Forage ratios or fresh grass. *Small Rumin. Res.* **2011**, *95*, 113–119. [CrossRef]
4. Papi, N.; Mostafa-Tehrani, A.; Amanlou, H.; Memarian, M. Effects of dietary forage-to-concentrate ratios on performance and carcass characteristics of growing fat-tailed lambs. *Anim. Feed Sci. Technol.* **2011**, *163*, 93–98. [CrossRef]
5. Issakowicz, J.; Bueno, M.S.; Sampaio, A.C.K.; Duarte, K.M.R. Effect of concentrate level and live yeast (Saccharomyces cerevisiae) supplementation on Texel lamb performance and carcass characteristics. *Livest. Sci.* **2013**, *155*, 44–52. [CrossRef]
6. Qiao, G.H.; Xiao, Z.G.; Li, Y.; Li, G.J.; Zhao, L.C.; Xie, T.M.; Wang, D.W. Effect of diet synchrony on rumen fermentation, production performance, immunity status and endocrine in Chinese Holstein cows. *Anim. Prod. Sci.* **2018**, *9*, 664–672. [CrossRef]
7. Cezar, M.F.; Sousa, W.H. *Carcaças Ovinas e Caprinas—Obtenção, Avaliação e Classificação*; Agropecuária Tropical: Uberaba, Brazil, 2007.
8. AMSA. *Research Guidelines for Cookery, Sensory Evaluation, and Instrumental Tenderness Measurements of Meat*; American Meat Science Association: Champaign, IL, USA, 2015.
9. Shackelford, S.D.; Wheeler, T.L.; Koohmaraie, M. Evaluation of slice shear force as an objective method of assessing beef Longissimus tenderness. *J. Anim. Sci.* **1999**, *77*, 2693–2699. [CrossRef]

10. Cañeque, V.; Sañudo, C. *Metodologia Para el Estúdio de la Calidad de la Canal y de la Carne en Ruminantes*; Instituto Nacional de Investigación y Tecnologia y Alimenticia: Madrid, Spain, 2000.

11. Miltenburg, G.A.J.; Wensing, T.H.; Smulders, F.J.M.; Breukink, H.J. Relationship between blood hemoglobin, plasma and tissue iron, muscle heme pigment, and carcass color of veal. *Sci. J. Anim. Sci.* **1992**, *70*, 2766–2772. [CrossRef]

12. Hunt, M.C.; King, A. *Meat Color Measurement Guidelines*; American Meat Science Association (AMSA): Champaign, IL, USA, 2012; Available online: https://meatscience.org/docs/default-source/publications-resources/hot-topics/2012_12_meat_clr_guide.pdf?sfvrsn=d818b8b3_0 (accessed on 10 January 2022).

13. Association of Official Analytical Chemists (AOAC). Humidity, Ash and Protein (crude) determination in animal feed (985.41; 920.15 e 928.08). In *Official Methods of Analysis*; AOAC: Gaithersburg, MD, USA, 2016.

14. Folch, J.; Lees, M.; Stanley, G.S. A simple method for the isolation and purification of total lipides from animal tissues. *J. Biol. Chem.* **1957**, *226*, 497–509. [CrossRef]

15. SAS-SAS Institute. *SAS User's Guide: Basics*; SAS Institute Inc.: Cary, NC, USA, 2014.

16. Salinas-Chavira, J.; Arzola, C.; García-Castillo, R.F.; Briseno, D.A. Approaches to the level and quality of forage in feedlot diets for lambs. *J. Dairy Vet. Anim.* **2017**, *5*, 96–97. [CrossRef]

17. Pinho, R.M.A.; Santos, E.M.; Oliveira, J.S.; Carvalho, G.G.P.; Silva, T.C.; Macêdo, A.J.D.S.; Corrêa, Y.R.; Zanine, A.M. Does the level of forage neutral detergent fiber affect the ruminal fermentation, digestibility and feeding behavior of goats fed cactus pear? *Anim. Sci. J.* **2018**, *89*, 1424–1431. [CrossRef]

18. Souza, A.F.N.; Araújo, G.G.L.; Santos, E.M.; Azevedo, P.S.; Oliveira, J.S.; Perazzo, A.F.; Pinho, R.M.A.; Zanine, A.M. Carcass traits and meat quality of lambs fed with cactus (*Opuntia fícus-indica* Mill) silage and subjected to an intermittent water supply. *PLoS ONE* **2020**, *15*, e0231191. [CrossRef]

19. Ríos-Rincón, F.G.; Estrada-Angulo, A.; Plascencia, A.; López-Soto, M.A.; Castro-Pérez, B.I.; Portillo-Loera, J.J.; Robles-Estrada, J.C.; Calderón-Cortes, J.F.; Dávila-Ramos, H. Influence of Protein and Energy Level in Finishing Diets for Feedlot Hair Lambs: Growth Performance, Dietary Energetics and Carcass Characteristics. *Asian-australas. J. Anim. Sci.* **2014**, *27*, 55–61. [CrossRef] [PubMed]

20. Sousa, W.H.; Cartaxo, F.Q.; Costa, R.G.; Cezar, M.F.; Cunha, M.G.G.; Pereira Filho, J.M.; Santos, N.M. Biological and economic performance of feedlot lambs feeding on diets with different energy densities. *Rev. Bras. Zootec.* **2012**, *41*, 1285–1291. [CrossRef]

21. Abbasi, S.A.; Vighio, M.A.; Soomro, S.A.; Kachiwal, A.B.; Gadahi, J.A.; Wang, G. Effect of different dietary energy levels on the growth performance of kamori goat kids. *Int. J. Agro Vet. Med. Sci.* **2012**, *6*, 473–479. [CrossRef]

22. Cui, K.; Qi, M.; Wang, S.; Diao, Q.; Zhang, N. Dietary energy and protein levels influenced the growth performance, ruminal morphology and fermentation and microbial diversity of lambs. *Sci. Rep.* **2019**, *9*, 16612. [CrossRef]

23. Brand, T.S.; Van der Westhuizen, E.J.; Van Der Merwe, D.A.; Hoffman, L.C. Analysis of carcass characteristics and fat deposition of Merino, South African Mutton Merino and Dorper lambs housed in a feedlot. *S. Afr. J. Anim. Sci.* **2018**, *48*, 477–478. [CrossRef]

24. Cartaxo, F.Q.; Sousa, W.H.; Costa, R.G.; Cezar, M.F.; Pereira Filho, J.M.; Cunha, M.G.G. Quantitative traits of carcass from lambs of different genotypes submitted to two diets. *R. Bras. Zootec.* **2011**, *40*, 2220–2227, (In Portuguese, with Abstract in English). [CrossRef]

25. Kannan, G.; Lee, J.H.; Kouakou, B. Chevon quality enhancement: Trends in pre-and post-slaughter techniques. *Small Rumin. Res.* **2014**, *121*, 80–88. [CrossRef]

26. Nute, G.R.; Richardson, R.I.; Wood, J.D.; Hughes, S.I.; Wilkinson, R.G.; Cooper, S.L.; Sinklar, L.A. Effect of dietary oil source on the flavour and the colour and lipid stability of lamb meat. *Meat Sci.* **2007**, *77*, 547–555. [CrossRef]

27. Lima, A.G.V.O.; Oliveira, R.L.; Silva, T.M.; Barbosa, A.M.; Nascimento, T.V.C.; Oliveira, V.D.S.; Ribeiro, R.D.X.; Pereira, E.S.; Bezerra, L.R. Feeding sunflower cake from biodiesel production to Santa Ines lambs: Physicochemical composition, fatty acid profile and sensory attributes of meat. *PLoS ONE.* **2018**, *13*, e0188648. [CrossRef]

28. Perlo, F.; Bonato, P.; Teira, G.; Tisocco, O.; Vicentin, J.; Pueyo, J.; Mansilla, A. Meat quality of lambs produced in the Mesopotamia region of Argentina finished on different diets. *Meat Sci.* **2008**, *79*, 576–581. [CrossRef] [PubMed]

29. Webb, E.C.; Casey, N.H.; Simela, L. Goat meat quality. *Small Rumin. Res.* **2005**, *60*, 153–166. [CrossRef]

30. Xazela, N.M.; Chimonyo, M.; Muchenje, V.; Marume, U. Effect of sunflower cake supplementation on meat quality of indigenous goat genotypes of South Africa. *Meat Sci.* **2012**, *90*, 204–208. [CrossRef]

31. Silva, T.M.; Medeiros, A.N.; Oliveira, R.L.; Neto, S.G.; Queiroga, R.C.R.E.; Ribeiro, R.D.X.; Leão, A.G.; Bezerra, L.R. Carcass traits and meat quality of crossbred Boer goats fed peanut cake as a substitute for soybean meal. *J. Anim. Sci.* **2016**, *94*, 2992–3002. [CrossRef] [PubMed]

32. Costa, J.B.; Oliveira, R.L.; Silva, T.M.; Ribeiro, R.D.X.; Silva, A.M.; Leão, A.G.; Bezerra, L.R.; Rocha, T.C. Intake, digestibility, nitrogen balance, performance, and carcass yield of lambs fed licuri cake. *J. Anim. Sci.* **2016**, *94*, 2973–2980. [CrossRef] [PubMed]

33. Wood, J.D. Consequences for meat quality of reducing carcass fatness. In *Reducing Fatness in Meat Animals*; Wood, J.D., Fisher, A.V., Eds.; Elsevier Applied Science: London, UK, 1990.

Article

Polycyclic Aromatic Hydrocarbons (PAHs) Occurrence in Traditionally Smoked Chicken, Turkey and Duck Meat

Cristian Ovidiu Coroian [1], Aurelia Coroian [2,*], Anca Becze [3], Adina Longodor [4], Oana Mastan [5] and Răzvan-Mihail Radu-Rusu [6,*]

[1] Department of Animal Nutrition and Feeding, Faculty of Animal Sciences and Biotechnology, University of Agricultural Sciences and Veterinary Medicine Cluj-Napoca, Calea Mănăștur 3–5, 400372 Cluj-Napoca, Romania

[2] Department of Toxicology, Faculty of Animal Sciences and Biotechnology, University of Agricultural Sciences and Veterinary Medicine Cluj-Napoca, Calea Mănăștur 3–5, 400372 Cluj-Napoca, Romania

[3] Research Institute for Analytical Instrumentation, National Institute of Research and Development for Optoelectronics INOE 2000, 67 Donath Street, 400293 Cluj-Napoca, Romania

[4] Department of Land Measurements and Exact Sciences, Faculty of Forestry and Cadastre, University of Agricultural Sciences and Veterinary Medicine Cluj-Napoca, Calea Mănăștur 3–5, 400372 Cluj-Napoca, Romania

[5] Department of Anatomy, Faculty of Veterinary Medicine, University of Agricultural Sciences and Veterinary Medicine Cluj-Napoca, Calea Mănăștur 3–5, 400372 Cluj-Napoca, Romania

[6] Department of Animal Resources and Technology, Faculty of Food and Animal Sciences, Ion Ionescu de la Brad University of Life Sciences, 8 Mihail Sadoveanu Alley, 700489 Iasi, Romania

* Correspondence: aurelia.coroian@usamvcluj.ro (A.C.); radurazvan@uaiasi.ro (R.-M.R.-R.)

Abstract: An increasingly high interest is given to the sensory, nutritional, and sanogenic qualities of meat. Considering that poultry meat is nowadays the main quantitatively demanded meat for human consumption, its quality is largely verified and monitored. Toxic compounds are trace markers to be monitored, as their health impacts often cause a high health risk for humans. We have evaluated how a traditional method of meat preservation—hot smoking with natural wood smoke—adds certain polycyclic aromatic hydrocarbons (PAHs) to chicken, duck, and turkey meat. One- vs two-day smoking period and three wood types for smoking (plum, cherry, and beech) have shown that the highest concentrations of PAHs were present in duck meat, irrespective of smoking time or wood type. A higher concentration overall of PAHs was quantified when beech wood was used, followed by cherry and plum woods. Fluorene associated with beech wood gave the highest values for day 1 and day 2, followed by duck and turkey meat, respectively. Very significant differences ($p < 0.001$) were usually observed for duck meat when compared with chicken and turkey meat, but it was also easy to notice absolute values for Anthracene, Phenanthrene, or Fluoranthene. As expected, two-day smoking contributed to higher concentrations of PAHs in meat.

Keywords: polycyclic aromatic hydrocarbons; meat; chicken; duck; turkey

Citation: Coroian, C.O.; Coroian, A.; Becze, A.; Longodor, A.; Mastan, O.; Radu-Rusu, R.-M. Polycyclic Aromatic Hydrocarbons (PAHs) Occurrence in Traditionally Smoked Chicken, Turkey and Duck Meat. *Agriculture* 2023, 13, 57. https://doi.org/10.3390/agriculture13010057

Academic Editor: Antonello Santini

Received: 18 November 2022
Revised: 20 December 2022
Accepted: 22 December 2022
Published: 24 December 2022

1. Introduction

Meat is one of the main protein sources in the human diet and an important component of a balanced diet [1]. Short-generation interval animal species, such as poultry and pig, provide the largest amount of meat globally. The previsions indicate these species as the main sources of meat for the upcoming period, poultry being among consumers' top preferences [2]. Worldwide, poultry is the second top meat produced as primary livestock and yields after pork, while turkey reaches 5–6% of chicken trades and duck meat is slightly above 4% [2]. While chicken meat production increased by about 50% throughout the past decade, turkey has increased slightly by about 8–9% and duck by about 40%. However, market demands for duck meat are far from being met, which is the same for turkey and even for chicken [2]. Duck requires the most particular rearing conditions and, therefore, has regional specificity. Both turkey and duck meats have different nutritional and sensory qualities as compared with chicken, and global demands have an ascending trend.

Besides production aspects, meat quality is an increasingly imperative requirement, especially among rich countries' consumers. The nutritional values seem to reach a stable level based on the applied farming technologies. Meat quality is a complex of features, of which the conditions of being safe, sanogenic, and from a trustworthy source have probably become the most important. Among multiple preserving and preparation methods, meat smoking imprints special sensory characteristics [3], rendering it very attractive for consumers. Such preservation was most likely begun in prehistoric times [4] and still represents a practice elsewhere. Smoking temperature may give more desirable aroma and flavour to meat [5]. Despite such benefits, wood smoking also brings into meat polycyclic aromatic hydrocarbons (PAHs) and their alkylated derivatives [6]. PAHs are carcinogenic, and thermal cooking processes add them to various foods [7]. Cooking and processing techniques, such as roasting, barbecuing, grilling, smoking, heating, drying, baking, ohmic-infrared cooking, etc., also contribute to PAHs formation. Various factors, such as the distance from the heat source, used fuel, level of processing, and cooking durations and methods, as well as processes, such as reuse, conching, concentration, crushing, and storage, all increase the concentration of PAHs in food [8]. Different cooking methods also contribute to the formation of heterocyclic aromatic amines (HCAs)—another mutagenic and/or carcinogenic compound naturally formed during the cooking of protein-rich foods, such as beef, poultry, and fish—which exhibit higher accumulation when the deep-fat frying method is used in food processing [9].

Specific legislation [10] limits the concentration of PAHs in foods. A wide-ranging debate is in full swing and will probably continue with the emergence of new data. This is due to associated toxicity and health issues that PAHs have been linked to [11]. Since the 1970s, a set of 16 PAHs have been monitored by the United States Environmental Protection Agency [12], while the European Food Safety Agency (EFSA) identified Σ8PAHs or Σ4PAHs as relevant toxicity indicators [12,13].

Not all types of wood produce the same amount of PAHs; large variation limits are described [14,15]. The time of smoking also dramatically contributes to final PAH concentrations in meat [13], as do the various smoke generating methods [16,17]. Although PAHs are naturally occurring in the environment, including in the air we breathe [18–20], the most acute toxic effect is given by direct dietary ingestion [21]. Of the many adverse effects of PAHs, those leading to serious illnesses are the most worrying [19,22–24].

Meat is a fully interesting subject of study when related to human health, regardless of dietary habits, quantities, or methods of preparation and consumption. A permanent comparison is the one about the way meat is consumed or the type of meat, i.e., species, anatomical region, white or red meat, aquatic originating or terrestrial, etc. Poultry seemed to be healthier than red or processed meat when colorectal cancer was a study subject [25], but consumed as refrigerated, frozen, and as minced meat. In pancreatic carcinoma, an inverse association was shown in poultry meat when compared with red meat [26]. What about when smoked meat is consumed? Traditional and long-lasting smoking of meat preserves its quality with antioxidant and antimicrobial properties, and also imparts a desirable colour, flavour, and aroma to smoked foods [27]. In fish, smoking either cold or hot has a long tradition [28], and PAH concentrations greatly vary among the producers, sometimes exceeding the regulation limits [29]; however, few associations with specific pathologies or genotoxic effects are documented [30]. In poultry, smoking methods vary, e.g., at least four were described at industrial scale [31], while traditional methods imply natural smoke generated from hardwood sawdust. Solutions are being sought for the optimization of technological ways to reduce the risk of PAHs depositing or synthesis in food [32], while smoke flavour is highly desirable in developing societies [33]. Nonetheless, a potential concern for consumers' health and for specific population groups can be easily seen in populations with traditions for consuming smoked food [34].

As a request to health concerns about PAH, European Commission experts are trying to standardize detection methods in tandem with new scientific discoveries [35], while the producers offer a wider range of varieties of smoked meat. Poultry meat is either traditionally or warm-flavour smoked, although liquid smoke could be a viable alternative to traditional methods sensitively appreciated [36], but even as an all-natural antimicrobial in preserving food [27]. In spite of more and more explicit methodology in PAHs detection, not much information is offered about smoking poultry meat, and even less is available for duck meat or turkey. Within this context, we aimed to assess PAH levels in traditional smoking hardwood sawdust applied (birch, plum, cherry wood) to poultry meat (chicken, duck, and turkey).

2. Materials and Methods

2.1. Biological Material

Chicken (*Gallus gallus domesticus*), turkey (*Meleagris gallopavo*), and ducks (*Anas platyrhynchos*) were traditionally bred and provided the meat used for the experiment. Farmers produce their own reproductive material each year; consequently, no pure-blood, certified breeds were raised. The fowl were reproduced in their own farm, specifically in Posmuș place (a village in Șieu rural community, Bistrița-Năsăud County, Transylvania, Romania, geo coordinates N lat. 46.9753, E long. 24.5814), and were slaughtered in Caraiman Slaughterhouse, Bistrița-Năsăud County, for commercial purposes, to be marketed as traditionally farmed poultry products. A semi-intensive farming system was applied, using locally available feedstuffs during warm season to formulate a maize-soybean diet, following the nutritional requirements of each species. Fowl (30 males from each species) were slaughtered at different ages and live weights: 16 weeks and 3.2 kg/head in chickens, 16 weeks and 3.0 kg/head in ducks, 21 weeks and 6.1 kg in turkey. Breast anatomic region has been used for smoking (*Pectoralis superficialis* and *Pectoralis profundis* muscles).

In terms of research bioethics, no animals were used for applying experimental factors on them within a farm-conducted protocol. They were raised for meat production and sold as eviscerated carcasses to farmers' market. Samples were taken from refrigerated carcasses. No ethical approval was necessary, as the biological material consisted of meat issued from marketed poultry, and the research did not interfere with fowl welfare and farming conditions.

2.2. Smoking Materials

Three types of wood were used for smoking: beech (*Fagus sylvatica*), plum (*Prunus domestica*), and cherry (*Prunus cerasus*). All types of wood had the same size of wood chunks (approximate dimensions: $6 \times 2 \times 0.3$ cm) while a continuous maintenance of the hot embers was carried out. Two time intervals were set, one-day smoke treatment (24 h) and two-day smoking treatment (48 h). A traditional wooden, handmade installation was used, and a natural velocity of smoke through a metal sieve covered pipe of 20×20 cm occurred.

2.3. Meat Proximate Composition

Breast pieces of 10 g were taken from all individuals, prior to smoking, then were minced and blended together to form a homogenous sample of 300 g/species, thereafter submitted to analytical chemistry investigation to assess the proximate composition. Most of the methods involved an initial gravimetric assessment of a crude sample using a KERN ABJ 220 4NM analytical scale (manufacturer KERN & SOHN GmbH, Stuttgart, Germany). AOAC standard protocols were used to assess proximate composition: dry matter (DM)—water content via AOAC 950.46 method [37], on a thermoregulated forced airflow MEMMERT UF110+ oven (manufacturer Memmert GmbH + Co., Schwabach, Germany); total minerals (TM) content via AOAC 920.153 method [38], on a Nabertherm muffle furnace L 9/13 (manufacturer Nabertherm GmbH, Lilienthal, Germany); total lipids (TL) content—ether extract via AOAC 960.39 method [39], on a Velp Scientific SER 148 Randall extractor (manufacturer VELP Scientifica SRL, Usmate, Italy); total nitrogen content and total protein (TP) content via AOAC 929.08 method [40], on a Velp Scientifica DK6 digester and a UDK127-distiller (manufacturer VELP Scientifica SRL, Usmate, Italy) and Biotrate 50 mL digital burette (manufacturer Sartorius Lab Instruments GmbH & Co., Goettingen, Germany). Total organic matter (TOM) was calculated by the difference between Dry Matter (DM) and Total minerals (TM) content (Relation (1)).

$$\text{TOM (g/100 g)} = \text{DM (g/100 g)} - \text{TM (g/100 g)} \tag{1}$$

Nitrogen-Free Extract (NFE) was calculated by the difference between Organic Matter content and Total lipids and Total protein content (Relation (2)).

$$\text{NFE (g/100 g)} = \text{DM (g/100 g)} - \text{TL (g/100 g)} - \text{TP (g/100 g)} \tag{2}$$

For each proximate compound and each derived computation, 6 analytical repetitions were run. All results were expressed as grams per 100 g crude sample.

2.4. HPLC Analysis

Extraction of polycyclic aromatic hydrocarbons (PAHs) from smoked, skinless chicken, turkey, and duck breast meat started from an average sample of 10 g/species, homogenized in a blender, and saponified in an ultrasound bath for 30 min at 60 °C, using 50 mL KOH 0.4 M solution in ethanol and water in a 9:1 ratio. Post saponification, the samples were passed through sodium sulphate and filtered, then a double extraction was performed using 15 mL cyclohexane. The supernatant was reunited and purified on a Florisil column, then evaporated to dryness under a continuous nitrogen atmosphere. Each sample was dissolved in acetonitrile (1 mL) and filtered through a 0.45 μm filter, then kept in the refrigerator until the beginning of chromatographic analysis. This was carried out on a Perkin-Elmer HPLC system (manufacturer Perkin-Elmer, equipped with a model 200 series binary pump delivery system (Perkin-Elmer), a model 200 series degasser (manufacturer Perkin-Elmer Inc., Boston, MA, USA), a Flexar autosampler (Perkin-Elmer), a Spark thermostat, and a fluorescence detector model 200 series (Perkin-Elmer) with a double monochromator. The identification of PAHs was carried out on a column of ZORBAX Eclipse PAH (5 μm, 15 cm × 4.6 mm) in a gradient consisting of acetonitrile and ultrapure water as the mobile phase.

The LOQ and LOD were calculated as 10 times and 3.3 times, respectively, the standard deviation of the response for the lowest feasible analytic concentration in the range and the calibration curve's slope. Calibration curve 0.05–10 ng/g. Limit of quantification 0.05 ng/g. Limit of detection 0.015 ng/g. Recovery between 88.3–97.2 %, calculated using spiked samples with a concentration of 5 ng/g.

Fifteen PAHs were aimed for detection: Benzo(a)pyrene, Naphthalene, Acenaphthene, Flourene, Phenanthrene, Anthracene, Fluoranthene, Pyrene, Benz(a)anthracene, Chrysene, Benzo(k)fluoranthene, Benzo(b)fluoranthene, Benzo[*ghi*]perylene, Dibenzo[a,h]anthracene, and Indeno [1,2,3-*cd*]pyrene. Eleven PAHs were utterly quantified among the 15. All acquired data resulted after 5 analytical replications for each PAH.

The sum of benzo(a)pyrene, benz(a)anthracene, benzo(b)fluoranthene, and chrysene (level of Σ4PAHs in smoked meat) was also quantified, representing an index for evaluating the toxicity induced by PAHs on meat quality, according to EU legislation.

2.5. Statistical Analyses

Analytical data were statistically processed via Graph Pad Prism 9.4.1 (manufacturer GraphPad Inc., Palo Alto, CA, USA) software to achieve the descriptive statistics values (mean, standard error of mean, coefficient of variation) and to analyse the variance using the ANOVA one-way algorithm, followed by Tukey post-hoc treatment, when 3 data groups were compared, or the unpaired 2-tailed *t* test when just 2 data groups were compared. The results of the analysis of variance were reported as *p* values, compared to *p* thresholds of 0.05, 0.01, and 0.001.

Within the same fowl species, meat samples were compared for differences in PAHs accumulation at 24 and 48 h post-smoking, in relation to smoking materials (burnt wood). Additionally, comparisons were carried out between the fowl species, within the same smoking type, to find out whether meat structure and its other particularities affect the accumulation of the same PAHs.

3. Results

The proximate composition of meat, prior to smoking, is displayed in Table 1. Fat content in meat, as the main accumulation matrix for the investigated PAH, varied from 1.88 g/100 g in chicken breast to 2.42 g/100 g in turkey and to 2.81 g/100 g in duck breast meat (Table 1).

Table 1. Proximate composition of the matured poultry meat prior to smoking (g/100 g).

Proximate Compound	Chicken		Turkey		Duck	
	Mean ± SEM	V%	Mean ± SEM	V%	Mean ± SEM	V%
Water	77.35 ± 0.55	1.75	73.28 ± 0.64	2.13	81.27 ± 0.67	2.01
Dry matter	22.65 ± 0.55	5.91	26.72 ± 0.64	5.86	18.73 ± 0.67	8.79
Total minerals	0.96 ± 0.02	5.12	1.31 ± 0.02	4.28	0.87 ± 0.02	4.81
Organic matters	21.69 ± 0.39	4.38	25.41 ± 0.48	4.66	17.86 ± 0.39	5.32
Total proteins	19.56 ± 0.27	3.44	22.62 ± 0.35	3.83	14.74 ± 0.25	4.09
Total lipids	1.88 ± 0.04	4.97	2.42 ± 0.05	5.08	2.81 ± 0.06	5.19
Nitrogen free extract	0.25 ± 0.01	5.63	0.37 ± 0.01	5.45	0.31 ± 0.01	5.81

These values of meat nutrients are generally expected when a traditional manner of farming is rolled on, in respect to feeding based mostly on locally available feedstuffs, such as corn and wheat crumbles (concentrated in carbohydrates and, subsequently, energy), and a relatively long period of fattening, compared to conventionally reared broilers, whose diets are based on a corn-soymeal mixture (better balanced diets in terms of energy/protein ratios).

Data on the accumulation of PAHs in chicken meat are presented in Table 2 with comparisons between the type of woods used in smoking and between the duration of exposure.

As expected, the two-day smoking period produced a higher accumulation of PAHs in meat; some of the PAHs even more than doubled in concentration (i.e., Benz(a)anthracene in beech wood smoking or Benzo(k)fluoranthene in cherry wood smoking). Benzo(k)fluoranthene has only been quantified in cherry wood smoking.

Data on the accumulation of PAHs in turkey meat are displayed in Table 3 with differences analysis related to the species of tree sawdust used in smoking and to the smoking length.

Table 2. Variance of total PAHs accumulation in chicken meat (μg PAH/kg meat, ppm).

Polycyclic Aromatic Hydrocarbons	Exposure Time	Beech Wood Smoking (B)		Plum Wood Smoking (P)		Cherry Wood Smoking (C)		Regulated Maximal Admitted Limit
		Mean ± SEM	V%	Mean ± SEM	V%	Mean ± SEM	V%	
Naphtalene	24 h	6.47^{a}_{x} ± 0.14	4.4	4.01^{da} ± 0.09	4.5	6.74^{ad}_{x} ± 0.24	7.1	At lowest as possible [10]
	48 h	8.24^{a}_{z} ± 0.21	5.8	4.66^{da} ± 0.05	2.2	9.07^{ad}_{w} ± 0.43	9.4	
Acenaphthene	24 h	8.95^{a} ± 0.15	3.3	6.46^{da} ± 0.30	9.4	10.24^{ad} ± 0.25	4.9	At lowest as possible [10]
	48 h	10.24^{a} ± 0.21	4.2	7.88^{da} ± 0.57	14.5	11.32^{ad} ± 0.22	3.8	
Fluorene	24 h	32.93^{a}_{x} ± 0.65	3.9	16.24^{da}_{x} ± 0.40	4.9	20.7^{dd}_{x} ± 0.32	3.1	At lowest as possible [10]
	48 h	37.02^{a}_{w} ± 0.31	1.7	18.56^{da}_{z} ± 0.17	1.9	22.53^{dd}_{y} ± 0.23	2.1	
Phenanthrene	24 h	54.85^{a}_{x} ± 0.59	2.2	42.45^{d}_{x} ± 1.09	5.1	43.68^{d}_{x} ± 0.52	2.4	At lowest as possible [10]
	48 h	59.60^{a}_{y} ± 0.99	3.3	64.53^{ca}_{w} ± 0.39	1.2	77.16^{dd}_{w} ± 1.07	2.8	
Anthracene	24 h	2.01^{a}_{x} ± 0.03	3.3	0.84^{d} ± 0.04	8.8	0.70^{d}_{x} ± 0.03	9.3	At lowest as possible [10]
	48 h	3.22^{a}_{w} ± 0.11	6.7	0.96^{d} ± 0.03	6.3	0.94^{d}_{y} ± 0.02	3.8	
Fluoranthene	24 h	6.05^{a}_{x} ± 0.08	2.6	4.09^{da}_{x} ± 0.20	9.9	2.51^{dd}_{x} ± 0.20	16.1	At lowest as possible [10]
	48 h	7.18^{a}_{w} ± 0.11	2.9	5.30^{da}_{w} ± 0.17	6.4	4.09^{dd}_{w} ± 0.12	5.9	
Pyrene	24 h	0.88^{a} ± 0.03	6.4	1.48^{ca} ± 0.07	8.9	2.18^{dd}_{x} ± 0.13	11.5	At lowest as possible [10]
	48 h	1.17^{a} ± 0.11	18.3	1.82^{da} ± 0.04	4.8	3.52^{dd}_{w} ± 0.12	6.8	
Benz(a)anthracene	24 h	0.32^{a}_{x} ± 0.02	15.1	0.68^{aa} ± 0.07	19.9	1.65^{dd} ± 0.13	16.2	At lowest as possible [10]
	48 h	0.77^{a}_{y} ± 0.05	12.1	0.95^{aa} ± 0.05	11.4	1.76^{dd} ± 0.15	16.9	
Chrysene	24 h	0.98^{a}_{x} ± 0.03	5.5	1.70^{da} ± 0.05	6.3	0.70^{ad}_{x} ± 0.06	17.4	Σ4PAHs < 30 μg/kg [10]
	24 h	1.59^{a}_{z} ± 0.125	15.3	2.08^{b} ± 0.12	11.3	1.69^{ab}_{w} ± 0.16	20.3	
Benzo(k)fluoranthene	24 h	-	-	-	-	0.75_{x} ± 0.07	18.4	At lowest as possible [10]
	48 h	-	-	-	-	1.88_{w} ± 0.03	2.9	

SEM = Standard error of mean; V% = coefficient of variation. Analysis of variance: per line, means with different superscripts on the same line differ significantly for: ^{ab} $p < 0.05$; ^{ac} $p < 0.01$; ^{ad} $p < 0.001$; per column (24 h vs. 48 h), means with different subscripts on the same column differ significantly for: $_{xy}$ $p < 0.05$; $_{xz}$ $p < 0.01$; $_{xw}$ $p < 0.001$. Σ4PAHs: Sum of benzo(a)pyrene, benz(a)anthracene, benzo(b)fluoranthene, and chrysene is <30 μg/kg in smoked meat and meat products [10].

Table 3. Variance of total PAHs accumulation in turkey meat (μg PAH/kg meat, ppm).

Polycyclic Aromatic Hydrocarbons	Exposure Time	Beech Wood Smoking (B)		Plum Wood Smoking (P)		Cherry Wood Smoking (C)		Regulated Maximal Admitted Limit
		Mean ± SEM	V%	Mean ± SEM	V%	Mean ± SEM	V%	
Naphtalene	24 h	4.85^{a} ± 0.14	5.82	3.37^{c}_{x} ± 0.26	15.19	4.14_{x} ± 0.33	15.86	At lowest as possible [10]
	48 h	5.37 ± 0.22	8.04	5.06_{w} ± 0.24	9.41	5.33_{y} ± 0.12	4.68	
Acenaphthene	24 h	7.93^{a}_{x} ± 0.29	7.36	6.19^{ba} ± 0.36	11.48	9.97^{cd}_{x} ± 0.53	10.56	At lowest as possible [10]
	48 h	9.60^{a}_{y} ± 0.30	6.27	7.74^{ba} ± 0.33	8.42	11.79^{cd}_{y} ± 0.39	6.59	
Fluorene	24 h	23.96^{a}_{x} ± 0.89	7.43	17.51^{d}_{x} ± 0.39	4.48	18.45^{d}_{x} ± 0.37	3.99	At lowest as possible [10]
	48 h	29.91^{a}_{w} ± 0.61	4.06	21.51^{d}_{z} ± 1.02	9.46	23.78^{d}_{w} ± 0.56	4.71	
Phenanthrene	24 h	46.45^{a} ± 1.23	5.32	32.82^{d}_{x} ± 1.16	7.05	46.70^{a}_{x} ± 3.00	12.85	At lowest as possible [10]
	48 h	53.58^{a} ± 1.21	4.52	44.50^{ba}_{z} ± 1.72	7.72	57.70^{ad}_{z} ± 2.25	7.79	
Anthracene	24 h	1.75^{a}_{x} ± 0.09	9.85	0.76^{d} ± 0.07	17.66	0.51^{d} ± 0.03	13.32	At lowest as possible [10]
	48 h	2.40^{a}_{y} ± 0.29	23.78	0.98^{d} ± 0.08	16.74	0.72^{d} ± 0.08	23.68	
Fluoranthene	24 h	2.64^{a} ± 0.23	17.39	4.27^{ca}_{x} ± 0.34	15.89	1.68^{cd} ± 0.18	21.36	At lowest as possible [10]
	48 h	3.82^{a} ± 0.49	25.92	5.57^{ca}_{y} ± 0.20	7.31	2.31^{bd} ± 0.13	10.88	
Pyrene	24 h	0.61^{a}_{x} ± 0.06	18.45	0.96^{a} ± 0.05	10.40	2.05^{d} ± 0.09	8.89	At lowest as possible [10]
	48 h	1.23^{a}_{z} ± 0.18	29.76	1.07^{a} ± 0.16	6.45	2.45^{d} ± 0.08	11.91	
Benz(a)anthracene	24 h	0.63 ± 0.04	11.91	0.69 ± 0.08	17.21	0.60_{x} ± 0.05	12.70	At lowest as possible [10]
	48 h	0.83 ± 0.05	12.70	0.80 ± 0.10	16.33	0.92_{y} ± 0.08	17.29	
Chrysene	24 h	0.71^{a} ± 0.06	17.29	1.01^{ba} ± 0.02	8.50	1.77^{dd}_{x} ± 0.08	10.40	Σ4PAHs < 30 μg/kg [10]
	48 h	0.89^{a} ± 0.05	10.40	1.06^{da} ± 0.05	8.69	2.19^{dd}_{z} ± 0.10	15.74	
Benzo(k)fluoranthene	24 h	-	-	-	-	0.38_{x} ± 0.16	32.76	At lowest as possible [10]
	48 h	-	-	-	-	0.71_{z} ± 0.17	19.57	
Benzo(a)pyrene	24 h	-	-	-	-	0.51 ± 0.04	15.74	< 5 μg/kg [10]
	48 h	-	-	-	-	0.59 ± 0.05	17.16	

SEM = Standard error of mean; V% = coefficient of variation. Analysis of variance: per line, means with different superscripts on the same line differ significantly for: ^{ab} $p < 0.05$; ^{ac} $p < 0.01$; ^{ad} $p < 0.001$; per column (24 h vs. 48 h), means with different subscripts on the same column differ significantly for: $_{xy}$ $p < 0.05$; $_{xz}$ $p < 0.01$; $_{xw}$ $p < 0.001$. Σ4PAHs: Sum of benzo(a)pyrene, benz(a)anthracene, benzo(b)fluoranthene, and chrysene is <30 μg/kg in smoked meat and meat products [10].

In turkey meat, the two-day smoking also quantified more PAHs, with Pyrene doubling the concentration in beech wood smoking from 0.61 µg after day 1 to 1.23 µg after two smoking days. Overall, the absolute values of the PAHs were lower in turkey meat as compared to chicken meat. The sum of Benz(a)anthracene and Chrysene (two of the set of four PAHs that were monitoring markers of PAHs toxicity accumulation in meat based on EU legislation) was similar after day 1 in chicken and turkey meat (2.35 µg vs. 2.37 µg, respectively), but the differences were slightly in favour of chicken meat after day 2 of smoking (3.45 µg vs 3.11 µg, respectively).

The dynamics of PAHs deposition and accumulation in duck meat in relation to the duration of smoking and the species of tree used in smoking are presented in Table 4.

Table 4. Variance of total PAHs accumulation in duck meat (µg PAH/kg meat, ppm).

Polycyclic Aromatic Hydrocarbons	Exposure Time	Beech Wood Smoking (B)		Plum Wood Smoking (P)		Cherry Wood Smoking (C)		Regulated Maximal Admitted Limit
		Mean ± SEM	V%	Mean ± SEM	V%	Mean ± SEM	V%	
Naphtalene	24 h	$8.55^{a} \pm 0.25$	5.93	$6.46^{da}_{x} \pm 0.23$	7.27	$9.42^{ad}_{x} \pm 0.26$	5.50	At lowest as possible [10]
	48 h	$9.53^{a} \pm 0.46$	9.66	$8.76^{aa}_{w} \pm 0.13$	2.89	$11.11^{cd}_{z} \pm 0.41$	7.42	
Acenaphthene	24 h	$9.10_{x} \pm 0.44$	9.71	8.94 ± 0.13	3.17	$10.24_{x} \pm 0.39$	7.68	At lowest as possible [10]
	48 h	$10.81^{a}_{y} \pm 0.21$	3.88	$10.40^{aa} \pm 0.30$	5.70	$12.26^{bc}_{z} \pm 0.42$	6.93	
Fluorene	24 h	27.12 ± 1.23	9.09	25.80 ± 0.52	4.02	24.60 ± 0.83	6.73	At lowest as possible [10]
	48 h	31.20 ± 1.36	8.72	29.26 ± 0.71	4.88	27.80 ± 0.86	6.19	
Phenanthrene	24 h	$84.74^{a} \pm 1.95$	4.59	$74.16^{ba}_{x} \pm 3.04$	8.21	$71.56^{ca}_{x} \pm 3.11$	8.70	At lowest as possible [10]
	48 h	93.24 ± 1.20	2.57	$94.00_{w} \pm 1.41$	3.01	$88.06_{w} \pm 2.12$	4.82	
Anthracene	24 h	$3.19^{a}_{x} \pm 0.21$	12.99	$0.92^{da}_{x} \pm 0.10$	20.36	$2.19^{dd}_{x} \pm 0.22$	20.44	At lowest as possible [10]
	48 h	$4.06^{a}_{z} \pm 0.14$	6.79	$1.62^{da}_{y} \pm 0.09$	11.39	$2.84^{dd}_{y} \pm 0.04$	2.86	
Fluoranthene	24 h	$8.84^{a} \pm 0.45$	5.09	$6.38^{d}_{x} \pm 0.67$	10.51	$5.40^{d} \pm 0.67$	12.45	At lowest as possible [10]
	48 h	$9.73^{a} \pm 0.22$	4.08	$7.67^{da}_{y} \pm 0.34$	8.81	$6.22^{db} \pm 0.34$	5.25	
Pyrene	24 h	$1.48^{a}_{x} \pm 0.14$	18.89	$2.10^{aa} \pm 0.14$	13.28	$2.56^{db}_{x} \pm 0.20$	15.74	At lowest as possible [10]
	48 h	$2.17^{a}_{y} \pm 0.14$	12.68	$2.50^{aa} \pm 0.18$	14.81	$3.26^{db}_{y} \pm 0.11$	6.87	
Benz(a)anthracene	24 h	$0.94^{a} \pm 0.08$	17.21	$1.08^{aa}_{x} \pm 0.15$	28.33	$1.68^{cb} \pm 0.09$	10.24	At lowest as possible [10]
	48 h	$1.38^{a} \pm 0.16$	22.88	$1.74^{aa}_{z} \pm 0.11$	12.90	$2.12^{ca} \pm 0.07$	6.37	
Chrysene	24 h	$1.88^{a} \pm 0.31$	22.05	$2.63^{ba} \pm 0.10$	7.37	$1.78^{ac}_{x} \pm 0.10$	11.40	Σ4PAHs < 30 µg/kg [10]
	48 h	$2.12^{a} \pm 0.16$	14.74	$3.10^{da} \pm 0.06$	3.98	$2.48^{ab}_{y} \pm 0.17$	13.59	
Benzo(k)fluoranthene	24 h	-	-	-	-	$0.89_{x} \pm 0.07$	16.54	At lowest as possible [10]
	48 h	-	-	-	-	$2.07_{w} \pm 0.09$	8.62	

SEM = Standard error of mean; V% = coefficient of variation. Analysis of variance: per line, means with different superscripts on the same line differ significantly for: ^{ab} $p < 0.05$; ^{ac} $p < 0.01$; ^{ad} $p < 0.001$; per column (24 h vs. 48 h), means with different subscripts on the same column differ significantly for: $_{xy}$ $p < 0.05$; $_{xz}$ $p < 0.01$; $_{xw}$ $p < 0.001$. Σ4PAHs: Sum of benzo(a)pyrene, benz(a)anthracene, benzo(b)fluoranthene, and chrysene is <30 µg/kg in smoked meat and meat products [10].

As shown in Table 3, the PAHs accumulation in duck meat increased from day 1 to day 2 in all smoking wood types, the highest being in cherry wood smoking for Benzo(k)fluoranthene, with more than doubling the concentration (from 0.89 µg to 2.07 µg). The most relevant finding is that the absolute values summarized for all PAHs are highest in duck meat when compared with chicken and turkey meat. In addition, the sum of Benz(a)anthracene and Chrysene was higher for duck meat, irrespective of the day of smoking, indicating similar values for day 1 with day 2 for chicken and turkey meat (3.46 µg in duck meat on day 1 vs. 3.45 µg for chicken meat and 3.11 µg for turkey meat, but the last two after two days of smoking). After two days of smoking, duck meat accumulated 4.6 µg of Benz(a)anthracene + Chrysene.

Table 5 reveals the influence of meat origin on the dynamics of PAHs accumulation, in relation to exposure time to smoking, within each type of fuel used to generate smoke.

Table 5. Analysis of variance of total PAHs accumulation in meat, under the influence of poultry species (arithmetic means, µg PAH/kg meat, ppm).

Polycyclic Aromatic Hydrocarbons	Exposure Time	Beech Wood Smoking (B)			Plum Wood Smoking (P)			Cherry Wood Smoking (C)			Regulated Maximal Admitted Limit
		Chicken	Turkey	Duck	Chicken	Turkey	Duck	Chicken	Turkey	Duck	
Naphtalene	24 h	6.47 [a]	4.85 [ca]	8.55 [dd]	4.01 [a]	3.37 [a]	6.46 [d]	6.74 [a]	4.14 [da]	9.42 [d]	At lowest as possible [10]
	48 h	8.24 [a]	5.37 [da]	9.53 [ad]	4.66 [a]	5.06 [a]	8.76 [d]	9.07 [a]	5.33 [da]	11.11 [d]	
Acenaphthene	24 h	8.95	7.93	9.10	6.46 [a]	6.19 [a]	8.94 [d]	10.24	9.97	10.24	At lowest as possible [10]
	48 h	10.24	9.60	10.81	7.88 [a]	7.74 [a]	10.40 [d]	11.32	11.79	12.26	
Fluorene	24 h	32.93 [a]	23.96 [d]	27.12 [d]	16.24 [a]	17.51 [a]	25.80 [d]	20.70 [a]	18.45 [aa]	24.60 [bd]	At lowest as possible [10]
	48 h	37.02 [a]	29.91 [d]	31.20 [d]	18.56 [a]	21.51 [a]	29.26 [d]	22.53 [a]	23.78 [aa]	27.80 [db]	
Phenanthrene	24 h	54.85 [a]	46.45 [a]	84.74 [d]	42.45 [a]	32.82 [ba]	74.16 [d]	43.68 [a]	46.70 [a]	71.56 [d]	At lowest as possible [10]
	48 h	59.60 [a]	53.58 [a]	93.24 [d]	64.53 [a]	44.50 [a]	94.00 [d]	77.16 [a]	57.70 [da]	88.06 [cd]	
Anthracene	24 h	2.01 [a]	1.75 [a]	3.19 [d]	0.84	0.76	0.92	0.70 [a]	0.51 [a]	2.19 [d]	At lowest as possible [10]
	48 h	3.22 [a]	2.40 [da]	4.06 [dd]	0.96 [a]	0.98 [a]	1.62 [b]	0.94 [a]	0.72 [a]	2.84 [d]	
Fluoranthene	24 h	6.05 [a]	2.64 [da]	8.84 [dd]	4.09 [a]	4.27 [aa]	6.38 [da]	2.51 [a]	1.68 [a]	5.40 [d]	At lowest as possible [10]
	48 h	7.18 [a]	3.82 [da]	9.73 [dd]	5.30 [a]	5.57 [a]	7.67 [d]	4.09 [a]	2.31 [da]	6.22 [d]	
Pyrene	24 h	0.88 [a]	0.61 [a]	1.48 [d]	1.48 [a]	0.96 [aa]	2.10 [ad]	2.18	2.05	2.56	At lowest as possible [10]
	48 h	1.17 [a]	1.23 [a]	2.17 [d]	1.82 [a]	1.07 [ca]	2.50 [bd]	3.52 [a]	2.45 [d]	3.26 [a]	
Benz(a)anthracene	24 h	0.32 [a]	0.63 [aa]	0.94 [ca]	0.68	0.69	1.08	1.65 [a]	0.60 [d]	1.68 [a]	At lowest as possible [10]
	48 h	0.77 [a]	0.83 [aa]	1.38 [cc]	0.95 [a]	0.80 [a]	1.74 [d]	1.76 [a]	0.92 [d]	2.12 [a]	
Chrysene	24 h	0.98 [a]	0.71 [a]	1.88 [d]	1.70 [a]	1.01 [ca]	2.63 [dd]	0.70 [a]	1.77 [d]	1.78 [d]	Σ4PAHs < 30 µg/kg [10]
	48 h	1.59 [a]	0.89 [ca]	2.12 [ad]	2.08 [a]	1.06 [da]	3.10 [dd]	1.69 [a]	2.19 [aa]	2.48 [da]	
Benzo(k)fluoranthene	24 h	-	-	-	-	-	-	0.75 [a]	0.38 [ca]	0.89 [ad]	At lowest as possible [10]
	48 h	-	-	-	-	-	-	1.88 [a]	0.71 [d]	2.07 [a]	
Benzo(a)pyrene	24 h	-	-	-	-	-	-	-	0.51	-	< 5 µg/kg [10]
	48 h	-	-	-	-	-	-	-	0.59	-	

Analysis of variance: per line, means with different superscripts on the same line differ significantly for: [ab] $p < 0.05$; [ac] $p < 0.01$; [ad] $p < 0.001$; Σ4PAHs: Sum of benzo(a)pyrene, benz(a)anthracene, benzo(b)fluoranthene, and chrysene is <30 µg/kg in smoked meat and meat products [10].

Table 5 shows comparatively how the PAHs varied between species, based on smoking wood and period of smoking. Overall, the duck meat has the higher PAHs accumulation, although there is an exception for Fluorene when beech wood was used. Chicken meat presented higher values for day 1 and day 2, followed by duck and turkey meat, respectively. Significant differences ($p < 0.001$) were usually observed for duck meat when compared with chicken and turkey meat, but also easy to notice as absolute values for Anthracene, Phenanthrene, or Fluoranthene.

4. Discussion

The concentration limits were found below the EU legislation [41] thresholds for benzo(a) pyrene (BaP), for the sum of benzo(a)pyrene, benz(a)anthracene, benzo(b)fluoranthene, and chrysene (level of Σ4PAHs in smoked meat should not exceed 12 µg/kg or 30 µg/kg for some of the member state countries since 13.12.2014 [10]) in all three poultry species. Those four PAHs are highly indicative and carefully monitored when present in food. The highest concentrations of PAHs occurred in duck meat, irrespective of time of smoking or wood type (Table 4), followed by chicken meat (Table 2) and turkey (Table 3). Duck meat also accumulated PAHs faster, as the percent of total PAHs at the end of the first day of smoking per total period.

In comparison, Peking roasted ducks in the hung oven roasted technique had a high BaP concentration of 8.7 µg/kg in the skin, but <3.0 µg/kg when other techniques, such as the closed oven procedure and electricity heating, were used [42]. Within the same study, ducks roasted by electrical heating presented lower concentrations of total PAHs. No BaP were detected in smoked duck meat, but higher Σ4PAHs were, followed by chicken and turkey meat. Duck meat reached the highest value for chrysene (one of the Σ4PAHs with legislative imposition limits) of 3.10 µg/kg when plum wood and two days of smoking treatment was used. T1hat represents 10.33% of the Σ4PAHs. When smoking time was used as reference, the duck meat accumulated the highest concentration of chrysene after only one day of smoking, with a value of 2.63 µg/kg (again in plum wood smoking), which was superior to all other meats after two days of smoking. This might correlate with the fat percentages of carcasses. Our samples had no skin at all. Generally, fatter meat accumulates higher PAH concentrations [43,44], while proteins and carbohydrates do not contribute to the formation of PAHs during thermal cooking [43]. Fat distribution inside the breast muscle differs among species, but usually a low percentage is found in chicken and higher in duck [45]. This fat is generally uniformly distributed within the muscle and surrounding muscle fibres.

The skin might significantly contribute to PAHs accumulation, but as a natural "barrier" too. In pork meat, skin has been suggested to protect the bacon from PAHs concentration [46]. Added fat in Portuguese traditional dry fermented sausages positively influenced PAHs contamination [47]. PAH compounds were found to be more abundant in high-fat doner samples than in low-fat groups when fat content was interrogated [48], while a positive linear correlation was observed for the PAHs' bio-accessibility and the fat contents of grilled meat [49]. That is also observed in various traditionally smoked products from Cyprus, where the highest PAH concentrations were found in samples with higher fat content [50]. The fat content of the meat products appears to favour PAHs accumulation in smoking products. Normally, smoked meat is left with the skin on, knowing that the fat gives it superior organoleptic qualities. There is also a small risk of dehydration of smoked meat when the skin is removed, but a short smoking period could help to avoid this phenomenon.

The tissue composition and tissue texture of the meat are other characteristics that could influence the smoking process. Tissue texture in poultry species varies depending on several factors [51], but genetic ones mostly differentiate the chemical and tissue composition of the meat from the three species. Meat texture and intramuscular connective tissue play an important role in meat quality [52], and texture affects bio-concentrating and PAHs accumulation properties [13]. In our sample data, turkey meat has the lowest total content of PAHs and also Σ4PAHs. Interestingly, benzo(a)pyrene was only quantified in low quantities (average of 0.512 µg/kg in the single-day smoking treatment and average of 0.592 µg/kg in the two-day smoking treatment) in turkey meat and only in cherry wood smoking. From this perspective, the values are far lower than the legislative imposition limit of 5 µg/kg. Anyway, the presence of a highly monitored PAH only in turkey meat is somehow surprising based on the chemical composition of the three types of meat used in the experiment. Duck meat has accumulated the highest concentrations overall of PAHs, but no benzo(a)pyrene. Possibly, the fat content was one of the most significant factors in PAHs accumulation, but there are certainly others to be interrogated. Duck meat has the highest fat content of the meat, while turkey exhibited the highest protein content (22.62% from dry matter vs 14.74% in duck meat in our experiment). Chicken meat had intermediate values with respect to both fat and protein content.

The type of wood used for smoke generated significant differences ($p < 0.01$) for the same smoking period (one day or two days). Irrespective of meat species, the highest concentration overall (average of the three poultry species) of PAHs was quantified when beech wood was used for smoking, followed by cherry and plum woods (Tables 1–3). Taking the poultry species separately, the same order was recorded (when averaging day one and day two). Beech wood is largely used in smoking meat and is part of Romanian tradition. It also generated higher concentrations of PAHs in various pork meat products, e.g., Frankfurter-type sausages and mini-salamis when compared with several other wood types in an experiment where a smoking chamber with a smoldering smoke generator was used [14,53].

Interestingly, only cherry wood generated benzo(k)fluoranthene in all species when the two-day smoking period occurred, and benzo(a)pyrene in duck meat, although in low concentrations (Tables 1–4). Bird-cherry wood has been shown to generate moderate to low concentrations of PAHs in pork meat smoked in a system of homemade smoking kiln, when 10 wood types and charcoal were used [54]. The same was not true for chicken and turkey meat, where plum wood utilization produced a significantly lower accumulation of PAHs when compared with both beech and cherry wood ($p < 0.05$).

Among the three species, duck recorded the fastest PAHs accumulation in meat, as the day one percentage from the final smoking time (83.95% as average for all three types of wood vs. 79.91% in turkey and 75.98% in poultry).

Surprisingly, the highest concentration of PAHs as the percent of accumulation from day one through day two was in chicken meat, although it has a low-fat percent in meat as compared with duck and turkey meat.

Benzo(a)pyrene, although present in turkey meat, was only a little concentrated ($p > 0.05$) with the doubling of the smoking time. The first smoking period has been shown to significantly accumulate higher PAH concentrations in grilled beef and pork meat, in direct correlation with fat percent and fat dripping from the meat samples onto the heat source during grilling [44]. In our experiment, fat dripping was excluded due to the smoking technique itself.

From PAHs, the higher concentration was observed for benz(a)anthracene, chrysene, and benzo(k)fluoranthene in chicken meat after 2 days of smoking. More than doubled values were generated by cherry wood for chrysene and benzo(k)fluoranthene and for benz(a)anthracene by smoking with beech wood.

In duck meat, the total average values of PAHs revealed non-significant differences for all types of wood after two days smoking treatment, suggesting a mild or any dependency of this parameter. This also might be interpreted as the high capacity of PAHs accumulation with increasing time. A longer smoking time entails a higher accumulation of PAHs, but a higher amount of fat seems to contribute even more to this accumulation in similar products [55]. In our experiment, the time factor is limited to only two days, but even so, in fatter duck meat, the rate shows uniformity in the accumulation of PAHs, regardless of the type of wood used in smoking.

Significant differences ($p < 0.001$) were recorded when the smoking time was the variable. The two-day smoking period dramatically increased the level of PAH contamination, especially in cherry wood.

ANOVA analysis based on wood type revealed a significant influence of the type of wood in PAH contamination, the most "polluting" being the cherry wood.

5. Conclusions

The highest concentrations of PAHs were present in duck meat, irrespective of smoking time (one vs. two days) or wood type (cherry, plum, and beech).

When wood type was interrogated, the higher concentration overall of PAHs was quantified when beech wood was used, followed by cherry and plum woods.

Two days of smoking contributed to higher concentrations of PAHs in meat, and the fastest PAHs accumulation was shown in duck meat, while turkey meat had the lowest total content of PAHs and also Σ4PAHs. The only meat that accumulated benzo(a)pyrene was turkey meat when cheery wood was used, but in low concentrations and far below the legally imposed limits.

As follow-up, the sensory analysis must complete the characterisation of poultry smoked breast. Moreover, lipid oxidation is a concern and should be approached through measuring the oxidative status of the meat and the likelihood of free radicals' occurrence after smoking.

Author Contributions: Conceptualization, C.O.C. and A.C.; methodology, C.O.C. and R.-M.R.-R.; software, C.O.C. and R.-M.R.-R.; validation, A.C., A.L., and A.B.; formal analysis, A.C., A.L., A.B., and O.M.; investigation, C.O.C., A.C., A.L., and O.M.; data curation, C.O.C. and R.-M.R.-R.; writing—original draft preparation, C.O.C. and A.C.; writing—review and editing, C.O.C. and R.-M.R.-R.; visualization, C.O.C., A.C., A.L., and O.M.; supervision, C.O.C. All authors have read and agreed to the published version of the manuscript.

Funding: This research received no external funding.

Institutional Review Board Statement: Not applicable. No animals were used for applying experimental factors on them. They were raised for meat production on a private farm and sold as eviscerated carcasses to farmers' market. Samples were taken from slaughtered refrigerated poultry, and the research did not interfere with poultry welfare and farming conditions throughout their raising period.

Informed Consent Statement: Not applicable.

Data Availability Statement: Data supporting reported results available, upon request, from the authors.

Conflicts of Interest: The authors declare no conflict of interest.

References

1. Pereira, P.M.d.C.C.; Vicente, A.F.d.R.B. Meat Nutritional Composition and Nutritive Role in the Human Diet. *Meat Sci.* **2013**, *93*, 586–592. [CrossRef]
2. sur les Additifs, C.d.C. 2005. Available online: https://shop.un.org/books/faostat-2005-fao-statistical-data-27775 (accessed on 11 August 2022).
3. Samant, S.S.; Crandall, P.G.; O'Bryan, C.A.; Lingbeck, J.M.; Martin, E.M.; Tokar, T.; Seo, H.-S. Effects of Smoking and Marination on the Sensory Characteristics of Cold-Cut Chicken Breast Filets: A Pilot Study. *Food Sci. Biotechnol.* **2016**, *25*, 1619–1625. [CrossRef]
4. Herring, J.L.; Smith, B.S. *Meat-Smoking Technology. Handbook of Meat and Meat Processing*, 2nd ed.; CRC Press Taylor and Francis Group: Boca Raton, FL, USA, 2012; pp. 547–556.
5. Agustinelli, S.; Yeannes, M. Sensorial Characterization and Consumer Preference Analysis of Smoked Mackerel (*Scomber Japonicus*) Fillets. *IFRJ* **2015**, *22*, 2010–2017.
6. Pogorzelska-Nowicka, E.; Kurek, M.; Hanula, M.; Wierzbicka, A.; Półtorak, A. Formation of carcinogens in processed meat and its measurement with the usage of artificial digestion—A review. *Molecules* **2002**, *27*(14), 4665.
7. Singh, L.; Agarwal, T.; Simal-Gandara, J. PAHs, Diet and Cancer Prevention: Cooking Process Driven-Strategies. *Trends Food Sci. Technol.* **2020**, *99*, 487–506. [CrossRef]
8. Singh, D.K.; Sharma, S.; Habib, G.; Gupta, T. Speciation of Atmospheric Polycyclic Aromatic Hydrocarbons (PAHs) Present during Fog Time Collected Submicron Particles. *Environ. Sci. Pollut. Res.* **2015**, *22*, 12458–12468. [CrossRef]
9. Oz, F.; Yuzer, M.O. The Effects of Different Cooking Methods on the Formation of Heterocyclic Aromatic Amines in Turkey Meat. *J. Food Process. Preserv.* **2017**, *41*, e13196. [CrossRef]
10. European Commission. *COMMISSION REGULATION (EU) No 1327/2014 of 12 December 2014 Amending Regulation (EC) No 1881/2006 as Regards Maximum Levels of Polycyclic Aromatic Hydrocarbons (PAHs) in Traditionally Smoked Meat and Meat Products and Traditionally Smoked Fish and Fishery Products*; European Commission: Brussels, Belgium, 2014.
11. Patel, A.B.; Shaikh, S.; Jain, K.R.; Desai, C.; Madamwar, D. Polycyclic Aromatic Hydrocarbons: Sources, Toxicity, and Remediation Approaches. *Front. Microbiol.* **2020**, *11*, 562813. [CrossRef]
12. Zelinkova, Z.; Wenzl, T. The Occurrence of 16 EPA PAHs in Food—A Review. *Polycycl. Aromat. Compd.* **2015**, *35*, 248–284. [CrossRef]
13. Singh, L.; Varshney, J.G.; Agarwal, T. Polycyclic Aromatic Hydrocarbons' Formation and Occurrence in Processed Food. *Food Chem.* **2016**, *199*, 768–781. [CrossRef]
14. Hitzel, A.; Pöhlmann, M.; Schwägele, F.; Speer, K.; Jira, W. Polycyclic Aromatic Hydrocarbons (PAH) and Phenolic Substances in Meat Products Smoked with Different Types of Wood and Smoking Spices. *Food Chem.* **2013**, *139*, 955–962. [CrossRef]
15. Avagyan, R.; Nyström, R.; Lindgren, R.; Boman, C.; Westerholm, R. Particulate Hydroxy-PAH Emissions from a Residential Wood Log Stove Using Different Fuels and Burning Conditions. *Atmos. Environ.* **2016**, *140*, 1–9. [CrossRef]
16. Varlet, V.; Serot, T.; Knockaert, C.; Cornet, J.; Cardinal, M.; Monteau, F.; Le Bizec, B.; Prost, C. Organoleptic Characterization and PAH Content of Salmon (Salmo Salar) Fillets Smoked According to Four Industrial Smoking Techniques. *J. Sci. Food Agric.* **2007**, *87*, 847–854. [CrossRef]
17. Pöhlmann, M.; Hitzel, A.; Schwägele, F.; Speer, K.; Jira, W. Polycyclic Aromatic Hydrocarbons (PAH) and Phenolic Substances in Smoked Frankfurter-Type Sausages Depending on Type of Casing and Fat Content. *Food Control* **2013**, *31*, 136–144. [CrossRef]
18. Kim, K.-H.; Jahan, S.A.; Kabir, E.; Brown, R.J.C. A Review of Airborne Polycyclic Aromatic Hydrocarbons (PAHs) and Their Human Health Effects. *Environ. Int.* **2013**, *60*, 71–80. [CrossRef]
19. Abdel-Shafy, H.I.; Mansour, M.S.M. A Review on Polycyclic Aromatic Hydrocarbons: Source, Environmental Impact, Effect on Human Health and Remediation. *Egypt. J. Pet.* **2016**, *25*, 107–123. [CrossRef]
20. WHO_2021_PAH from Air on Human Health Effects. Available online: https://www.who.int/europe/publications/i/item/9789289056533 (accessed on 11 August 2022).
21. Ramesh, A.; Walker, S.A.; Hood, D.B.; Guillén, M.D.; Schneider, K.; Weyand, E.H. Bioavailability and Risk Assessment of Orally Ingested Polycyclic Aromatic Hydrocarbons. *Int. J. Toxicol.* **2004**, *23*, 301–333. [CrossRef]

22. Agency for Toxic Substances and Disease Registry_ATSDR_2012_Health Effects of PAHs. Available online: https://www.atsdr.cdc.gov/csem/pah/docs/pah.pdf (accessed on 11 August 2022).
23. Holme, J.A.; Brinchmann, B.C.; Refsnes, M.; Låg, M.; Øvrevik, J. Potential Role of Polycyclic Aromatic Hydrocarbons as Mediators of Cardiovascular Effects from Combustion Particles. *Env. Health* **2019**, *18*, 74. [CrossRef]
24. Sopian, N.A.; Jalaludin, J.; Abu Bakar, S.; Hamedon, T.R.; Latif, M.T. Exposure to Particulate PAHs on Potential Genotoxicity and Cancer Risk among School Children Living Near the Petrochemical Industry. *IJERPH* **2021**, *18*, 2575. [CrossRef]
25. Norat, T.; Bingham, S.; Ferrari, P.; Slimani, N.; Jenab, M.; Mazuir, M.; Overvad, K.; Olsen, A.; Tjønneland, A.; Clavel, F. Meat, Fish, and Colorectal Cancer Risk: The European Prospective Investigation into Cancer and Nutrition. *J. Natl. Cancer Inst.* **2005**, *97*, 906–916. [CrossRef]
26. Larsson, S.C.; Håkanson, N.; Permert, J.; Wolk, A. Meat, Fish, Poultry and Egg Consumption in Relation to Risk of Pancreatic Cancer: A Prospective Study. *Int. J. Cancer* **2006**, *118*, 2866–2870. [CrossRef]
27. Lingbeck, J.M.; Cordero, P.; O'Bryan, C.A.; Johnson, M.G.; Ricke, S.C.; Crandall, P.G. Functionality of Liquid Smoke as an All-Natural Antimicrobial in Food Preservation. *Meat Sci.* **2014**, *97*, 197–206. [CrossRef]
28. Stołyhwo, A.; Sikorski, Z.E. Polycyclic Aromatic Hydrocarbons in Smoked Fish–a Critical Review. *Food Chem.* **2005**, *91*, 303–311. [CrossRef]
29. Wretling, S.; Eriksson, A.; Eskhult, G.; Larsson, B. Polycyclic Aromatic Hydrocarbons (PAHs) in Swedish Smoked Meat and Fish. *J. Food Compos. Anal.* **2010**, *23*, 264–272. [CrossRef]
30. Erhunmwunse, N.O.; Ekaye, S.A. Carcinogenic and Genotoxicity of some PAHs in commonly consumed smoked fish (Parachanna obscura and Ethmalosa fimbriata). *J. Appl. Sci. Environ. Manag.* **2019**, *23*, 1349–1352. [CrossRef]
31. Alan, R.; Sams, P.D. Preslaughter factors affecting poultry meat quality. In *Poultry Meat Processing*; CRC Press: Boca Raton, FL, USA, 2001; p. 5.
32. Onopiuk, A.; Kolodziejczak, K.; Szpicer, A.; Wojtasik-Kalinowska, I.; Wierzbicka, A.; Poltorak, A. Analysis of Factors That Influence the PAH Profile and Amount in Meat Products Subjected to Thermal Processing. *Trends Food Sci. Technol.* **2021**, *115*, 366–379. [CrossRef]
33. McGee, H. *McGee on Food & Cooking: An Encyclopedia of Kitchen Science, History and Culture*; Hodder & Stoughton: London, UK, 2004; ISBN 0-340-83149-9.
34. Rozentale, I.; Zacs, D.; Bartkiene, E.; Bartkevics, V. Polycyclic Aromatic Hydrocarbons in Traditionally Smoked Meat Products from the Baltic States. *Food Addit. Contam. Part B* **2018**, *11*, 138–145. [CrossRef]
35. European Commission. Commission Regulation (EC) No 333/2007 of 28 March 2007 Laying down the Methods of Sampling and Analysis for the Official Control of the Levels of Lead, Cadmium, Mercury, Inorganic Tin, 3-MCPD and Benzo (a) Pyrene in Foodstuffs. *Off. J. Eur. Union* **2007**, *L 88*, 29.
36. Hattula, T.; Elfving, K.; Mroueh, U.-M.; Luoma, T. Use of Liquid Smoke Flavouring as an Alternative to Traditional Flue Gas Smoking of Rainbow Trout Fillets (Oncorhynchus Mykiss). *LWT-Food Sci. Technol.* **2001**, *34*, 521–525. [CrossRef]
37. AOAC International: AOAC 950.46-1950 Loss on Drying (Moisture) in Meat. AOAC Official Method. 2015. Available online: http://www.aoacofficialmethod.org/index.php?main_page=product_info&products_id=2998 (accessed on 12 July 2022).
38. AOAC International: AOAC, 920.153-1920, Ash of meat. AOAC Official Method. 2015. Available online: http://www.aoacofficialmethod.org/index.php?main_page=product_info&products_id=1694, (accessed on 12 July 2022).
39. AOAC International: AOAC 960.39-1960, Fat (crude) or ether extract in meat. AOAC Official Method. 2015. Available online: http://www.aoacofficialmethod.org/index.php?main_page=product_info&products_id=605, (accessed on 12 July 2022).
40. AOAC International: AOAC 928.08-1974, Nitrogen in meat. Kjeldahl method. AOAC Official Method. 2015. Available online: http://www.aoacofficialmethod.org/index.php?main_page=product_info&products_id=2823, (accessed on 12 July 2022).
41. European Commission. Commission Regulation (EU) No 836/2011 of 19 August 2011 amending Regulation (EC) No 333/2007 laying down the methods of sampling and analysis for the official control of the levels of lead, cadmium, mercury, inorganic tin, 3-MCPD and benzo (a) pyrene in foodstuffs. *Off. J. Eur. Union* **2011**, *215.9*, 1–8.
42. Lin, G.; Weigel, S.; Tang, B.; Schulz, C.; Shen, J. The Occurrence of Polycyclic Aromatic Hydrocarbons in Peking Duck: Relevance to Food Safety Assessment. *Food Chem.* **2011**, *129*, 524–527. [CrossRef]
43. Saito, E.; Tanaka, N.; Miyazaki, A.; Tsuzaki, M. Concentration and Particle Size Distribution of Polycyclic Aromatic Hydrocarbons Formed by Thermal Cooking. *Food Chem.* **2014**, *153*, 285–291. [CrossRef]
44. Lee, J.-G.; Kim, S.-Y.; Moon, J.-S.; Kim, S.-H.; Kang, D.-H.; Yoon, H.-J. Effects of Grilling Procedures on Levels of Polycyclic Aromatic Hydrocarbons in Grilled Meats. *Food Chem.* **2016**, *199*, 632–638. [CrossRef]
45. Gong, Y.; Parker, R.S.; Richards, M.P. Factors Affecting Lipid Oxidation in Breast and Thigh Muscle from Chicken, Turkey and Duck: Poultry Muscle and Lipid Oxidation. *J. Food Biochem.* **2010**, *34*, 869–885. [CrossRef]
46. Djinovic, J.; Popovic, A.; Jira, W. Polycyclic Aromatic Hydrocarbons (PAHs) in Different Types of Smoked Meat Products from Serbia. *Meat Sci.* **2008**, *80*, 449–456. [CrossRef]
47. Gomes, A.; Santos, C.; Almeida, J.; Elias, M.; Roseiro, L.C. Effect of Fat Content, Casing Type and Smoking Procedures on PAHs Contents of Portuguese Traditional Dry Fermented Sausages. *Food Chem. Toxicol.* **2013**, *58*, 369–374. [CrossRef]
48. Karslıoğlu, B.; Kolsarıcı, N. The Effects of Fat Content and Cooking Procedures on the PAH Content of Beef Doner Kebabs. *Polycycl. Aromat. Compd.* **2022**, 1–14. [CrossRef]

49. Hamidi, E.N.; Hajeb, P.; Selamat, J.; Lee, S.Y.; Abdull Razis, A.F. Bioaccessibility of Polycyclic Aromatic Hydrocarbons (PAHs) in Grilled Meat: The Effects of Meat Doneness and Fat Content. *IJERPH* **2022**, *19*, 736. [CrossRef]
50. Kafouris, D.; Koukkidou, A.; Christou, E.; Hadjigeorgiou, M.; Yiannopoulos, S. Determination of Polycyclic Aromatic Hydrocarbons in Traditionally Smoked Meat Products and Charcoal Grilled Meat in Cyprus. *Meat Sci.* **2020**, *164*, 108088. [CrossRef]
51. Fletcher, D.L. Poultry Meat Quality. *World's Poult. Sci. J.* **2002**, *58*, 131–145. [CrossRef]
52. Purslow, P.P. Intramuscular Connective Tissue and Its Role in Meat Quality. *Meat Sci.* **2005**, *70*, 435–447. [CrossRef] [PubMed]
53. Malarut, J.; Vangnai, K. Influence of Wood Types on Quality and Carcinogenic Polycyclic Aromatic Hydrocarbons (PAHs) of Smoked Sausages. *Food Control* **2018**, *85*, 98–106. [CrossRef]
54. Stumpe-Vīksna, I.; Bartkevičs, V.; Kukāre, A.; Morozovs, A. Polycyclic Aromatic Hydrocarbons in Meat Smoked with Different Types of Wood. *Food Chem.* **2008**, *110*, 794–797. [CrossRef]
55. Santos, C.; Gomes, A.; Roseiro, L.C. Polycyclic Aromatic Hydrocarbons Incidence in Portuguese Traditional Smoked Meat Products. *Food Chem. Toxicol.* **2011**, *49*, 2343–2347. [CrossRef] [PubMed]

 *agriculture*

Article

Effect of *Saccharomyces cerevisiae* Supplementation on Reproductive Performance and Ruminal Digestibility of Queue Fine de l'Ouest Adult Rams Fed a Wheat Straw-Based Diet

Samia Ben Saïd [1,*], Jihene Jabri [2], Sihem Amiri [1], Mohamed Aroua [1], Amel Najjar [3], Sana Khaldi [2], Zied Maalaoui [4], Mohamed Kammoun [2] and Mokhtar Mahouachi [1]

[1] Ecole Supérieure d'Agriculture du Kef, Université de Jendouba, Jendouba 8189, Tunisia
[2] Ecole Nationale de Médecine Vétérinaire, Université de Mannouba, Mannouba 2010, Tunisia
[3] Institut Nationale Agronomique de Tunisie, Université de Carthage, Carthage 1054, Tunisia
[4] Arm and Hammer Animal and Food Production, Ewing, NJ 08628, USA
*   Correspondence: bensaid.samia@esakef.u-jendouba.tn

Abstract: This study aimed to investigate the effect of supplementing a wheat straw-based diet with *Saccharomyces cerevisiae* (S.C.) on feed intake, nutrient digestibility, nitrogen balance, body weight and reproduction performance. The experiment was conducted on 14 Queue Fine de l'Ouest rams between 3 and 4 years of age (body weight (B.W.): 54.7 ± 2.03 kg; body condition score (B.C.S.): 3.5 ± 0.5), for 80 days during the breeding season. The rams were divided into two homogenous groups (n = 7), housed individually in floor pens, and allocated to two dietary treatments. The control group was offered a basal diet of 1 kg/d of wheat straw and 700 g of concentrate. The experimental group (yeast) received the same basal diet supplemented with 10 g of S.C./head/day. The results indicated that the S.C. supplementation had no significant effect on the animal's body weight, volume and concentration of semen, dry matter intake, crude protein digestibility and nitrogen balance. Compared to the control group, the S.C. addition improved ($p < 0.05$) the digestibility of dry matter by 7.3%, organic matter by 11.9% and crude fiber by 24%. In addition, the mass motility score increased for the yeast group compared to the control (3.7 ± 0.24 vs. 1.9 ± 0.27, $p < 0.05$). The total number of dead and abnormal spermatozoa decreased for the yeast group in contrast to the control group (9.28 ± 0.95 vs. $26.8 \pm 3.85\%$ and 25.5 ± 3.33 vs. $59.2 \pm 2.78\%$, respectively; $p < 0.05$). These results showed that adding S.C. to Queue Fine de l'Ouest ram's diet during breeding season could improve nutrient digestibility and reproductive performance.

Keywords: ram; sperm quality; *Saccharomyces cerevisiae*; apparent digestibility

Citation: Ben Saïd, S.; Jabri, J.; Amiri, S.; Aroua, M.; Najjar, A.; Khaldi, S.; Maalaoui, Z.; Kammoun, M.; Mahouachi, M. Effect of *Saccharomyces cerevisiae* Supplementation on Reproductive Performance and Ruminal Digestibility of Queue Fine de l'Ouest Adult Rams Fed a Wheat Straw-Based Diet. *Agriculture* **2022**, *12*, 1268. https://doi.org/10.3390/agriculture12081268

Academic Editor: Daniel Simeanu

Received: 15 June 2022
Accepted: 14 August 2022
Published: 20 August 2022

Publisher's Note: MDPI stays neutral with regard to jurisdictional claims in published maps and institutional affiliations.

1. Introduction

Animal feeding is vital for maintaining animal health, including reproductive performance, and producing safe animal products [1]. Puberty onset, ovulation rate, embryo survival, depth of anestrus and the response to the male effect were all proven to be affected by dietary changes in ewes [2]. High-energy diets increase the libido, testes size, seminiferous tubule diameter and spermatogenesis for rams [3] and promote the onset of puberty in young males [4,5]. Furthermore, malnutrition was reported to reduce the scrotal circumference and inhibit the intensity of the goat male's sexual behavior [6].

Despite their poor digestibility and their negative impact on animal productivity [7], forages and agricultural by-products continue to be the primary feeding source for ruminants.

This has created an interest among nutritionists in manipulating the rumen ecosystem and fermentation characteristics of livestock in order to improve feed utilization, animal productivity and health [8,9].

Currently, probiotics that are live and non-pathogenic microbes, such as live yeast (*Saccharomyces cerevisiae*), are widely used as feed additives for ruminants. Probiotic supplementation has proved its effectiveness for enhancing intestinal function by maintaining a healthy gastrointestinal environment, thus improving nutrient intake and digestibility, production performance and feed efficiency [10–13]. Further, Saccharomyces cerevisiae (S.C.) application to the ruminant diet has been shown to reduce methane emissions from anaerobic fermentation [14], decrease the number of pathogenic intestinal bacteria [15] and repopulate gut microflora in diarrheal cases [16]. However, far too little attention has been paid to the effect of S.C. supplementation on ruminant's reproductive performance. Recently, Ahmadzadeh [17] proved that S.C. supplementation to the ewe in the breeding season could enhance the reproductive performance by increasing fertility and the twinning rate. Hence, this study aimed to evaluate the effect of S.C. supplementation on reproductive performance, intake and ruminal digestibility of native Tunisian rams fed a wheat straw and concentrate diet during the breeding season.

2. Materials and Methods

The study was carried out on the experimental farm of the Higher School of Agricultural of Kef, located in the northwestern part and the semi-arid area of Tunisia (latitude 35.7° N, longitude 8.83° W). The experiment meets the ethical guidelines and adheres to Tunisian legal requirements (The Livestock Law No. 2005-95 of 18 October 2005) and was conducted by trained staff strictly following good animal practices as defined by national authorities.

2.1. Experimental Design

A total of 14 Queue Fine de l'Ouest rams (also called "Bergui" in the local language) were used during the traditional mating season in Tunisia for 80 days from March to May. Animals were aged between 3 and 4 years, with a mean body weight of 54.75 ± 2.03 kg and a mean body condition score (B.C.S.) of 3.5 ± 0.5. The rams were checked for the integrity of their sexual tracts before the experiment, then assigned into two homogenous groups. For the control group (n = 7), rams received 1 kg of wheat straw and a fixed amount of concentrate (700 g) according to their metabolic weight. The concentrate was composed of 9% soyabean meal, 87% barley and a 4% mineral and vitamin complement. For the experimental group (yeast, n = 7), the same diet was supplemented with 10 g/head/day of Saccharomyces cerevisiae yeast culture Celmanax® (Arm & Hammer Animal and Food Production, Ewing, NJ, USA). The chemical composition of the diet ingredients (concentrate and wheat straw) is detailed in Table 1. Animals were allowed free access to fresh and clean water and received two equal meals at 09:00 a.m. and 15:00 p.m. The rams were housed individually in cleaned experimental pens of 2 m² (1 m × 2 m) and raised under the same management and nutritional conditions. They were exposed to the same normal seasonal photoperiod without any contact with ewes. All rams were vaccinated against enterotoxemia (Coglavax®, CEVA, Libourne, France, 2 mL/animal) and received an antiparasitic treatment (IVOMEC®, Boehringer-Ingelheim, Reims, France, 1 mL/50 kg B.W. by subcutaneous rout).

Table 1. Ingredients of the experimental diets and their chemical composition (g/kg fresh matter).

	Dry Matter	Mineral	Crude Protein	Crude Fiber
Wheat straw	890	54	42	430
Concentrate	910	50	182.9	70

2.2. Measurements
2.2.1. Feed Intake, Live Weight, Body Score Condition and Testicular Diameter

The daily feed intake of concentrate and wheat straw (the difference between offered feeds and refusals) for each animal was recorded throughout the trial. Samples of feed offered and refused were removed and weighed before every morning feeding and pooled over the week for each ram. Live body weight, body score condition (B.S.C., noted on a scale from 0 to 5, [18]) and testicular diameter were measured every two weeks before distributing the morning's food. The testicular diameter was assessed using a clipper with the ram in a standing position.

2.2.2. Semen Collection and Evaluation

Sperm collection was performed every 10 days. Rams were put individually in the collection room in the presence of a teaser female that was previously induced by inserting a progestogen-impregnated vaginal sponge for 10 days [19]. After collection, sperm volume was immediately recorded by direct reading from a graduated glass tube. The general appearance of the semen was visually assessed [20]. Samples were immediately placed in a water bath at 35 °C [19]. Mass activity (wave motion or motility score) in undiluted semen and individual motility in diluted semen were assessed by examining a drop of semen under a warm stage using a phase-contrast microscope (score, 0–5, [21]). The sperm concentration was assessed by a hemocytometer slide (Malassez slide, Marienfeld, Germany) [21]. Percentages of dead and abnormal spermatozoa were studied using the eosin/nigrosin staining technique described by Baril et al. [21].

2.2.3. Digestibility Trial

Over the last two weeks of the trial, rams were placed in an individual metallic box with a wire-mesh floor (1.2 m × 0.6 m) to evaluate the diet's (in vivo) digestibility and nitrogen balance (7 days for adaptation and 7 days for measurements) [22]. The crates were specifically designed to separate feces, urine and refusals. The offered feed, refusals, total urine and feces excreted during 24 h by each ram were collected and weighed daily before the morning feeding. A sample of 10% of raw material from the feed refusals and total feces of each ram was pooled daily (during the 7 days of measurements). The fecal samples were stored at −20 °C until further analysis. The total urine was collected daily into plastic buckets and preserved with 50 mL of 10% sulfuric acid (0.1 N) solution. Representative aliquots of each animal's urine (10%) were immediately stored in a freezer (−20 °C) until analysis.

For each ram, samples of offered diets, refusals and feces were dried in a forced air oven at 55 °C until a constant weight [23]. The samples were subsequently ground through a 1 mm screen using a Wiley mill. Dry matter, ash, crude fiber and crude protein contents were determined according to the Association of Official Analytical Chemists [23]. Moreover, the nitrogen content of urine was determined according to the Kjeldahl method [23]. All chemical analyses were performed in triplicate for each sample.

2.3. Statistical Analysis

Mixed models with a random animal effect were run for repeated data on body weight, testicular diameter and semen parameters (volume, concentration, massal motility and individual motility, percentages of dead and morphologically abnormal spermatozoa) using the MIXED model's procedure (S.A.S. Version 9.1; S.A.S. Inst. Inc., Cary, NC, USA). We included the fixed effect of the treatment, the rank of the sperm collection (week of collection) and their interaction in the analysis. The random variable was the ram within the treatment. An ANOVA was performed to study the effect of treatment on dry matter, digestibility parameters and nitrogen balance. Duncan's test was used to compare variables between groups. Data on dry matter intake, digestibility parameters and nitrogen balance were compared using the analysis of variance (ANOVA) [24]. Each ram was regarded as the experimental unit. The significance level was set at 0.05, and trends were discussed for *p*-values between 0.05 and 0.10.

3. Results

3.1. Live Body Weight, Body Condition Score and Testicular Diameter

The live body weight and B.C.S. of rams did not vary between the experimental and control groups ($p > 0.05$; Table 2). Regarding the testicular diameter, results showed no variation between the treated and the control groups ($p > 0.05$).

Table 2. Live weight, body score condition and testicular diameter of control and yeast-treated rams (means ± S.E.M.).

	Beginning *		End *	
	Control	**Yeast**	**Control**	**Yeast**
L.B.W. (kg)	55.7 ± 4.0	54.8 ± 3.8	56.0 ± 1.9	55.05 ± 2.7
B.S.C.	3.1 ± 0.32	3.2 ± 0.27	3.4 ± 0.21	3.5 ± 0.15
T.D. (cm)	4.95 ± 0.20	4.90 ± 0.31	4.93 ± 0.18	4.83 ± 0.12

S.E.M: standard error of the mean; L.B.W.: live body weight, B.S.C.: body score condition, T.D.: testicular diameter. * Beginning: mean data relative to the first sample collection during the trial. End: mean data of the last sample collection during the trial.

3.2. Semen Characteristics

Data for semen characteristics such as ejaculation volume, sperm concentration, motility scores and percentage of dead and abnormal spermatozoa are represented in Figure 1. There was no significant difference between the two groups regarding the ejaculation volume, mass motility, individual motility, sperm concentration and the percentage of abnormal spermatozoa at the beginning of the trial (Table 3; $p > 0.05$).

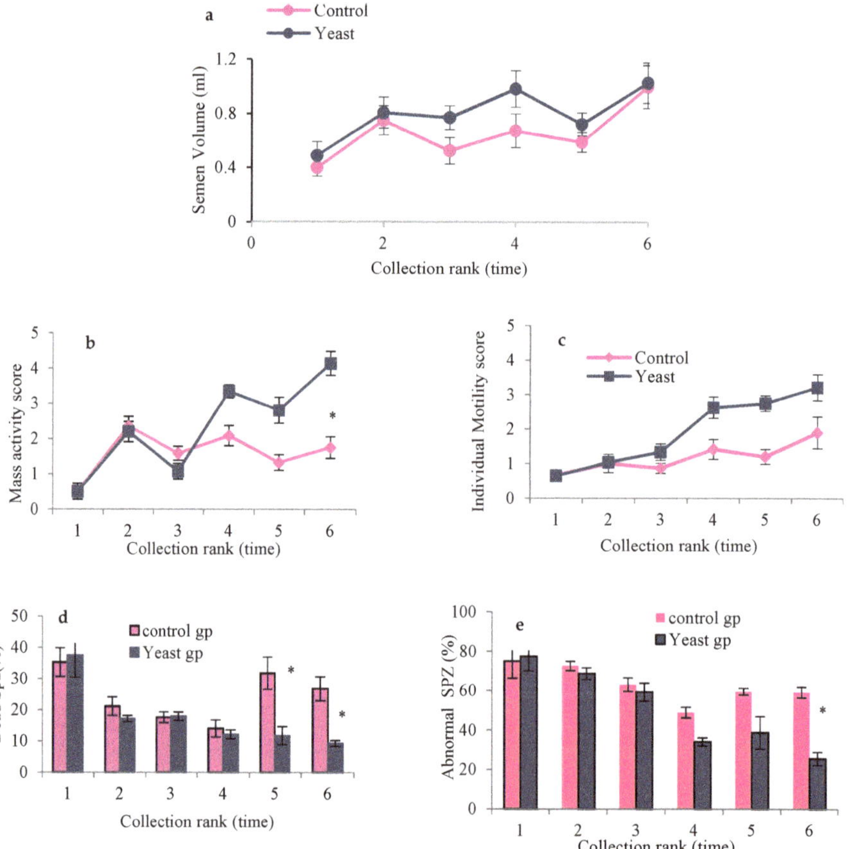

Figure 1. Variation in (**a**) ejaculate volume, (**b**) mass motility score, (**c**) individual motility, (**d**) dead spermatozoa rate and (**e**) abnormal spermatozoa rate in yeast-treated and control rams. Effect of supplementation is significant, * $p < 0.05$.

Table 3. Sperm quality traits of control and yeast-treated Queue Fine de l'Ouest rams (means ± S.E.M.).

	Beginning *		End *	
	Control	Yeast	Control	Yeast
Volume (mL)	0.4 ± 0.06	0.48 ± 0.1	1 ± 0.15	1.02 ± 0.14
MM	0.77 ± 0.12	0.75 ± 0.18	1.91 ± 0.27 [a]	3.70 ± 0.24 [b]
MI	0.66 ± 0.08	0.64 ± 0.14	1.90 ± 0.40	3.20 ± 0.30
CC (10^6 spz/mL)	1512.29 ± 86.91	1840.7 ± 246	2036.47 ± 127 [a]	2296.23 ± 132 [b]
Dead spz (%)	35.25 ± 4.59	37.5 ± 7.13	26.8 ± 3.85 [a]	9.28 ± 0.95 [b]
Abnormal spz (%)	75 ± 8.62	77.4 ± 7.38	59.2 ± 2.78 [a]	25.5 ± 3.33 [b]

MM: mass motility score; MI: individual motility score, CC: sperm concentration, spz: spermatozoa. a,b: different letters within the same row (different diets) differ significantly ($p < 0.05$). S.E.M.: standard error of the mean. * Beginning: mean data relative to the first sample collection during the trial. End: mean data of the last sample collection during the trial.

Regardless of the treatment, the mean ejaculation volume was 0.73 ± 0.2 mL (0.66 ± 0.2 and 0.80 ± 0.1 for control and yeast-treated rams, respectively, $p > 0.05$; Table 3). However, the ejaculate volume was significantly affected by the rank of sperm collection ($p < 0.05$). The volume increased during the weeks of the trial (Figure 1a) and the sixth sperm collection showed the highest sperm volume (v = 1 mL, $p < 0.05$). The interaction between the treatment and sperm collection had no significant effect on sperm volume ($p > 0.05$). At the end of the experiment, S.C. supplementation improved the mass motility (Table 3) compared with the control group (3.7 ± 0.24 vs. 1.91 ± 0.17, $p < 0.05$). Statistical analysis also showed a significant effect on the collection rank ($p < 0.05$) and on the interaction between the treatment and the collection rank ($p < 0.05$; Figure 1b). In the same way, statistical analysis showed the effect of the collection rank ($p < 0.05$) on individual motility. However, there was no effect of S.C. supplementation on this parameter (Figure 1c).

S.C. supplementation slightly improved sperm concentration ($p = 0.056$; Table 3). Mean sperm concentration during the whole period of the experiment was $2240 \pm 201 \times 10^6$ spz/mL and $2534 \pm 190 \times 10^6$ spz/mL for control and yeast-treated rams, respectively. In addition, an effect of the collection rank was observed ($p < 0.05$).

The percentage of dead spermatozoa decreased ($p < 0.05$) during the trial (Table 3 and Figure 1d). At the end of the experiment, the improvement in this parameter was more pronounced ($p < 0.05$) with the S.C. diet than in the control one ($9.28 \pm 0.95\%$ and 26.8 ± 3.85, respectively). The interaction between the treatment and the collection rank was significant ($p < 0.05$).

The mean of the abnormal spermatozoa percentage followed the same trend as the dead spermatozoa since it decreased significantly ($p < 0.05$) during the trial for both diets. However, this decrease was more significant in the yeast-supplemented group than in the control group (59.2 ± 2.78 vs. $25.5 \pm 3.33\%$, respectively, $p < 0.05$; Table 3).

3.3. Nutrient Intake and Diet Digestibility

The results revealed no significant effect ($p > 0.05$) of the diet supplementation with S.C. on the daily feed intake (g kg^{-1} body weight per day) (Table 4). However, for the nutrient's digestibility (Table 5), the addition of S.C. to the rams' diet improved the dry matter, organic matter and fiber digestibility by 7.3% ($p = 0.02$), 11.9% ($p = 0.01$) and 24% ($p = 0.02$), respectively, as compared with the control. Likewise, protein digestibility tended to increase by 10.5% ($p = 0.08$). Moreover, the N intake, excreted nitrogen (urinary nitrogen (U.R.) and fecal nitrogen (F.N.)) and nitrogen retention (Table 5) were unaffected by the addition of S.C. to the diet ($p > 0.05$).

Table 4. Effect of adding S.C. supplementation on nutrient intake (g/kg B.W.$^{0.75}$/day) of rams (means ± S.E.M.).

	Control	Yeast
Dry matter	63.3 ± 1.00	68.5 ± 1.10
Crude protein	6.4 ± 0.04	6.8 ± 0.11
Crude fiber	14.4 ± 0.38	15.6 ± 0.31
Organic matter	60.3± 0.95	65.3 ± 1.05

S.E.M.: standard error of the mean.

Table 5. Apparent digestibility (%) and nitrogen balance (g/d) parameters (means ± S.E.M.).

	Control	Yeast
Dry matter digestibility (D.M.D.)	70.19 ± 0.90 [a]	75.30 ± 0.96 [b]
Organic matter digestibility (O.M.D.)	58.73 ± 1.22 [a]	65.72 ± 1.25 [b]
Crude protein digestibility (C.P.D.)	62.07 ± 1.20	68.61 ± 2.06
Crude fiber digestibility (C.F.D.)	40.44 ± 2.51 [a]	50.10 ± 1.61 [b]
Nitrogen intake (g/d)	22.45 ± 0.91	21.72 ± 0.77
Fecal nitrogen (g/d)	13.32 ± 0.48	12.4 ± 0.74
Urinary nitrogen (g/d)	4.22 ± 0.28	5.36 ± 0.80
Retained nitrogen (g/d)	4.91 ± 0.86	3.96 ± 0.73

a,b: different letters within the same row (different diets) differ significantly ($p < 0.05$). S.E.M.: standard error of the mean.

4. Discussion

In this study, adding 10 g/day of S.C. during the experimental period did not affect body live weight, testicular diameter, ejaculation volume and concentration. This may be explained, in part, by the fact that there was not any measured difference in the feed intake between both dietary groups. In addition, the animals are already mature, thus they do not have additional needs. Although there is a general agreement that prebiotic supplementation to ruminant feeding increases the feed intake [25], the S.C. addition to the rams' diet did not affect feed intakes for dry matter, crude proteins, crude fiber and organic matter in our study. In agreement with our finding, Haddad and Goussous [26] and Titi et al. [27] reported that supplying yeast culture at increasing addition rates of 0, 3, 6 and 12.6 g/day to Awassi lambs' diets had no significant effect on nutrient intake. In another meta-analysis of 141 yeast treatment studies, Desnoyers et al. [28] reported that the effect of yeast supplementation on ruminant diets is related to some interfering factors. Indeed, the positive effect on intake could be affected by the proportion of concentrate in the diet but not be influenced by the neutral detergent fiber or ruminant species. Issakowicz et al. [29] reported a significant increase in dry matter intake and the feed conversion ratio by adding S.C. to lamb-fed diets with a greater concentrate proportion (80%).

Several studies reported the interrelationship between energy intake and reproductive performance in adult rams [30–32]. In this regard, Tufarelli et al. [33] reported that a dietary level with a higher concentrate of supplementation would improve body weight gain, feed intake and semen characteristics (volume and concentration) of rams. In the Tunisian Queue Fine de l'Ouest breed, alternating feeding levels from high (1.6 of the metabolizable energy for maintenance (M.E.M.) requirement) to low (1.0 M.E.M.) levels was associated with a reduction in the volume of ejaculate, but did not reduce the sperm concentration in rams [9]. The current study showed that both quantitative and qualitative parameters of rams' semen were improved during the trial period independently of the diet. These results are similar to those reported by Mahouachi and Khaldi [34], suggesting a resumption of sexual activity of the local breed at the beginning of the traditional mating season even if the photoperiod is unfavorable. Therefore, there would be interference between the two factors of diet and time, as shown by the significant interactions. Compared to the control group, S.C. supplementation had no apparent effect on ejaculation volume. However, it increased the mass motility significantly after 60 days of its introduction to the diet.

Furthermore, the percentage of dead spermatozoa and abnormal spermatozoa were reduced with the yeast supplementation compared to the control group. These results suggest that yeast supplementation's main positive effect is improving sperm quality. In this regard, Aboul-Ela et al. [35] reported that S.C. reduces primary sperm abnormalities in rats. Furthermore, Emmanuel et al. [36] showed that an S.C.-based diet in bucks improved motility and live sperm, tubule diameter, epididymal volume, the volume fraction of the duct and total duct volume. They also reported that the incidence of head and tail sperm abnormalities decreased significantly in bucks fed S.C.-based diets compared with the control diet.

For humans, Sahar-Eissa et al. [37] found that the baker's yeast extraction of S.C. at the different doses of 10, 20 and 50 mg/mL improved sperm viability and motility.

For female ruminants, it was proved that dietary supplementation with fermented yeast culture induced an early onset of the first post-partum heat, improved the conception rate and reduced the number of inseminations per conception [38].

In the same way, our results indicate that rams fed with S.C. had higher sperm motility and viability and a lower number of abnormal sperm. Based on these findings and those proved by [39,40], we could hypothesize that the antioxidant properties of S.C. could stimulate the improvement of semen characteristics. Dietary antioxidants are crucial to control and counteract the harmful effect of oxidative stress. Therefore, a sufficient antioxidant intake contributes to a lower risk of oxidative stress-mediated diseases and infertility. For that reason, yeast has been shown to contain significant amounts of antioxidants [41]. In addition, S.C. supplementation in dairy cows' diet increased total antioxidant capacity, glutathione peroxidase and superoxide dismutase in serum while decreasing malondialdehyde [42].

Furthermore, this improvement in sperm quality could be explained, in part, by the positive effect of S.C. on diet digestibility. One of the limitations that influences the use of cereal crop residues in ruminant diets, such as wheat straw as used in this study, is its low digestibility and protein deficiency [43]. Our results showed that for nutrient digestibility, the S.C. yeast supplementation tended to increase the crude protein digestibility and significantly improved the digestibility of dry matter, organic matter and crude fiber. Similar results were reported earlier for lambs [44], dairy heifers [45] and buffalos [46]. This improvement in nutrient digestibility could be due to different possible yeast action modes in a ruminal environment. It is admitted that the presence of a yeast source in the rumen improves the release of various proteolytic, glycolytic or lipolytic enzymes, which leads to better digestion of the organic matter [47] and promotes the ruminal ecosystem through the stimulation of the growth and the function of ruminal microbiota [48]. This stimulation is ensured by releasing amino acids and monosaccharides through the cell wall [49].

Our results showed a significant increase in fiber digestibility with the addition of yeast to a wheat straw-based diet for rams, which is in agreement with the findings of other authors such as Ghazanfar et al. [45] and Mallekahi et al. [44] who reported a significant positive effect of yeast on neutral detergent fiber, acid detergent fiber and cellulose digestibility. It was also shown by Ding et al. [50] that S.C. increased the number of the total ruminal bacteria, fungi and protozoa, especially the Ruminococci, by two- to four-fold [12] in fibrolytic protozoa and *Fibrobacter succinogenes* feed colonization [51]. Furthermore, in one of the few papers that studied the efficiency of yeast addition on gene expression, Durand et al. [51] reported an increase in the expression of carbohydrate-active enzymes GH5 and GH43 in the rumen of lambs receiving yeast supplementation. This suggests the ability of this feed additive to improve the fibrolytic potential of the rumen microbiota.

However, findings regarding crude protein digestibility are inconsistent. Similar to our findings, Osita et al. [52] showed that crude protein digestibility improved by 24.5% by adding 1.5 g yeast/kg to the sheep's diet. Likewise, Malekkhahi et al. [44] found that crude protein digestibility improved by 7.8% by supplementing the wheat straw-based lambs' diet with 4 g yeast/animal/day. In another study on dairy heifers, Ghazanfar et al. [45] recorded an improvement of 6% in crude protein digestibility by adding 5 g of yeast/animal/day on a wheat straw-based diet. However, some other studies found that yeast supplementation does not significantly affect crude protein digestibility [27,46]. Therefore, as proved by the meta-analysis study of Desnoyers et al. [28], all ruminal fermentation characteristics influenced by yeast addition are significantly dependent on at least one of the dietary characteristics, such as the concentrate proportion.

5. Conclusions

The present study has shown that the supplementation of S.C. into the diet of rams improves the semen quality and the diet digestibility of rams when fed low-quality straw and used in mating season. These findings suggest the importance of using prebiotics in animal feeding to improve animal performance when raised under hard conditions and fed with low-quality forage.

Author Contributions: S.B.S.: conceived and designed the experiment, investigation, formal analysis, visualization, and writing—original draft; J.J. and S.A.: writing, software and formal analysis; A.N. and S.K.: writing—review; Z.M.: contributed new reagents (Celmanax) and methodology; M.A. and M.K.: analytical tools; M.M.: manuscript revision and funding. All authors have read and agreed to the published version of the manuscript.

Funding: This research was funded by the Arm & Hammer Animal and Food Production.

Institutional Review Board Statement: All procedures employed in this study meet ethical guidelines and adhere to Tunisian legal requirements (The Livestock Law No. 2005-95 of 18 October 2005).

Informed Consent Statement: Not applicable.

Data Availability Statement: The data presented in this study are available on request from the corresponding author.

Acknowledgments: We are deeply grateful to the staff of the experimental farm of ESAKef for the feeding experiment servicing and Arm & Hammer Animal and Food Production for the financial support.

Conflicts of Interest: The authors declare no conflict of interest.

References

1. Robinson, J.J.; Ashworth, C.J.; Rooke, J.A.; Mitchell, L.M.; McEvoy, T.G. Nutrition and fertility in ruminant livestock. *Anim. Feed Sci. Technol.* **2006**, *126*, 259–276. [CrossRef]
2. Forcada, F.; Abecia, J.-A. The effect of nutrition on the seasonality of reproduction in ewes. *Reprod. Nutr. Dev.* **2006**, *46*, 355–365. [CrossRef] [PubMed]
3. Martin, G.B.; Blache, D.; Miller, D.W.; Vercoe, P.E. Interactions between nutrition and reproduction in managing the mature male ruminant. *Animal* **2010**, *4*, 1214–1226. [CrossRef] [PubMed]
4. Kenny, D.A.; Byrne, C.J. Review: The effect of nutrition on timing of pubertal onset and subsequent fertility in the bull. *Animal* **2018**, *12*, s36–s44. [CrossRef]
5. Guan, Y.; Liang, G.; Hawken, P.A.; Meachem, S.J.; Malecki, I.A.; Ham, S.; Stewart, T.; Martin, G.B. Nutrition affects Sertoli cell function but not Sertoli cell numbers in sexually mature male sheep. *Reprod. Fertil. Dev.* **2016**, *28*, 1152–1163. [CrossRef]
6. Delgadillo, J.A.; Sifuentes, P.I.; Flores, M.J.; Espinoza-Flores, L.A.; Andrade-Esparza, J.D.; Hernández, H.; Keller, M.; Chemineau, P. Nutritional supplementation improves the sexual response of bucks exposed to long days in semi-extensive management and their ability to stimulate reproduction in goats. *Animal* **2021**, *15*, 100114. [CrossRef]
7. Serrapica, F.; Masucci, F.; Romano, R.; Napolitano, F.; Sabia, E.; Aiello, A.; Di Francia, A. Effects of chickpea in substitution of Soybean meal on milk production, blood profile and reproductive response of primiparous buffaloes in early lactation. *Animals* **2020**, *10*, 515. [CrossRef]
8. Marrero, Y.; Castillo, Y.; Ruiz, O.; Burrola, E.; Angulo, C. Feeding of yeast (*Candida* spp.) improves in vitro ruminal fermentation of fibrous substrates. *J. Integr. Agric.* **2015**, *14*, 514–519. [CrossRef]
9. Rekik, M.; Lassoued, N.; Salem, H.B.; Mahouachi, M. Interactions between nutrition and reproduction in sheep and goats with particular reference to the use of alternative feed sources. *Options Méditerranéennes* **2007**, *74*, 375–383.
10. Francesca, G.; Francesca, M.; Tania, G.; Marina, B.; Maurizio, S.; Alessandra, M. Effect of different glucose concentrations on proteome of *Saccharomyces cerevisiae*. *Biochim. Et Biophys. Acta-Proteins Proteom.* **2010**, *1804*, 1516–1525. [CrossRef]
11. Kulkarni, P.; Bhattacharya, S.; Achuthan, S.; Behal, A.; Jolly, M.K.; Kotnala, S.; Mohanty, A.; Rangarajan, G.; Salgia, R.; Uversky, V. Intrinsically Disordered Proteins: Critical Components of the Wetware. *Chem. Rev.* **2022**, *122*, 6614–6633. [CrossRef] [PubMed]

12. Mosconi, P.; Chaucheyras-Durand, F.; Béra-Maillet, C.; Forano, E. Quantification by real-time PCR of cellulolytic bacteria in the rumen of sheep after supplementation of a forage diet with readily fermentable carbohydrates: Effect of a yeast additive. *J. Appl. Microbiol.* **2007**, *103*, 2676–2685. [CrossRef]

13. Poppy, G.; Rabiee, A.; Lean, I.; Sanchez, W.; Dorton, K.; Morley, P. A meta-analysis of the effects of feeding yeast culture produced by anaerobic fermentation of *Saccharomyces cerevisiae* on milk production of lactating dairy cows. *J. Dairy Sci.* **2012**, *95*, 6027–6041. [CrossRef] [PubMed]

14. Preston, T.; Leng, R.; Garcia, Y.; Binh, P.T.; Inthapanya, S.; Gomez, M.E. Yeast (*Saccharomyces cerevisiae*) fermentation of polished rice or cassava root produces a feed supplement with the capacity to modify rumen fermentation, reduce emissions of methane and improve growth rate and feed conversion. *Development* **2021**, *33*, 5.

15. Elghandour, M.; Tan, Z.; Abu Hafsa, S.; Adegbeye, M.; Greiner, R.; Ugbogu, E.; Cedillo Monroy, J.; Salem, A. *Saccharomyces cerevisiae* as a probiotic feed additive to non and pseudo-ruminant feeding: A review. *J. Appl. Microbiol.* **2020**, *128*, 658–674. [CrossRef]

16. Cheng, G.; Hao, H.; Xie, S.; Whang, X.; Dai, M.; Huang, L.; Yuan, Z. Antibiotic alternatives: The substitution of antibiotics in animal husbandry. *Front. Microbiol.* **2014**, *5*, 217. [CrossRef]

17. Ahmadzadeh, L.; Hosseinkhani, A.; Kia, H.D. Effect of supplementing a diet with monensin sodium and Saccharomyces Cerevisiae on reproductive performance of Ghezel ewes. *Anim. Reprod. Sci.* **2018**, *188*, 93–100. [CrossRef]

18. Russel, A.; Doney, J.; Gunn, R. Subjective assessment of body fat in live sheep. *J. Agric. Sci.* **1969**, *72*, 451–454. [CrossRef]

19. Rouatbi, M.; Gharbi, M.; Rjeibi, M.R.; Salem, I.B.; Akkari, H.; Lassoued, N.; Rekik, M. Effect of the infection with the nematode Haemonchus contortus (Strongylida: Trichostrongylidae) on the haematological, biochemical, clinical and reproductive traits in rams. *Onderstepoort J. Vet. Res.* **2016**, *83*, 1–8. [CrossRef]

20. Cortell, J.; Baril, G.; Leboeuf, B. Development and application of artificial insemination with deep frozen semen and out-of-season breeding of goats in France. In Proceedings of the IV International Conference on Goats, Brasilia, Brazil, 8–13 March 1987; pp. 523–547.

21. Baril, G.; Chemineau, P.; Cognie, Y. *Manuel de Formation Pour L'insémination Artificielle Chez les Ovins et les Caprins*; Food & Agriculture Organisation: Rome, Italy, 1993.

22. Abid, K.; Jabri, J.; Ammar, H.; Said, S.B.; Yaich, H.; Malek, A.; Rakhis, J.; López, S.; Kamoun, M. Effect of treating olive cake with fibrinolytic enzymes on feed intake, digestibility and performance in growing lambs. *Anim. Feed Sci. Technol.* **2020**, *261*, 114405. [CrossRef]

23. AOAC. *Association of Official Analytical Chemists, Official Methods of Analysis*, 15th ed.; Association of Official Analytical Chemists: Arlington, VA, USA, 1990.

24. S.A.S. *User's Guide*; Version 9.1; S.A.S. Institute, Inc.: Cary, NC, USA, 2005.

25. Mohammed, S.F.; Mahmood, F.A.; Abas, E.R. A review on effects of yeast (*Saccharomyces cerevisiae*) as feed additives in ruminants performance. *J. Entomol. Zool. Stud* **2018**, *6*, 629–635.

26. Haddad, S.; Goussous, S. Effect of yeast culture supplementation on nutrient intake, digestibility and growth performance of Awassi lambs. *Anim. Feed Sci. Technol.* **2005**, *118*, 343–348. [CrossRef]

27. Titi, H.; Dmour, R.; Abdullah, A. Growth performance and carcass characteristics of Awassi lambs and Shami goat kids fed yeast culture in their finishing diet. *Anim. Feed Sci. Technol.* **2008**, *142*, 33–43. [CrossRef]

28. Desnoyers, M.; Giger-Reverdin, S.; Bertin, G.; Duvaux-Ponter, C.; Sauvant, D. Meta-analysis of the influence of *Saccharomyces cerevisiae* supplementation on ruminal parameters and milk production of ruminants. *J. Dairy Sci.* **2009**, *92*, 1620–1632. [CrossRef] [PubMed]

29. Issakowicz, J.; Bueno, M.; Sampaio, A.; Duarte, K. Effect of concentrate level and live yeast (*Saccharomyces cerevisiae*) supplementation on Texel lamb performance and carcass characteristics. *Livest. Sci.* **2013**, *155*, 44–52. [CrossRef]

30. Braden, A.; Turnbull, K.; Mattner, P.; Moule, G. Effect of protein and energy content of the diet on the rate of sperm production in rams. *Aust. J. Biol. Sci.* **1974**, *27*, 67–74. [CrossRef]

31. Rowe, J.; Murray, P. Production characteristics of rams given supplements containing different levels of protein and metabolisable energy. In Proceedings of the Australian Society of Animal Production, Armidale, NSW, Australia, February 1984; pp. 565–568.

32. Murray, P.J.; Rowe, J.; Pethick, D.; Adams, N. The effect of nutrition on testicular growth in the Merino ram. *Aust. J. Agric. Res.* **1990**, *41*, 185–195. [CrossRef]

33. Tufarelli, V.; Lacalandra, G.; Aiudi, G.; Binetti, F.; Laudadio, V. Influence of feeding level on live body weight and semen characteristics of Sardinian rams reared under intensive conditions. *Trop. Anim. Health Prod.* **2011**, *43*, 339–345. [CrossRef] [PubMed]

34. Mahouachi, M.; Khaldi, G. Variations saisonnières de la production spermatique chez les béliers de races Barbarine et Noire de Thibar. *Ann. De L'INRAT* **1987**, *60*, 11–28.

35. Aboul-Ela, A.; Abdel-Hamid Mabrouk, E.; Atef Helmy, N.; Ragab Mohamed, S.J. Effects of some dietary supplements on the reproductive and productive performances in male rats. *J. Vet. Med. Res.* **2018**, *25*, 199–212. [CrossRef]

36. Emmanuel, D.; Amaka, A.; Okezie, E.; Sunday, U.; Ethelbert, O. Epididymal sperm characteristics, testicular morphometric traits and growth parameters of rabbit bucks fed dietary Saccharomyces cerevisiae and/or zinc oxide. *Braz. J. Poult. Sci.* **2019**, *21*, 1–10. [CrossRef]

37. Sahar-Eissa, A.; Saad, A.; Enan, G. Yeast cells as antioxidant-producing probiotics: Can they emulate vitamin E on in-vitro protection of human spermatozoa against oxidative stress? *Adv. Environ. Biol.* **2016**, *10*, 60–69.
38. Ahmad Para, I.; Ahmad Shah, M.; Punetha, M.; Hussain Dar, A.; Rautela, A.; Gupta, D.; Singh, M.; Ahmad Naik, M.; Rayees, M.; Sikander Dar, P. Feed fortification of periparturient Murrah buffaloes with dietary yeast (*Saccharomyces cerevisiae*) elevates metabolic and fertility indices under field conditions. *Biol. Rhythm Res.* **2020**, *51*, 858–868. [CrossRef]
39. Bayat, E.; Moosavi-Nasab, M.; Fazaeli, M.; Majdinasab, M.; Mirzapour-Kouhdasht, A.; Garcia-Vaquero, M. Wheat Germ Fermentation with Saccharomyces cerevisiae and Lactobacillus Plantarum: Process Optimization for Enhanced Composition and Antioxidant Properties In Vitro. *Foods* **2022**, *11*, 1125. [CrossRef]
40. Zhang, D.; Tan, B.; Zhang, Y.; Ye, Y.; Gao, K. Improved nutritional and antioxidant properties of hulless barley following solid-state fermentation with Saccharomyces cerevisiae and Lactobacillus Plantarum. *J. Food Processing Preserv.* **2022**, *46*, e16245. [CrossRef]
41. Al-Farsi, M.A.; Lee, C.Y. Nutritional and functional properties of dates: A review. *Crit. Rev. Food Sci. Nutr.* **2008**, *48*, 877–887. [CrossRef]
42. Du, D.; Feng, L.; Chen, P.; Jiang, W.; Zhang, Y.; Liu, W.; Zhai, R.; Hu, Z. Effects of Saccharomyces Cerevisiae Cultures on Performance and Immune Performance of Dairy Cows During Heat Stress. *Front. Vet. Sci.* **2022**, *9*, 1–11. [CrossRef]
43. Bhandari, B.; Bahadur. *Crop Residue as Animal Feed*; 2019. [CrossRef]
44. Malekkhahi, M.; Tahmasbi, A.M.; Naserian, A.A.; Danesh Mesgaran, M.; Kleen, J.; Parand, A. Effects of essential oils, yeast culture and malate on rumen fermentation, blood metabolites, growth performance and nutrient digestibility of Baluchi lambs fed high-concentrate diets. *J. Anim. Physiol. Anim. Nutr.* **2015**, *99*, 221–229. [CrossRef]
45. Ghazanfar, S.; Anjum, M.; Azim, A.; Ahmed, I. Effects of dietary supplementation of yeast (*Saccharomyces cerevisiae*) culture on growth performance, blood parameters, nutrient digestibility and fecal flora of dairy heifers. *J. Anim. Plant Sci.* **2015**, *25*, 53–59.
46. Anjum, M.; Javaid, S.; Ansar, M.; Ghaffar, A. Effects of yeast (*Saccharomyces cerevisiae*) supplementation on intake, digestibility, rumen fermentation and milk yield in Nili-Ravi buffaloes. *Iran. J. Vet. Res.* **2018**, *19*, 96.
47. Shurson, G. Yeast and yeast derivatives in feed additives and ingredients: Sources, characteristics, animal responses, and quantification methods. *Anim. Feed Sci. Technol.* **2018**, *235*, 60–76. [CrossRef]
48. Ghazanfar, S.; Khalid, N.; Ahmed, I.; Imran, M. Probiotic yeast: Mode of action and its effects on ruminant nutrition. In *Yeast Industrial Applications*; IntechOpen: London, UK, 2017; pp. 179–202.
49. Samuel, B. *Medical Microbiology*, 4th ed.; University of Texas Medical Branch at Galveston: Galveston, TX, USA, 1996.
50. Ding, G.; Chang, Y.; Zhao, L.; Zhou, Z.; Ren, L.; Meng, Q. Effect of Saccharomyces cerevisiae on alfalfa nutrient degradation characteristics and rumen microbial populations of steers fed diets with different concentrate-to-forage ratios. *J. Anim. Sci. Biotechnol.* **2014**, *5*, 24. [CrossRef] [PubMed]
51. Durand, F.; Ameilbonne, A.; Auffret, P.; Bernard, M.; Mialon, M.-M.; Dunière, L.; Forano, E. Supplementation of live yeast-based feed additive in early life promotes rumen microbial colonization and fibrinolytic potential in lambs. *J. Appl. Microbiol.* **2019**, *9*, 19216. [CrossRef]
52. Osita, C.; Ani, A.; Ikeh, N.; Oyeagu, C.; Akuru, E.; Ezemagu, I.; Udeh, V. Growth performance and nutrient digestibility of West African dwarf sheep fed high roughage diet containing Saccharomyces cerevisiae. *Agro-Science* **2019**, *18*, 25–28. [CrossRef]

 *agriculture*

Article

Influence of Supplemental Feeding on Body Condition Score and Reproductive Performance Dynamics in Botosani Karakul Sheep

Ionică Nechifor [1], Marian Alexandru Florea [1], Răzvan-Mihail Radu-Rusu [2] and Constantin Pascal [2,*]

[1] Research and Development Station for Sheep and Goat Breeding, Popăuți-Botoșani, 710004 Botoșani, Romania
[2] Department of Animal Resources and Technologies, Faculty of Food and Animal Sciences, Iasi University of Life Sciences, 700490 Iași, Romania
* Correspondence: pascalc@uaiasi.ro

Abstract: The aim of this research was to study the impact of supplementary feeding on reproductive traits in sheep. Two groups, L1 (control) and L2 (experimental treatment), of adult females aged between two and six years belonging to the Botosani Karakul sheep breed were formed. The experimental treatment group (L2) received supplementary feeding 25 days prior to mating. Improvements in body condition and significant increases in live weight occurred by the time of mating in those ewes that had been flushed (L2) ($p < 0.01$ vs. L1). The number of lambs per individual parturition was influenced by the body condition score (BCS), especially in females with a BCS of 2.0. The total number of weaned lambs in females with a BCS of 2.0 differed in comparison to that of females with a BCS of 2.5 or 3.0. All results highlighted that supplementary feeding applied to ewes prior to mating affected their reproductive and economic performance, translating to an increased live weight of the litter at weaning in the L2 group ($p < 0.01$ in lambs from BCS 2.0 ewes and $p < 0.001$ in lambs from BCS 2.5 to 3.5 ewes).

Keywords: body condition score; ewes; reproductive traits; flushing

Citation: Nechifor, I.; Florea, M.A.; Radu-Rusu, R.-M.; Pascal, C. Influence of Supplemental Feeding on Body Condition Score and Reproductive Performance Dynamics in Botosani Karakul Sheep. *Agriculture* **2022**, *12*, 2006. https://doi.org/10.3390/agriculture12122006

Academic Editor: Maria Grazia Cappai

Received: 12 October 2022
Accepted: 19 November 2022
Published: 25 November 2022

Publisher's Note: MDPI stays neutral with regard to jurisdictional claims in published maps and institutional affiliations.

1. Introduction

Northeastern Romania is a geographical region in which traditional sheep farming technologies are mostly applied and the rearing of the Botosani Karakul sheep breed represents a priority activity. The breed flock comprises more than 250,000 heads, of which 65% represent reproductive females [1]. Climate change, in the form of drought and high temperatures, induces alterations in the botanical composition and nutritional value of pastures, affecting the body conditions of ewes before the reproductive season and, subsequently, their level of productivity. The research aim was to evaluate the impact of supplementary feeding on the main reproductive and productive features of the Botosani Karakul sheep breed.

To establish the research objectives, we started from the idea that the management of sheep farms cannot be competitive without the satisfaction of certain minimal demands related to the biological state of flocks, i.e., their body condition and live weight throughout the essential periods of reproduction and production [2]. Bringing the breeders to an optimal body condition is important, due to the strong and complex connection between their metabolic functioning and their reproductive control system [3].

In reproductive ewes, an appropriate body condition is important. Hence, metabolic factors have intense manifestations and exercise variable effects, from yield increases to reproductive performance improvements, when the conditions become favourable or could induce the total inhibition of reproduction when the circumstances are adverse [4–6]. Supplementary feeding could be an efficient means for improvement in this regard when correctly applied and must comprise nutrient-rich feedstuffs. Intake and digestibility can be reduced or unaffected by supplementation with energy, while, in some cases, it has been proven that lower levels of energetic supplements increase the intake of grazed fodders [7]. Supplementary feeding is also useful because it covers ewes' nutritional demands throughout the developmental stages, providing a solution to the scarcity of specific nutrients and contributing to the formation of body reserves and the improvement of the growth rate and fertility, as well as meat, milk, and wool yields and quality [8]. Ewes' nutritional needs vary by age, body development, gestational status, gestation phase, and suckling period. The application of supplementary feeding is beneficial, preventing live weight losses in ewes and facilitating body weight maintenance at optimal levels [9].

The nutritional status of ewes can affect both their reproductive performance and their main productive functions, with yields decreasing below their genetic potential due to many limiting factors, such as malnutrition throughout critical physiological stages (i.e., gestation and lactation) [10].

Scoring the body development of sheep highlights quite well the general body condition of the herd, and farmers can intervene to correct nutritional uptake in relation to demand, which in the grazing period is very high, depending on the quality and quantity of the consumed grass [11]. Continuously updating the procedures and criteria used in the body condition score (BCS) calculation offers efficient technical support for the use of this tool in informing important farming decisions, with direct effects on the level of future production, as well as on the fodder management mode [12]. Body condition is determined by growth peculiarities and by farm management activities [13]. Evaluation based on point-scoring the body condition represents a valuable instrument for the assessment and management of ewes' welfare in farms [13,14].

Body condition assessment is highly relevant, along with live weight and size measurements, and could mitigate the errors that appear during certain periods (gestation or lactation) [15,16]. The results of the body condition scoring method are not affected by temporary mass variations caused by digestive tract filling or by wool length and moisture content [16–21]. The evaluation technique of the BCS was proposed by Jefferies (1961), in order to obtain accurate information on aspects affecting ewes' body development state and nutrition so that the available nutritional resources could be utilised more efficiently [17]. This technique allows the evaluation of the body condition in the absence of certain concrete aspects that are not visible through the analysis of the body's appearance (external traits). When correctly applied, the farmer can observe whether a major decline in body condition has occurred and can intervene to eliminate/prevent the apparition of certain negative effects [22].

The achievement of favourable economic results is determined by the provision of appropriate welfare conditions, including supplemental feeding for the restoration or improvement of body condition, especially during certain critical periods of the production cycle, due to external factors (the poor nutritional value of feed, drought, etc.) or internal factors (special physiological statuses and metabolic conditions of animals) [23]. The application of stimulating feeding has a positive influence on static characteristics (live weight at mating) as well as on dynamic traits, such as variations in body weight across different periods [20,23].

Within the context of natural feed resource depletion and degradation due to climate change, i.e., the deterioration of the floristic composition and nutritional value of pastures, it is important to study the supplemental feeding of ewes prior to mating in order to assess the direct effect of nutritional boosting on their productive and reproductive performance. Many farmers rear Botosani Karakul sheep using inherited traditions and facilities, and such findings could be of substantial benefit by offering an example of effective practices for improving farm management. Additionally, this study could provide a technical example to other research facilities, based on which they might modify their fodder resources and nutritional optimisation process in order to use locally available feedstuffs and raise the economic performance of their herds.

2. Materials and Methods

2.1. Study Area

The rearing area of the Botosani Karakul sheep breed is situated in the North-east part of Romania (Figure 1) between 25°02′ and 28°07′ Eastern longitude and between 46°37′ and 48°15′ Northern latitude.

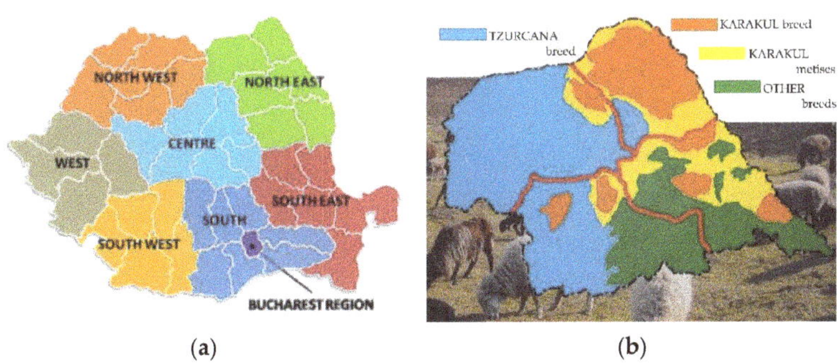

(a) (b)

Figure 1. Distribution map of Romania's regions (**a**) and representation of rearing area of Botosani Karakul sheep breed (**b**).

Temperatures have continued to increase throughout the last decades, progressively till July and August, and sometimes during September. Drought lasts more than 60 days. When higher temperatures are associated with less rainfall, the pastures' floristic composition modifies, and drought-resistant plants increase as proportion. They have poor nutritional value, unable to provide the optimal nutrients for ewes' body development prior to the reproduction season.

2.2. Animals and Data Management

The research was carried out within the Research and Development Station for Sheep and Goat Rearing Popauti-Botosani and took place on a complete reproduction-production cycle. Biological material was represented by adult females belonging to the Botosani Karakul sheep breed, representative of the Northeast Region of Romania (it counts up to 30% of the whole area sheep population) [24].

To study the impact of supplementary feeding on body condition (BC), reproductive and productive traits, two groups (L1 and L2) were formed, each comprising 100 adult females, aged 2 and 6 years (Figure 2). Each group benefited from the same experimental conditions, maintained throughout the whole research period in a local traditional system, i.e. indoors (barn) throughout the cold season (December–May) and on pasture during the warm season (May–November).

(a) (b)

Figure 2. Females exclusively grazed (**a**) and females which benefited from supplementary feeding (**b**).

During the cold season, a diet composed of a unique mix of gramineae (grasses) hay, alfalfa hay, and rough fodders was fed to ewes, and they were grazed exclusively on natural pasture, in the warm season.

The experimental factor that differentiated the Groups was the supplementary feeding provided to Group L2: 25 days prior to mating season onset, a mix of concentrated feedstuffs (cracked corn, sunflower meal, grains of barley, and oat). Their nutritional values and usage rates are presented in Table 1. Supplementary feeding in Group L2 aimed to improve the females' BC prior to reproduction season onset (150 g of the supplemental mixture was fed daily to each female), thanks to the 15% higher dietary energy, in comparison with the regular diet, fed to Group L1.

Table 1. Structure of diet throughout experimental period.

Feedstuffs Assortment	MU	Diet Provided before Mating and during the 1st Part of Gestation (First 3 Months)	Flushing Diet	Diet during the 2nd Part of Gestation		Diet during the Lactation Period		
				4th Gestational Month	5th Gestational Month	1st–2nd Lactation Months	3rd Lactation Month	4th Lactation Month
Pasture	Kg	7.5	6	7.5	-	-	-	5.833
Alfalfa hay	Kg	-	-	-	0.2	0.44	0.711	-
Grasses hay	Kg	-	-	-	0.5	0.5	0.915	-
Wheat straw	Kg	-	-	-	0.33	0.611	0.3	-
Barley	Kg	-	0.04	-	0.36	0.05	-	-
Oat	Kg	-	0.03	-		0.05	-	-
Corn	Kg	-	0.05	-		0.1	-	-
Sunflower meal	Kg	-	0.03	-	0.113	0.2	-	-
Salt	Kg	-	-	-	0.001	0.001	0.001	-
Nutritional value								
Dry matter	G	1077	1200	1200	1300	1687	1657	1225
Crude protein	G	199.81	225	225	160	215	195	175
NEM (Net energy for milk production)	MJ	7.47	8.59	8.23	7.0	8.2	7.71	7.65
Gross fibre	G	225	177	225	304	481	537	292
Ca	G	6.09	7.5	7.5	8.137	12.662	18.746	8.167
P	G	3.65	3.75	3.75	4.348	5.215	4.0	4.083
Ca/P Ratio		1.67:1	2:1	2:1	1.87:1	2.43:1	4.69:1	2:1
Na	g	1.18	1.42	1.425	1500	1.5	1.5	1.458
Mg	g	1.5	1.5	1.5	2.716	3.862	3.551	1.75

Water and salt were provided *ad libitum*. Reproduction season began in September and ended in October. Conducted natural mating was used, providing a ratio of 1 male to 25 females. Lambing season lasted between March and April and lambs were weaned 70 days post-partum.

To assess the economic efficacy of the supplemental feeding, average daily costs generated by the diets have been calculated for each experimental version, using the real pricing for producing or procuring the feedstuffs at the moment of the experiment.

2.3. Traits

The research aimed to highlight the impact of supplementary feeding on body condition (BC) dynamics, reported as body condition score (BCS), and on certain reproductive and productive traits of the Botosani Karakul sheep breed.

Body development condition was evaluated by palpation of muscular masses and fat deposits on the episome (topside trunk, loin, and rump) and was graded from 1 (slim ewes) to 5 (very fat ewes) with subrankings of 0.5 using a method developed by Russel [17]. BC was evaluated by two experienced graders who scored the ewes consensually. Live weight (LW) was measured with an electronic scale, with an accuracy of ± 100 g.

The impact of supplementary feeding on reproduction performances was evaluated through certain specific parameters, in accordance with a pattern presented by Landais and Sissoko, 1986 [25]:

Lambing rate (%) = (number of lambing ewes/number of mated ewes) $\times$ 100;

Fertility rate (%) = (number of pregnant ewes/number of mated ewes) $\times$ 100;

Litter size (prolificacy) = (number of lambs/number of lambing ewes) $\times$ 100;

Weaned rate (%) = (number of weaned lambs/number of born lambs) $\times$ 100;

Fecundity rate (%) = (number of live births/number of mated ewes) $\times$ 100;

Abortion rate (%) = (number of abortions/number of mated ewes) $\times$ 100;

Pregnancy rate (%) = (number of pregnant ewes/total mated ewes) $\times$ 100.

Other specific investigated traits were:
LLW-lambs live weight at birth (kg);
WLW-lambs live weight at weaning (kg);
LLW/LE-lambs live weight at birth(kg)/LE-number of lambing ewes (n);
WLW/LE-lambs live weight at weaning (kg)/LE-number of lambing ewes (n);
WLW/MWE-lambs live weight at weaning (kg)/MWE-metabolic weight of ewes (kg$^{-0.75}$).

2.4. Statistical Data Processing

Data have been assembled in a database using Microsoft Excel 2016 software and statistically processed via the GraphPad Prism 9.4.1., software (manufacturer, GraphPad Ltd., Palo Alto, CA, USA), to obtain the statistical descriptors (mean, standard error of the mean—SEM, coefficient of variation). Analysis of variance for the difference of average performances between studied groups was performed, using the ANOVA single factor algorithm followed by Tukey post-hoc for multiple comparisons correction using the statistical hypothesis testing.

In the quantifiable studied traits, such as LW at lambing and weaning moments (both for ewes and lambs), there were run comparisons between groups when at least three individual different values were present on the data class. Each data table displays the number of individuals that were compared.

Those traits with single output values, such as the reproduction indices, i.e., fecundity, prolificacy, fertility, abortion, pregnancy, lambing, and weaning rates, were expressed as percentages out of the main group and they were not compared statistically, because no repeated data were available.

3. Results

3.1. Impact of Supplementary Feeding on Body Condition and Live Weight

The obtained results show the effect of supplementary feeding, due to the 15% energetic dietary surplus, on the BC, through the restoration of body reserves as well as through depositing nutritional reserves that animals are supposed to metabolise further. The ewes BCS increased by 12% in L2 vs. L1 at mating and by 16% at weaning, while the LW values were 5.1 to 5.2% higher in L2 vs. L1, due to supplementary feeding usage ($p < 0.01$) (Table 2).

Table 2. Descriptive statistics for BCS (points) and ewe live weight (kg).

Assessment Moment	Trait	L1		L2		*p*-Value
		Mean	±St. Dev	Mean	±St. Dev	
Mating	BCS (points)	2.48	0.059	2.79	0.054	0.1960
	LW (kg)	46.61 [a]	5.484	49.02 [c]	5.049	0.0014
Weaning	BCS (points)	2.21	0.054	2.57	0.063	0.7713
	LW (kg)	45.09 [a]	5.666	47.44 [c]	5.514	0.0032

BCS = Body Condition Score; LW = Live Weight. ANOVA: means with different superscripts per row are statistically different: [ac] for $p < 0.01$.

Supplementary feeding determined more ewes to achieve better BCS categories in group L2 (Figure 3) due to LW increasing, due to muscular masses restoration, and due to adipose tissue deposition.

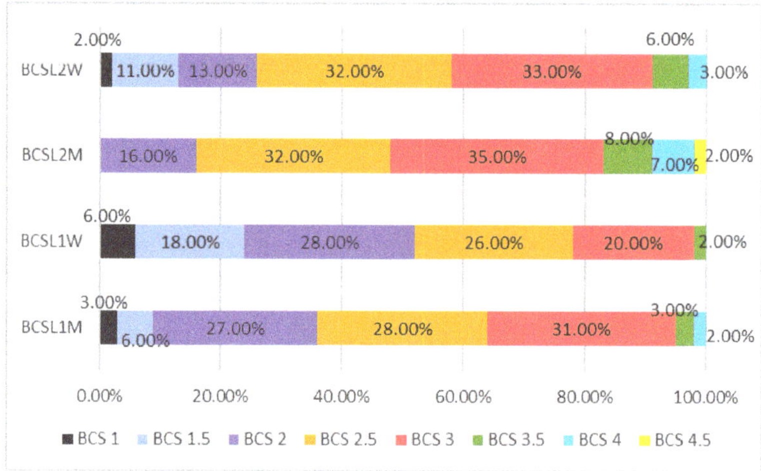

Figure 3. Females' rate (%) in relation to BCS (points) assessed at the mating moment and lambs' weaning moment. Notes: BCSL1M and BCSL2M: body condition score at the mating moment; BCSL1W and BCSL2W: body condition score at lambs weaning moment. ABWL1M and ABWL2M; average body weight at mating moment; ABWL1W and ABWL2W average body weight at lambs weaning moment.

3.2. Impact of Supplementary Feeding on Ovulation Rate and on Some Reproduction Characters

In the L1 group, ewes presented a narrow body frame and observable bone angles. After finishing mating in L1, 55.55% of females with BCS ≤ 1.5 did not become pregnant even if heats were observable. Female fertility was affected in a direct way by both BC and LW.

The positive impact of supplementary feeding is highlighted by the reproduction performances in L2 ewes, where just 7% of all ewes needed the third mating service, and less than 2% experienced abortions, in comparison with L1 ewes which had worse BC and, consequently poorer reproductive performance (12% of females with third mating service and 5% abortion rate) (Table 3).

After finalizing all lambing, better reproductive performances were recorded in the group that was supplementarily fed prior to mating. The number of lambs/ewes was influenced by ewes BC and LW at mating time.

Table 3. Effect of body condition on reproduction and production performances.

BCS	Pregnancy Rate, Lambing, and Weaning (n)						
	Total (%)	1st Service	2nd Service	3nd Service and More	Abortion	Lambing	Weaning
Group L1							
1	3	1	1	1	2	1	0
1.5	6	2	2	2	3	3	2
2	27	19	5	3	0	30	22
2.5	28	23	3	2	0	32	27
3	31	20	8	3	0	35	28
3.5	3	1	1	1	0	2	2
4	2	1	1	0	0	2	0
4.5	0	0	0	0	0	0	0
Total (n)	100	67	21	12	5	103	97
Group L2							
1	0	0	0	0	0	0	0
1.5	0	0	0	0	0	0	0
2	16	12	2	2	1	20	17
2.5	33	28	3	2	0	41	38
3	36	32	3	1	0	43	43
3.5	8	5	2	1	0	12	12
4	7	5	1	1	1	6	5
4.5	2	1	1	0	0	1	1
Total (n)	100	82	11	7	2	123	116

Impact of Supplementary Feeding on Reproduction Traits

Supplementary feeding had a positive impact on prolificacy (Table 4); in L2 group, there were 15.47% more lambs obtained in comparison with L1. The nutritional surplus meant not only energy intake but higher quantities of digestible proteins prior to mating, stimulating, most likely, the ovulatory rate.

Table 4. Effect of supplementary feeding on reproduction indices.

Groups	Fecundity, %	Prolificacy, %	Fertility Rate, %	Lambing Rate, %	Weaning Rate, %	Abortion, %	Pregnancy, %
L1	95	104	95	96	94.1	3	97
L2	97	123	97	98	94.3	2	98

Reduced differences were recorded for fecundity and fertility rates, due to the fact that those characters are genetically well consolidated and their different manifestation is relatively low even when ewes did not have an optimal BC or LW at mating.

When traditional rearing technologies are applied under the influence of climatic changes or of the quantity and quality of grazed forage, supplementary feeding prior to the mating period could improve reproduction traits and it is advisable to become a current technique for all farmers rearing Botosani Karakul sheep.

3.3. Evaluation of Relation between Ewes' Body Condition and Live Weight of Lambs at Lambing and Weaning

Even if weight at lambing was close among groups and no significant differences occurred (Table 5), by reporting the total LW of lambs to the total number of females which carried out the gestation till the end (LLW/LE), higher values were achieved by females L2 with favourable BCS (3.0 and 3.5).

Table 5. Descriptive statistics (mean $\pm$ SEM) and the ratio between the lambs' weight at lambing and at weaning in relation to the mothers' BCS.

Group	Characters	Body Condition Score (Points)				
		2	2.5	3	3.5	4
L1	LLW	4.57 ± 0.06	4.80 ± 0.07	5.08 ± 0.06	4.90 ± 0.10	0
	WLW	$20.78^{a} \pm 0.26$	$21.09^{a} \pm 0.20$	$21.53^{a} \pm 0.17^{a}$	$21.77^{a} \pm 0.12$	0
	LLW/LE	5.13	5.53	5.89	4.90	0
	WLW/LE	16.93	20.33	19.44	21.77	0
	WLW/MWE	1.30	1.23	1.24	1.22	0
L2	LLW	4.62 ± 0.06	4.84 ± 0.06	5.21 ± 0.06	5.03 ± 0.09	5.10 ± 0.13
	WLW	$22.28^{c} \pm 0.14$ $p = 0.0051$	$22.53^{d} \pm 0.08$ $p = 2.7 \times 10^{-5}$	$23.06^{d} \pm 0.07$ $p = 0.0001$	$23.37^{d} \pm 0.15$ $p = 0.0099$	22.66 ± 0.13
	LLW/LE	5.71	5.96	6.22	7.54	4.37
	WLW/LE	23.67	25.94	27.54	35.05	16.18
	WLW/MWE	1.36	1.30	1.32	1.25	1.19

ANOVA: statistical significance of differences between live weights of lambs at weaning, per classes of ewes Body Condition Scores (means with different superscripts per column are statistically different: [ac] for $p < 0.01$; [ad] for $p < 0.001$). LLW—lambing live weight (kg); WLW—weaning live weight (kg); LLW/LE—lambing live weight (kg)/LE—lambing ewes (n); WLW/LE—weaning live weight (kg)/LE—lambing ewes(n); WLW/MWE—weaning live weight (kg)/MWE—metabolic weight of ewes (kg$^{-0.75}$).

Supplementary feeding applied to group L2 prior to mating did not have a relevant effect on lamb weight at lambing because the resulting differences were not significant, and the lambs obtained from ewes with a mediocre BC had a reduced survival rate. Neonatal mortality exceeded 6% in ewes with BCS below 2.0. Lambs obtained from ewes in group L2 had a better survival rate and greater LW at weaning ($p < 0.001$). In group L2, ewes with BCS 2.0 and BCS 3.5 had better body development, higher prolificacy, and a better rate between lambs LW at weaning and the number of ewes that lambed (WLW/LE).

L2 ewes with BCS = 2.0 and BCS = 2.5 during mating raised the lambs easier till weaning, but the tendency was to have lower LW values, compared with the ewes with BCS $\geq$ 3.0. The ratio between the lambs' mean total weight at weaning and metabolic weight of ewes indicates certain variations, with higher values in females with BCS = 2.

3.4. Effect of Supplementary Feeding on the Economical Results

Supplementary feeding contributed to achieving differentiation in economic results, between groups (Table 6).

Thus, in the L2 group, the daily supplementary feeding costs exceeded by 1.3% the regular feeding in L1. At birth, the value of the lambs' LW was 23.1% higher in L2, compared to L1, while after weaning, the value gap became more intense, reaching 27.4%, in the favour of lambs issued from supplementary fed mothers (L2). Therefore, supplementary feeding probably generated higher ovulation rates, with positive effects on prolificacy, leading to better reproduction and higher economical performances.

Table 6. Economical outcomes generated by supplementary feeding.

Economical Item	n	Live Weight /Lamb (kg)	Live Weight /Group (kg)	Total Value/Group (Euro)
		L1		
Total born lambs (n)	103	4.81	495.43	1694.37
Total weaned lambs (n)	97	21.37	2072.89	7089.28
Feeding costs (euro/group/day)			189.21	
		L2		
Total born lambs (n)	123	4.96	610.08	2086.47
Total weaned lambs (n)	116	22.76	2640.16	9029.35
Feeding costs (euro/group/day)			191.72	

4. Discussions

4.1. Effect of Supplementary Feeding on Ewes' Body Condition and Live Weight

Supplementary feeding should be a common practice in farms, especially in arid areas where climate negatively affects the floral composition of pastures. It would be ideal that female breeders have fulfilled their nutritional requirements, in order to reach the appropriate BC and LW prior to mating, and to become able to express well the production and reproduction characteristics. Close and positive relations are between BCS and ewes' LW (LW) at mating and weaning, which improve many basic features. Additionally, stimulating feeding applied to Group L2 improved both BC and LW. The 25 days of stimulating feeding in L2 ewes produced moderate improvement of BCS and a significant increase of LW ($p < 0.01$), compared to L1, in mating and weaning moments. Supplementary feeding had a positive influence on the L2 group. More than 84% of ewes achieved a BCS ≥ 2.5 points, out of which 20.23% with a BCS between 3.5 and 4.0 points. In Group L1 (no supplementarily fed), around 9% of females achieved a BCS ≤ 1.5 (palpation revealed weakly developed muscle masses and lack of fat layer).

According to our results, the existence of quite a high proportion of females with BCS ≤ 2 could due to the extremely high temperatures and poor rainfalls throughout the months preceding mating (July and August), which affected not only pastures' productivity but also the grass quality and its nutritional value. From all climatic variables, fluctuations of ambient temperatures had the highest impact on animals' yields and welfare [26] but also on reproductive characteristics [2], and compromise the productivity of ewes with suckling lambs [27]. The combined effects of low feed intake and of higher energy demands, under conditions of thermal stress, induce the mobilisation of body-stored fats and of labile protein reserves to provide amino acids for protein neo-synthesis and carbon sources for gluconeogenesis [28]. Longer periods of exposure of animals to thermal stress are not desirable due to the generated negative effects: compromised growth, decreased milk and meat yields, and disrupted reproductive capacity [29].

The relation between BC and LW is dynamic due to the oscillations occurring throughout a production cycle. BCS values were consistent with ewes LW, and influenced ewes' reproductive function and yields. The impact of supplementary feeding was positive, and the BC was improved, in accordance with other studies (LW increases by 3.1 kg for each BCS unit) [30]. When nutritional requirements cannot be covered by grazing alone, decreasing in ewes' LW can occur, due to the demobilisation of body nutrients stored reserves. One study reported LW significant decrease throughout ewes' physiological stages, in relation to fodder availability, with the gestation stage ($p < 0.05$), but not with the pregnancy rank ($p > 0.05$) [31]. Experimental data suggest a linear relationship between LW and BCS. A study on the Aragonesa sheep breed [32] reported a curvilinear relation between LW and BCS, i.e., increasing in LW by 7, 10, 12, and 16 kg is necessary to improve BCS with 1.0 units within the BCS ranges of 1.0–2.0, 2.0–3.0, 3.0–4.0, and 4.0–5.0, respectively. The correlation between LW and BCS could be influenced by breed, age, or physiological status. A Turkish study indicated different regression equations that could be used to predict LW on the BCS basis and in accordance with ewes' physiological status: LW = 28.716 + 6.962 × BCS (reproduction period); LW = 39.977 + 6.771 × BCS (lambing period); LW = 33.444 + 7.074 × BCS (weaning period) [11]. Other authors [33] stated that the absence of a feed stimulating program (throughout gestation and milking) can induce metabolic dysfunctions, translating to some variations of LW, depreciation of BC, and modifications of certain blood serum parameters.

4.2. Effect of Supplementary Feeding on Sexual Cycle and Ovulation Rate

Ewes' mating period was placed during the regular—natural season for mating in the study area (September and October), to eliminate any abnormal seasonal interference on reproduction function. The results revealed the beneficial influence of the appropriate body condition on the manifestation of the sexual cycle and ovulation rate. Ewes in group L2 performed better till the end of mating season, suggesting that supplementary feeding improved the ovulatory rate. These females had a 93% ovulation rate and better fecundity values after two heats cycles. Ewes in group L1 needed the 3rd mating in quite high proportion because poor body condition at mating influences the number of sexual cycles required for pregnancy onset, inducing economic loss as well. Scottish ewes, Blackface, with lower BCS, had a late onset of the mating season [34], while at the end of the reproduction season, Masham females with higher BCS had a higher probability to manifest oestrus [35]. Aragonesa sheep with higher BCS had longer reproduction season, due to a tardive onset of seasonal anoestrus [36] and to a shorter total seasonal anoestrus [37]. Therefore, higher BCS could induce longer reproductive seasons. However, BCS effects on the duration of the reproduction season are not relevant in the majority of studies and are less probable to significantly change the moment or lastingness of the reproduction season, by improving the BCS of breeder ewes [38].

Static and dynamic effects of nutrition on ovulation rate are very well defined in sheep; ewes with higher LW and fed supplementary prior to mating could produce more lambs [39,40]. The relation between LW and ovulation rate is curvilinear. Each LW supplementary kg corresponds to a relative increase in ovulation rates, but to a point. After a certain level, exceeding LW does not lead to ovulation rate improvement, even if supplementary feeding is provided prior to mating [39]. Above a certain BCS threshold (3.5 points), the ovulation rate can decrease and ewes respond less to supplementary feeding, in comparison with ewes with average BCS values (2.0–3.0). Covering the nutritional requirements of ewes leads to positive effects in achieving a favourable BCS, an appropriate LW, and in reaching better values for reproduction indices, suggesting that a flushing diet has a positive impact on reproductive function [41,42].

The management of sheep reproductive activity influences farm economics and artificial selection activity. The number of lambs per lambing or weaned from each ewe, as well as their LW in both moments, affect profitability [43,44]. Therefore, breeder ewes must benefit from adequate feeding, knowing that the metabolic energy balance of one ewe represents an important factor in determining how many lambs are weaned and their LW [40]. Ewes with low BCS can have diminished reproductive performances in comparison with the ones with higher BCS; a study carried out on Lori-Bakhtiari sheep [30] reported that BCS values up to three at mating exert positive effects on the number of lambs weaned/ewe. Individual ovulatory rate establishes the potential for the number of lambs weaned per ewe and per year.

4.3. Effect of Supplementary Feeding on Reproductive Indices

BCS effect on the reproductive performance of ewes reared in Romania was not studied till now, and the current research brings new findings in this respect.

All reproductive traits were improved in the L2 group, probably due to stimulating feeding, compared to L1 (-33% abortion rate; $+2.1\%$ fecundity, fertility and lambing rates; $+18.3\%$ prolificacy rate; $+0.21\%$ weaning rate). In group L1, some females manifested signs of sub-nutrition, because grazed-only ewes did not have the ability to cover their daily nutritional needs (36% of the group achieved a BCS ≤ 2.0). After a long-lasting period of poor feeding, the miscarriage rate reached 5% in this group. A daily decrease in nutrient uptake induces sub-nutrition, diminishing the endometrial sensitivity to progesterone and affecting embryo survival [45].

Studies on other sheep breeds revealed that the rate of oocyte loss could be higher and reproductive performances lower even if ewes have good levels of body condition [41,46,47]. It is recommended to score the ewes for body condition 6–8 weeks prior to mating, to allow the farmer to take adequate measures to bring the breeding stock to an optimal state (BCS of 2.5–3) at the moment of mating. Other studies recommend to assess the BCS of adult ewes at least once a trimester [48], to keep it within the optimal range while, on the contrary, other authors [49] reported insignificant effects of BCS on ovulation rate, pregnancy rate, and oocyte loss. On the other hand, in ewes with poor BC at the onset of reproductive activity, deletion or attenuation of oestrus can occur [50]. Effects of supplementary feeding on prolificacy are positive and relevant, knowing this trait has a reduced heritability, hence a longer time needed for improvement via artificial selection. Genetic control of ovulation rates at many breeds, even at the prolific ones, is hard to manage, because litter size is a trait subjected to polygenic influence, implying a great number of genes with small individual effects [51].

Botosani Karakul sheep breed was formed and selected for pelts production. Therefore, among the selection criteria included in the breeding synthesis program, technical activities to support prolificacy improvement were not included. Obtaining more products per lambing led to less developed new born lambs, and to decreased individual pelts surface. However, in the current research, it was noticed that supplementary feeding led to an increase in litter size, with better values in L2 ewes presenting BCS between 2.0 and 2.5 prior to mating.

Ewes in the L2 group deposited body reserves efficiently when their BCS was between 2 and 2.5 and, subsequently, it can be hypothesised that the concentration of hormones and metabolites which appear during chronological modifications was optimal. In pregnant ewes, prolongation of the sub-nutrition state induces certain modifications and increasing of endometrial sensitivity to steroid hormones throughout the gestational beginning stages, with a negative impact on the intrauterine environment and on embryo survival, as well [52,53]. This can explain why L1 ewes with BCS ≤ 1.5 points were not fertile or why they experienced abortions.

4.4. Effect of Body Condition on Lambs Weight at Lambing and Weaning

Ewes' LW and BC influence the postpartum lambs' survival rate. Providing balanced diets prior to mating is essential to allow ewes to reach a BC that will guarantee reproductive performance. Additionally, it is recommended to provide well-balanced feeding throughout the whole gestation period. The current research confirms that next to reproductive performance, ewes in L2 with higher BCS (2, 2.5, and 3) produced more lambs, with better LW in both lambing and weaning moments.

Weight at birth is a determinant of the lambs' immediate and further survival rate. Usually, lambs born below 3 kg LW have a higher risk of sudden death, regardless of the ranking received at birth or the age and BC of the mother. Litter weight was higher in ewes with better BSC values at mating (Table 5). Ewes with BCS ≥ 3.0 produced lambs above 5.0 kg LW. Ewes with BCS ≤ 2.5 gave birth to lambs with close mean LW, with very small differences between groups ($p > 0.05$). Both LW and BCS of mothers had a great influence on the postpartum adaptation of lambs to the extra-uterine environment. Supplementary feeding aimed to build up ewes' energy reserves, used during gestation to positively influence the foetal development and lambs' LW at birth. In both groups, ewes with BCS $\geq$ 3.0 produced lambs with LW values between close limits (4.90 $\pm$ 0.10 kg till 5.21 $\pm$ 0.06 kg).

Statistically significant differences appeared between groups, on the lambs' LW at weaning: for $p < 0.01$ in females with BCS 2; for $p < 0.001$ in those having BCS $\geq$ 2.5. The ratio between lambs' total weight at weaning (WLW) and the total number of females which lambed (LE) indicated better values in the flushed group (25.94 in ewes with BCS 2.5 and in 35.05 in ewes with BCS 3.5).

For the same BC, the LW of ewes was different, therefore we considered that the assessment of the ratio between lambs' LW at weaning (WLW) and metabolic weight of ewes (MWE) will be more relevant. Maximal values of 1.30 were calculated in group L1 and higher in group L2, except the value of this ratio in group L2 ewes with BCS 3.5. In this case, the value was lower, influenced by lower litter size and by higher LW of the ewes.

The performance related to lambs' total weight at lambing, lambs' weight at weaning moment (WLW) and litter size was considered the most representative evaluation, because it mixed multiple traits, such as conception rate, fecundity, prolificacy, lambs survival rate till weaning, weight at weaning, suckle capacity, and maintenance conditions provided to ewes and lambs during suckle period. The ratio between lambs' LW and the number of ewes that lambed (LLW/LE) had higher values in females with BCS 3.0 (6.22 kg/ewe) and in ewes with BCS 3.5 (7.54 kg/ewe). This rate can be improved only in two situations: when the litter size is greater, or when lambs' LW is higher at lambing. A study carried out on the Afshari breed [54] confirmed that BCS had a significant effect on litter size. Females with BCS 3 had better reproductive performances and bore heavier lambs, while the lambing rate at ewes with BCS > 3.5 decreased. Other research [55] has shown that ewes' BCS at the lambing moment did not have any effect on the LW of lambs at birth or at the age of 30, 60, 90, and 120 days. On the contrary, Sezenler et al. [56] noticed that ewes with higher BCS values at lambing had lambs with better LW. Other studies reported the direct impact of ewes' BCS on reproductive performances [56,57], on colostrum production [58], with a direct and significant impact on lambs' mortality reduction throughout the incipient neonatal period [59]. According to other authors, flushing diet and BCS value at mating did not have any effect on lambs' LW or on apparent colostrum intake [21]. When ewes are malnourished throughout gestation, negative effects can occur, especially on Maternal Behaviour Score (MBS) and on the postpartum moment when lambs make their first suckling [60] with further negative consequences on their development.

In accordance with BCS and by reporting the lambs' LW at weaning (WLW) to the total numbers of females which lambed (LE), better values were observed in group L2. Ewes' BCS had a significant effect on productivity and on some features with high economic relevance (number of lambs at lambing, lambs' total weight at lambing and at weaning). In both groups, females with BCS $\leq$ 3.0 had lower performances for the same parameters, proving that in advanced gestation, ewes' nutritional requirements were not covered at an optimal level. The current study results are in concordance with other published data on other breeds from semi-arid zones, such as the Ossimi ewes [50] and Malpura breed [61] which proved that ewes' BCS has a curvilinear correlation with productive traits.

The tendency of decreasing performance (lambs' LW at birth and weaning) in females with BCS lower than 3–3.5 could be attributed to poor feeding of mothers throughout the advanced gestation stage [49]. The decreasing tendency of performance in ewes' with BCS above 3.5 can probably be attributed to higher requirements for maintenance because ewes with BCS 4 were heavier than the ones with a BCS 3–3.5. On the other hand, the litter size and the number of weaned lambs in ewes with BCS 3.0 was higher, compared to ewes with lower BCS values.

Better rates between lambs' LW at weaning (WLW) and ewes' metabolic weight (MW) in Group L2 can be due to the supplementary feeding, that facilitated the achievement of a better body condition and induced the constitution of some energetic deposits as backup for the increased nutritional requirements throughout both advanced gestation stage and whole suckling period. Reaching an optimal BC has a positive effect on milk yield [62,63], knowing that approx. 33% of the yielded milk is based on the mobilization of fat and proteins previously deposited by females [64,65].

Besides these, reproductive performances depend on the ewes' age. Studies of other sheep reared in drought areas, indicate that the older the ewes were, the higher were prolificacy and weaning rates [66]. When adequately fed, ewes can cover the metabolic demands for gestation and milk secretion, regardless of their age and litter size.

5. Conclusions

All the original results highlight the positive effect of a supplementary diet on breeder ewes, prior to mating.

Botosani Karakul sheep provide better performances in ewes having a BCS between 2.5 and 3.5, therefore it advisable for farmers to identify those technical-economic ways to bring and keep their female breeders within this BCS interval.

Supplementary feeding positively influenced the reproductive performance and, generated better economical results (23–27% better live weight in lambs from flushed ewes, with only 1.3% supplemental dietary costs), supporting thus the idea of its applicability into farm conditions.

Author Contributions: Conceptualization, C.P. and I.N.; methodology, C.P. and R.-M.R.-R.; validation, C.P., R.-M.R.-R. and M.A.F.; formal analysis, I.N. and M.A.F.; investigation, I.N. and M.A.F.; resources, I.N. and M.A.F.; data curation, C.P. and M.A.F.; writing—original draft preparation, C.P. and R.-M.R.-R.; writing—review and editing, C.P. and R.-M.R.-R.; visualization, I.N. and C.P.; supervision, C.P. All authors have read and agreed to the published version of the manuscript.

Funding: This study was realised via partial funding from the Romanian Ministry of Agriculture and Rural Development research program ADER 2019–2022, grant no. 814.

Institutional Review Board Statement: The animal study protocol was approved by the Institutional Committee for Animal Ethics of the Research and Development Station for Sheep and Goat Breeding, Popauti—Botosani, Romania, on 10 August 2021, as specified in the Statement on Bioethics no. 1008 per 10 October 2022.

Informed Consent Statement: Not applicable.

Data Availability Statement: Detailed data that generated the results presented in this study are available on request from the corresponding author.

Acknowledgments: The kind contribution of the research team from the Animal Resources and Technologies Department within the Faculty of Food and Animal Sciences, Iasi University of Life Sciences is kindly recognised.

Conflicts of Interest: The authors declare no conflict of interest. The funders had no role in the design of the study; in the collection, analyses, or interpretation of data; in the writing of the manuscript; or in the decision to publish the results.

References

1. Florea, M.A.; Nechifor, I.; Crîșmaru, A.; Pascal, C. Influence of some external factors on specific characteristics of seminal material for Karakul de Botoșani breed rams. *Sci. Pap. Ser. D-Anim. Sci.* **2021**, *64*, 247–252.
2. Florea, M.A.; Pascal, C.; Nechifor, I. Influence of atmospheric temperature on heating release at Karakul de Botoșani sheep breed. *Sci. Pap. D-Anim. Sci. Ser.* **2017**, *68*, 151–158.
3. Araújo, A.R.; Rodriguez, N.M.; Rogério, M.C.P.; Borges, I.; Saliba, E.O.S.; Santos, S.A.; Pompeu, R.C.F.F.; Fernandes, F.E.P.; Monteiro, J.P.; Muir, J.P. Nutritional evaluation and productivity of supplemented sheep grazing in semiarid rangeland of northeastern Brazil. *Trop. Anim. Health Prod.* **2018**, *51*, 957–966. [CrossRef]
4. Martin, G.; Walkden-Brown, S. Nutritional influences on reproduction in mature male sheep and goats. *J. Reprod. Fertil. Suppl.* **1995**, *49*, 437–449. [CrossRef]
5. Rekik, M.; Lassoued, N.; Ben Salem, H.; Mahouachi, M. Interactions between nutrition and reproduction in sheep and goats with particular reference to the use of alternative feed sources. *Options Méditerranéennes* **2007**, *74*, 375–383.
6. Caton, J.S.; Dhuyvetter, D.V. Influence of energy supplementation on grazing ruminants: Requirements and responses. *J. Anim. Sci.* **1997**, *75*, 533–542. [CrossRef]

7.	Martin, G.B.; Blache, D.; Williams, I.H. Allocation of resources to reproduction. *Resour. Alloc. Theory Appl. Farm Anim. Prod.* **2008**, *10*, 169–191. [CrossRef]

8.	Aitchison, E.M. Cereal straw and stubble for sheep feed. *J. Dep. Agric. West. Aust.* **2008**, *29*, 7.

9.	Brand, T.; Brundyn, L. Effect of supplementary feeding to ewes and suckling lambs on ewe and lamb live weights while grazing wheat stubble. *S. Afr. J. Anim. Sci.* **2015**, *45*, 89. [CrossRef]

10.	Alharthi, A.S.; Alobre, M.M.; Abdelrahman, M.M.; Al-Baadani, H.H.; Swelum, A.A.; Khan, R.U.; Alhidary, I.A. The Effects of Different Levels of Sunflower Hulls on Reproductive Performance of Yearly Ewes Fed with Pelleted Complete Diets. *Agriculture* **2021**, *11*, 959. [CrossRef]

11.	Sezenler, T.; Özder, M.; Yıldırır, M.; Ceyhan, A.; Yüksel, M.A. The relationship between body weight and body condition score some indigenous sheep breeds in Turkey. *J. Anim. Plant Sci.* **2011**, *21*, 443–447.

12.	Walkom, S.F.; Brien, F.D.; Hebart, M.L.; Pitchford, W.S. The impact of selecting for increased ewe fat level on reproduction and its potential to reduce supplementary feeding in a commercial composite flock. *Anim. Prod. Sci.* **2016**, *56*, 698. [CrossRef]

13.	Phythian, C.J.; Michalopoulou, E.; Jones, P.H.; Winter, A.C.; Clarkson, M.J.; Stubbings, L.A.; Grove-White, D.; Cripps, P.J.; Duncan, J.S. Validating indicators of sheep welfare through a consensus of expert opinion. *Animal* **2011**, *5*, 943–952. [CrossRef]

14.	Morgan-Davies, C.; Waterhouse, A.; Pollock, M.L.; Milner, J.M. Body condition score as an indicator of ewe survival under extensive conditions. *Anim. Welf.* **2008**, *17*, 71–77.

15.	Hernández, J.C.; Schilling, S.R.; Arias, M.A.V.; Pérez, R.A.E.; Castelán-Ortega, O.A.; Pérez, A.H.R.; Ronquillo, M.G. Effect of live weight pre- and post-lambing on milk production of East Friesian sheep. *Ital. J. Anim. Sci.* **2017**, *17*, 184–194. [CrossRef]

16.	Oldham, C.M.; Thompson, A.; Ferguson, M.B.; Gordon, D.J.; Kearney, G.A.; Paganoni, B.L. The birthweight and survival of Merino lambs can be predicted from the profile of liveweight change of their mothers during pregnancy. *Anim. Prod. Sci.* **2011**, *51*, 776–783. [CrossRef]

17.	Jefferies, B.C. Body condition scoring and its use in management. *Tasman. J. Agric.* **1961**, *32*, 19–21.

18.	Adalsteinsson, S. The independent effects of live weight and body condition on fecundity and productivity of Icelandic ewes. *Anim. Sci.* **1979**, *28*, 13–23. [CrossRef]

19.	Russel, A.J.F. Means of assessing the adequacy of nutrition of pregnant ewes. *Livest. Prod. Sci.* **1984**, *11*, 429–436. [CrossRef]

20.	Gonzalez, R.E.; Labuonora, D.; Russel, A.J.E. The effects of ewe live weight and body condition score around mating on production from four sheep breeds in extensive grazing systems in Uruguay. *Anim. Sci.* **1997**, *64*, 139–145. [CrossRef]

21.	Kenyon, P.R.; Maloney, S.K.; Blache, D. Review of sheep body condition score in relation to production characteristics. *N. Z. J. Agric. Res.* **2014**, *57*, 38–64. [CrossRef]

22.	Florea, M.A.; Pascal, C.; Nechifor, I. Research concerning the influence of stimulative feeding over reproduction activity of Karakul of Botosani breed. *Sci. Pap. -Anim. Sci. Ser. Lucr. Ştiinţifice Ser. Zooteh.* **2019**, *71*, 1–7.

23.	Coop, I.E. Effect of flushing on reproductive performance of ewes. *J. Agric. Sci.* **1966**, *67*, 305–323. [CrossRef]

24.	Pascal, C. *Treaty for Sheep and Goat Breeding*; Ion Ionescu de la Brad Publishing House: Iaşi, Romania, 2015.

25.	Landais, E.; Sissoko, M.M. Methodological bases for measuring animal performances. *IEMVT/ISRA* **1986**, *20*, 433–485.

26.	Joy, A.; Dunshea, F.R.; Leury, B.J.; Clarke, I.J.; Digiacomo, K.; Chauhan, S.S. Resilience of Small Ruminants to Climate Change and Increased Environmental Temperature: A Review. *Animals* **2020**, *10*, 867. [CrossRef]

27.	Sevi, A.; Caroprese, M. Impact of heat stress on milk production, immunity and udder health in sheep: A critical review. *Small Rumin. Res.* **2012**, *107*, 1–7. [CrossRef]

28.	Rhoads, R.; La Noce, A.; Wheelock, J.; Baumgard, L. Short communication: Alterations in expression of gluconeogenic genes during heat stress and exogenous bovine somatotropin administration. *J. Dairy Sci.* **2011**, *94*, 1917–1921. [CrossRef]

29.	Berihulay, H.; Abied, A.; He, X.; Jiang, L.; Ma, Y. Adaptation Mechanisms of Small Ruminants to Environmental Heat Stress. *Animals* **2019**, *9*, 75. [CrossRef]

30.	Vatankhah, M.; Talebi, M.; Zamani, F. Relationship between ewe body condition score (BCS) at mating and reproductive and productive traits in Lori-Bakhtiari sheep. *Small Rumin. Res.* **2012**, *106*, 105–109. [CrossRef]

31.	Semakula, J.; Corner-Thomas, R.; Morris, S.; Blair, H.; Kenyon, P. The Effect of Herbage Availability, Pregnancy Stage and Rank on the Rate of Liveweight Loss during Fasting in Ewes. *Agriculture* **2021**, *11*, 543. [CrossRef]

32.	Teixeira, A.; Delfa, R.; Colomer-Rocher, F. Relationships between fat depots and body condition score or tail fatness in the Rasa Aragonesa breed. *Anim. Sci.* **1989**, *49*, 275–280. [CrossRef]

33.	Yıldırır, M.; Çakır, D.; Yurtman, I.Y. Effects of restricted nutrition and flushing on reproductive performance and metabolic profiles in sheep. *Livest. Sci.* **2022**, *258*, 104870. [CrossRef]

34.	Gunn, R.G.; Doney, J.M. The interaction of nutrition and body condition at mating on ovulation rate and early embryo mortality in Scottish Blackface ewes. *J. Agric. Sci.* **1975**, *85*, 465–470. [CrossRef]

35.	Newton, J.E.; Betts, J.E.; Wilde, R. The effect of body condition and time of mating on the reproductive performance of Masham ewes. *Anim. Sci.* **1980**, *30*, 253–260. [CrossRef]

36.	Forcada, F.; Abecia, J.; Sierra, I. Seasonal changes in oestrus activity and ovulation rate in Rasa Aragonesa ewes maintained at two different body condition levels. *Small Rumin. Res.* **1992**, *8*, 313–324. [CrossRef]

37.	Rondon, Z.; Forcada, F.; Zarazaga, L.; Abecia, J.; Lozano, J. Oestrous activity, ovulation rate and plasma melatonin concentrations in Rasa Aragonesa ewes maintained at two different and constant body condition score levels and implanted or reimplanted with melatonin. *Anim. Reprod. Sci.* **1996**, *41*, 225–236. [CrossRef]

38. Celik, H.T.; Aslan, F.A.; Arıcı, Y.K.; Kahveci, M.E.; Kiper, I. Determining the factors affecting the gestational length in sheep. *Arch. Anim. Breed.* **2021**, *64*, 83–89. [CrossRef]

39. Smith, J.F. A Review of recent developments on the effect of nutrition on ovulation rate (the flushing) effect with particular reference to research at Ruakura. *Proc. N. Z. Soc. Anim. Prod.* **1991**, *51*, 15–23.

40. Scaramuzzi, R.J.; Campbell, B.K.; Downing, J.A.; Kendall, N.; Khalid, M.; Muñoz-Gutiérrez, M.; Somchit, N. A review of the effects of supplementary nutrition in the ewe on the concentrations of reproductive and metabolic hormones and the mechanisms that regulate folliculogenesis and ovulation rate. *Reprod. Nutr. Dev.* **2006**, *46*, 339–354. [CrossRef]

41. Gunn, R.G. The influence of nutrition on the reproductive performance of ewes. In *Sheep Production*; Haresign, W., Ed.; Butterworths: London, UK, 1983; pp. 99–110.

42. Gunn, R.G.; Sim, D.A.; Hunter, E.A. Effects of nutrition *in utero* and in early life on the subsequent lifetime reproductive performance of Scottish Blackface ewes in two management systems. *Anim. Sci.* **1995**, *60*, 223–230. [CrossRef]

43. Morel, P.C.H.; Kenyon, P.R. Sensitivity analysis of weaner lamb production in New Zealand. *Proc. N. Z. Soc. Anim. Prod.* **2006**, *66*, 377–381.

44. Young, J.M.; Thompson, A.N.; Kennedy, A.J. Bioeconomic modelling to identify the relative importance of a range of critical control points for prime lamb production systems in south-west Victoria. *Anim. Prod. Sci.* **2010**, *50*, 748–756. [CrossRef]

45. Abecia, J.-A.; Sosa, C.; Forcada, F.; Meikle, A. The effect of undernutrition on the establishment of pregnancy in the ewe. *Reprod. Nutr. Dev.* **2006**, *46*, 367–378. [CrossRef] [PubMed]

46. Rhind, S.M.; Gunn, R.G.; Doney, J.M.; Leslie, I.D. A note on the reproductive performance of greyface ewes in moderately fat and very fat condition at mating. *Anim. Sci.* **1984**, *38*, 305–307. [CrossRef]

47. Merrell, B.G. The effect of duration of flushing period and stocking rate on the reproductive performance of Scottish Blackface ewes. *BSAP Occas. Publ.* **1990**, *14*, 138–141. [CrossRef]

48. Mosenfechtel, S. Good fertiltiy with a good body condition. *Milchpraxis* **2004**, *42*, 12–15.

49. Rhind, S.; Rae, M.; Brooks, A. Effects of nutrition and environmental factors on the fetal programming of the reproductive axis. *Reproduction* **2001**, *122*, 205–214. [CrossRef]

50. Abdel-Mageed, I. Body condition scoring of local ssimi ewes at mating and its impact on fertility and prolificacy. *Egypt. J. Sheep Goat Sci.* **2009**, *4*, 37–44.

51. Fhamy, M.H.; Mason, I.L. Less known and rare breeds. In *Prolific Sheep*; Fahmy, M.H., Ed.; CAB International: Willingford, UK, 1996; pp. 178–186.

52. Grazul-Bilska, A.T.; Borowicz, P.P.; Johnson, M.L.; Minten, M.A.; Bilski, J.J.; Wroblewski, R.; Redmer, D.A.; Reynolds, L.P. Placental development during early pregnancy in sheep: Vascular growth and expression of angiogenic factors in maternal placenta. *Reproduction* **2010**, *140*, 165–174. [CrossRef]

53. Bairagi, S.; Grazul-Bilska, A.T.; Borowicz, P.P.; Reyaz, A.; Valkov, V.; Reynolds, L.P. Placental development during early pregnancy in sheep: Progesterone and estrogen receptor protein expression. *Theriogenology* **2018**, *114*, 273–284. [CrossRef]

54. Aliyari, D.; Moei, M.M.; Sh, M.H.; Sirjan, M.A. Effect of Body Condition Score, Live Weight and Age on Reproductive Performance of Afshari Ewes. *Asian J. Anim. Vet. Adv.* **2012**, *7*, 904–909. [CrossRef]

55. Karakuş, F.; Atmaca, M. The effect of ewe body condition at lambing on growth of lambs and colostral specific gravity. *Arch. Anim. Breed.* **2016**, *59*, 107–112. [CrossRef]

56. Sezenler, T.; Köycü, E.; Özder, M. The effect of Body Condition Score in Karacabey Merino at lambing on the lamb growth. *J. Tekirdag Agric. Fac.* **2008**, *5*, 45–52.

57. Yilmaz, M.; Altin, T.; Karaca, O.; Cemal, I.; Bardakcioglu, H.E.; Yilmaz, O.; Taskin, T.; Yılmaz, M. Effect of body condition score at mating on the reproductive performance of Kivircik sheep under an extensive production system. *Trop. Anim. Health Prod.* **2011**, *43*, 1555–1560. [CrossRef] [PubMed]

58. Jalilian, M.T.; Moeini, M.M. The effect of body condition score and body weight of Sanjabi ewes on immune system, productive and reproductive performance. *Acta Agric. Slov.* **2013**, *102*, 99. [CrossRef]

59. Corner-Thomas, R.A.; Back, P.J.; Kenyon, P.R.; Hickson, R.E.; Ridler, A.L.; Stafford, K.J.; Morris, S.T. Ad libitum Pasture Feeding in Late Pregnancy Does Not Improve the Performance of Twin-bearing Ewes and Their Lambs. *Asian-Australas. J. Anim. Sci.* **2015**, *28*, 360–368. [CrossRef]

60. Gronqvist, G.V.; Corner-Thomas, R.A.; Kenyon, P.R.; Stafford, K.J.; Morris, S.T.; Hickson, R.E. The effect of nutrition and body condition of triplet-bearing ewes during late pregnancy on the behaviour of ewes and lambs. *Asian-Australas. J. Anim. Sci.* **2018**, *31*, 1991–2000. [CrossRef]

61. Sejian, V.; Maurya, V.P.; Naqvi, S.M.K.; Kumar, D.; Joshi, A. Effect of induced body condition score differences on physiological response, productive and reproductive performance of Malpura ewes kept in a hot, semi-arid environment. *J. Anim. Physiol. Anim. Nutr.* **2010**, *94*, 154–161. [CrossRef] [PubMed]

62. Hossamo, H.E.; Owen, J.B.; Farid, M.F.A. Body condition score and production in fat-tailed Awassi sheep under range conditions. *Res. Dev. Agric.* **1996**, *3*, 99–104.

63. Gibb, M.J.; Treacher, T.T. The effect of body condition and nutrition during late pregnancy on the performance of grazing ewes during lactation. *Anim. Sci.* **1982**, *34*, 123–129. [CrossRef]

64. Cannas, A. Feeding of lactating ewes. In *Dairy Sheep Feeding and Nutrition*; Pulina, G., Ed.; CABI Publishing: Wallingford, UK, 2002; pp. 123–166.

65. Pascal, C.T.; Nechifor, I.; Florea, A.M.; Cristian, C. Heritability determination for reproduction characters in the new milk population formed in the North-East part of Romania. *Sci. Pap.-Anim. Sci. Ser.* **2019**, *72*, 3.
66. Koyuncu, M. Reproductive Performance of Kivircik Ewes on Accelerated Lambing Management. *Pak. J. Biol. Sci.* **2005**, *8*, 1499–1502. [CrossRef]

Article

The Effect of Increasing Dietary Manganese from an Organic Source on the Reproductive Performance of Sows

Clint E. Edmunds [1,2,*], Alyssa S. Cornelison [3], Chantale Farmer [4], Christof Rapp [5], Valerie E. Ryman [2], Wes P. Schweer [3], Mark E. Wilson [6] and C. Robert Dove [2,*]

1 Department of Biology, 2000 Clayton State Boulevard, Clayton State University, 30260 Morrow, Georgia
2 Department of Animal and Dairy Science, 425 River Rd, University of Georgia, 30602 Athens, Georgia
3 Zinpro Corporation, 10400 Viking Drive, Suite 240, Eden Prairie, MN 55344, USA
4 Agriculture and Agri-Food Canada, Sherbrooke R&D Centre, 2000 College, Sherbrooke, QC J1M 0C8, Canada
5 Zinpro Corporation, Akkerdistel 2E, 5831 PJ Boxmeer, The Netherlands
6 FeedworksUSA, 2407 Ashland Avenue, Cinncinati, OH 45206, USA
* Correspondence: cedmunds@clayton.edu (C.E.E.); crdove@uga.edu (C.R.D.)

Abstract: The objective of this study was to determine the effect of dietary manganese on the reproductive performance of sows. Sows ($n = 39$; 231 ± 8 kg) were randomly assigned to one of three dietary levels of supplemented Mn (CON: 0 ppm Mn; PRO20: 20 ppm Mn; PRO40: 40 ppm Mn). The experimental treatments were initiated at breeding and continued through two parities. The sows were blocked by parity within each farrowing group. The data were analyzed as a randomized complete block design using the MIXED procedure of SAS with diet as a fixed effect and block as a random effect. The lactation feed intake increased in the PRO20 sows compared to the CON and PRO40 sows ($p < 0.05$). The PRO20 and PRO40 sows farrowed piglets with improved average daily gain from birth to weaning (CON 214 g/day; PRO20 237 g/day; 220 g/day; $p < 0.05$) compared to the CON sows. The milk fat content was lower in the PRO20 (5.5%) and PRO40 sows (6.1%; $p < 0.05$) compared to the CON sows (7.8%), possibly due to increased milk demand. Supplementary dietary Mn throughout two gestation and lactation cycles led to improved birth weights and pre-weaning growth of piglets.

Keywords: lactation; manganese; reproductive performance; sows

Citation: Edmunds, C.E.; Cornelison, A.S.; Farmer, C.; Rapp, C.; Ryman, V.E.; Schweer, W.P.; Wilson, M.E.; Dove, C.R. The Effect of Increasing Dietary Manganese from an Organic Source on the Reproductive Performance of Sows. *Agriculture* **2022**, *12*, 2168. https://doi.org/10.3390/agriculture12122168

Academic Editor: Daniel Simeanu

Received: 1 November 2022
Accepted: 14 December 2022
Published: 17 December 2022

Publisher's Note: MDPI stays neutral with regard to jurisdictional claims in published maps and institutional affiliations.

1. Introduction

Reproductive efficiency is an important aspect of the swine industry. Genetics, nutrition, and the environment are contributing factors to a sow's reproductive efficiency [1], but nutrition is one of the easiest factors for producers to control. Adequate nutrition for gestating and lactating sows ensures a balance between energy expenditure and investment as her body and metabolism shift to accommodate the developing offspring and mammary tissue [1,2]. Neonates experience innate nutritional deficiencies that must be corrected via the formulation of the sow's diet because suckling piglets obtain the majority of their required nutrients from the sow via colostrum and milk until weaning [3,4]. Maximizing the performance of sows and their litters during the lactation period of reproduction is therefore a major focus for swine nutritionists [5].

Manganese (Mn) is an important inorganic dietary component found in low concentrations in most feedstuffs [6]. The basal levels of Mn in feedstuffs alone are not sufficient for optimal growth and are of unknown availability to the animal [6]. Therefore, Mn must be supplemented in the diets of pigs. It is well established that Mn plays a role in development, digestion, reproduction, antioxidant defense, and immune function in multiple species [7–12]. Leibholz et al. established that feeding piglets 0.4 mg/kg Mn is sufficient for normal growth, and no toxicity symptoms were noted in pigs supplemented with 4000 mg/kg [10]. Plumlee et al. demonstrated that feeding a Mn-deficient diet to gilts results in bone growth abnormalities, irregular estrous cycles, fetal resorption, and the birth of small and weak neonates [13]. Supplemental Mn in the diets of swine, poultry, and other species is necessary to prevent growth abnormalities, reproductive failure, and overall negative health concerns [6,7,13]. The form in which trace minerals, such as Mn, are supplemented (organic vs. inorganic) can affect their efficacy in improving growth, utilization, and decreasing mineral excretion [14,15]. It is believed that organic mineral sources are more available to the animal because the organic mineral complex can pass through the stomach intact and arrive in the small intestine to be absorbed more readily [15]. However, the literature is varied in the supposed benefits of using organic mineral sources compared to inorganic mineral sources. The objective of this research project was to determine the effect of increasing dietary Mn from an organic source on the reproductive performance of sows and the antioxidant status of their offspring. It was hypothesized that the supplementation of Mn would improve the reproductive performance of sows and the antioxidant status of their offspring.

2. Materials and Methods

The animal care, handling, and processing procedures were approved by the University of Georgia Institutional Animal Care and Use Committee (AUP #A2018 11-014-R1).

2.1. Sow Management

Sows (Choice Genetics Line CG32; $n = 39$; 231 ± 8 kg (at first breeding)) were blocked by parity (ranged 3–6 parities) within farrowing groups at breeding, with all dietary treatments being represented within each block. Sow diets were formulated with three levels of supplemental Mn (ProPath® Mn, Zinpro Corp, Eden Prairie, MN, USA): Parity 1 on treatment ($n = 39$): Control ($n = 13$; CON; no supplemental Mn, Table 1), 20 ppm Mn ($n = 13$; PRO20), or 40 ppm Mn ($n = 13$; PRO40); Parity 2 on treatment ($n = 35$): CON ($n = 11$), PRO20 ($n = 11$), PRO40 ($n = 13$). The current NRC (2012) requirement for Mn in gestating and lactating sows is 25 ppm. Sows began their respective dietary treatments on the day of breeding and remained on dietary treatment until pregnancy determination after the third breeding cycle. All the sows were on a common diet for at least 90 d prior to the start of this study. During gestation, the experimental diet was mixed to contain 320 ppm Mn and then fed at 1/16 (20 ppm) or 1/8 (40 ppm) of the daily intake with the remainder being the control diet. This blending of the experimental and control diets allowed the use of the dual hopper system in the electronic sow feeder (AP®, AGCO; Duluth, Georgia). The sows were fed to maintain a body condition score of three during gestation [16]. The low Mn gestation diet and the CON lactation diet were both 42 ppm Mn, which met the NRC requirement and was not Mn-deficient (Table 1). Although 42 ppm Mn was in excess of the NRC recommendation, basal levels of Mn in corn, soybean meal, DDGS, and other ingredients are of unknown availability to the animal. During breeding and gestation, the sows were housed in a temperature-controlled (21.1 ± 2.8 °C) barn at the University of Georgia Double Bridges Swine Unit (Oglethorpe County, Georgia).

The pregnant sows were transported to an environmentally controlled (21.1 ± 2.8 °C) farrowing room (LARU; University of Georgia, Athens, Georgia) approximately d 110 ± 1 of gestation. The sows were restrict-fed upon arrival (2.27 kg/day) until 1 d post-farrowing with ad libitum access to water. The sows consumed the lactation feed and water ad libitum after 1 d post-farrowing until weaning (Table 1). The sows remained on the same dietary treatment during the gestation and lactation phases. Diet samples were collected throughout the study and were sent to the University of Georgia Feed, Water, and Soil Laboratory (Athens, Georgia) for proximate analysis and mineral concentration via inductively coupled plasma (ICP) analysis (Table 1). The sows were weighed upon entering the farrowing unit, within 24 h of farrowing, and at weaning (Mosdal Scale Systems; Broadview, MT, USA). The weekly feed intake was recorded during the lactation period. The data were collected through two full breeding cycles and lactations. Only those sows that completed parity 1 on the treatment were included in parity 2 data collection.

Table 1. Dietary composition and analysis on an as-fed basis.

Dietary Treatment	Gestation		Lactation		
	Low Mn	High Mn	CON	PRO20	PRO40
Ingredient, %					
Corn	54.370	54.160	54.410	54.400	54.380
Corn DDGS	40.000	40.000	20.000	20.000	20.000
Soybean meal, 47.5%	1.700	1.700	21.620	21.620	21.620
L-Lysine	0.210	0.210	0.220	0.220	0.220
Dicalcium phosphate	0.870	0.870	1.260	1.260	1.260
Limestone	1.600	1.600	1.240	1.240	1.240
Salt	0.250	0.250	0.250	0.250	0.250
Vitamin premix [1]	0.250	0.250	0.250	0.250	0.250
Sow Add Pack- Vit [2]	0.250	0.250	0.250	0.250	0.250
Mineral premix [3]	0.500	0.500	0.500	0.500	0.500
ProPath® Mn [4]	0.000	0.210	0.000	0.014	0.027
Analysis [5]					
ME [6,7], Mcal/kg	3303.000	3303.000	3297.000	3297.000	3297.000
Crude protein, %	18.400	17.600	20.000	19.900	19.900
Lysine [6], %	0.520	0.520	0.970	0.970	0.970
Crude fat, %	3.700	3.600	3.800	4.100	3.800
Ash, %	5.900	5.900	5.500	5.600	6.000
Crude fiber, %	5.400	5.100	3.700	4.000	3.900
Phosphorus (total), %	0.600	0.600	0.700	0.700	0.700
Phosphorus (avail) [6], %	0.400	0.420	0.390	0.390	0.390
Calcium, %	1.000	0.900	0.900	0.900	0.900
Potassium, %	0.800	0.740	0.920	0.940	0.920
Magnesium, %	0.210	0.180	0.190	0.210	0.210
Sulfur, %	0.070	0.080	0.080	0.080	0.080
Manganese, ppm	42.000	310.000	42.000	73.000	81.000
Iron, ppm	243.000	176.000	464.000	576.000	479.000
Copper, ppm	34.000	30.000	48.000	40.000	35.000
Zinc, ppm	181.000	113.000	225.000	255.000	251.000

1.Vitamin Premix: supplied per kg of diet: vitamin A (4,134 IU); vitamin D (1,653 IU); vitamin E (66 IU); vitamin K (3.3 mg); riboflavin (8.27 mg); niacin (49.6 mg); vitamin B12 (0.033 mg); pantothenic acid (27.6 mg); ADM Alliance Nutrition, Quincy, IL 62305. 2.Sow Add Pack: supplied per kg of diet: vitamin A (4,134 IU); vitamin E (33 IU); pyridoxine (0.992 mg); folic acid (2.205 mg); biotin (0.2205 mg); choline (551.25 mg); carnitine (49.6 mg); ADM Alliance Nutrition, Quincy, IL 62305. 3.Mineral Premix: supplied per kg of diet: Copper (10 ppm Cu as $CuSO_4$; 10 ppm Cu as ProPath® Cu, Zinpro); Zinc (50 ppm Zn as ZnO and 50 ppm Zn as ProPath® Zn, Zinpro); Iron (100 ppm Fe as $FeSO_4$); Iodine (1 ppm iodine as KIO_3); Selenium (0.3 ppm Se as Na_2SeO_3). 4.Zinpro, Eden Prairie, MN. 5.Analysis performed at University of Georgia Feed, Water, and Soil Laboratory (Athens, GA). 6.Metabolizable energy. 7.Calculated value.

2.2. Piglet Handling and Care

The piglets were processed and weighed (Ohaus Corporation; Parsippany, NJ, USA) within 24 h of birth and at 21 ± 3 d of age (weaning). Pre-weaning mortality, number of live piglets, number of stillborn piglets, and number of mummies were recorded. The males were castrated at 7–10 d of age. Pre-weaning survivability was calculated on a per litter basis. The number of litters on treatment from the LARU were: Parity 1 (n = 39): CON (n = 13), PRO20 (n = 13), PRO40 (n = 13); Parity 2 (n = 35): CON (n = 11), PRO20 (n = 11), PRO40 (n = 13). The piglets were not cross-fostered or allowed access to creep feed during the course of this study.

2.3. Blood Collection and Storage

Sow blood samples were obtained 10 d ± 1 post-breeding and at 3 d ± 1 of lactation via syringe using the jugular vein. The blood samples were transferred into heparinized blood tubes (BD Vacutainer®, Franklin Lakes, NJ, USA), inverted several times, and placed on ice until arriving at the laboratory within an hour. The samples were centrifuged (2000 × g, 10 min, 4 °C) and plasma was aliquoted and stored at −80 °C for subsequent analyses.

The samples obtained 10 d post-breeding were analyzed for progesterone concentration. The samples obtained 3 d into lactation were analyzed for prolactin and immune marker concentrations.

Piglets were chosen for blood sampling by selecting an average-sized piglet from each litter based on the average weaning weight for that litter. The piglet blood samples were obtained 5 d ± 1 post-weaning via the orbital sinus [17] and collected into heparinized tubes (BD Vacutainer®, Franklin Lakes, NJ, USA). The samples were inverted several times after collection and placed on ice until arriving at the laboratory within an hour. The samples were then centrifuged (2000 × g, 10 min, 4 °C), and plasma was aliquoted and stored at −80 °C for subsequent analysis of immune marker concentrations. All the piglets were on a common phase 1 nursery diet containing 12 ppm Mn (supplemented) post-weaning.

2.4. Prolactin and Progesterone Assays

A previously described radioimmunoassay (RIA) was used to determine the concentration of prolactin [18] with the modification that 100 μL of plasma sample was used. The radio-inert prolactin and the first antibody to porcine prolactin were purchased from A.F. Parlow (U.S. National Hormone and Peptide Program, Harbor UCLA Medical Centre, Torrance, CA, USA). The parallelism of a pooled sample from lactating sows was 98.4%. The average recovery calculated by the addition of various doses of the radio-inert prolactin to 50 μL of a pooled sample was 96.3%. The sensitivity of the assay was 1.5 ng/mL. The intra- and inter-assay CV were 1.57% and 3.16%, respectively. Progesterone was measured with a RIA commercial kit (Progesterone CT, ICN Pharmaceuticals Inc., Costa Mesa, CA, USA). The validation showed a parallelism of 105.3% and an average recovery of 94.4%. Intra- and inter-assay CV were 2.96% and 0.65%, respectively.

2.5. Cytokine Analyses

Cytokines were measured using Luminex xMAP technology for multiplexed quantification of 13 porcine cytokines, chemokines, and growth factors. The multiplexing analysis was performed using the Luminex™ 200 system (Luminex, Austin, TX, USA) by Eve Technologies Corp. (Calgary, Alberta). Thirteen markers were measured simultaneously using Eve Technologies' Porcine Cytokine 13-Plex Discovery Assay® (MilliporeSigma, Burlington, MA, USA) according to the manufacturer's protocol. The 13-plex consisted of GM-CSF (granulocyte-macrophage colony-stimulating factor), IFNγ, IL-1α, IL-1ra, IL-1β, IL-2, IL-4, IL-6, IL-8, IL-10, IL-12, IL-18, and TNF-α. The assay sensitivities of these markers ranged from 5 to 42 pg/mL for the 13-plex. The individual analyte sensitivity values are available in the MilliporeSigma MILLIPLEX® MAP protocol.

2.6. Tissue Collection and Storage

Piglet tissue samples were collected at weaning (21 ± 3 d of age) during the second parity. An average-sized female piglet was chosen at weaning for tissue sampling based on the average weaning weight of each litter. The piglets were euthanized with CO_2 and the ileum, heart, and liver were removed. The ileum was flushed with 1X phosphate buffered saline solution, and the heart and liver were rinsed with the same solution. Approximately 45–65 cm of the ileum was removed immediately prior to the ileocecal junction. The left ventricle of the heart and the right medial lobe of the liver were removed. The gall bladder was removed from the liver before tissue was homogenized and frozen. The tissues (1.0 ± 0.05 g) were homogenized in 4 mL of homogenizing buffer according to Marklund and Marklund [19]. The resulting tissue homogenates were frozen at −80 °C for subsequent analysis of MnSOD activity. Additional ileum, heart, and liver tissues (5.0 ± 0.1 g) were sent to the University of Georgia Feed, Water, and Soil Laboratory (Athens, Georgia) to determine the tissue mineral concentrations via ICP analysis. The samples were kept at −20 °C until analysis.

2.7. MnSOD Analysis

The tissue MnSOD specific activity (EC 1.15.1.1) was determined according to the protocol outlined by Marklund and Marklund [19] with slight modifications. The tissue samples were collected and homogenized as previously stated. The tissue homogenates were then kept at $-80\,^{\circ}\mathrm{C}$ until analysis, according to the cited protocol.

In order to inactivate the Cu/Zn-dependent SOD, 1 mmol potassium cyanide (KCN) was added to the reaction buffer (50 mM Tris-HCl, 1.0 mM diethylenetriamine pentaacetic acid, pH 8.2) and was used to measure the relative MnSOD activity [19]. The tissue homogenate (0.5 mL) was added to 1.1 mL of reaction buffer and centrifuged ($2000 \times g$, 15 min, $4\,^{\circ}\mathrm{C}$). The resulting supernatant was diluted in a reaction buffer (1:10). An assay control with no added sample (900 µL reaction buffer + 50 µL of 10 mM sodium azide (NaN_3)), 200 µL (of diluted homogenate + 700 µL reaction buffer + 50 µL NaN_3), and 400 µL (of diluted homogenate + 500 µL reaction buffer + 50 µL NaN_3) were plated in triplicate on a 12-well microcuvette plate (VWR® Tissue Culture Plates, Radnor, PA, USA), with each well having a volume of 950 µL. The reaction was initiated when 50 µL of 4 mM pyrogallol in 10 mM HCl was added to each well and the plate was quickly mixed. The reaction was monitored at 320 nm for 3 min using the kinetic reading program of a spectrophotometer (Biotek® µQuant, 2006). Deionized water (1.0 mL) was used to blank the spectrophotometer. The amount of supernatant that resulted in the 50% inhibition of the autooxidation of pyrogallol was the equivalent of one unit of MnSOD activity (IU). The Lowry protein determination [20] was performed on the tissue homogenate to determine the milligram of soluble protein. Specific activity was defined by the internationally recognized unit, IU mg/soluble protein.

2.8. Milk Collection and Component Analysis

Milk samples were obtained from sows on d 1, 7, and 14 of their individual lactation. The success of adequate milk collection was dependent on sow temperament and safety of collection. The piglets were removed 45 min to 1 hour prior to milking. The sow's udder was manually stimulated in order to induce milk letdown. In cases where milk letdown would not occur after manual stimulation, an injection of 0.25 mL of oxytocin (20 USP oxytocin/mL) was administered to accelerate milk letdown. Several minutes post-injection, milk was collected from the sow into labelled, 15-mL conical tubes. Milk was collected from as many of the sow's functional teats (those teats that expressed milk) as possible to ensure a representative sample. The samples were frozen at $-20\,^{\circ}\mathrm{C}$ until milk component analysis.

Protein concentration of colostrum and milk was determined with the Lowry method [20]. The total fat percent of colostrum and milk was determined using a modification of the method described by Folch et al. [21]. The milk samples were thawed, mixed thoroughly, and one mL of sample was placed in a 25 mL glass screw top tube. Ten milliliters of chloroform and methanol (1:2) were added to the tube, which was then capped and vortexed, and left at room temperature for at least 10 min. Then, 5 mL of chloroform and 5 mL of saline (9.0 g/L NaCl) were added, the tubes were capped, vortexed, and the phases were allowed to separate overnight. The next morning, samples were centrifuged ($800 \times g$; $37\,^{\circ}\mathrm{C}$; 10 min) to further separate the phases. The upper methanol and water layer was aspirated, and the protein precipitate at the interface was discarded. Aluminum weigh pans were accurately weighed using gloves and forceps, and 6 mL of chloroform were then transferred from the tube to its corresponding weigh pan. The weigh pans with chloroform were left under the fume hood to allow the chloroform to evaporate (1 h minimum) and were placed in a 100 °C drying oven for an hour to allow any additional water to evaporate. The pans with dried lipid were weighed.

The milk samples were submitted to the University of Georgia Feed, Water, and Soil Laboratory (Athens, Georgia) for determination of the mineral concentration via the method of ICP analysis.

2.9. Statistical Analysis

All the analyses were performed using sow or litter as the experimental unit and a farrowing group based on farrowing dates as the block. There were five groups of 8, 8, 8, 8, and 7 sows, respectively. The sow performance data were analyzed as a randomized complete block design. The dietary treatment served as a fixed effect, and the block served as a random effect. All the models were analyzed using the MIXED procedure of SAS 9.4 (SAS Enterprise, Cary, NC, USA). The dietary treatment within the study parity (first or second) was included to detect the differences between the diets within each parity. The lactation length was utilized as a covariate for sow feed intake, total number weaned, and survivability. The piglet birthweights, weaning weights, and average daily gains were analyzed on an individual piglet basis. For the total litter weaning weight, the of number piglets weaned was used as a covariate in addition to the lactation length. The cytokine data was not normally distributed, and thus a $\log_{10}$ transformation of the cytokine concentrations was analyzed with the MIXED procedure of SAS as described above. Pairwise comparisons between the least squares means of the Mn level comparisons were computed using the PDIFF option of the LSMEANS statement. Statistical significance was declared at $p < 0.05$ and tendencies were considered at $0.05 \leq p < 0.10$.

3. Results

3.1. Sow Performance

The sow body weights during lactation were not affected by treatment ($p > 0.10$; Table 2). There was a dietary treatment within parity effect on relative weight change between d 110 of gestation and d 1 of lactation ($p = 0.024$). The PRO20 sows lost less weight from d 110 until d 1 of lactation in the second parity compared to their first ($p < 0.05$). There were no differences between treatments in sow body weight loss during lactation or from d 110 until weaning ($p > 0.10$). The gestation length was not affected by dietary treatment ($p > 0.10$). The weekly sow feed intake was affected by dietary treatment during all three weeks of lactation ($p < 0.05$). In either parity, feed intake during week 1 of lactation was higher for the PRO20 sows than the CON sows ($p < 0.05$) and did not differ from that of the PRO40 sows ($p > 0.10$). The feed intake during week 2 of the second parity was increased by 0.5–1.0 kg when compared to the first parity across all dietary treatments ($p < 0.01$). In either parity, the feed intake during week 3 of lactation was greater for the PRO20 sows than the CON sows ($p < 0.05$) and did not differ from that of the PRO40 sows ($p > 0.10$). Overall, average daily feed intake was affected by dietary treatment. In the first parity, the PRO20 sows ate significantly more feed ($p < 0.05$) when compared to either the CON or PRO40 sows. During parity 2, the PRO20 sows ate significantly more feed than the PRO40 sows ($p < 0.05$), while feed intake did not differ from the CON ($p > 0.10$).

Table 2. The effect of supplemental dietary Mn (0, 20, 40 ppm) on the reproductive performance of sows over two parities.

Dietary Treatment	Parity 1 [1]			Parity 2 [1]			SEM	*p*–Values [2]		
	CON	PRO20	PRO40	CON	PRO20	PRO40		Mn	Lin Mn	Mn (Parity)
Sow body weight, kg (N)	13	13	13	11	11	13				
d 110 ± 1 Gestation	233.0	226.4	233.1	222.0	225.4	233.5	8.0	0.588	0.450	0.807
d 1 ± 1 Lactation	224.6	217.7	224.4	216.9	221.9	222.9	10.9	0.879	0.707	0.907
d 21 ± 1 Lactation	233.4	223.6	230.6	230.2	228.8	236.5	9.3	0.612	0.822	0.918
Relative weight change, kg										
d 110–d 1 Lactation	−15.3 [a]	−11.0 [a]	−8.9 [ab]	−7.2 [ab]	−3.6 [b]	−16.2 [a]	3.6	0.118	0.689	0.024
d 1 Lactation–d 21 Lact	11.2	7.0	6.3	16.8	6.6	15.6	3.9	0.148	0.376	0.260
d 110–d 21 Lact	−4.2	−0.3	0.3	8.7	2.2	1.6	4.5	0.932	0.748	0.273
Gestation length, d	114.7	115.2	114.8	115.3	115.1	115.0	0.5	0.878	0.899	0.695
Feed intake, kg/sow/day (N)	13	13	13	11	11	13				
Week 1	5.15 [b]	6.26 [a]	5.72 [ab]	5.63 [b]	6.68 [a]	6.20 [ab]	0.43	0.012	0.128	0.726
Week 2	6.38 [c]	7.73 [a]	6.38 [c]	7.87 [ab]	8.01 [ab]	7.50 [ab]	0.38	0.039	0.621	0.003
Week 3	6.36 [bc]	6.58 [bc]	6.04 [cd]	7.63 [a]	8.16 [a]	6.41 [b]	0.56	0.025	0.060	0.131

Table 2. *Cont.*

Dietary Treatment	Parity 1 [1]			Parity 2 [1]				*p*–Values [2]		
	CON	PRO20	PRO40	CON	PRO20	PRO40	SEM	Mn	Lin Mn	Mn (Parity)
ADFI	5.92 [d]	7.05[a]	6.04 [bd]	6.96 [abc]	7.52 [a]	6.73 [bc]	0.35	0.006	0.848	0.150
Lactation length, d	18.5	16.1	18.8	17.9	17.9	18.8	3.0			
Litter performance (N)	13	13	13	11	11	13				
Total number born	16.2 [a]	12.9 [b]	15.2 [ab]	15.2 [ab]	13.3[b]	14.6 [ab]	0.9	0.021	0.396	0.817
Total live born	13.6 [a]	12.0 [ab]	13.2 [a]	12.6 [ab]	10.6 [b]	12.4 [ab]	0.7	0.035	0.713	0.524
Stillborn	1.7	0.5	1.3	2.3	2.3	2.1	0.6	0.541	0.594	0.083
Mummies	0.9	0.6	0.6	0.2	0.4	0.1	0.3	0.750	0.466	0.348
Total number weaned	10.7	10.0	10.5	9.8	9.2	9.3	0.7	0.686	0.652	0.355
Survival, %	79.8	82.3	82.3	78.9	88.4	77.0	4.3	0.282	0.937	0.592
Avg piglet birthweight, kg[3]	1.23 [c]	1.59 [a]	1.35[b]	1.22 [c]	1.55 [a]	1.45 [b]	0.04	0.001	0.001	0.150
Avg piglet weaning wt, kg[3]	5.18 [b]	5.44 [ab]	5.74[a]	5.11 [b]	5.82 [a]	5.24 [b]	0.26	0.001	0.010	0.023
Avg litter weaning wt, kg	52.58 [ab]	59.08 [a]	58.09[ab]	51.66 [ab]	58.54 [ab]	51.40 [b]	2.7	0.049	0.310	0.318
ADG, g/pig/day[3]	216 [b]	232 [ab]	236[a]	211 [b]	241 [a]	204[b]	11	0.001	0.298	0.014

a–dLS Means within a row that do not share a letter superscript differ significantly ($p < 0.05$). 1. This refers to Parity 1 or 2 on treatment; this applies throughout the text and in tables. 2. *p*-values reported are for the main effect of manganese (Mn), the preplanned linear orthogonal contrast (Lin Mn), and the manganese treatment within parity (Mn(Parity)). 3. Variables with a quadratic *p*-value of $p < 0.01$ as a result of a quadratic orthogonal contrast statement.

3.2. Litter Performance

The total number of piglets born, the total number of piglets born alive, the number of stillborn piglets, and the number of mummies were not affected by dietary treatment ($p > 0.10$; Table 2). There was a significant effect of Mn on the total number of piglets born ($p < 0.05$) and the total number of piglets born alive ($p < 0.05$). The numbers of total piglets and total live piglets were greater for the CON sows compared to the PRO20 or PRO40 sows, while values were lowest for the PRO20 sows. The PRO40 sows were intermediate between the CON and PRO20 sows. The number of stillborn piglets between Mn treatments within parity tended to be different ($p = 0.08$). The PRO20 sows tended to have lower numbers of stillborn piglets in the first parity. There was no effect of Mn and parity on the previously listed variables: the total number of piglets, the total born alive, the stillborn, and the mummies. The total number of piglets weaned and litter survivability were not affected by dietary treatment ($p > 0.23$). The average piglet birthweight increased linearly in response to increasing the concentration of dietary Mn ($p < 0.01$). The PRO20 piglets weighed more at birth (1.57 kg; $p < 0.05$) than the CON piglets (1.23 kg) and the PRO40 piglets (1.40 kg), and the PRO40 piglets weighed more at birth than the CON piglets ($p < 0.05$). The weaning weights differed in response to dietary treatment ($p < 0.01$). The PRO20 and PRO40 piglets weighed more at weaning than the CON piglets ($p < 0.05$), the PRO20 piglets (5.6 kg) had similar weights to the PRO40 piglets (5.5 kg) at weaning ($p > 0.10$). The dietary treatment did have a significant effect on litter weaning weight ($p < 0.05$; Table 2). Looking at parity 1 and parity 2 separately, there were no significant differences in litter weaning weight ($p > 0.10$). However, there was a significant difference ($p < 0.05$) in the weaning weights of the PRO20 litters from parity 1 (59.08 kg) and the PRO40 litters from parity 2 (51.40 kg). It does not appear that litter weaning weights followed any particular biological pattern in response to the maternal dietary treatment. When averaged across parity, the piglet average daily gain (ADG) was affected by dietary treatment ($p < 0.01$). The PRO20 piglets gained, on average across both parities, 23 g more per day (237 g/pig/day) than the CON piglets (214 g/pig/day; $p < 0.05$), while it was similar for the PRO20 and the PRO40 (220 g/pig/day; $p > 0.10$). In the second parity, the PRO40 piglets had decreased ADG compared to that in the first parity.

3.3. Sow Immune Marker and Plasma Hormone Concentrations

In sows, plasma concentrations (log transformed) of GM-CSF, IFN-γ, IL-1β, IL-1rα, IL-6, IL-8, and TNF-α did not differ in response to increasing the dietary Mn concentration (Table 3) or effect of the dietary treatment within parity ($p > 0.10$). There was an effect of the dietary treatment on the log concentrations of IL-1α, IL-2, IL-4, IL-10, IL-12, and IL-18 ($p < 0.05$; Table 3), as well as a linear effect on the log concentrations of these same markers ($p < 0.05$). There was a linear effect of Mn on IL-1β and IL-6 ($p < 0.05$), even though the Mn effect was not significant itself. Nevertheless, mean separations for these two markers are presented in Table 3. In the first parity, concentrations of IL-1α, IL-1β, IL-2, IL-4, IL-6, IL-10, IL-12, and IL-18 were not significantly different from one another in response to the dietary treatment. During the second parity, IL-1α, IL-1β, IL-4, IL-6, IL-10, and IL-12 had similar patterns of log concentrations. The PRO40 sows had significantly decreased log concentrations compared to the CON sows ($p < 0.05$), while the PRO20 sows had an intermediate log concentration that did not differ from that of the CON or PRO40 sows ($p > 0.10$). During the second parity, the CON sow log concentrations of IL-2 were significantly increased compared to the log concentrations in the PRO20 and PRO40 sows ($p < 0.05$). In addition, the PRO20 sows had a significantly increased log concentration of IL-2 compared to the PRO40 sows ($p < 0.05$). During the second parity, the CON and PRO20 sow log concentrations of IL-18 were significantly increased compared to the PRO40 sows ($p < 0.05$).

Table 3. The effect of supplemental dietary Mn (0, 20, 40 ppm) on the log concentration of plasma immune markers in sows 3 d into lactation over two parities.

| | Parity 1 | | | Parity 2 | | | | p-Values [1] | | |
Dietary Treatment	CON	PRO20	PRO40	CON	PRO20	PRO40	SEM	Mn	Lin Mn	Mn (Parity)
Log$_{10}$Conc, pg/mL (*N*)	13	13	13	11	11	13				
GM-CSF	2.38	2.57	2.37	2.33	2.57	2.47	0.15	0.24	0.61	0.94
IFN-γ	2.94	3.35	3.29	3.01	3.22	3.09	0.30	0.38	0.30	0.86
IL-1α	2.03 [ab]	1.95 [ab]	1.89[b]	2.32 [a]	2.08 [ab]	1.82[b]	0.15	0.05	0.01	0.45
IL-1β	3.10 [ab]	2.96 [ab]	2.94 [ab]	3.38 [a]	3.18 [ab]	2.90[b]	0.18	0.08	0.02	0.52
IL-1rα	3.00	3.12	3.03	3.37	3.15	2.99	0.14	0.26	0.12	0.24
IL-2	3.03 [abc]	2.96 [abc]	2.86 [bc]	3.33 [a]	3.11 [b]	2.72 [c]	0.17	0.02	0.01	0.38
IL-4	3.53 [ab]	3.36 [ab]	3.30 [b]	3.84 [a]	3.48 [ab]	3.21 [b]	0.19	0.03	0.01	0.61
IL-6	2.43 [ab]	2.41 [ab]	2.31 [b]	2.79 [a]	2.55 [ab]	2.34[b]	0.17	0.11	0.04	0.45
IL-8	1.47	1.04	1.52	1.25	1.30	1.17	0.18	0.50	0.93	0.24
IL-10	3.29 [ab]	3.31 [ab]	3.07 [b]	3.62 [a]	3.35 [ab]	2.95 [b]	0.19	0.01	0.01	0.52
IL-12	2.89 [ab]	2.90 [ab]	2.83 [b]	3.07 [a]	2.96 [ab]	2.76 [b]	0.09	0.04	0.02	0.38
IL-18	3.51 [ab]	3.52 [ab]	3.35[b]	3.82 [a]	3.62 [a]	3.25 [b]	0.16	0.01	0.01	0.43
TNF-α	2.51	1.81	2.21	2.43	2.09	1.96	0.28	0.20	0.10	0.69
Prolactin[2], ng/mL	30.95	28.88	30.52	32.03	30.28	28.23	3.55	0.75	0.47	0.91
(*N*)	13	13	13	11	11	13				
Progesterone[3], ng/mL	25.83	26.86	23.04	21.56	25.45	21.12	2.40	0.09	0.36	0.60
(*N*)	13	13	13	11	11	13				

a-bLS Means within a row that do not share a letter superscript differ significantly ($p < 0.05$).1P-values reported are for the main effect of manganese (Mn), the preplanned linear orthogonal contrast (Lin Mn), and the effect of Mn within parity (Mn(Parity)). 2. Prolactin was measured from plasma obtained on d 3 ± 1 of lactation. 3. Progesterone was measured from plasma obtained on 10 ± 1 d post breeding.

The prolactin concentrations in sows did not differ due to increasing dietary Mn ($p > 0.10$; Table 3). There was a tendency for the progesterone concentration to differ in response to the dietary Mn level ($p < 0.10$; Table 3), with concentrations being greater in the PRO20 sows compared to the CON and PRO40 sows.

3.4. Piglet Immune Marker Concentrations

In piglets, the log plasma concentrations of all the tested immune markers did not differ in response to increasing the maternal dietary Mn concentrations ($p > 0.10$; Table 4).

3.5. Tissue MnSOD Activity and Mineral Composition in Piglets

The dietary treatment had no effect on the MnSOD activity in the ileal or cardiac tissue of piglets ($p > 0.10$; Table 4). The hepatic MnSOD activity decreased in a linear manner across increasing levels of the maternal dietary Mn supplementation ($p = 0.03$; Table 4). There was a tendency for the Mn to affect the cardiac and hepatic zinc concentrations ($p = 0.09$, $p = 0.077$, respectively; Table 5). None of the other analyzed mineral concentrations differed across the dietary treatments ($p > 0.10$; Table 5).

Table 4. The effect of maternal supplemental dietary manganese on the log concentration of plasma immune markers in piglets 5 d post-weaning and MnSOD tissue activity at weaning.

Dietary Treatment	Parity 1			Parity 2				*p*–Values [1]		
	CON	PRO20	PRO40	CON	PRO20	PRO40	SEM	Mn	Lin Mn	Mn (Parity)
$Log_{10}Conc^2$, pg/mL (*N*)	13	13	13	11	11	13				
GM-CSF	1.21	0.97	1.35	1.36	1.67	0.88	0.24	0.57	0.42	0.06
IFN-γ	2.18	2.40	2.13	2.03	2.24	2.34	0.20	0.50	0.45	0.68
IL-1α	0.80	0.62	0.96	0.86	0.89	0.79	0.23	0.83	0.80	0.72
IL-1β	2.38	2.31	2.42	2.31	2.38	2.30	0.14	0.99	0.91	0.87
IL-1rα	3.17	3.02	3.15	3.24	3.18	3.17	0.12	0.58	0.64	0.76
IL-2	1.80	1.76	1.89	1.64	1.87	1.81	0.24	0.78	0.51	0.90
IL-4	2.25	1.81	2.04	2.17	2.06	2.12	0.20	0.28	0.41	0.76
IL-6	1.57	1.33	1.52	1.57	1.44	1.49	0.12	0.13	0.50	0.87
IL-8	1.46	1.30	1.38	1.78	1.42	1.66	0.17	0.12	0.44	0.44
IL-10	2.30	2.17	2.13	2.27	2.42	2.13	0.14	0.23	0.14	0.57
IL-12	3.19	3.15	3.13	3.13	3.18	3.17	0.06	0.97	0.88	0.71
IL-18	2.93	2.83	2.92	2.89	3.07	2.85	0.12	0.69	0.70	0.17
TNF-α	1.74	1.84	1.82	1.85	1.63	1.49	0.15	0.51	0.25	0.16
MnSOD [3,4], IU/mg (*N*)				11	11	13				
Ileum	.	.	.	6.95	6.36	6.89	0.81	0.67	0.93	
Heart	.	.	.	8.79	5.73	6.77	1.25	0.21	0.23	
Liver	.	.	.	10.02 [a]	7.46 [b]	7.87 [ab]	0.71	0.03	0.03	

a-bLS Means within a row that do not share a letter superscript differ significantly ($p < 0.05$). 1.P-values reported are for the main effect of manganese (Mn), the preplanning linear orthogonal contrast (Lin Mn), and the effect of Mn within parity (Mn(Parity)). 2.This data is derived from plasma obtained from piglets on d 5 ± 1 post-weaning. One average sized piglet was bled from each litter, based on the average weaning weight for their litter. 3.MnSOD activity is expressed in the internationally recognized unit for enzymatic activity, IU/mg soluble protein. In addition, tissue samples were only collected from piglets during the second lactation of the study. 4. *p*–values reported (from left to right) are for the main effect of manganese (Mn), the preplanned linear orthogonal contrast (Lin Mn), and the effect of tissue on MnSOD activity. There was no Mn x Tissue interaction ($p = 0.653$).

Table 5. The effect of supplemental maternal dietary manganese on tissue mineral concentrations of piglets at weaning.

Dietary Treatment		CON	PRO20	PRO40	SEM	*p*-Value
Mineral Concentration [1,2]	*N*	11	11	13		Mn
Ileum						
Phosphorus, %		1.385	1.320	1.427	0.060	0.249
Calcium, %		0.041	0.046	0.051	0.004	0.114
Manganese, ppm		5.316	5.000	5.556	0.235	0.227
Iron, ppm		116.000	114.000	118.000	11.000	0.937
Copper, ppm		6.000	15.000	6.000	5.000	0.364
Zinc, ppm		124.000	118.000	119.000	5.000	0.728
Heart						
Phosphorus, %		1.015	0.970	1.013	0.010	0.117
Calcium, %		0.023	0.024	0.031	0.004	0.226
Manganese, ppm		5.704	5.410	5.670	0.190	0.479
Iron, ppm		238.000	222.000	219.000	16.000	0.589
Copper, ppm		7.000	7.000	16.000	4.000	0.215
Zinc, ppm		89.000	87.000	91.000	2.000	0.091
Liver						
Phosphorus, %		0.908	0.912	0.922	0.060	0.970
Calcium, %		0.022	0.023	0.023	0.002	0.692
Manganese, ppm		6.023	6.441	5.907	0.446	0.591
Iron, ppm		845.000	1152.000	863.000	204.000	0.478
Copper, ppm		172.000	159.000	148.000	27.000	0.662
Zinc, ppm		262.000	265.000	199.000	24.000	0.077

1.Mineral concentration analysis performed via inductively coupled plasma (ICP) analysis. 2.Tissue samples were only collected from the second parity of the study.

3.6. Sow Milk Composition

There was no interaction between the Mn and the parity for any of the measured milk components, therefore, averages were presented across the dietary treatments (Table 6).

There was no day effect between d 7 and d 14 milk samples for any measured component, therefore, averages were presented and an orthogonal contrast between colostrum (d 1) and milk (average of d 7 and d 14) was reported (Table 6). The protein percentage decreased ($p < 0.01$) between the colostrum and milk, while there was a tendency for the milk protein to decrease in response to the supplemented dietary Mn ($p = 0.08$). The colostral fat content differed in response to dietary treatment ($p < 0.05$), with The PRO20 sows having higher colostral fat than the PRO40 sows ($p < 0.05$), but similar values to the CON sows ($p > 0.10$). The CON sows had the highest percent milk fat when compared to the PRO20 and PRO40 sows ($p < 0.05$), whereas the PRO20 sows and the PRO40 sows had similar values ($p > 0.10$). The increased dietary supplementation of Mn did not affect the mineral composition of the colostrum or milk ($p > 0.10$; Table 6). The calcium content increased ($p < 0.01$) from the colostrum to the milk, while the copper and zinc concentrations decreased ($p < 0.01$).

Table 6. The effect of supplemental dietary manganese on the colostrum (d 1) and milk (d 7 and 14) composition of lactating sows.

Dietary Treatment	CON	PRO20	PRO40	*p*–Values	
Milk composition (as-received)				Mn	Col v. Milk
Protein[1], %					
Colostrum (*N*)	15.4 (19)	16.0 (18)	13.5 (21)	0.39	0.01
Milk [2] (*N*)	9.2 (16)	8.7 (15)	7.7 (20)	0.08	
Fat [1], %					
Colostrum (*N*)	5.9 [ab] (19)	6.2 [a] (18)	4.4 [b] (17)	0.05	0.02
Milk [2] (*N*)	7.8 [a] (15)	5.5 [b] (12)	6.1 [b] (16)	0.01	
Mineral Concentration [1,3] (as-received)					
Colostrum (*N*)	19	17	17		
Milk [2] (*N*)	15	11	15		
Phosphorus, %					
Colostrum	0.11	0.12	0.12	0.21	0.85
Milk [2]	0.12	0.12	0.12	0.20	
Calcium, %					
Colostrum	0.09	0.11	0.12	0.22	0.01
Milk [2]	0.16	0.17	0.17	0.71	
Manganese, ppm					
Colostrum	0.25	0.25	0.26	0.32	0.13
Milk [2]	0.26	0.25	0.30	0.12	
Iron, ppm					
Colostrum	1.69	1.90	1.53	0.68	0.52
Milk [2]	1.68	1.84	2.17	0.63	
Copper, ppm					
Colostrum	2.59	2.60	2.21	0.52	0.01
Milk [2]	1.29	1.03	1.10	0.30	
Zinc, ppm					
Colostrum	9.05	10.34	8.66	0.48	0.01
Milk [2]	5.34	5.74	5.58	0.80	

a-bLS Means within a row that do not share a letter superscript differ significantly ($p < 0.05$). 1.There was no Mn x parity effect, so data were averaged over both lactation periods for protein and fat percentages and mineral composition. 2. There was no day effect ($p > 0.10$) between d 7 and d 14 samples for any component, therefore those samples were combined and an orthogonal contrast between colostrum and milk was reported. 3.Mineral analysis performed via inductively coupled plasma (ICP) analysis.

4. Discussion

Sow body weight is an important variable to monitor during gestation and lactation as it can affect subsequent reproductive performance and longevity [22]. Before farrowing, feed intake is restricted to prevent unnecessary weight gain and the onset of constipation in the days leading up to parturition [22]. In general, sows lose body weight after farrowing and throughout lactation until weaning [22,23]. The dietary treatment had no effect on sow body weight before farrowing, after farrowing, or after weaning. There was an effect of the Mn level within parity on the relative body weight change from d 110 to birth ($p < 0.02$), but there was no clear or logical pattern of change. This statistically significant observation does not have a significant impact in the larger frame of the study, based on the absence of the observed differences in the other relative weight change variables.

The lactation feed intake across all the treatments in this study approximated the industry mean of 6.5 kg/sow/day [24]. The increased lactation feed intake in the sows fed 20 ppm of Mn is in agreement with Tsai et al. [25] in regards to organic minerals improving sow growth characteristics. Nevertheless, Peters and Mahan [26] have reported no change in sow feed intake between mineral sources. Feed intake may be impacted by the type of mineral, but the previous literature on this subject is not consistent from one study to the next. Trace minerals are involved in many physiological pathways and affect many cellular processes, including hormone synthesis and distribution.

Progesterone is the pregnancy and conceptus maintenance hormone and is necessary for fetal growth [27,28]. Manganese is a cofactor for enzymes related to squalene synthesis, a precursor for steroid hormones like progesterone [29,30]; however, dietary treatment did not impact plasma progesterone concentrations in the current study. Current findings indicate that the dietary levels of Mn provided sufficient support of progesterone synthesis and did not differ based on Mn supplementation.

Prolactin stimulates the production of milk in mammals and has key roles in mammary development [31]. Increasing prolactin secretion in late gestation [32] or during lactation [33] led to greater sow milk yield. Prolactin is known to be a peripheral marker of Mn toxicity in rats and could also serve as a sensitive biomarker of cumulative exposure to Mn [34]. More specifically, Mn stimulates dopamine depletion, thereby increasing prolactin secretion and circulating concentrations [34]. In this study, prolactin concentrations were not affected by feeding increasing amounts of Mn from an organic source, even though organic minerals have better absorption and body retention compared to the inorganic form [35].

Efficient reproductive performance is a key component in swine husbandry. Increasing litter size while minimizing labor costs is a goal in piglet production, but there are many routes that may lead to this goal [3,4]. Plumlee et al. demonstrated that sows fed Mn-deficient diets gave birth to weak and poorly structured piglets [13]. Therefore, it was anticipated that sows fed increasing amounts of supplemental Mn would have improved Mn utilization for mineral deposition and bone development in the conceptus during gestation. It was predicted in the current study that more piglets would be born to the PRO20 and PRO40 sows and would be heavier in comparison to the CON piglets due to improved dam nutrient intake and deposition in fetal piglets. Increased maternal Mn supplementation may lead to improved Mn utilization and, as a result, lead to more piglets being born alive by the improvement of the embryonic survivability and the oxidative defense of the sow.

The sows in the current study had litter size characteristics (total number, total live born, total number weaned) slightly below the US industry average [24]. The piglet weights at birth and pre-weaning survivability were similar to the industry means. There was a numerical increase in the pre-weaning survivability when comparing the CON to PRO20 litters (79.4% vs. 85.4% over two parities, respectively). This is likely linked to the greater birthweight of the PRO20 piglets. Increased birthweight has been shown to improve pre-weaning survivability [36,37]. Improved piglet birthweights, weaning weights, and pre-weaning average daily rates of gain are important factors for efficient piglet production [3,4]. Piglets that are born heavier have a reduced risk of pre-weaning mortality and generally will gain more weight during the pre-weaning period, which was the case in the present study. Heavier piglets at weaning also typically result in improved average daily gain during the grow-finish phase of production [38]. Although the PRO20 sows did have significantly less total born and total live born piglets, there was not a feasible explanation for the heavier piglet birthweights for those particular sows.

Maternal immunity is essential for suckling piglets because there is no placental transfer of immunoglobulins to the developing offspring in swine. As is the case with most mammals, immunity must be passed from the dam to the neonate via colostrum in order for the newborn piglet to fight off pathogenic organisms [39–41]. Sow colostrum and milk contain a variety of immunomodulatory agents: prolactin, nucleotides that enhance the activity of natural killer (NK) cells, macrophages, T helper cells, and cytokines [39]. These immune markers and agents are used by the animal and can usually be found circulating in the plasma [40]. It was anticipated that increasing Mn in sow diets may impact immune markers due to improved Mn utilization by the immune system and the reduction of oxidative stress. In a recent study, plasma cytokine profiles of sows during early gestation and the second half of pregnancy were characterized by the increased production of IL-1α and IL-4 and the reduction of the production of IFN-γ [42]. Following parturition, sows experience metabolic stress, and pro-inflammatory cytokine concentrations increase after farrowing [42].

When looking at the immune status of sows post-farrowing in the present experiment, clear patterns emerge. For markers that did respond significantly to the dietary treatment, the CON sows generally had increased concentrations compared to the PRO20 sows, while the PRO40 sows had lower concentrations compared to the CON and PRO20 sows. It may be that the CON sows had a more primed and capable immune response when faced with the metabolic stress that accompanies parturition. On the other hand, the PRO40 sows may have experienced a suppression or deficiency in their immune response based on the lower plasma concentrations of various immune markers. It is also possible that the PRO40 sows overcame the metabolic stress associated with farrowing more quickly or had less metabolic stress to begin with, and the plasma immune marker concentrations had already begun to decrease. It would be of interest to obtain multiple blood samples before and after farrowing and throughout lactation to draw meaningful conclusions on the impact of feeding supplementary organic Mn on the immune status of sows.

Manganese has been linked to nutritional immunity, which is the idea that the body sequesters trace nutrients to impair or prevent the growth of certain pathogens [43]. In addition, trace elements can play messenger roles in immune system cascades [43]. The reduced plasma concentrations (log transformed) of IL-1α in the PRO40 sows compared to the CON and PRO20 sows are of importance because IL-1α is produced by activated macrophages and plays an important role in the regulation of immune responses [39,44]. It is an intermediary cellular signal in the pathway activating the pro-inflammatory cytokine, tumor necrosis factor- α (TNF-α). However, log plasma concentrations of the TNF-α were not affected by the dietary treatment. It has been determined that TNF-α concentrations peak 24–36 h following parturition in sows of differing immune status [45]. The plasma concentrations (log transformed) of the pro-inflammatory cytokines, IL-1β, IL-2, and IL-6 and the anti-inflammatory cytokine, IL-4, were also lower in the PRO40 compared with the CON and PRO20 sows, indicating altered immune status. IL-1β is released by macrophages and monocytes during cell injury, infection, invasion, and inflammation [39,44]. IL-2 is a signaling molecule that regulates the activities of white blood cells that are responsible for immune status, while the anti-inflammatory cytokine IL-4 induces the differentiation of naïve helper T-cells and reduces pro-inflammatory responses [39,44]. IL-6 is a pro-inflammatory cytokine that has been shown to suppress feed intake and stimulate the acute phase immune response [44]. There was no effect of Mn on the plasma concentration of IL-6. The plasma concentrations of IL-10, IL-12, and IL-18 were also lower in the PRO40 sows compared with the CON and PRO20 sows, especially in parity 2. IL-1rα is secreted by various cell types (for example: epithelial and adipocytes) and is a natural inhibitor of the pro-inflammatory effects of IL-1α and IL-1β [44]. IL-1rα may have suppressed the inflammatory effects of the IL-1α and IL-1β in this study. IL-10 is an anti-inflammatory cytokine with multiple, pleiotropic effects in immune regulation and inflammation, while IL-18 is a pro-inflammatory cytokine that facilitates type 1 responses along with IL-12, to induce cell-mediated immunity following infection [39,44]. IL-12 stimulates the production of IFN-γ, TNF-α, and NK cells [44]. Supplemental Mn concentrations exceeding 20 ppm may have a role in the disruption of secondary messenger cascades and, as a result, a reduction in immune marker expression.

Overall, plasma concentrations of immune markers that did change in response to dietary treatment seemed to do so in similar patterns, showing that the PRO40 sows had reduced immune marker concentrations (log transformed) compared to the CON and PRO20 sows, especially in parity 2. A reduction in concentrations in response to the Mn supplementation is not necessarily a negative result. As previously mentioned, an explanation may be that the PRO20 and PRO40 sows overcame the metabolic stress associated with parturition more quickly than the CON sows. The significant responses in log concentrations of these markers to the dietary Mn occurred in parity 2, and it is possible that feeding these diets for a longer period of time could have longer term effects on the sow's immune system. It is important to understand that these are log concentrations at a specific moment in time in the farrowing room; therefore, more definitive conclusions could be made about the immune status of the animal if samples were taken at various points in time. Looking at a variety of acute phase proteins like serum amyloid A, haptoglobin, albumin, and others could aid to better understand the immune responses observed in this study. The concentrations and activity of these acute phase proteins can provide a more complete understanding of the immune system around parturition in sows of differing immune status [45].

There were no significant differences observed in the log concentrations of immune markers in piglets at 5 d post-weaning. It is known that weaning induces the expression of pro-inflammatory cytokines in piglets, with peak expression primarily occurring 1 d post-weaning [46,47]. Based on the current data, it can be concluded that there were no long-term effects of maternal Mn supplementation on the immune status of their piglets at 5 d post-weaning.

The dietary treatment did not affect tissue mineral concentrations in the present study. In a related study, some tissues, such as the liver, showed similar trends in mineral concentration compared to the present study, independent of dietary treatment [48,49]. There is limited research on the impact of supplementing maternal diets with organic minerals on the mineral status of offspring. It was reported that maternal supplementation of chelated organic mineral sources (Cu, Fe, Zn and Mn) did not impact whole body tissue analysis in piglets when compared to inorganic sources [50]. Furthermore, as described by Papadopolous et al. [50], there was no difference in tissue concentrations of Mn in response to the maternal dietary addition of Mn, the element of interest in this study.

Weaning for piglets is a time of nutritional, immunological, social, and oxidative stress [47]. Oxidative stress results from the formation of reactive oxygen species (ROS), a byproduct of oxygen metabolism [12]. If left unchecked by antioxidant regulators, the accumulation of ROS can cause damage to membrane lipids, proteins, and DNA [51,52]. The tissue types of interest for the analysis of MnSOD activity in the present study were chosen based on the increased cellular energy requirement of hepatic, cardiac, and ileal tissues and localization of MnSOD to the mitochondria [12]. It was expected that increasing maternal Mn supplementation would increase MnSOD activity in tissues post-weaning. In weanling pigs fed the increasing dietary levels of Mn (0.24 to 32 ppm Mn), the hepatic MnSOD concentrations averaged 4.89 IU/mg and the cardiac concentrations averaged 10.0 IU/mg irrespective of Mn dose [53]. Weanling pigs fed 12 mg/kg had the increased red blood cell MnSOD activity 7 d post-weaning compared to those with no supplemented Mn [54]. The MnSOD values in the current study are higher than those previously reported. These differences are likely attributed to the supplemental Mn post-weaning rather than the maternal diet supplementation. In the current study, the liver MnSOD activity showed a reduction as the maternal Mn supplementation increased. This could be a result of suppressed oxidative stress in the PRO20 and PRO40 piglets at weaning. Cardiac tissue MnSOD activity, though not significantly different, showed a similar numerical reduction in activity from the CON to the PRO20 and PRO40 piglets.

Sow Mn supplementation influenced percent fat of both colostrum and milk and tended to change milk protein content. Mineral composition did not change due to supplemental Mn. The PRO20 sows had increased colostrum fat when compared to the PRO40 sows ($p < 0.05$), while colostrum fat in the PRO20 sows did not differ from the CON sows ($p > 0.05$). The CON sows had the highest milk fat percentage when compared to the PRO20 and PRO40 sows. Decreased percent fat of colostrum in response to increased dietary Mn may be due to increased milk demand from heavier the PRO20 and PRO40 piglets. Piglets from the PRO20 and PRO40 sows had significantly increased ADG when compared to CON piglets, suggesting consumption of increased amounts of milk. In conjunction, the PRO20 sows had a significantly increased overall ADFI. It is known that litter size can have an impact on sow milk yield [55]. The milk fat in thePRO20 and PRO40 may have been reduced or diluted, due to presumably higher milk yield. However, based on this data, there was no conclusive evidence that this was the case. The milk and colostral percent fat values from the current study are in line with those previously reported [56–58].

5. Conclusions

Supplementing gestating and lactating sow diets with Mn increased piglet birth-weights and improved pre-weaning growth. The driver of increased pre-weaning piglet weight gain seems to be increased milk production, driven by increased sow lactation feed intake. Immune markers in lactating sows and their offspring at weaning responded inconsistently to Mn supplementation, suggesting that dietary Mn beyond what is provided by feedstuffs does not significantly impact either sow or piglet immune status at 3 d into lactation. Manganese supplementation to sow diets is critical to ensure increased piglet birthweights and pre-weaning growth rates. These improvements can have compounding effects as pigs continue to grow and progress through the nursery and grow-finish phases of production.

Author Contributions: Conceptualization, C.R.D.; methodology, C.R.D.; software, C.E.E.; validation, C.R.D.; formal analysis, C.E.E.; investigation, C.E.E. and C.F.; resources, A.S.C., C.R., W.P.S. and C.R.D.; data curation, C.E.E.; writing—original draft preparation, C.E.E.; writing—review and editing, A.S.C., C.F., C.R., V.E.R., W.P.S., M.E.W. and C.R.D.; visualization, C.E.E. and C.R.D.; supervision, C.R.D.; project administration, C.R.D.; funding acquisition, C.R.D. All authors have read and agreed to the published version of the manuscript.

Funding: Zinpro® Corporation based in Eden Prairie, MN provided the financial support for the conduct of this research project and preparation of the article.

Institutional Review Board Statement: See statement at the beginning of Materials and Methods section.

Informed Consent Statement: Not applicable.

Data Availability Statement: Not applicable.

Acknowledgments: Zinpro® was involved in the study design, interpretation of data, editing of the report, and the decision to submit the article for publication.

Conflicts of Interest: The authors declare no conflict of interest.

References

1. Kyriazakis, I.; Whittemore, C.T. *Whittemore's Science and Practice of Pig Production*, 3rd ed.; Blackwell Publishing: Hoboken, NJ, USA, 2006.
2. Schoknecht, P.A. Swine Nutrition: Nutrient Usage during Pregnancy and Early Postnatal Growth, An Introduction. *J. Anim. Sci.* **1997**, *75*, 2705–2707. [CrossRef] [PubMed]
3. Knox, R.V. Impact of Swine Reproductive Technologies on Pig and Global Food Production. In *Advances in Experimental Medicine and Biology*; Lamb, G.C., DiLorenzo, N., Eds.; Springer: New York, NY, USA, 2014; Volume 752, pp. 131–160.
4. Knox, R.V.; Dziuk, P.; Hollis, G.R. *The Changing Nature of Pig Reproduction Livestock Trail*; The University of Illinois: Champaign, IL, USA, 2005; Available online: http://livestocktrail.illinois.edu/swinerepronet/paperDisplay.cfm?ContentID=7588 (accessed on 1 April 2020).
5. Derouchey, J.M.; Hancock, J.D.; Hines, R.H.; Cummings, K.R.; Lee, D.J.; Maloney, C.A.; Dean, D.W.; Park, J.S.; Cao, H. Effects of dietary electrolyte balance on the chemistry of blood and urine in lactating sows and sow litter performance. *J. Anim. Sci.* **2003**, *81*, 3067–3074. [CrossRef] [PubMed]

6. Berta, E.; Andrasofszky, E.; Bersenyu, A.; Glavits, R.; Gaspardy, A.; Fekete, S.G. Effect of Inorganic and Organic Manganese Supplementation on the Performance and Tissue Manganese Content of Broiler Chicks. *Acta Vet. Hung.* **2004**, *52*, 199–209. [CrossRef] [PubMed]

7. Hansen, S.L.; Spears, J.W.; Lloyd, K.E.; Whisnant, C.S. Feeding a Low Manganese Diet to Heifers During Gestation Impairs Fetal Growth and Development. *J. Dairy Sci.* **2006**, *89*, 4305–4311. [CrossRef] [PubMed]

8. McDowell, L.R. *Minerals in Animal and Human Nutrition*, 2nd ed.; Elsevier Science B. V.: Amsterdam, The Netherlands, 2003.

9. Leach, R.M.; Muenster, A. Studies on the Role of Manganese in Bone Formation I. Effect upon the Mucopolysaccharide Content of Chick Bone. *J. Nutr.* **1962**, *78*, 51–56. [CrossRef] [PubMed]

10. Leibholz, J.M.; Speer, V.C.; Hays, V.W. Effect of Dietary Manganese on Baby Pig Performance and Tissue Manganese Levels. *J. Anim. Sci.* **1962**, *21*, 772–776. [CrossRef]

11. Chen, P.; Bornhorst, J.; Aschner, M. Manganese metabolism in humans. *Front. Biosci. Landmark Ed.* **2018**, *23*, 1655–1679. [CrossRef]

12. Holley, A.K.; Bakthavatchalu, V.; Velez-Roman, J.M.; Clair, D.K.S. Manganese superoxide dismutase: Guardian of the powerhouse. *Int. J. Mol. Sci.* **2011**, *12*, 7114–7162. [CrossRef]

13. Plumlee, M.P.; Thrasher, D.M.; Beeson, W.M.; Andrews, F.N.; Parker, H.E. The Effects of a Manganese Deficiency Upon the Growth, Development, and Reproduction of Swine. *J. Anim. Sci.* **1956**, *15*, 352–368. [CrossRef]

14. Veum, T.L.; Carlson, M.S.; Wu, C.W.; Bollinger, D.W.; Ellersieck, M.R. Copper proteinate in weanling pig diets for enhancing growth performance and reducing fecal copper excretion when compared with copper sulfate. *J. Anim. Sci.* **2004**, *82*, 1062–1070. [CrossRef]

15. Burkett, L.J.; Stalder, K.J.; Powers, W.J.; Bregendahl, K.; Pierce, J.L.; Bass, T.J.; Bailey, T.; Shafer, B.L. Effect of inorganic and organic trace mineral supplementation on the performance, carcass characteristics, and fecal mineral excretion of phase-fed, grow-finish swine. *Asian Australas. J. Anim. Sci.* **2009**, *22*, 1279–1287. [CrossRef]

16. Knauer, M.; Baitinger, D.J. The sow body condition caliper. *Appl. Eng. Agric.* **2015**, *31*, 175–178.

17. Dove, C.R.; Alworth, L.C. Blood collection from the orbital sinus of swine. *Lab Anim.* **2015**, *44*, 383–384. [CrossRef] [PubMed]

18. Robert, S.; De Passile, A.-M.B.; St-Pierre, N.; Pelletier, G.; Petitclerc, D.; Dubreuil, P.; Brazeau, P. Effect of the stress of injections on the serum concentration of cortisol, prolactin, and growth hormone in gilts and lactating sows. *Can. J. Anim. Sci.* **1989**, *69*, 663–672. [CrossRef]

19. Marklund, S.; Marklund, G. Involvement of the Superoxide Anion Radical in the Autoxidation of Pyrogallol and a Convenient Assay for Superoxide Dismutase. *Eur. J. Biochem.* **1974**, *47*, 469–474. [CrossRef]

20. Lowry, O.H.; Rosebrough, N.J.; Farr, A.L.; Randall, R.J. Protein measurement with the Folin phenol reagent. *J. Biol. Chem.* **1951**, *193*, 265–275. [CrossRef]

21. Folch, J.; Lees, M.; Sloane, G.H. A simple method for isolation and purification of total lipids from animal tissues. *J. Biol. Chem.* **1957**, *226*, 497. [CrossRef]

22. Kim, J.S.; Yang, X.; Baidoo, S.K. Relationship between body weight of primiparous sows during late gestation and subsequent reproductive efficiency over six parities. *Asian Australas. J. Anim. Sci.* **2016**, *29*, 768–774. [CrossRef]

23. Close, W.H.; Noblet, J.; Heavens, R.P. The partition of body-weight gain in the pregnant sow. *Livest. Prod. Sci.* **1984**, *11*, 517–527. [CrossRef]

24. Knauer, M.T.; Hostetler, C.E. US swine industry productivity analysis, 2005 to 2010. *J. Swine Health Prod.* **2013**, *21*, 248–252.

25. Tsai, T.; Apgar, G.A.; Estienne, M.J.; Wilson, M.; Maxwell, C.V. A cooperative study assessing reproductive performance in sows fed diets supplemented with organic or inorganic sources of trace minerals. *Transl. Anim. Sci.* **2020**, *4*, 59–66. [CrossRef] [PubMed]

26. Peters, J.C.; Mahan, D.C. Effects of dietary organic and inorganic trace mineral levels on sow reproductive performances and daily mineral intakes over six parities. *J. Anim. Sci.* **2008**, *86*, 2247–2260. [CrossRef] [PubMed]

27. Spencer, T.E.; Bazer, F.W. Conceptus signals for establishing and maintenance of pregnancy. *Reprod. Biol. Endocrinol.* **2004**, *2*, 49. [CrossRef] [PubMed]

28. Spencer, T.E.; Johnson, G.A.; Burghardt, R.C.; Bazer, F.W. Progesterone and placental hormone actions on the uterus: Insights from domestic animals. *Biol. Reprod.* **2004**, *71*, 2–10. [CrossRef]

29. Curran, G.L. Effect of Certain transition elements on cholsterol biosynthesis. *J. Biol. Chem.* **1961**, *20*, 109–111.

30. Xie, J.; Tian, C.; Zhu, Y.; Zhang, L.; Lu, L.; Luo, X. Physiology, endocrinology, and reproduction:effects of inorganic and organic manganese supplementation on gonadotropin-releasing hormone-i and follicle-stimulating hormone expression and reproductive performance of broiler breeder hens. *Poult. Sci.* **2014**, *93*, 959–969. [CrossRef]

31. Farmer, C. Altering prolactin concentrations in sows. *Domest. Anim. Endocrinol.* **2016**, *56*, S155–S164. [CrossRef]

32. VanKlompenberg, M.K.; Manjarin, R.; Trott, J.F.; McMicking, H.F.; Hovey, R.C. Late gestational hyperprolactinemia accelerates mammary epithelial cell differentiation that leads to increased milk yield. *J. Anim. Sci.* **2013**, *91*, 1102–1111. [CrossRef]

33. Farmer, C.; Palin, M.F. Hyperprolactinemia using domperidone in prepubertal gilts: Effects on hormonal status, mammary development and mammary and pituitary gene expression. *Domest. Anim. Endocrinol.* **2021**, *76*, 106630. [CrossRef]

34. Santos, A.P.M.d.; Santos, M.; Batoreu, C.; Aschner, M. Prolactin is a peripheral marker of manganese neurotoxicity. *Brain Res.* **2011**, *1382*, 282–290. [CrossRef]

35. Liu, Y.; Ma, Y.L.; Zhao, J.M.; Vazquez-Añón, M.; Stein, H.H. Digestibility and retention of zinc, copper, manganese, iron, calcium, and phosphorus in pigs fed diets containing inorganic or organic minerals. *J. Anim. Sci.* **2014**, *92*, 3407–3415. [CrossRef] [PubMed]

36. Fix, J.S.; Cassady, J.P.; Holl, J.W.; Herring, W.O.; Culbertson, M.S.; See, M.T. Effect of piglet birth weight on survival and quality of commercial market swine. *Livest. Sci.* **2010**, *132*, 98–106. [CrossRef]

37. Feldpausch, J.A.; Jourqin, J.; Bergstorm, J.R.; Bargen, J.L.; Bokenkroger, C.D.; Davis, D.L.; Gonzalez, J.M.; Nelssen, J.L.; Puls, C.L.; Trout, W.E.; et al. Birth weight threshold for identifying piglets at risk for preweaning mortality. *Transl. Anim. Sci.* **2019**, *3*, 633–640. [CrossRef] [PubMed]

38. Cabrera, R.A.; Boyd, R.D.; Jungst, S.B.; Wilson, E.R.; Johnston, M.E.; Vignes, J.L.; Odle, J. Impact of lactation length and piglet weaning weight on long-term growth and viability of progeny. *J. Anim. Sci.* **2010**, *88*, 2265–2276. [CrossRef] [PubMed]

39. Salmon, H.; Berri, M.; Gerdts, V.; Meurens, F. Humoral and cellular factors of maternal immunity in swine. *Dev. Comp. Immunol.* **2009**, *33*, 384–393. [CrossRef]

40. Šinkora, M.; Butler, J.E. The ontogeny of the porcine immune system. *Dev. Comp. Immunol.* **2009**, *33*, 273–283. [CrossRef]

41. Sinkora, J.; Rehakova, Z.; Sinkora, M.; Cukrowska, B.; Tlaskalova-Hogenova, H. Early development of immune system in pigs. *Vet. Immunol. Immunopathol.* **2002**, *87*, 301–306. [CrossRef]

42. Brigadirov, Y.N.; Kotsarev, V.N.; Shaposhnikov, I.T.; Volkova, I.V.; Lobanov, A.E.; Falkova, Y.O. Cytokine Profile of Sows under Effects of Inflammatory Processes in Reproductive Organs. *Russ. Agric. Sci.* **2018**, *44*, 365–368. [CrossRef]

43. Haase, H. Innate Immune Cells Speak Manganese. *Immunity* **2018**, *48*, 616–618. [CrossRef]

44. Zhang, J.-M.; An, J. Cytokines, Inflammation, and Pain. *Int. Anesth. Clin.* **2007**, *45*, 27–37. [CrossRef]

45. Kaiser, M.; Jacobson, M.; Andersen, P.H.; Bækbo, P.; Ceron, J.J.; Dahl, J.; Escribano, D.; Jacobsen, S. Inflammatory markers before and after farrowing in healthy sows and in sows affected with postpartum dysgalactia syndrome. *BMC Vet. Res.* **2018**, *14*, 83.

46. Pié, S.; Lallès, J.P.; Blazy, F.; Laffitte, J.; Sève, B.; Oswald, I.P. Weaning Is Associated with an Upregulation of Expression of Inflamatory Cytokines in the Intestine of Piglets. *J. Nutr.* **2004**, *134*, 641–647. [CrossRef] [PubMed]

47. Cao, S.T.; Wang, C.C.; Wu, H.; Zhang, Q.H.; Jiao, L.F.; Hu, C.H. Weaning disrupts intestinal antioxidant status, impairs intestinal barrier and mitochondrial function, and triggers mitophagy in piglets. *J. Anim. Sci.* **2018**, *96*, 1073–1083. [CrossRef] [PubMed]

48. Martin, R.E.; Mahan, D.C.; Hill, G.M.; Link, J.E.; Jolliff, J.S. Effect of dietary organic microminerals on starter pig performance, tissue mineral concentrations, and liver and plasma enzyme activities. *J. Anim. Sci.* **2011**, *89*, 1042–1055. [CrossRef] [PubMed]

49. Ma, Y.L.; Lindemann, M.D.; Webb, S.F.; Rentfrow, G. Evaluation of trace mineral source and preharvest deletion of trace minerals from finishing diets on tissue mineral status in pigs. *Asian Australas. J. Anim. Sci.* **2018**, *31*, 252–262. [CrossRef]

50. Papadopoulos, G.A.; Maes, D.G.D.; Janssens, G.P.J. Mineral accretion in nursing piglets in relation to sow performance and mineral source. *Vet. Med.* **2009**, *54*, 41–46. [CrossRef]

51. Lubos, E.; Loscalzo, J.; Handy, D.E. Glutathione Peroxidase-1 in Health and Disease: From Molecular Mechanisms to Therapeutic Opportunities. *Antioxid. Redox Signal.* **2011**, *15*, 1957–1997. [CrossRef]

52. Schwarz, C.; Ebner, K.M.; Furtner, F.; Duller, S.; Wetscherek, W.; Wernert, W.; Kandler, W.; Schedle, K. Influence of high inorganic selenium and manganese diets for fattening pigs on oxidative stability and pork quality parameters. *Animal* **2017**, *11*, 345–353. [CrossRef]

53. Pallauf, J.; Kauer, C.; Most, E.; Habicht, S.D.; Moch, J. Impact of dietary manganese concentration on status criteria to determine manganese requirement in piglets. *J. Anim. Physiol. Anim. Nutr.* **2012**, *96*, 993–1002. [CrossRef]

54. Edmunds, C.E.; Seidel, D.S.; Welch, C.B.; Lee, E.A.; Azain, M.J.; Callaway, T.R.; Dove, C.R. The effect of varying dietary manganese and selenium levels on the growth performance and manganese-superoxide dismutase activity in nursery pigs. *Livest. Sci.* **2022**, *265*, 105100. [CrossRef]

55. Kim, S.W.; Osaka, I.; Hurley, W.L.; Easter, R.A. *Litter Size Affects Mammary Gland Growth in Lactating Sows*; Department of Animal Sciences, University of Illinois at Urbana-Champaign: Champaign, IL, USA, 1998.

56. Farmer, C. *The Gestating and Lactating Sow*; Wageningen Academic Publishers: Wageningen, The Netherlands, 2015.

57. Hu, P.; Yang, H.; Lv, B.; Zhao, D.; Wang, J.; Zhu, W. Dynamic changes of fatty acids and minerals in sow milk during lactation. *J. Anim. Physiol. Anim. Nutr.* **2018**, *103*, 603–611. [CrossRef] [PubMed]

58. Csapó, J.; Martin, T.G.; Csapó-Kiss, Z.S.; Házas, Z. Protein, fats, vitamin and mineral concentrations in porcine colostrum and milk from parturition to 60 days. *Int. Dairy J.* **1996**, *6*, 881–902. [CrossRef]

Article

Nutritive Value of *Ajuga iva* as a Pastoral Plant for Ruminants: Plant Phytochemicals and In Vitro Gas Production and Digestibility

Hajer Ammar [1,*,†], Ahmed Eid Kholif [2,†], Yosra Ahmed Soltan [3], Mohammad Isam Almadani [4], Walid Soufan [5], Amr Salah Morsy [6], Saloua Ouerghemmi [7], Mireille Chahine [8], Mario E. de Haro Marti [9], Sawsan Hassan [10], Houcine Selmi [11], Egon Henrique Horst [12,13] and Secundino Lopez [13]

1. Laboratoire de Systèmes de Production Agricole et Développement Durable "SPADD", UCAR, Ecole Supérieure d'Agriculture de Mograne, Mograne Zaghouan 1121, Tunisia
2. Dairy Science Department, National Research Centre, 33 Bohouth St. Dokki, Giza 12622, Egypt
3. Animal and Fish Production Department, Faculty of Agriculture, Alexandria University, Alexandria 21516, Egypt
4. Thünen Institute of Farm Economics, Bundesallee 63, 38116 Braunschweig, Germany
5. Plant Production Department, Faculty of Food and Agriculture Sciences, King Saud University, Riyadh 11451, Saudi Arabia
6. Livestock Research Department, Arid Lands Cultivation Research Institute, City of Scientific Research and Technological Applications, Alexandria 21934, Egypt
7. Laboratoire de Biotechnologie des Plantes, Banque Nationale des Gènes (BNG), ZI Charguia 1, Tunis 1080, Tunisia
8. Department of Animal, Veterinary and Food Sciences, University of Idaho, Moscow, ID 83844, USA
9. Gooding County Extension, College of Agricultural and Life Sciences, University of Idaho, 203 Lucy Lane, Gooding, ID 83330, USA
10. International Center for Agricultural Research in Dry Areas (ICARDA), Amman 11185, Jordan
11. Laboratoire des Ressources Sylvo-Pastorales, Institut Sylvo-Pastoral de Tabarka, Tabarka 8110, Tunisia
12. Department of Veterinary Medicine, Universidade Estadual do Oeste do Paraná, Campus de Cascavel, Guarapuava 85040-167, Paraná, Brazil
13. Departamento de Producción Animal, Universidad de León, 24007 León, Spain
* Correspondence: hjr.mmr@gmail.com or hajer.ammar@esamo.ucar.tn; Tel.: +216-96-721-222
† These authors contributed equally to this work.

Citation: Ammar, H.; Kholif, A.E.; Soltan, Y.A.; Almadani, M.I.; Soufan, W.; Morsy, A.S.; Ouerghemmi, S.; Chahine, M.; de Haro Marti, M.E.; Hassan, S.; et al. Nutritive Value of *Ajuga iva* as a Pastoral Plant for Ruminants: Plant Phytochemicals and In Vitro Gas Production and Digestibility. *Agriculture* **2022**, *12*, 1199. https://doi.org/10.3390/agriculture12081199

Academic Editor: Daniel Simeanu

Received: 10 July 2022
Accepted: 5 August 2022
Published: 11 August 2022

Publisher's Note: MDPI stays neutral with regard to jurisdictional claims in published maps and institutional affiliations.

Abstract: This study aims to evaluate the nutritive value of Ajuga iva (*A. iva*) harvested from three distinct altitude regions in Tunisia (Dougga, Mograne, and Nabeul). The chemical composition, phenolic concentration, gas production, and in vitro dry matter (DM) digestibility were determined. The highest concentrations of neutral detergent fiber (NDF) and acid detergent fiber (ADF) were for *A. iva* cultivated in Nabeul. In contrast, the highest crude protein (CP) concentration was observed in that cultivated in Mograne, and the lowest ($p < 0.01$) CP concentration was noted in that cultivated in Dougga. Additionally, the cultivation regions affected the concentrations of free-radical scavenging activity, total flavonoids, and total polyphenols ($p < 0.01$). The highest free-radical scavenging activity was observed with *A. iva* cultivated in Dougga and Mograne. The highest ($p < 0.05$) gas production rate and lag time were observed in *A. iva* cultivated in Mograne and Nabeul regions. DM digestibility differed between regions and methods of determination. The highest ($p < 0.01$) DM degradability, determined by the method of Tilley and Terry and the method of Van Soest et al., was for *A. iva* cultivated in Mograne and Dougga, while the lowest ($p < 0.01$) value was recorded for that cultivated in the Nabeul region. Likewise, metabolizable energy (ME) and protein digestibility values were higher for *A. iva* collected from Mograne region than that collected from the other sampling areas. In conclusion, the nutritive value of *A. iva* differed between regions. Therefore, care should be taken when developing recommendations for using *A. iva* in an entire region. Season- and region-specific feeding strategies for feeding *A. iva* are recommended.

Keywords: *Ajuga iva*; chemical composition; nutritive value; unconventional feeds; phenolic; growing conditions

---

1. Introduction

Inadequate feed supply is one of the significant challenges facing ruminant livestock producers, making exploring new feeds a premium issue for successful animal production [1]. Evaluating the nutritive value of unconventional feeds is essential before feeding them to animals. However, the nutritive value of plants depends on many factors and may differ for the same plant under different conditions. Reasons for that variability can be classified as intrinsic (variety, chemical composition) or extrinsic factors (growing conditions, storage, etc.) [2]. In addition, soil types, environmental conditions, geographical areas, and many more characteristics affect the nutritive value of feeds [3].

Plants contain secondary metabolites, including flavonoids, phytosterols, tannins, saponins, alkaloids, terpenoids, cyanogenic glycosides, etc., with multiple biological activities [4]. The concentration of secondary metabolites in plants depends on growth stage, soil type, etc. Soil type plays a vital role in determining the concentration and type of plant secondary metabolites. It is the matrix through which potential secondary metabolites are adsorbed and pass [5]. The activities of plant secondary metabolites in the soil are strongly linked with the soil's physical, chemical, biological, and physicochemical properties, which in turn affect adsorption and degradation [6]. *Ajuga iva* (L.) Schreber (Lamiaceae) (*A. iva*) is a plant that has been used in traditional medicine due to its anti-inflammatory, antifungal, antimicrobial, antifebrile, and anthelmintic activity [7]. *A. iva* contains polyphenolic compounds with antioxidant properties [8]. Its extract has been used traditionally as a diuretic, cardiac tonic, hypoglycemic, or a cure for fever. It exhibits a high stimulating effect on animal protein synthesis [9,10]. Chemical studies on *A. iva* have revealed the presence of several flavonoids, tannins, terpenes, and steroids [11]. The natural presence of bioactive compounds in the plant suggests the possibility of its use in animal feed to alter ruminal fermentation [12,13]. Recently, Bouyahya et al. [14] compared the volatile compounds of *A. iva* essential oils at three developmental periods, and noted that phenological stages significantly affected the volatile compounds resulting in different biological properties. They identified 28 volatile compounds in *A. iva* essential oils at the three developmental periods, with carvacrol, methyl chavicol, and octadecane among the significant compounds with different concentrations in each developmental period.

Additionally, the antioxidant, antibacterial, and antifungal properties of *Ajuga* were significantly affected by the concentrations of total phenolics and flavonoids [9]. Thus, the hypothesis of the present study depends on the possibility of a relationship between the nutritional value of *A. iva* feed materials and the sites of cultivation, as well as their content of phenolic substances, which may have an impact on their in vitro gas production and digestibility. Therefore, the present trial was undertaken to compare in vitro the nutritive value of *A. iva* at different sites (Dougga, Mograne, and Nabeul) in Tunisia.

2. Materials and Methods

2.1. Sampling Source of Ajuga iva (A. iva)

Approximately 1000 g of naturally cultivated mature *A. iva* parts (leaves and small stems) were randomly collected in Spring 2018 from three different sites in Tunisia: Nabeul (latitude 36°22′556″ N longitude 11°40′4581″ E), Dougga (latitude 36°25′94″ N; longitude 9°13′05″ E), and Mograne (latitude 36°40′920″ N and longitude 10°27′918″ E). These sites were selected to represent the majority of Mediterranean conditions in Tunisia. The texture of the soil in Nabeul was sandy, it was silty in Mograne, and vertisol (very fertile and rich in clay) in Dougga. The three regions are situated in semi-arid areas (precipitation ranges between 400–600 mm/year). The plant samples were air-dried at room temperature ($40 \pm 2\ °C$) for one week, ground by a Retsch blender mill (Normandie-Labo, 7210, type ZM1, Lintot, France), and sieved through a 0.5 mm mesh screen to obtain a uniform particle size. The ground substrates were bagged and stored at room temperature until the chemical analysis and in vitro experiments.

2.2. Chemical Analysis

All *A. iva* samples were analyzed in triplicate for dry matter (DM, method ID 934.01), ash (method ID 942.05), ether extract (EE, method ID 920.30), and crude protein (CP, method ID 984.13) content following the methods of AOAC [15]. Neutral detergent fiber (NDF), acid detergent fiber (ADF), and acid detergent lignin (ADL) were determined using an ANKOM2000 fiber analyzer [16] (ANKOM 2000, ANKOM Technology, Macedon, NY, USA) with the reagents described by Van Soest et al. [17]. Sodium sulfite, but not β-amylase, was added to the solution for NDF determination.

2.3. Phytochemical Analysis

For a detailed analysis of the bioactive components in *A. iva*, triplicate samples (1 g) were extracted with 20 mL of hydro-ethanolic solution (700 mL/L) according to the method described by Neffati et al. [18]. Extractions were carried out using maceration at room temperature for 24 h. The mixture was then filtered through Wattman No.1 filter paper (Bärenstein, Germany) and micro filter paper (Wattman, 0.45 μm). The resulting solutions were evaporated under vacuum at 40 °C using a rotavapor (Buchi Corporation R-210, New-Castle, DE, USA), and the yield (%) of extraction was determined. Samples were stored at 4 °C until use. The extract yield (%) was determined according to the below equation:

$$\text{Yield } (\%) = \frac{weight\ of\ dried\ extract\ (mg)}{weight\ of\ dried\ plant\ material\ (mg)} \times 100 \tag{1}$$

The total phenolic (TP) content was determined by the Folin–Ciocalteu colorimetric method [19] with some modifications, using gallic acid as the standard. The method is based on reducing phosphotungstate–phosphomolybdate complex to blue reaction products [19]. The modified method is described briefly: *A. iva* leaf extract and the chosen standard (gallic acid) were dissolved at different concentrations, and 0.1 mL of each solution was mixed with 1 mL of Folin–Ciocalteu reagent (10%). The mixture was incubated for 5 min before adding 1 mL of 10% (w/v) Na_2CO_3. Prepared solutions were then diluted with 8.4 mL of deionized water and incubated in the dark at room temperature for 90 min. The absorbance of each sample and of the standard mixture was measured at 760 nm against the appropriate blank using a spectrophotometer (Jenway spectrophotometer monofaisceau UV/visible model 7315). The TP concentration was expressed as gallic acid equivalents per milligram of dry extract (mg GAE/mg DE).

Total flavonoid content (TF) was measured using a colorimetric method based on the formation of flavonoid [20]. Each diluted sample extract (0.25 mL) was added to 0.075 mL of $NaNO_2$ solution (7%) and mixed for 6 min before adding 0.15 mL of freshly prepared $AlCl_3$ solution ($6H_2O$, 10%). Catechin was used as a standard. After 5 min, 0.5 mL of 1 mol/L NaOH solution was added. The final volume was adjusted with distilled water to 2.5 mL, and thoroughly mixed; the absorbance of the mixture was determined at 510 nm. TF of dried *A. iva* leaf extract was estimated according to the calibration curve obtained by a series of concentrations of the catechin standard (0 to 700 μg/mL range). Samples were analyzed in triplicate, and results were expressed as catechin equivalents per mg dry extract (mg CE/mg DE).

2.4. Antioxidant Activity

The anti-radical activity (ARSA) of *A. iva* leaves was evaluated as the scavenging of the free anionic 2,2-diphenyl-1-picrylhydrazyl (DPPH) radical. At different concentrations, the sample solution (50 μL) or standard Trolox solution was added to 1 mL of 40 μM DPPH in methanol. The mixture was then shaken vigorously. After incubation (1 h), changes in color (from deep violet to dark yellow) were measured at 517 nm. The radical scavenging activity for DPPH was evaluated by calculating the percentage of inhibition (PI) using the control and sample recorded absorbance (Abs):

$$\text{PI } (\%) = \frac{Abs\ control - Abs\ sampl}{Abs\ control} \times 100 \tag{2}$$

The half-maximal inhibitory concentration (IC50) was calculated from the linear equation of the curve obtained by projection of PI versus the respective sample concentrations.

2.5. In Vitro Assays

In vitro trials were carried out using two different methods: the in vitro dry matter digestibility, a gravimetrical method, and the in vitro gas production technique. In vitro dry matter digestibility analysis was performed in Spain (University of León, León, Spain) according to the technique proposed by Tilley and Terry [21] or by Van Soest et al. [22]. The in vitro gas production analysis was carried out in Tunisia at the Sylvo-Pastoral Institute of Tabarka.

2.5.1. In Vitro Dry Matter Digestibility (IVDMD)

Four mature Merino sheep with 49.4 $\pm$ 4.2 kg body weight (mean $\pm$ standard error) and fitted with a permanent ruminal cannula were used as inoculum donors to carry out in vitro incubations of the plant material. Animals were allowed 1 kg of Lucerne (Medicago sativa) hay (the traditional Mediterranean forage) once per day supported with ground maize and soybean meal (0.7 kg/100 kg live weight, 156 g CP/kg), and had free access to a mineral premix and fresh water. Sheep were cared for and handled by trained personnel in accordance with the Spanish guidelines for experimental animal protection (Spanish Royal Decree 53/2013 on the protection of animals used for experimentation or other scientific purposes). The experimental protocols were approved by the Institutional Ethics Committee on Animal Experimentation (ULE_014_2016) of Universidad de León and the Junta de Castilla y León (León, Spain). Ruminal liquid and solid parts were collected separately from each animal before morning feeding. The liquid part was collected by a stainless steel probe (2.5 mm screen) attached to a large-capacity syringe. The solid part was collected from the dorsal rumen sac through the cannula, and squeezed by hand. Liquid and solid parts were placed separately under anaerobic surroundings into pre-warmed thermo containers (39 °C) and were carried immediately to the laboratory. The two parts were blended at 1:1 (v/v) for 10 s, squeezed through four layers of cheesecloth, and retained in a water bath (39 °C) flushed under CO_2 until the inoculation took place.

The in vitro dry matter digestibility (IVDMD) was determined using the Ankom Daisy procedure [16], to which two different approaches were applied as proposed by Tilley and Terry, and Van Soest et al. [22]. Both techniques were carried out separately in different trials. A culture medium containing macro and micro mineral solutions, resazurin, and a bicarbonate buffer solution was prepared as described by Van Soest et al. [22]. The medium was kept at 39 °C and saturated with CO_2. Oxygen in the medium was reduced by adding a solution containing cysteine–HCl and Na_2S, as Van Soest et al. [22] described. Rumen fluid was then diluted into the medium at a proportion of 1:5 (v/v). *A. iva* samples (250 mg) were weighed into artificial fiber bags (size 5 cm $\times$ 5 cm, pore size 20 μm), sealed with heat, and placed in incubation jars. Each jar was a 5 L glass recipient with a plastic lid provided with a single-way valve to avoid the accumulation of fermentation gases. Each incubation jar was filled with 2 L of the buffered rumen fluid transferred anaerobically, closed with the lid, and the contents mixed thoroughly. The jars were then placed in a revolving incubator (Ankom Daisy Incubator, ANKOM Technology Corp, Macedon, NY, USA) at 39 °C, with continuous rotation to facilitate the effective immersion of the bags in the rumen fluid. After 48 h of incubation in buffered rumen fluid, samples were either subject to 48 h pepsin–HCl digestion as described by Tilley and Terry [21], or gently rinsed in cold water followed by extraction with a neutral detergent solution at 100 °C for 1 h as described by Van Soest et al. [22]. According to Van Soest [23], the original method of Tilley and Terry is a measurement of the apparent in vitro digestibility (AIVD).

Treatment with the neutral detergent solution removed bacterial cell walls and other endogenous products and, therefore, the residuals can be considered a determination of the true in vitro digestibility (TIVD) of dry matter. The first stage of ruminal incubation (48 h) following the Goering and Van Soest technique corresponds to the determination of dry matter degradability (IVdeg). Each technique was performed in duplicate (two bags per sample) and repeated in three runs in different weeks, giving six observations per sample.

2.5.2. Kinetics of Gas Production

Rumen fluid was extracted from four mature slaughtered Queue Fine de l'Ouest sheep (48.5 ± 4.3 kg body weight), collected in a thermos, and transported immediately to the laboratory where it was strained through various layers of cheesecloth and kept at 39 °C under a CO_2 atmosphere. A culture medium was prepared as described previously and the rumen fluid was diluted in the culture medium at the proportion 1:2 (*v:v*). Plant material samples (300 mg) were weighed in a glass syringe (capacity 100 mL), added to 30 mL of the culture medium, and incubated in a water bath. The volume of gas produced in the syringes was measured every 2 h (from 0 to 72 h). Data were fitted to the model proposed by France et al. [24]:

$$G = A\,[1 - e - c(t - L)] \tag{3}$$

where G (mL) denotes the cumulative gas production (GP) at time t; A (mL) the asymptotic gas production; c (h − 1) the fractional rate of gas production and L (h) is the lag time. Effective degradability (ED, g DM degraded/g DM ingested) for a given rate of passage (k, h^{-1}) was estimated following the approach derived by France et al. [24]. To calculate ED, a mean retention time of digesta in the rumen of 30 h was assumed, giving a rate of passage of $0.033\ h^{-1}$ (which can be found in sheep fed on a forage diet at maintenance level). The partitioning factor (PF) was calculated as the ratio between net GP and the degradation of the organic matter during 24 h, and was used as an indicator of microbial protein syntheses [25].

2.6. Calculations

Metabolizable energy (ME, MJ/kg DM) content was estimated using CP and EE contents (g/kg DM) and the volume of gas measured after 24 h of incubation (G24 in mL per 300 mg DM incubated), as described by Menke and Steingass [26]:

$$ME = 2.43 + 0.1206 \times G24 + 0.0069 \times CP + 0.0187 \times EE \tag{4}$$

The digested organic matter (DOM), protein values [dietary protein undegraded in the rumen (PDIA), true protein degraded in the small intestine (PDIN), and true protein absorbable in the small intestine (PDIE)], and net energy status (in terms of forage units for lactation (UFL) or meat production (UFV)) of Ajuga foliage were assessed according to the INRA [27] feed evaluation system. These were estimated from the feed characteristics (chemical composition and in vitro digestibility parameters) obtained in our study using the INRAtion software (V5, RUMIN'AL, Paris, France).

The partitioning factor (PF) was calculated as mg DM digested potential of degradability (D144)/mL gas production (A) [25], as an indicator of the efficiency of ruminal microbial protein synthesis.

2.7. Statistical Analysis

All data were analyzed by Tukey's test according to a split-plot design, with the whole plots arranged in a randomized block design. Statistics were carried out using the PROC GLM procedure of SAS (v. 9.2; SAS Institute Inc., Cary, NC, USA). The mean values of each parameter and the pooled standard error of the mean (S.E.M.) are reported in the tables. Differences between treatments were considered significant at $p < 0.05$ using Duncan's test.

3. Results

3.1. Chemical Composition and Phytochemical Contents

Table 1 shows the chemical composition of *A. iva* cultivated in different Tunisian regions. *A. iva* cultivated in Nabeul provided the highest ($p < 0.001$) values of Ash, NDF, ADF, and EE compared with the other regions, while that collected in Mograne had the highest ($p < 0.001$) CP value compared with the other two regions. On the other hand, the lowest ($p < 0.001$) values of CP and EE were observed for *A. iva* cultivated in the Dougga region.

Table 1. Chemical composition, active phytochemicals (g/kg dry matter), and anti-radical scavenging activity (µg/mL) of *A. iva* leaves cultivated in three different Tunisian regions.

Items	Regions			S.E.M.	*p*-Value
	Dougga	Mograne	Nabeul		
DM	892	898	905	2.94	0.060
Ash	165 [b]	155 [b]	244 [a]	3.07	<0.001
CP	81.7 [c]	134.5 [a]	102.4 [b]	2.27	<0.001
NDF	279 [b]	262 [b]	332 [a]	3.90	<0.001
ADF	212 [b]	202 [b]	274 [a]	4.34	<0.001
ADL	50.1	51.6	46.2	2.03	0.229
EE	10.8 [c]	11.2 [b]	12.2 [a]	0.072	<0.001
TF (mg CE [1]/mg DM [2])	0.34 [b]	0.17 [c]	0.93 [a]	0.020	<0.001
TP (mg GAE [3]/mg DM)	0.79 [a]	0.61 [c]	0.72 [b]	0.013	<0.001
ARSA (µg/mL)	485 [a]	343 [b]	71.2 [c]	6.283	<0.001

DM = Dry matter, CP = Crude protein, NDF = Neutral detergent fibre, ADF = Acid detergent fibre, ADL = Acid detergent lignin, EE = Ether extract, TF = Total flavonoid, TP = Total phenolic, ARSA = Anti-radical scavenging activity. [a,b,c] = letters within the same row, mean values not sharing a common superscript represent significant differences ($p < 0.05$), S.E.M. = Standard error of the mean. [1] CE = Catechin equivalents, [2] DM = Dry mater, [3] GAE = Gallic acid equivalents.

Results of TF, TP, and anti-radical scavenging activity "ARSA" of *A. iva* samples collected from different regions of Tunisia generally showed that Ajuga is a rich phytochemical plant (Table 1). Highly significant values ($p < 0.001$) of TP and ARSA were recorded in *A. iva* from Dougga, while those collected from Nabeul had the highest ($p < 0.001$) TF compared with *A. iva* from other regions. Leaves collected from Mograne had the lowest TF and TP, while samples collected from Nabeul resulted in the lowest ($p < 0.001$) ARSA compared with other plant samples.

3.2. In Vitro DM Digestibility (IVDMD)

As shown in Table 2, the highest values ($p < 0.05$) of in vitro DM digestibility (IVDMD) measured either by the Tilley and Terry or Van Soest et al. methods [23] were observed in *A. iva* collected from Mograne and Dougga, while leaves collected from Nabeul had the lowest ($p < 0.01$) nutrient digestibility values.

Table 2. In vitro dry matter degradability of *Ajuga. iva* leaves cultivated from three different regions in Tunisia.

Parameters	Regions			S.E.M.	*p*-Value
	Dougga	Mograne	Nabeul		
AIVD (g/kg)	699 [a,b]	741 [a]	631 [b]	1.58	0.008
IVdeg (g/kg)	577 [a]	599 [a]	491 [b]	1.69	0.009
TIVD (g/kg)	776 [a]	793 [a]	679 [b]	1.04	<0.001
ED	451 [a]	492 [a]	392 [b]	1.18	0.003

AIVD = Apparent in vitro dry matter digestibility, IVdeg = in vitro degradability of dry matter, TIVD = true in vitro digestibility, ED = effective degradability. [a,b] = letters within the same line, mean values not sharing a common superscript represent significant differences ($p < 0.05$), S.E.M. = Standard error of the mean.

3.3. Gas Production Kinetics and Energy Status

The gas emitted from leaves of *A. iva* collected from three different regions incubated at different times (0–72 h) is illustrated in Table 3. *A. iva* collected from Nabeul had the lowest ($p < 0.05$) GP at 8, 16, and 24 h incubation times, the lowest yield of GP calculated at 24 h (GY24), and lowest average rate (AR) of GP compared with those collected from Dougga and Mograne regions.

Table 3. In vitro cumulative gas production (mL/g DM) and gas kinetics through 72 h of incubation of *Ajuga. iva* cultivated in three different regions in Tunisia.

Items	Regions			S.E.M.	*p*-Value
	Dougga	Mograne	Nabeul		
8 h	112 [a]	113 [a]	100 [b]	2.3	0.012
16 h	163 [a]	167 [a]	147 [b]	1.9	<0.001
24 h	226 [a]	210 [b]	199 [b]	2.9	0.002
36 h	286	250	256	10.9	0.119
48 h	312	277	279	11.4	0.122
72 h	323	292	291	12.3	0.189
A	344	295	311	17.3	0.209
c	0.046	0.054	0.046	0.003	0.202
GY24	231 [a]	215 [a,b]	204 [b]	4.1	0.012
AR	11.5 [a]	11.6 [a]	10.2 [b]	0.18	0.002
PF	2.26	2.69	2.22	0.125	0.069

A = Asymptotic gas production (mL/g DM incubated), c = Fractional rate of fermentation (/h), GY24 = Volume of fermentation gas produced at 24 h incubation (mL/g DM incubated), AR = Average gas production rate (mL/g DM per h), PF = Partitioning factor. [a,b] = letters within the same row, mean values not sharing a common superscript represent significant differences ($p < 0.05$); S.E.M. = Standard error of the mean.

Table 4 shows the ME, DOM, net energetic values (UFL and UFV), and protein values (PDIA, PDIN, PDIE) of *A. iva* leaves cultivated from three different regions. *A. iva* cultivated in Nabeul showed the lowest ($p < 0.001$) ME, UFL, and UFV values compared with samples collected in Dougga and Mograne. However, the highest ($p < 0.001$) PDIA (41.7 g/kg DM), PDIN (84.7 g/kg DM), and PDIE (87.7 g/kg DM) were recorded in *A. iva* cultivated in Mograne, compared with those cultivated in Nabeul and Dougga. As compared with those cultivated in other regions, samples of *A. iva* cultivated in Mograne tended to have the highest PF values ($p = 0.06$).

Table 4. Metabolizable energy (MJ/kg DM), digested organic matter (g/kg DM), energetic value (kg DM), and protein value (g/kg DM) of *A. iva* leaves cultivated from three different regions in Tunisia.

Parameters	Regions			S.E.M.	*p*-Value
	Dougga	Mograne	Nabeul		
ME (MJ/kg DM)	8.59 [a]	8.51 [a]	8.08 [b]	0.064	0.003
DOM (g/kg)	673	670	664	0.44	0.449
UFL (MJ/kg DM)	0.71 [a]	0.70 [a]	0.61 [b]	0.006	<0.001
UFV (MJ/kg DM)	0.64 [a]	0.63 [a]	0.54 [b]	0.008	<0.001
Protein values (g/kg DM)					
PDIA	25.7 [c]	41.7 [a]	31.7 [b]	0.81	<0.001
PDIN	51.3 [c]	84.7 [a]	65.0 [b]	1.44	<0.001
PDIE	73.0 [b]	87.7 [a]	73.3 [b]	0.64	<0.001

ME = Metabolizable energy, DOM = Digested organic matter, UFL = Net energy for lactation, UFV = Net energy for meat production, PDIA = Dietary protein undegraded in the rumen, PDIN = True protein digested in small intestine, PDIE = True protein absorbable in the small intestine. [a,b,c] = letters within the same row, mean values not sharing a common superscript represent significant differences ($p < 0.05$), S.E.M. = Standard error of the mean.

4. Discussion

4.1. Chemical Composition

The CP concentration of *A. iva* ranged between 8.2 and 13.5%, which is within the acceptable range reported for different foliage plants [28], and was above the minimum threshold of 80 g/kg DM required for rumen microbial growth and activity [29]. *A. iva* cultivated in Dougga had the lowest CP concentration ($p < 0.05$), while that cultivated in Mograne had the highest CP (13.5%). Irrespective of the region, based on CP, it appears that leaves of *A. iva* were at least comparable in value to most traditional Mediterranean legume forages such as lucerne hay [30]. Therefore, the significant contribution of such pastoral plants would suggest their potential for overcoming feed limitations for ruminant livestock in Mediterranean regions, especially during the drought season, justifying their use to complement poor-quality pastures and crop residues [30]. However, it is supposed that some nitrogenous compounds are encrusted in the cell wall structure [31], and consequently, the utilization of CP by animals may not be as high as expected. Thus, the chemical composition of these browse species should not be the sole criterion for judging the relative importance of a particular species. Concerning cell wall fractions (NDF and ADF), *A. iva* cultivated in Mograne had the lowest values ($p < 0.05$), at 26% and 20% for NDF and ADF, respectively. The variability observed between cultivated regions could be due to differences in climatic conditions, soil types, soil fertility, agronomical management, and other environmental factors [30,32,33]. In this context, Mountousis et al. [34] reported that NDF and ADF content of forages were affected by the altitudinal zone and the season.

4.2. Bioactive Phytochemicals and Antioxidant Activity of A. iva

Our present study shows that secondary metabolites (types and concentrations) varied widely with the site of *A. iva* cultivation. The environmental conditions across the three collection sites are the most probable causes of variations in the plant phytochemicals [35]. In the present experiment, these differences also resulted in variations of antioxidant activity. The differences between regions are related to many factors including differences in meters above sea level, soil type, soil chemical composition, erosion status, management systems, and other related aspects [33,36]. Moreover, the differences in AA between regions and extraction methods could be associated with differences in active ingredients due to different concentrations of phenolic compounds in *A. iva* [14].

Free-radical scavenging is one known mechanism by which antioxidants inhibit lipid oxidation [9]. In the present study, the AA differed between regions and from other studies examining the leaves of *A. iva* harvested in Tunisia [9]. In the current study, we used ethanolic extraction; thus, the solvent extraction method seems to be the main reason for the differences between studies [37]. Extracts with high polarities, such as ethanol, give better results than weakly polar solvents such as petroleum ether or methanol. The primary function of plant secondary metabolites is defense against different environmental threats. Therefore, concentrations of plant secondary metabolites are expected to differ between cultivation zones. Extraction is the foremost step for recovering and isolating phytochemicals from plant materials, and the concentration of phytochemicals in plants depends on plant samples' physical properties and the solvent's polarity [34,35]. Extraction efficiency is affected by the chemical nature of phytochemicals, the extraction method, and the solvent used [5,38]. The sensitivity of the chemical method used to quantify the phenolic compounds and the nature of the standard can affect concentrations in the same sample. Makni et al. [9] observed that the extraction yield of *A. iva* differed between methanol, aqueous, hexane, and chloroform extractions. For *A. iva*, Bendif et al. [39] observed different concentrations of total phenolics and free-radical scavenging activity with different extraction methods (acetone, ethanol, and water). Ouerghemmi et al. [40] compared the phenolic composition and antioxidant properties of methanol and ethyl acetate extracts from leaves of *Rosa canina*, *Rosa sempervirens*, and *Rosa moschata* collected from different Tunisian regions and observed differed yields. Higher phenolic compounds indicate higher antioxidant activity (i.e., low free-radical scavenging activity).

Phenolic compounds are critical components in plant samples, and their ability to scavenge free radicals is due to their hydroxyl groups [35]. The highest free-radical scavenging activity was observed for *A. iva* cultivated in Dougga and Mograne. It has been proven that levels of total phenols and flavonoids are high when the living environment of the plant is not appropriate. In this case, the plant promotes the synthesis of secondary metabolites to adapt and survive.

4.3. In Vitro DM Digestibility and Kinetics of Gas Production

DM degradability differed among plant samples cultivated in different regions. The in vitro DM digestibility measured by the Van Soest et al. method [22] (TIVD) or by Tilley and Terry's method [21] (AIVD) for *A. iva* from different regions was within the range (36 to 69%) of in vitro DM digestibility observed for most browse plants [41]. The digestibility of DM determined using Goering and Van Soest's method was high for *A. iva* cultivated in Mograne, while the lowest value was recorded for that cultivated in the Nabeul region. The different results obtained by different methods of DM digestibility determination could be related to the conditions of each determination method. In vitro methods such as in vitro digestibility and gas production measurements are more reliable for detecting inhibitory compounds in feeds, because these compounds are likely to affect the activity of rumen microbes in a closed system [42,43]. As previously observed by Ammar et al. [44], using the in vitro gas production technique is preferred to other in vitro methods for estimating digestibility [26]. Moreover, in vitro gas production is very suitable for assessing the biological activity of tannins and other anti-nutritional factors affecting the digestibility of browse plants [44,45]. In the present experiment, *A. iva* cultivated in Nabeul showed the lowest values of AIVD, TIVD, degradability potential, and effective degradability compared with samples cultivated in Dougga and Mograne.

The kinetics of the in vitro GP differed between regions. Gas production is a good indicator of the ruminal fermentability of feeds [46,47]. It depends mainly on the degradability of soluble components in the incubated substrates, and the partitioning of fermented substrates to volatile fatty acids and microbial biomass production [25,48]. During the first 24 h of incubation, *A. iva* from Dougga and Mograne regions produced higher gas levels, a higher average rate of gas production, and higher gas yields at 24 h; however, the asymptote and the rate of gas production were not significantly affected, indicating different fermentability between *A. iva* from different regions. Differences could be due to variations in the chemical composition and nutrient degradability [49] of the *Ajuga* cultivated in different zones. In the present experiment, the asymptotic gas production followed the same trend as OM and CP content and in vitro digestibility, conversely to the fiber content in *A. iva*, which confirms the results obtained by Ammar et al. [44,50]. They observed significant positive correlations between in vitro digestibility, GP parameters, and CP content, and negative correlations with NDF, ADF, and lignin contents. Furthermore, other factors including non-soluble carbohydrate fractions and phytochemicals affect the production of gases [51]. In the present experiment, the insignificantly different partitioning factor indicates similar efficiency of ruminal microbial protein synthesis [25].

The higher NDF and ADF concentrations in *A. iva* collected from Nabeul may be the main reason for the low degradability revealed in the GP experiment. The observed greater ED of *A. iva* cultivated in Dougga and Mograne, compared to that in Nabeul, is an indicator of how well it can be utilized by ruminants. Differences in ED may be attributed to chemical composition, particularly the structural and non-structural protein and carbohydrate fractions [52–54].

The low values of UFL and UFV indicate low energy availability for milk and meat production for animals consuming *A. iva* cultivated in Nabeul, compared to those in the Dougga and Mograne regions [27]. Moreover, the measured parameters of protein value indicate that *A. iva* cultivated in Dougga and Nabeul had lower nutritive protein value compared to that cultivated in the Mograne region [27]. Greater concentrations of protein undegraded in the rumen but truly digestible in the small intestine, as well as true protein absorbable in the small intestine when rumen fermentable energy is limited, are good indicators of high nutritive value and are important from a nutritional view as lower degradability at the beginning of incubation indicates greater bypass protein that can be utilized in the duodenum [55,56]. Microorganisms could more easily attach to better degradable protein in the rumen and reflect greater protein solubility [57].

5. Conclusions

Based on the chemical composition and the in vitro digestibility results, it seems that *A. iva* could be successfully used to complement protein deficiencies in the diet of ruminants during periods of feed scarcity. The nutritive value of *A. iva* greatly varies between geographical zones, suggesting a need for season- and region-specific feeding strategies. Further studies are needed to evaluate its palatability and demonstrate its efficacy in vivo. Studies are ongoing examining other biochemical activities of *A. iva* to demonstrate its medicinal properties.

Author Contributions: Conceptualization, H.A.; methodology, H.A., A.E.K., H.S. and S.L.; software, H.A.; validation, H.A., A.E.K., Y.A.S., A.S.M. and M.I.A.; formal analysis, A.E.K., S.O. and S.L.; investigation, H.A., A.E.K., Y.A.S. and A.S.M.; resources, H.A.; data curation, H.A., A.E.K., Y.A.S., S.O., S.H., E.H.H. and S.L.; writing—original draft preparation, H.A. and A.E.K.; writing—review and editing, H.A., A.E.K., Y.A.S., A.S.M., S.H., M.C. and M.E.d.H.M.; visualization, H.A., A.E.K., M.C., M.E.d.H.M., W.S. and S.L.; supervision, H.A. and M.I.A.; project administration, H.A.; funding acquisition, W.S. All authors have read and agreed to the published version of the manuscript.

Funding: This research received no external funding.

Institutional Review Board Statement: The study was conducted in accordance with the Declaration of Helsinki, and approved by the Institutional Review Board (or Ethics Committee) of University of Leon, Spain. The protocol was designed and all practices were performed according to the Directives 2010/63/EU of the European Parliament and of the Council of 22 September 2010 on the protection of the animals used for scientific purposes.

Informed Consent Statement: Not applicable.

Data Availability Statement: The data presented in this study are available on request from the corresponding author.

Acknowledgments: The authors want to thank the Researchers Supporting Project number (RSP2021/390), King Saud University, Riyadh, Saudi Arabia.

Conflicts of Interest: The authors declare no conflict of interest.

References

1. Kholif, A.E.; Gouda, G.A.; Morsy, T.A.; Salem, A.Z.M.; Lopez, S.; Kholif, A.M. *Moringa oleifera* Leaf Meal as a Protein Source in Lactating Goat's Diets: Feed Intake, Digestibility, Ruminal Fermentation, Milk Yield and Composition, and Its Fatty Acids Profile. *Small Rumin. Res.* **2015**, *129*, 129–137. [CrossRef]
2. Gutiérrez-Alamo, A.; Pérez De Ayala, P.; Verstegen, M.W.A.; Den Hartog, L.A.; Villamide, M.J. Variability in Wheat: Factors Affecting Its Nutritional Value. *World's Poult. Sci. J.* **2008**, *64*, 20–39. [CrossRef]
3. Onyango, A.A.; Dickhoefer, U.; Rufino, M.C.; Butterbach-Bahl, K.; Goopy, J.P. Temporal and Spatial Variability in the Nutritive Value of Pasture Vegetation and Supplement Feedstuffs for Domestic Ruminants in Western Kenya. *Asian-Australas. J. Anim. Sci.* **2019**, *32*, 637–647. [CrossRef]
4. Khalaj, M.; Amiri, M.; Azimi, M.H. Allelopathy: Physiological and Sustainable Agriculture Important Aspects. *Int. J. Agron. Plant Prod.* **2013**, *4*, 950–962.
5. Dallali, S.; Rouz, S.; Aichi, H.; Hassine, H.B. Phenolic Content and Allelopathic Potential of Leaves and Rhizosphere Soil aqueous Extracts of White Horehound (*Maribum Vulgare* L.). *J. New Sci. Agric. Biotechnol.* **2017**, *39*, 2106–2120.
6. Gulzar, A.; Siddiqui, M.B. Root-Mediated Allelopathic Interference of Bhringraj (*Eclipta Alba* L.) Hassk. on Peanut (*Arachis Hypogaea*) and Mung Bean (*Vigna Radiata*). *Appl. Soil Ecol.* **2015**, *87*, 72–80. [CrossRef]
7. Bondì, M.L.; Al-Hillo, M.R.Y.; Lamara, K.; Ladjel, S.; Bruno, M.; Piozzi, F.; Simmonds, M.S.J. Occurrence of the Antifeedant 14,15-Dihydroajugapitin in the Aerial Parts of *Ajuga iva* from Algeria. *Biochem. Syst. Ecol.* **2000**, *28*, 1023–1025. [CrossRef]
8. Chenni, A.; Yahia, D.A.; Boukortt, F.O.; Prost, J.; Lacaille-Dubois, M.A.; Bouchenak, M. Effect of Aqueous Extract of *Ajuga iva* Supplementation on Plasma Lipid Profile and Tissue Antioxidant Status in Rats Fed a High-Cholesterol Diet. *J. Ethnopharmacol.* **2007**, *109*, 207–213. [CrossRef]
9. Makni, M.; Haddar, A.; Kriaa, W.; Zeghal, N. Antioxidant, Free Radical Scavenging, and Antimicrobial Activities of *Ajuga Iva* Leaf Extracts. *Int. J. Food Prop.* **2013**, *16*, 756–765. [CrossRef]
10. Israili, Z.H.; Lyoussi, B. Ethnopharmacology of the Plants of Genus Ajuga. *Pak. J. Pharm. Sci.* **2009**, *22*, 425–462. [PubMed]
11. Houghton, P.J.; Raman, A. *Laboratory Handbook for the Fractionation of Natural Extracts*; Chapman and Hall: London, UK, 1998.
12. Kholif, A.E.; Gouda, G.A.; Anele, U.Y.; Galyean, M.L. Extract of *Moringa oleifera* Leaves Improves Feed Utilization of Lactating Nubian Goats. *Small Rumin. Res.* **2018**, *158*, 69–75. [CrossRef]
13. Elghandour, M.M.Y.; Kholif, A.E.; Salem, A.Z.M.; Olafadehan, O.A.; Kholif, A.M. Sustainable Anaerobic Rumen Methane and Carbon Dioxide Productions from Prickly Pear Cactus Flour by Organic Acid Salts Addition. *J. Clean. Prod.* **2016**, *139*, 1362–1369. [CrossRef]
14. Bouyahya, A.; El Omari, N.; Belmehdi, O.; Lagrouh, F.; El Jemli, M.; Marmouzi, I.; Faouzi, M.E.A.; Taha, D.; Bourais, I.; Zengin, G.; et al. Pharmacological Investigation of *Ajuga iva* Essential Oils Collected at Three Phenological Stages. *Flavour Fragr. J.* **2021**, *36*, 75–83. [CrossRef]
15. AOAC. *Official Methods of Analysis of AOAC International*, 16th ed.; AOAC International: Washington, DC, USA, 2005; ISBN 0935584544.
16. Ammar, H.; López, S.; Bochi-Brum, O.; García, R.; Ranilla, M. Composition and in Vitro Digestibility of Leaves and Stems of Grasses and Legumes Harvested from Permanent Mountain Meadows at Different Stages of Maturity. *J. Anim. Feed Sci.* **1999**, *8*, 599–610. [CrossRef]
17. Van Soest, P.J.; Robertson, J.B.; Lewis, B.A. Methods for Dietary Fiber, Neutral Detergent Fiber, and Nonstarch Polysaccharides in Relation to Animal Nutrition. *J. Dairy Sci.* **1991**, *74*, 3583–3597. [CrossRef]
18. Neffati, N.; Aloui, Z.; Karoui, H.; Guizani, I.; Boussaid, M.; Zaouali, Y. Phytochemical Composition and Antioxidant Activity of Medicinal Plants Collected from the Tunisian Flora. *Nat. Prod. Res.* **2017**, *31*, 1583–1588. [CrossRef]
19. Singleton, V.L.; Orthofer, R.; Lamuela-Raventós, R.M. Analysis of Total Phenols and Other Oxidation Substrates and Antioxidants by Means of Folin-Ciocalteu Reagent. In *Methods in Enzymology*; Academic Press Inc.: Cambridge, MA, USA, 1999; Volume 299, pp. 152–178.
20. Dewanto, V.; Wu, X.; Adom, K.K.; Liu, R.H. Thermal Processing Enhances the Nutritional Value of Tomatoes by Increasing Total Antioxidant Activity. *J. Agric. Food Chem.* **2002**, *50*, 3010–3014. [CrossRef]
21. Tilley, J.M.A.; Terry, R.A. A Two-Stage Technique for the in Vitro Digestion of Forage Crops. *Grass Forage Sci.* **1963**, *18*, 104–111. [CrossRef]
22. Van Soest, P.J.; Wine, R.H.; Moore, L.A. Estimation of the True Digestibility of Forages by the in Vitro Digestion of Cell Walls. In Proceedings of the 10th International Grassland Congress, Helsinki, Finland, 7–16 July 1966; pp. 438–441.
23. Van Soest, P.J. *Nutritional Ecology of the Ruminant*; Cornell University Press: Ithaca, NY, USA, 2019; ISBN 9781501732355.
24. France, J.; Dijkstra, J.; Dhanoa, M.S.; Lopez, S.; Bannink, A. Estimating the Extent of Degradation of Ruminant Feeds from a Description of Their Gas Production Profiles Observed in Vitro: Derivation of Models and Other Mathematical Considerations. *Br. J. Nutr.* **2000**, *83*, 143–150. [CrossRef]
25. Blümmel, M.; Makkar, H.P.S.; Becker, K. In Vitro Gas Production: A Technique Revisited. *J. Anim. Physiol. Anim. Nutr.* **1997**, *77*, 24–34. [CrossRef]

26. Menke, K.H.; Steingass, H. Estimation of the Energetic Feed Value Obtained from Chemical Analysis and in Vitro Gas Production Using Rumen Fluid. *Anim. Res. Dev.* **1988**, *28*, 7–55.

27. INRA. *INRA Feeding System for Ruminants*; Wageningen Academic Publishers: Wageningen, The Netherlands, 2018; ISBN 978-90-8686-292-4.

28. Lee, M.A. *A Global Comparison of the Nutritive Values of Forage Plants Grown in Contrasting Environments*; Springer: Tokyo, Japan, 2018; Volume 131, pp. 641–654.

29. Soliva, C.R.; Amelchanka, S.L.; Kreuzer, M. The Requirements for Rumen-Degradable Protein per Unit of Fermentable Organic Matter Differ between Fibrous Feed Sources. *Front. Microbiol.* **2015**, *6*, 715. [CrossRef] [PubMed]

30. Soltan, Y.A.; Morsy, A.S.; Lucas, R.C.; Abdalla, A.L. Potential of Mimosine of *Leucaena leucocephala* for Modulating Ruminal Nutrient Degradability and Methanogenesis. *Anim. Feed Sci. Technol.* **2017**, *223*, 30–41. [CrossRef]

31. Shayo, C.M.; Udén, P. Nutritional Uniformity of Neutral Detergent Solubles in Some Tropical Browse Leaf and Pod Diets. *Anim. Feed Sci. Technol.* **1999**, *82*, 63–73. [CrossRef]

32. Dönmez, H.B.; Uzun, F. Geographical Variation in Nutrient Composition of *Lotus Tenuis* (Walds.&Kit.) Populations from Seeds Collected from Different Locations. *Türkiye Tarımsal Araştırmalar Derg.* **2015**, *3*, 30–36. [CrossRef]

33. Metoui, M.; Essid, A.; Bouzoumita, A.; Ferchichi, A. Chemical Composition, Antioxidant and Antibacterial Activity of Tunisian Date Palm Seed. *Pol. J. Environ. Stud.* **2019**, *28*, 267–274. [CrossRef]

34. Mountousis, I.; Dotas, V.; Stanogias, G.; Papanikolaou, K.; Roukos, C.H.; Liamadis, D. Altitudinal and Seasonal Variation in Herbage Composition and Energy and Protein Content of Grasslands on Mt Varnoudas, NW Greece. *Anim. Feed Sci. Technol.* **2011**, *164*, 174–183. [CrossRef]

35. Kholif, A.E.; Olafadehan, O.A. Essential Oils and Phytogenic Feed Additives in Ruminant Diet: Chemistry, Ruminal Microbiota and Fermentation, Feed Utilization and Productive Performance. *Phytochem. Rev.* **2021**, *20*, 1087–1108. [CrossRef]

36. Yisehak, K.; Janssens, G.P.J. Evaluation of Nutritive Value of Leaves of Tropical Tanniferous Trees and Shrubs. *Livest. Res. Rural Dev.* **2013**, *25*, 28.

37. Senhaji, S.; Lamchouri, F.; Bouabid, K.; Assem, N.; El Haouari, M.; Bargach, K.; Toufik, H. Phenolic Contents and Antioxidant Properties of Aqueous and Organic Extracts of a Moroccan *Ajuga iva Subsp.* Pseudoiva. *J. Herbs Spices Med. Plants* **2020**, *26*, 248–266. [CrossRef]

38. Ksouri, R.; Megdiche, W.; Falleh, H.; Trabelsi, N.; Boulaaba, M.; Smaoui, A.; Abdelly, C. Influence of Biological, Environmental and Technical Factors on Phenolic Content and Antioxidant Activities of Tunisian Halophytes. *Comptes Rendus Biol.* **2008**, *331*, 865–873. [CrossRef] [PubMed]

39. Bendif, H.; Lazali, M.; Harir, M.; Miara, M.D.; Boudjeniba, M.; Venskutonis, P.R. Biological Screening of *Ajuga Iva* Extracts Obtained by Supercritical Carbon Dioxide and Pressurized Liquid Extraction. *J. Med. Bot.* **2017**, *1*, 33. [CrossRef]

40. Ouerghemmi, S.; Sebei, H.; Siracusa, L.; Ruberto, G.; Saija, A.; Cimino, F.; Cristani, M. Comparative Study of Phenolic Composition and Antioxidant Activity of Leaf Extracts from Three *Wild rosa* Species Grown in Different Tunisia Regions: *Rosa Canina* L., *Rosa Moschata Herrm.* and *Rosa Sempervirens* L. *Ind. Crop. Prod.* **2016**, *94*, 167–177. [CrossRef]

41. Sawe, J.J.; Tuitoek, J.K.; Ottaro, J.M. Evaluation of Common Tree Leaves or Pods as Supplements for Goats on Range Area of Kenya. *Small Rumin. Res.* **1998**, *28*, 31–37. [CrossRef]

42. Khazaal, K.; Boza, J.; Ørskov, E.R. Assessment of Phenolics-Related Antinutritive Effects in Mediterranean Browse: A Comparison between the Use of the in Vitro Gas Production Technique with or without Insoluble Polyvinylpolypyrrolidone or Nylon Bag. *Anim. Feed Sci. Technol.* **1994**, *49*, 133–149. [CrossRef]

43. El Hassan, S.M.; Lahlou Kassi, A.; Newbold, C.J.; Wallace, R.J. Chemical Composition and Degradation Characteristics of Foliage of Some African Multipurpose Trees. *Anim. Feed Sci. Technol.* **2000**, *86*, 27–37. [CrossRef]

44. Ammar, H.; López, S.; González, J.S.; Ranilla, M.J. Chemical Composition and in Vitro Digestibility of Some Spanish Browse Plant Species. *J. Sci. Food Agric.* **2004**, *84*, 197–204. [CrossRef]

45. Ammar, H.; López, S.; González, J.S.; Ranilla, M.J. Comparison between Analytical Methods and Biological Assays for the Assessment of Tannin-Related Antinutritive Effects in Some Spanish Browse Species. *J. Sci. Food Agric.* **2004**, *84*, 1349–1356. [CrossRef]

46. Cedillo, J.; Vázquez-Armijo, J.F.; González-Reyna, A.; Salem, A.Z.M.; Kholif, A.E.; Hernández-Meléndez, J.; Martínez-González, J.C.; de Oca Jiménez, R.M.; Rivero, N.; López, D. Effects of Different Doses of *Salix babylonica* Extract on Growth Performance and Diet in Vitro Gas Production in Pelibuey Growing Lambs. *Ital. J. Anim. Sci.* **2014**, *13*, 609–613. [CrossRef]

47. Salem, A.Z.M.; Kholif, A.E.; Elghandour, M.M.Y.; Hernandez, S.R.; Domínguez-Vara, I.A.; Mellado, M. Effect of Increasing Levels of Seven Tree Species Extracts Added to a High Concentrate Diet on in Vitro Rumen Gas Output. *Anim. Sci. J.* **2014**, *85*, 853–860. [CrossRef]

48. Elghandour, M.M.Y.; Kholif, A.E.; Hernández, A.; Salem, A.Z.M.; Mellado, M.; Odongo, N.E. Effects of Organic Acid Salts on Ruminal Biogas Production and Fermentation Kinetics of Total Mixed Rations with Different Maize Silage to Concentrate Ratios. *J. Clean. Prod.* **2017**, *147*, 523–530. [CrossRef]

49. Elghandour, M.M.Y.; Kholif, A.E.; Salem, A.Z.M.; Montes de Oca, R.; Barbabosa, A.; Mariezcurrena, M.; Olafadehan, O.A. Addressing Sustainable Ruminal Methane and Carbon Dioxide Emissions of Soybean Hulls by Organic Acid Salts. *J. Clean. Prod.* **2016**, *135*, 194–200. [CrossRef]

50. Ammar, H.; López, S.; González, J.S.; Ranilla, M.J. Seasonal Variations in the Chemical Composition and in Vitro Digestibility of Some Spanish Leguminous Shrub Species. *Anim. Feed Sci. Technol.* **2004**, *115*, 327–340. [CrossRef]
51. Bureenok, S.; Langsoumechai, S.; Pitiwittayakul, N.; Yuangklang, C.; Vasupen, K.; Saenmahayak, B.; Schonewille, J.T. Effects of Fibrolytic Enzymes and Lactic Acid Bacteria on Fermentation Quality and in Vitro Digestibility of Napier Grass Silage. *Ital. J. Anim. Sci.* **2019**, *18*, 1438–1444. [CrossRef]
52. Yang, S.H. Pyridinium Hydrobromide Perbromide: A Versatile Reagent in Organic Synthesis. *Synlett* **2009**, *58*, 1351–1352. [CrossRef]
53. Gusha, J.; Ngongoni, N.T.; Halimani, T.E. Nutritional Composition and Effective Degradability of Four Forage Trees Grown for Protein Supplementation. *J. Anim. Feed Res. Sci.* **2013**, *3*, 170–175.
54. Jabri, J.; Ammar, H.; Abid, K.; Beckers, Y.; Yaich, H.; Malek, A.; Rekhis, J.; Morsy, A.S.; Soltan, Y.A.; Soufan, W.; et al. Effect of Exogenous Fibrolytic Enzymes Supplementation or Functional Feed Additives on In Vitro Ruminal Fermentation of Chemically Pre-Treated Sunflower Heads. *Agriculture* **2022**, *12*, 696. [CrossRef]
55. Barry, T.N.; Manley, T.R. Interrelationships between the Concentrations of Total Condensed Tannin, Free Condensed Tannin and Lignin in Lotus Sp. and Their Possible Consequences in Ruminant Nutrition. *J. Sci. Food Agric.* **1986**, *37*, 248–254. [CrossRef]
56. Kumar, R.; D'mello, J.P.F.; Devendra, C. Anti-Nutritional Factors in Forage Legumes. In *Tropical Legumes in Animal Nutrition*; D'Mello, F.J.P., Devendra, C., Eds.; CAB International: Wallingford, UK, 1995; pp. 95–133.
57. Ebeid, H.M.; Kholif, A.E.; Chrenkova, M.; Anele, U.Y. Ruminal Fermentation Kinetics of *Moringa oleifera* Leaf and Seed as Protein Feeds in Dairy Cow Diets: In Sacco Degradability and Protein and Fiber Fractions Assessed by the CNCPS Method. *Agrofor. Syst.* **2020**, *94*, 905–915. [CrossRef]

Review

Antimicrobial and Digestive Effects of *Yucca schidigera* Extracts Related to Production and Environment Implications of Ruminant and Non-Ruminant Animals: A Review

Aracely Zúñiga-Serrano [1], Hugo B. Barrios-García [1], Robin C. Anderson [2], Michael E. Hume [2], Miguel Ruiz-Albarrán [1], Yuridia Bautista-Martínez [1], Nadia A. Sánchez-Guerra [1], José Vázquez-Villanueva [1], Fidel Infante-Rodríguez [1] and Jaime Salinas-Chavira [1],*

[1] College of Veterinary Medicine and Animal Science, Universidad Autónoma de Tamaulipas, Ciudad Victoria 87000, Mexico
[2] Food and Feed Safety Research Unit, Southern Plains Agricultural Research Center, Agricultural Research Service, United States Department of Agriculture, College Station, TX 77845, USA
* Correspondence: jsalinasc@hotmail.com

Abstract: Plant extracts have been used over time in traditional medicine, mainly for their antimicrobial activity as well as for their medicinal effects. Plant-derived products contain secondary metabolites that prevent pathogenic microbial growth similar to conventional medicines. These secondary metabolites can enhance animal health and production in a more natural or organic manner and may contribute to the reduction in the use of pharmacological drugs in animal feed, which is of great concern for emerging microbial resistance. Plant secondary metabolites can be cost effective, while improving the production efficiency of ruminants, non-ruminants, and aquatic food animals. Among the plant-derived products is the *Yucca schidigera* extract (YSE), containing steroidal saponins as their main active component. YSE has multiple biological effects, including inhibition of some pathogenic bacteria, protozoa, and nematodes. YSE is used to control odor and ammonia and consistently enhance poultry production by enhancing intestinal health and function. In pigs, results are as yet inconclusive. In ruminants, YSE works against protozoa, has selective action against bacteria, and reduces the archaea populations; all these effects are reflected in the reduction in emissions of polluting gases, mainly methane, although the effects are not observed in all feeding conditions. These effects of YSE are discussed in this review. YSE has potential as a natural feed additive for sustainable animal production while contributing to the mitigation of contaminant gas emissions.

Keywords: *Yucca schidigera*; antimicrobial; secondary metabolites; sustainability; pollution; production; food animals

Citation: Zúñiga-Serrano, A.; Barrios-García, H.B.; Anderson, R.C.; Hume, M.E.; Ruiz-Albarrán, M.; Bautista-Martínez, Y.; Sánchez-Guerra, N.A.; Vázquez-Villanueva, J.; Infante-Rodríguez, F.; Salinas-Chavira, J. Antimicrobial and Digestive Effects of *Yucca schidigera* Extracts Related to Production and Environment Implications of Ruminant and Non-Ruminant Animals: A Review. *Agriculture* **2022**, *12*, 1198. https://doi.org/10.3390/agriculture12081198

Academic Editor: Daniel Simeanu

Received: 13 July 2022
Accepted: 8 August 2022
Published: 11 August 2022

Publisher's Note: MDPI stays neutral with regard to jurisdictional claims in published maps and institutional affiliations.

1. Introduction

For several years, antibacterial drugs have been routinely used in diets to enhance food animal production efficiency in intensive systems; nevertheless, there is concern about increased bacterial resistance to these drugs [1,2]. As a consequence, alternative products are sought to prevent the proliferation of pathogenic microorganisms, as well as those that can enhance food animal production. Among these alternatives are plant extracts that may offer fewer adverse effects than conventional drugs and have similar or superior effects against pathogenic microorganisms to those of traditional drugs [3–5].

Much of the research on natural products to prevent or control infections is primarily geared towards plant extracts and essential oils, several of them being related to phenolic compounds, terpenes, saponins, tannins, flavonoids, organic acids, complex carbohydrates, and non-protein amino acids [6–8]. Antimicrobial effects of plant secondary metabolites include cell membrane disruption, enzyme inhibition, substrate deprivation, and prevention of bacterial colonization. Plant extracts can contain many different secondary metabolites that can have a synergistic biological effect when used in combination with other plant extracts or with other feed additives used to promote growth in livestock, such as probiotics and organic acids [9,10]. The antimicrobial compounds obtained from secondary metabolites of plants have potential use in ruminants, non-ruminants, and aquatic organisms [11] and are a real alternative to traditional antibiotics to improve the health, productive performance, and quality of food animals.

Different research works are being conducted for agricultural systems sustainability [12,13]. In this aspect, diet formulations must provide nutrients for optimal growth of animals, and the inclusion of additives extracted from plants has potential as a common practice to enhance animal production efficiency [14]. Among these extracts is the one obtained from *Yucca schidigera* (YSE), which has antibacterial, antiprotozoal, and antifungal effects [15–17]. YSE contains saponins and polyphenols; however, the main biological effects are associated with the steroidal fraction of saponins [18–23]. The saponins trap ammonia-nitrogen, reducing odor and ammonia emissions in pig and poultry production [10,24,25], as well as demonstrating influence on cat and dog fecal metabolites and microorganisms [26] and in reducing ammonia accumulations in aquaculture environments [27–30].

In ruminants, feed supplementation with YSE reduces gas emissions that pollute the environment. The positive effects of YSE on methane and ammonia mitigation in livestock are well documented [31]. The saponins in YSE reduce the methanogenic archaea (methanogens) in the rumen, mainly reducing the protozoa associated with methanogens [32]. Additionally, YSE also inhibits the growth of some rumen bacteria and fungi [15].

There is information available regarding the effects of YSE on the different enteric microorganisms in ruminants and in non-ruminants [25,33–35]; also, there are some scientific reports about the use of YSE to enhance animal production efficiency and to reduce ambient pollution [36,37]. All of this information must be integrated and organized into an approach to maximize the potential of YSE for sustainable animal production, reduce conventional drug use as feed additives, and reduce polluting gas emissions. It is also important to show those areas that need more research because some aspects of efficacy and application are not conclusive. The objective of the present study is to review the antimicrobial and digestive effects of *Yucca schidigera* extracts related to the production and environmental implications of ruminant and non-ruminant food animals.

2. *Yucca schidigera*, Distribution, and General Characteristics

Yucca schidigera is a native plant in the states of Baja California and Sonora, Mexico; its habitat extends from the arid region of northwestern Mexico to the Mojave Desert in the southwestern United States. *Yucca schidigera*, with the common names Mojave yucca, Mohave yucca, and Spanish dagger, is a shrub or small tree that belongs to the Agavaceae family. It is a slow-growing plant that thrives in a climate characteristic of high temperature and low rainy precipitation [38]. *Yucca schidigera* has medicinal properties and was used by Native Americans to treat various diseases, one of them being arthritis [39]. The plant has straight leaves and white, bell-shaped, edible flowers and bears edible fruits [40]. The *Yucca schidigera* products are approved for human use by the FDA [39]. To obtain extracts, the plants are harvested and transported to the factory, where the trunks are mechanically processed to obtain the extracts [41]. Currently, the plant is used in animal feed with multiple benefits enhancing animal production efficiency and acting against multiple harmful organisms such as protozoa, nematodes, and bacteria (mainly Gram-positive). *Yucca schidigera* contributes to reducing ammonia and methane emissions in confined animals. In addition, it has other uses for commercial purposes such as in beverages and agriculture [39]. Yucca is used to control nematodes and fungi in agricultural crops [42].

2.1. Saponin in Yucca schidigera Extracts

Saponin is the main active compound in YSE. The word saponin comes from Latin "sapo" that in English is "soap" because it forms foams like soaps with water; saponins have surfactant properties due to their amphiphile nature. In the chemical structure of saponins are observed two components: one is the hydrophobic aglycone (sapogenin) bounded to the second part that is the hydrophilic component composed of carbohydrates such as glucose, fructose, galactose, and arabinose [43]. According to Francis et al. [43] and Tamura et al. [44], the aglycone component is used for the classification of saponins into two groups: steroidal (steroid saponins) or triterpenoid (triterpene saponins). The steroidal saponins may contain smilagenin and sarsasapogenin. Cheeke [41] recognized that the main plants used for industrial extraction of saponins are *Yucca schidigera* (from Mexico) and *Quillaja Saponaria* (from Chile), which have steroidal (Yucca) and triterpenoid (Quillaja) structures, respectively; however, saponins are also found in many other plants [43].

Both steroid and triterpene saponins have been explored for many years with commercial applications. Due to their detergent properties, saponins have exhibited several biological activities of economic importance with applications in the pharmaceutical industry, foods, cosmetics, and beverages [43,44]. In animal production, the commercial applications of saponins in YSE are in the cereal grain processing known as tempering [45,46] and for odor and pollution control from poultry and pigs' feces. Matusiak et al. [25] observed in poultry manure that YSE had effective control of odorous and volatile compounds such as ammonia, hydrogen sulfide, dimethylamine, trimethylamine, and isobutyric acid; they observed greater effect when combining YSE with other microbial treatments.

Extracts from the *Yucca schidigera* plant show antimicrobial effects against yeasts and dermatophytes, as well as antibacterial and antiprotozoal properties, and have been used for many years in food production and cosmetics [47]. However, it has been reported that YSE has no effect on the concentration of microorganisms under specific conditions [48], and the growth of microorganisms depends on the substrate and the concentration of microbiota present. Thus, the implications for effects on the gut microbiota differ according to their ecological niche [33].

The presence of saponins in plants appears to be linked to defensive functions against pathogenic bacteria and fungi [9,11,49]. In animals, saponins inhibit microbial growth by adsorption to the microbial cell membranes, disruption of cytoplasmic membranes, and cell leakage [33,50]. However, these inhibitory effects of saponins are seen at lower population densities and may not be effective with higher densities [33]. Some bacteria, including *Fusobacterium necrophorum* and *Clostridium perfringens*, have reduced growth when exposed to culture medium containing 10% YSE. In addition, YSE has reduced rumen protozoa numbers in cattle in in vitro and in vivo studies [15,17,22,51,52]. Extracts of *Yucca schidigera* eliminated *Giardia* trophozoites in in vitro studies; however, these extracts, when included in diets of gerbils and lambs, did not alter the course of experimentally induced giardiasis but did diminish the excretion of cysts [17].

2.2. Influence of Yucca schidigera Extracts on Enteric Microorganisms

The influence of YSE on microbial growth in different conditions of study is shown in summary in Table 1. According to this table, YSE has different effects on microbial growth. Katsunuma et al. [34], in in vitro studies with 20 strains of bacteria isolated from the intestinal tract of animals, observed that *Yucca schidigera* extract did not inhibit the growth of seven strains of bacteria (*Lactobacillus plantarum, Lactobacillus rhamnosus, Enterococcus hirae, Escherichia coli, Bifidobacterium thermophilum, Bifidobacterium longum*, and *Streptococcus bovis*); however, the other 13 strains (*Bacteroides fragilis, Fusobacterium varium, Fusobacterium necrophorum, Clostridium perfringens, Clostridium innocuum, Clostridium sporogenes, Veillonella parvula, Propionibacterium acnes, Eubacterium aerofaciens, Selenomonas ruminantium, Peptococcus asaccharolyticus, Ruminococcus productus*, and *Megasphaera elsdenii*) were inhibited by YSE. It is interesting to note that the antimicrobial effects of saponins from *Yucca schidigera* are due to the compounds sarsasapogenin and smilagenin, present in the butanol-extractable fraction. Conversely, the aqueous non-butanol-extractable fraction of *Y. schidigera* saponins influences various effects on in vivo nitrogen metabolism [33,35].

Matusiak et al. [25] observed that 5% YSE reduced most populations of several pathogenic microorganisms, including *E. coli.*, *Listeria monocytogenes*, *Salmonella Typhimurium*, and *Enterococcus faecalis*. In addition, 15% YSE reduced all potentially pathogenic microorganisms. The potentially beneficial lactic acid bacteria *Leuconostoc mesenteroides* and *Lactiplantibacillus plantarum* were not influenced by YSE.

2.3. Influence of Yucca schidigera on the Growth Microorganisms of Poultry

Yucca schidigera extract has the potential to decrease the number of pathogenic microorganisms or those that may reduce the productive performance of poultry. In addition, YSE may not influence the growth of beneficial bacteria such as those producing lactic acid. More research is warranted in this aspect, mainly using YSE plus probiotics and prebiotics. Ayoub et al. [53] supplemented YSE to broiler chickens in drinking water and observed reductions of intestinal bacteria for total bacteria counts and *E. coli*; however, lactic-acid-producing bacteria numbers were not influenced. Alghirani et al. [54] tested the oxytetracycline antibiotic or *Yucca schidigera* at different levels in broiler feed. In all treatments, they observed in cecal contents the presence of *Escherichia coli* and *Bacillus* sp.; however, *Enterococcus faecalis* growth was inhibited by the antibiotics and by *Yucca schidigera*. Bafundo et al. [55] observed reductions in mortality of broiler chickens, reduced numbers of fecal Clostridia, and decreased percentages of broiler chickens with *Salmonella* when treated with a product obtained from *Quillaja saponaria* trees and *Yucca schidigera* plants. In another study, *Yucca schidigera* additions to the litter of broiler chickens did not influence total colony counts of Enterobacteriaceae, yeast, mold, pH, moisture, or ammonia-N [56]. A blend of caprylic acid plus *Yucca schidigera* extract consistently reduced *E. coli* and had no effect on *Lactobacillus* in laying hens [57] or broiler chickens [58] but had no effect on the growth of *Clostridium perfringens* and Bifidobacteria.

2.4. Influence of Yucca schidigera on the Growth Microorganisms of Pigs

Weaning is critical due to the stress of piglets. In this stage, a decrease in feed intake, an increase in morbidity, and enteric infections with diarrhea are observed [59]. Diarrhea is of economic importance in pig production [60]. As mentioned, YSE may reduce or inhibit the growth of enteric microorganisms with pathogenic potential. In this sense, Yang et al. [61] supplemented weaning piglet diets with YSE and observed reduced diarrhea rate and mortality and increased diversity and abundance of cecal microflora. At the family level, the authors saw reduced numbers of Bacteroidales, Clostridiaceae, Veillonellaceae, Erysipelotrichaceae, Acidaminococcaceae, Streptococcaceae, Campylobacteraceae, and Streptomycetaceae. The authors also mentioned that YSE reduced the growth of bacteria associated with ammonia production in the intestine. Katsunuma et al. [62] observed with pig YSE-supplemented diets that there was no change in total fecal counts of viable microorganisms. However, Bifidobacteria, eubacteria, and staphylococci were more abundant, while the *Veillonella* number was lower in the feces of pigs supplemented with YSE. More research is warranted on effect of YSE on other pathogenic microorganisms such as those of the Enterobacteriaceae family, as well as on normal or beneficial microorganisms found in the gastrointestinal tract of healthy pigs.

2.5. Influence of Yucca schidigera on Microbial and Fermentation in Ruminants

The possible use of YSE to decrease gas emissions that pollute the environment and that are produced by livestock was reviewed by Adegbeye et al. [63]. YSE mainly reduced methane and nitrous oxide, as well as urinary and fecal nitrogen excretion. The authors documented that saponins as the main active compounds in YSE reduced ruminal cellulolytic bacteria and fungi that produce the contaminant gases. In addition, Jayanegara et al. [64], in a meta-analysis study, established that the increase in saponin levels reduced methane and acetate levels and protozoal counts; however, propionate was increased with increased saponin levels. They also observed greater reductions in methane levels for yucca than for tea or quillaja as sources of saponins. Sun et al. [31] recognized the beneficial effects of YSE on methane and ammonia mitigation in in vitro studies; however, in some studies, YSE did not influence methane or ammonia production, attributing those differences to different experimental conditions.

The influence of YSE on microbial and ruminal fermentation characteristics is shown in summary in Table 2. In this table, the different factors that influence the effect of YSE on rumen fermentation can be observed. Hristov et al. [36], in beef heifers supplemented with YSE, did not observe an influence on ruminal pH or degradability of peptides; amino acids; microbial protein synthesis; or digestibility of crude protein (CP), dry matter (DM), or neutral detergent fiber (NDF). However, they did observe reduced ruminal ammonia, protozoa numbers, and an increase in propionate concentration with YSE supplementation. Consistently, Liu and Li [37], in ruminal cannulated lambs fed a basal diet supplemented with YSE, did not observe changes in ruminal pH, but ruminal ammonia concentrations and protozoa populations were decreased, and they saw increased ruminal propionate concentrations. In addition, the authors observed reduced acetate levels and improved digestibility of DM, CP, and NDF with YSE supplementation.

Eryavuz and Dehority [48] found that in lambs, YSE supplementation did not influence bacteria (total and cellulolytic) and fungal concentrations. Lambs given YSE at 30 g/head/day had higher protozoal numbers and pH in rumen contents. With different results, Wang et al. [15], in in vitro studies with ruminal microorganisms, observed that saponin extracted from *Yucca schidigera* had different effects on the growth of amylolytic bacteria, decreasing the numbers of *Prevotella bryantii*, *Ruminobacter amylophilus*, and *Streptococcus bovis* but increased *Selenomonas ruminantium*. The authors also observed that saponins from *Yucca schidigera* inhibited the growth of cellulolytic bacteria and fungi.

Narvaez et al. [65], using in vitro fermentations of a barley-based diet supplemented with monensin alone or combined with extracts of hops and YSE, found that all treatments reduced CH4, NH3-N, microbial protein, and propionate. In addition, YSE increased total VFA production. Monensin reduced *Ruminococcus flavefaciens* and increased *Selenomonas ruminantium*, while both microorganisms increased with only YSE supplementation, as did *Fibrobacter succinogenes* and *Ruminococcus albus*. The methanogenic archaea were decreased by all treatments. YSE has multiple effects in rumen related to methane reductions (Figure 1). The influence of YSE on other ruminal changes, such as on metabolites and on different ruminal microorganisms, is not complete, and research on these themes is warranted.

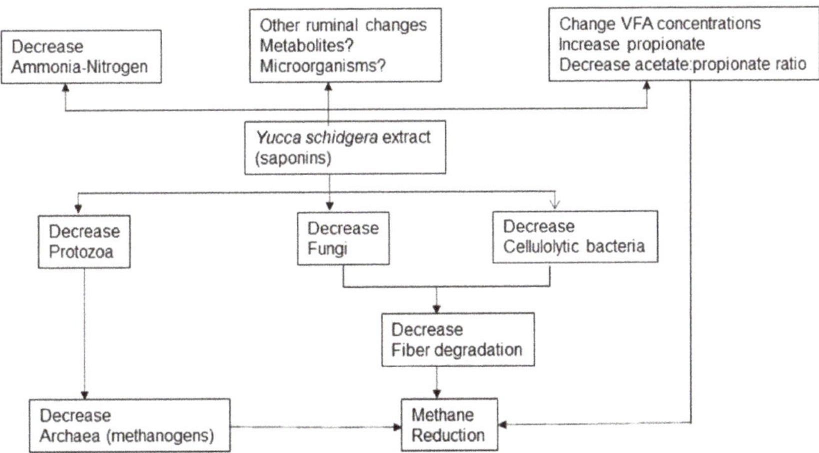

Figure 1. Proposed modes of action for ruminal methane reduction with *Yucca schidigera* extracts.

The beneficial effects of YSE on rumen fermentation are not always observed. Canul-Solis et al. [66], in lambs fed *Pennisetum purpureum* grass, did not find effects of saponins from *Yucca schidigera* on DM intake and digestibility of DM, OM, NDF, or production of VFA or CH4. Similarly, Galindo et al. [67], using in vitro ruminal incubations with star grass *(Cynodon nlemfuensis)* as substrate and saponin extract from *Sapindus saponaria*, observed reductions in protozoa numbers but increased methanogen numbers and methane production. Wang et al. [68] observed that YSE improved in vitro fermentation of barley grain but not with alfalfa hay. In addition, the fermentation of barley grain resulted in increased propionic acid and VFA; however, ammonia production was decreased. These results may indicate that the beneficial influence of YSE could be expected with high grain-based diets. Furthermore, Guyader et al. [69] reported that saponins caused reductions in protozoa populations and methane production in in vitro studies with ground cereal grains as substrate. However, in in vivo studies with dairy cows fed a diet with 60% forage, the same saponin did not mitigate methane production. Because several factors account for methane and ruminal fermentation characteristics, more research is warranted to elucidate the role of saponins or YSE on methane production and feed efficiency in in vivo studies under different feeding conditions.

Table 1. Effect of *Yucca schidigera* extract (YSE) on microbial growth in different conditions of research.

Microorganism	Conditions of the Study	Reference
Reports Where YSE Inhibited Microbial Growth		
Enterococcus faecalis	Inhibited by antibiotics and by YSE. Study in broiler chickens	Alghirani et al. [54]
Bacteroides fragilis; Fusobacterium varium; Fusobacterium necrophorum; Clostridium perfringens; Clostridium innocuum; Clostridium sporogenes; Veillonella párvula; Propionibacterium acnés; Eubacterium aerofaciens; Selenomonas ruminantium; Peptococcus asaccharolyticus; Ruminococcus productus; Megasphaera elsdenii	In vitro study	Katsunuma et al. [34]
E. coli; Listeria monocytogenes; Salmonella Typhimurium; Enterococcus Faecalis	15% YSE reduced all potentially pathogenic microorganisms	Matusiak et al. [25]
Candida famata; Pichia carsonii; Pichia nakazawae; Dermatomyces hansenii; Zygosaccharomyces	In vitro study	Miyakoshi et al. [16]
Reports where YSE reduced microbial growth		
E. coli	A blend of caprylic acid plus YSE	Wang et al. [57]
E. coli	Cecal microbiota in chickens	Begum et al. [58]
Neocallimastix frontalis; Piromyces rhizinflata	In vitro study. Ruminal fungi	Wang et al. [68]
Prevotella bryantii; Ruminobacter amylophilus; Streptococcus bovis	In vitro study. Cellulolytic ruminal bacteria	Wang et al. [68]
Bacteroidales; Clostridiaceae; Veillonellaceae; Erysipelotrichaceae; Acidaminococcaceae; Streptococcaceae; Campylobacteraceae; Streptomycetaceae	Bacterial Family. Study with piglets	Yang et al. [61]
Salmonella	Reductions in mortality of broiler chickens	Bafundo et al. [55]
Giardia trophozoite	Trophozoite in the ileum	McAllister et al. [17]
Escherichia coli	Study in broiler chickens (intestinal bacteria)	Ayoub et al. [53]
Clostridia spp.	bacterial fecal	Bafundo et al. [55]
Veillonella	Study with pig feces	Katsunuma et al. [62]

Table 1. *Cont.*

Reports Where YSE Inhibited Microbial Growth		
Reports where YSE had no effect on microbial growth		
Lactobacillus plantarum; Lactobacillus rhamnosus Enterococcus hirae; Escherichia coli; Bifidobacterium thermophilum; Bifidobacterium longum; Streptococcus bovis	In vitro study	Katsunuma et al. [34]
Leuconostoc mesenteroides; Lactiplantibacillus plantarum	Lactic acid bacteria important in nutrition	Matusiak et al. [25]
Lactobacillus	Laying hens	Wang et al. [57]
Lactobacillus; Bifidobacteria; Clostridium perfringens	Broiler chickens	Begum et al. [58]
Lactobasullus; Streptococcus; Staphylococcus	Study with pig feces	Katsunuma et al. [62]
Enterobacteriaceae E. coli	Study in litter of broiler chickens In vitro study	Onbasilar et al. [56] Killeen et al. [33]

Table 2. Effects of *Yucca schidigera* extracts (YSE) on microbial and ruminal fermentation characteristics.

Effect of *Yucca schidigera* Extracts	Study Characteristics	Reference
YSE reduced ruminal ammonia (33.5%) and protozoa numbers (20.3%). Increased propionate concentration (18.2%).	Beef heifers fed a diet with 61% barley grain and 39% alfalfa silage (DM basis)	Hristov et al. [36]
YSE reduced ruminal ammonia (14.42%), protozoa (17.3%), and acetate (17.5%). Improved propionate (16.9%) and digestibility of DM, CP, and NDF.	Lambs fed a diet of 50:50 forage to concentrate	Liu and Li [37]
YSE increased protozoal numbers (23.3%) and pH (6.5%) in rumen. YSE did not influence total or cellulolytic bacteria and fungi.	Lambs fed a diet with 69.5% concentrate and 30.5% alfalfa meal (DM basis)	Eryavuz and Dehority [48]
YSE had different effects on amylolytic bacteria. Decreased numbers of *Prevotella bryantii, Ruminobacter amylophilus,* and *Streptococcus bovis* but increased *Selenomonas ruminantium*. YSE inhibited the growth of cellulolytic bacteria and fungi.	Pure culture in in vitro studies	Wang et al. [15]
YSE reduced methane (55.3%), ammonia (11.9%), microbial protein (39.9%), acetate (16.8%), total VFA, and methanogenic archaea (23.9%). YSE increased propionate (67%), as well as the microorganisms *Ruminococcus flavefaciens, Selenomonas ruminantium, Fibrobacter succinogenes,* and *Ruminococcus albus.*	In vitro study with barley grain as substrate	Narvaez et al. [65]
YSE had no effect on DM intake, digestibility of DM, OM, NDF, or on the production of VFA or methane	In lambs fed *Pennisetum purpureum* grass	Canul-Solis et al. [66]
YSE improved in vitro fermentation of barley grain but not with alfalfa hay. Fermentation of barley grain resulted in increased propionic acid and VFA; however, ammonia production was decreased.	In vitro fermentation	Wang et al. [68]
Saponins caused reductions in protozoa populations and methane production. However, in in vivo studies with dairy cows fed a diet with 60% forage, the same saponin did not mitigate methane production.	In in vitro studies with ground cereal grains as substrate.	Guyader et al. [69]

2.6. Effects of Yucca schidigera Supplementation on Production Behavior and Carcass Traits of Broiler Chickens

The enhanced production parameters of broiler chickens supplemented with YSE are well documented (Table 3), revealing improved carcass characteristics and meat composition. Other beneficial effects of YSE supplementation were improved antioxidative and immune capability, as well as improved nutrient digestibility. Sahoo et al. [70] reported that YSE supplementation in broiler chickens improved average weight gain (9.7%), feed conversion ratio (9%), survivability (3.5%), eviscerated weight yield (8.1%), breast yield (6.2%), and thigh yield (11.4%). All of these improvements were reflected in higher economic profit with YSE supplementation.

Table 3. Effects of *Yucca schidigera* supplementation on production behavior and carcass traits of broiler chickens.

Effect of YSE on Broiler Productive Variables	Conditions of the Study	Reference
YSE improved average weight gain (9.7%), feed conversion ratio (9%), survivability (3.5%), eviscerated weight yield (8.1%), breast yield (6.2%), and thigh yield (11.4%).	Broiler chickens in winter season	Sahoo et al. [70]
YSE had no effect on weight gain. It improved feed conversion ratio (13.5%), protein efficiency (19.5%), and reduced feed intake (11.4%).	YSE in drinking water of broiler chickens in finisher phase	Ayoub et al. [53]
YSE enhanced feed efficiency in the whole experimental period.	Broiler chickens	Sun et al. [31]
The product of Quillaja and Yucca at 42 d enhanced FCR (5.4%) and reduced mortality (79%).	Broiler chickens with a product of *Quillaja saponaria* and *Yucca schidigera*.	Bafundo et al. [55]
At 42 d, YSE improved body weight gain (15.2%), FCR (9.4%), carcass weight (7.1%), and breast weight (2.05%).	Broiler chickens in a tropical environment	Alghirani et al. [54]
At 35 d, YSE plus caprylic acid improved body weight gain (10.7%) and FCR (5.6%). The effect was small on carcass characteristics.	*Yucca schidigera* extract combined with caprylic acid	Begum et al. [58]

Ayoub et al. [53], supplementing YSE in the drinking water of broiler chickens, observed enhanced feed conversion, antioxidative biomarkers, and immunoglobulin levels; in addition, litter nitrogen and water concentration were reduced. Sun et al. [31], with YSE supplementation, saw in broiler chickens enhanced production performance during the finisher period at grow out, as well as improved liver antioxidative capability.

Bafundo et al. [55] examined disease response in broiler chickens given a product of *Quillaja saponaria* trees and *Yucca schidigera* plants. They observed reductions in mortality of birds and reductions in numbers of fecal Clostridia and *Salmonella* colonization. In broiler chickens produced in a tropical environment, Alghirani et al. [54] observed that YSE improved production parameters, digestibility of nutrients, gut health, carcass yields of main cuts, and meat chemical composition through increased CP, reduced crude fat, and saturated fatty acids. Begum et al. [58] showed that *Yucca schidigera* extract combined with caprylic acid improved the growth performance of broilers. The improved production variables were evident during the finisher phase and at grow out. The effect was small on carcass characteristics. In broiler chickens supplemented with probiotics plus YSE, Benamirouche et al. [71] observed improved meat chemical composition. They saw increased pH, crude protein, and reduced crude fat levels. In addition, saturated fatty acids were reduced, and polyunsaturated fatty acids were increased with the supplements.

There is limited information about the effects of YSE on the coccidia of broiler chickens. Rodríguez et al. [72] reported that broiler chickens challenged with coccidia and supplemented with YSE saponin and *Trigonella foenum-graecum* showed reduced oocyst counts and intestinal lesions, and these improvements were reflected in enhanced production performance of the birds. These results agree with Bafundo et al. [73] using saponins of quillaja plus yucca. Alfaro et al. [74] reported the synergic influence of YSE and coccidiosis vaccine on the productive performance of broiler chickens; however, YSE did not have the same positive effect when combined with the coccidiostat monensin.

2.7. Effect of Yucca schidigera Supplementation on Production Behavior and Carcass Traits of Pigs

Yang et al. [61] observed in weaned piglets supplemented YSE or *Candida utilis* that YSE improved production performance and reduced diarrhea. There was also enhanced immunity, antioxidant function, and intestinal health. The improved production performance with YSE and *Candida utilis* was quite similar, although no synergic effect was detected.

In agreement, Fan et al. [75], in weaned piglets, also reported that YSE improved feed efficiency and reduced hindgut NH3-N production with increased nutrient digestibility and reduced blood urea concentrations. Espinosa-Muñoz et al. [14], in growing and finishing pigs supplemented with YSE in diets, also found decreased blood urea, triglycerides, and cholesterol. In contrast, Gebhardt et al. [76], in growing finishing pigs, found a very marginal advantage of YSE combined with chromium propionate on the production performance. The treatments did not influence carcass characteristics. These differences may be related to the initial body weights of the animals, and other factors related to management; the available information is limited as to document these effects, and research is warranted on these themes.

3. Conclusions

In addition to odor and ammonia control, YSE can be used to enhance animal production efficiency and contributes to diminishing ambient pollution mainly in confined animals. YSE may decrease or inhibit the growth of some pathogenic microorganisms that reduce animal production efficiency. The most consistent effect of YSE is in poultry production enhancement. In pig production, more research is warranted to identify the stage of production (from weaning to finishing and for animals in reproduction) where YSE supplementation could be more effective to improve the health and production of animals. More information is required for poultry and pig production regarding the effect of YSE combined with other feed additives such as probiotics or prebiotics. In ruminants, the effects of YSE can be influenced by diet type; however, YSE has the potential to reduce methane and contribute to reduced ambient contamination. YSE is a natural feed additive that may have the potential to improve animal production.

Author Contributions: Conceptualization and investigation, A.Z.-S., H.B.B.-G. and J.S.-C.; Writing (original draft preparation, review and editing), A.Z.-S., R.C.A., M.E.H., H.B.B.-G., M.R.-A., Y.B.-M., N.A.S.-G., J.V.-V., F.I.-R. and J.S.-C. All authors have read and agreed to the published version of the manuscript.

Funding: This research received no external funding.

Institutional Review Board Statement: Not applicable.

Informed Consent Statement: Not applicable.

Data Availability Statement: Not applicable.

Conflicts of Interest: The authors declare no conflict of interest.

References

1. Cabrera-Cao, Y.; Fadragas-Fernández, A.; Guerrero-Guerrero, L.G. Antibióticos naturales. Mito o realidad. *Rev. Cuba. Med. Gen. Integr.* **2005**, *21*. Available online: http://scielo.sld.cu/scielo.php?script=sci_arttext&pid=S0864-21252005000300025&lng=es&tlng=es (accessed on 28 May 2022).
2. Carrique-Mas, J.J.; Rushton, J. Integrated interventions to tackle antimicrobial usage in animal production systems: The ViParc project in Vietnam. *Front Microbiol.* **2017**, *8*, 1062. [CrossRef] [PubMed]
3. Barbosa, E.; Calzada, F.; Campos, R. Antigiardial activity of methanolic extracts from *Helianthemum glomeratum* Lag. And *Rubus coriifolius* Focke in suckling mice CD-1. *J. Ethnopharmacol.* **2006**, *108*, 395–397. [CrossRef] [PubMed]
4. Annan, K.; Houghton, P.J. Antibacterial, antioxidant and fibroblast growth stimulation of aqueous extracts of *Ficus asperifolia* Miq. and *Gossypium arboreum* L., wound-healing plants of Ghana. *J. Ethnopharmacol.* **2008**, *119*, 141–144. [CrossRef] [PubMed]
5. Puvaca, N.; Tufarelli, V.; Giannenas, I. Essential oils in broiler chicken production, immunity and meat quality: Review of *Thymus vulgaris*, *Origanum vulgare*, and *Rosmarinus officinalis*. *Agriculture* **2022**, *12*, 874. [CrossRef]
6. Molina-Salinas, G.; Pérez-López, A.; Becerril-Montes, P.; Salazar-Aranda, R.; Said-Fernández, S.; de Torres, N.W. Evaluation of the flora of northern Mexico for in vitro antimicrobial and antituberculosis activity. *J. Ethnopharmacol.* **2007**, *12*, 435–441. [CrossRef]
7. Miguel, M.G. Antioxidant and anti-inflammatory activities of essential oils: A short review. *Molecules* **2010**, *15*, 9252–9287. [CrossRef]
8. Rosas-Piñón, Y.; Mejía, A.; Díaz-Ruiz, G.; Aguilar, M.I.; Sánchez-Nieto, S.; Rivero-Cruz, J.F. Ethnobotanical survey and antibacterial activity of plants used in the Altiplane region of Mexico for the treatment of oral cavity infections. *J. Ethnopharmacol.* **2012**, *14*, 860–865. [CrossRef]

9.   Wink, M. Modes of action of herbal medicines and plant secondary metabolites. *Medicines* **2015**, *2*, 251–286. [CrossRef]
10.   Windisch, W.; Schedle, K.; Plitzner, C.; Kroismayr, A. Use of phytogenic products as feed additives for swine and poultry. *J. Anim. Sci.* **2008**, *86*, E140–E148. [CrossRef]
11.   Reddy, P.R.K.; Elghandour, M.; Salem, A.; Yasaswini, D.; Reddy, P.; Reddy, A.N.; Hyder, I. Plant secondary metabolites as feed additives in calves for antimicrobial stewardship. *Anim. Feed Sci. Technol.* **2020**, *264*, 114469. [CrossRef]
12.   Sabia, E.; Claps, S.; Morone, G.; Bruno, A.; Sepe, L.; Aleandri, R. Field inoculation of arbuscular mycorrhiza on maize (*Zea mays* L.) under low inputs: Preliminary study on quantitative and qualitative aspects. *Ital. J. Agron.* **2015**, *10*, 30–33. [CrossRef]
13.   Serrapica, F.; Masucci, F.; Romano, R.; Napolitano, F.; Sabia, E.; Aiello, A.; Di Francia, A. Effects of chickpea in substitution of soybean meal on milk production, blood profile and reproductive response of primiparous buffaloes in early lactation. *Animals* **2020**, *10*, 515. [CrossRef] [PubMed]
14.   Espinosa-Muñoz, V.; García-Contreras, A.D.C.; Herrera-Haro, J.G.; Álvarez-Macías, A.G.; Estrada-Barrón, S.G.; Meza-Cortes, M. Effects of *Yucca schidigera* extract on biochemical and hematological profiles of growing and fattening pigs. *Rev. Cient.* **2008**, *18*, 51–58. Available online: http://ve.scielo.org/scielo.php?pid=S0798-22592008000100009&script=sci_abstract&tlng=en (accessed on 28 May 2022).
15.   Wang, Y.; McAllister, T.A.; Yanke, L.J.; Cheeke, P.R. Effect of steroidal saponin from *Yucca schidigera* extract on ruminal microbes. *J. Appl. Microbiol.* **2000**, *88*, 887–896. [CrossRef] [PubMed]
16.   Miyakoshi, M.; Tamura, Y.; Masuda, H.; Mizutani, K.; Tanaka, O.; Ikeda, T.; Ohtani, K.; Kasai, R.; Yamasaki, K. Antiyeast steroidal saponins from *Yucca schidigera* (*Mohave yucca*), a new anti-food-deteriorating agent. *J. Nat. Prod.* **2000**, *63*, 332–338. [CrossRef] [PubMed]
17.   McAllister, T.; Annett, C.; Cockwill, C.; Olson, M.; Wang, Y.; Cheeke, P. Studies on the use of *Yucca schidigera* to control giardiosis. *Vet. Parasitol.* **2001**, *97*, 85–99. [CrossRef]
18.   Cheeke, P.R. Biological Effects of Feed and Forage Saponins and Their Impacts on Animal Production. In *Saponins Used in Food and Agriculture*, 1st ed.; Waller, G.R., Yamasaki, K., Eds.; Springer: Boston, MA, USA, 1996; pp. 377–385.
19.   Oleszek, W.; Sitek, M.; Stochmal, A.; Piacente, S.; Pizza, C.; Cheeke, P. Steroidal saponins of *Yucca schidigera* Roezl. *J. Agric. Food Chem.* **2001**, *49*, 4392–4396. [CrossRef]
20.   Oleszek, W.; Sitek, M.; Stochmal, A.; Piacente, S.; Pizza, A.C.; Cheeke, P. Resveratrol and other phenolics from the bark of *Yucca schidigera* Roezl. *J. Agric. Food Chem.* **2001**, *49*, 747–752. [CrossRef]
21.   Piacente, S.; Pizza, C.; Oleszek, W. Saponins and phenolics of *Yucca schidigera* Roezl: Chemistry and bioactivity. *Phytochem. Rev.* **2005**, *4*, 177–190. [CrossRef]
22.   Patra, A.K.; Saxena, J. The effect and mode of action of saponins on the microbial populations and fermentation in the rumen and ruminant production. *Nutr. Res. Rev.* **2009**, *22*, 204–219. [CrossRef] [PubMed]
23.   Njagi, G.W.; Lee, S.; Won, S.; Hong, J.; Hamidoghli, A.; Bai, S.C. Effects of dietary Yucca meal on growth, haematology, non-specific immune responses and disease resistance of juvenile Nile tilapia, *Oreochromis niloticus* (Linnaeus, 1758). *Aquac. Res.* **2017**, *48*, 4399–4408. [CrossRef]
24.   Biagi, G.; Cipollini, I.; Paulicks, B.R.; Roth, F.X. Effect of tannins on growth performance and intestinal ecosystem in weaned piglets. *Arch. Anim. Nutr.* **2010**, *64*, 121–135. [CrossRef] [PubMed]
25.   Matusiak, K.; Oleksy, M.; Borowski, S.; Nowak, A.; Korczynski, M.; Dobrzanski, Z.; Gutarowska, B. The use of *Yucca schidigera* and microbial preparation for poultry manure deodorization and hygienization. *J. Environ. Manag.* **2016**, *170*, 50–59. [CrossRef] [PubMed]
26.   Pinna, C.; Vecchiato, C.G.; Cardenia, V.; Rodriguez-Estrada, M.T.; Stefanelli, C.; Grandi, M.; Gatta, P.P.; Biagi, G. An *in vitro* evaluation of the effects of a *Yucca schidigera* extract and chestnut tannins on composition and metabolic profiles of canine and feline faecal microbiota. *Arch. Anim. Nutr.* **2017**, *71*, 395–412. [CrossRef]
27.   Santacruz-Reyes, R.A.; Chien, Y.-H. The potential of *Yucca schidigera* extract to reduce the ammonia pollution from shrimp farming. *Bioresour. Technol.* **2012**, *113*, 311–314. [CrossRef]
28.   Yang, Q.-H.; Tan, B.-P.; Dong, X.-H.; Chi, S.-Y.; Liu, H.-Y. Effects of different levels of *Yucca schidigera* extract on the growth and nonspecific immunity of Pacific white shrimp (*Litopenaeus vannamei*) and on culture water quality. *Aquaculture* **2015**, *439*, 39–44. [CrossRef]
29.   Elumalai, P.; Kurian, A.; Lakshmi, S.; Faggio, C.; Esteban, M.A.; Ringø, E. Herbal immunomodulators in aquaculture. *Rev. Fish. Sci. Aquac.* **2020**, *29*, 33–57. [CrossRef]
30.   Elbialy, Z.I.; Salah, A.S.; Elsheshtawy, A.; Rizk, M.; Abualreesh, M.H.; Abdel-Daim, M.M.; Salem, S.M.R.; El Askary, A.; Assar, D.H. Exploring the multimodal role of *Yucca schidigera* extract in protection against chronic ammonia exposure targeting: Growth, metabolic, stress and inflammatory responses in Nile Tilapia (*Oreochromis niloticus* L.). *Animals* **2021**, *11*, 2072. [CrossRef]
31.   Sun, D.; Jin, X.; Shi, B.; Su, J.; Tong, M.; Yan, S. Dietary *Yucca schidigera* extract improved growth performance and liver antioxidative function in broilers. *Ital. J. Anim. Sci.* **2017**, *16*, 677–684. [CrossRef]
32.   Patra, A.K.; Saxena, J. A new perspective on the use of plant secondary metabolites to inhibit methanogenesis in the rumen. *Phytochemistry* **2010**, *71*, 1198–1222. [CrossRef] [PubMed]
33.   Killeen, G.F.; Madigan, C.A.; Connolly, C.R.; Walsh, G.A.; Clark, C.; Hynes, M.J.; Timmins, B.F.; James, P.; Headon, D.R.; Power, R.F. Antimicrobial saponins of *Yucca schidigera* and the implications of their in vitro properties for their in vivo impact. *J. Agric. Food Chem.* **1998**, *46*, 3178–3186. [CrossRef]

34. Katsunuma, Y.; Nakamura, Y.; Toyoda, A.; Minato, H. Effect of *Yucca shidigera* extract and saponins on growth of bacteria isolated from animal intestinal tract. *Nihon Chikusan Gakkaiho.* **2000**, *71*, 164–170. [CrossRef]
35. Killeen, G.F.; Connolly, C.R.; Walsh, G.A.; Duffy, C.F.; Headon, D.R.; Power, R.F. The effects of dietary supplementation with *Yucca schidigera* extract or fractions thereof on nitrogen metabolism and gastrointestinal fermentation processes in the rat. *J. Sci. Food Agric.* **1998**, *76*, 91–99. [CrossRef]
36. Hristov, A.N.; McAllister, T.A.; Van Herk, F.H.; Cheng, K.-J.; Newbold, C.J.; Cheeke, P.R. Effect of *Yucca schidigera* on ruminal fermentation and nutrient digestion in heifers. *J. Anim. Sci.* **1999**, *77*, 2554–2563. [CrossRef] [PubMed]
37. Liu, C.L.; Li, Z.Q. Effect of Levels of Yucca Schidigera Extract on Ruminal Fermentation Parameters, Digestibility of Nutrients and Growth Performance in Sheep. In *Advanced Materials Research*; Wang, D., Ed.; Trans Tech Publications Ltd.: Zurich, Switzerland, 2012; Volume 343, pp. 655–660. Available online: https://doi.org/10.4028/www.scientific.net/AMR.343-344.655 (accessed on 8 August 2022).
38. Gucker, C.L. *Yucca schidigera* . In *Fire Effects Information System*; U.S. Department of Agriculture, Forest Service, Rocky Mountain Research Station, Fire Sciences Laboratory (Producer): Missoula, MT, USA, 2006. Available online: https://www.fs.fed.us/database/feis/plants/shrub/yucsch/all.html (accessed on 26 May 2022).
39. Cheeke, P.R.; Piacente, S.; Oleszek, W. Anti-inflammatory and anti-arthritic effects of *Yucca schidigera*: A review. *J. Inflamm.* **2006**, *3*, 6. [CrossRef]
40. Mielke, J. *Native Plants for Southwestern Landscape*, 1st ed.; University of Texas Press: Austin, TX, USA, 1993; pp. 272–275.
41. Cheeke, P.R. Actual and Potential Applications of *Yucca Schidigera* and *Quillaja Saponaria* Saponins in Human and Animal Nutrition. In *Saponins in Food, Feedstuffs and Medicinal Plants*; Proceedings of the Phythochemical Society of Europe, Volume 45; Oleszek, W., Marston, A., Eds.; Springer: Dordrecht, The Netherlands, 2000; pp. 241–254. [CrossRef]
42. Ruiz-Romero, P.; Valdez-Salas, B.; Gonzalez-Mendoza, D.; Mendez-Trujillo, V. Antifungal Effects of Silver Phytonanoparticles from *Yucca shilerifera* Against Strawberry Soil-Borne Pathogens: *Fusarium Solani Macrophomina Phaseolina*. *Mycobiology* **2018**, *46*, 47–51. [CrossRef]
43. Francis, G.; Kerem, Z.; Makkar, H.P.S.; Becker, K. The biological action of saponins in animal systems: A review. *Br. J. Nutr.* **2002**, *88*, 587–605. [CrossRef]
44. Tamura, Y.; Miyakoshi, M.; Yamamoto, M. Application of Saponin-Containing Plants in Foods and Cosmetics. In *Alternative Medicine*; Sakagami, H., Ed.; IntechOpen: London, UK, 2012; pp. 85–101. [CrossRef]
45. McDonough, C.; Anderson, B.J.; Acosta-Zuleta, H.; Rooney, L.W. Effect of conditioning agents on the structure of tempered and steam-flaked sorghum. *Cereal Chem.* **1998**, *75*, 58–63. [CrossRef]
46. Zinn, R.A.; Alverez, E.G.; Montano, M.; Salinas-Chavira, J. Influence of dry-rolling and tempering agent addition during the steam-flaking of sorghum grain on its feeding value for feedlot cattle. *J. Anim. Sci.* **2008**, *86*, 916–922. [CrossRef]
47. Wallace, R.J.; Arthaud, L.; Newbold, C.J. Influence of *Yucca-shidigera* extract on ruminal ammonia concentrations and ruminal microorganisms. *Appl. Environ. Microbiol.* **1994**, *60*, 1762–1767. [CrossRef] [PubMed]
48. Eryavuz, A.; Dehority, B.A. Effect of *Yucca schidigera* extract on the concentration of rumen microorganisms in sheep. *Anim. Feed Sci. Technol.* **2004**, *117*, 215–222. [CrossRef]
49. Díaz-Puentes, L.N. Molecular Interactions between Plants and Microorganisms: Saponins as Plant Chemical Defenses and Their Microbial Tolerance—A Review. *Rev. Estud. Transdiscipl.* **2009**, *1*, 32–55. Available online: Chrome-extension://efaidnbmnnnibpcajpcglclefindmkaj/https://www.uv.mx/personal/tcarmona/files/2010/08/Diaz-2009.pdf (accessed on 26 May 2022).
50. Dong, S.X.; Yang, X.S.; Zhao, L.; Zhang, F.X.; Hou, Z.H.; Xue, P. Antibacterial activity and mechanism of action saponins from Chenopodium quinoa Willd. husks against foodborne pathogenic bacteria. *Ind. Crops. Prod.* **2020**, *149*, 112350. [CrossRef]
51. Wallace, R.J. Antimicrobial properties of plant secondary metabolites. *Proc. Nutr. Soc.* **2004**, *63*, 621–629. [CrossRef]
52. Holtshausen, L.; Chaves, A.V.; Beauchemin, K.A.; McGinn, S.M.; McAllister, T.A.; Odongo, N.E.; Cheeke, P.R.; Benchaar, C. Feeding saponin-containing *Yucca schidigera* and *Quillaja Saponaria* to decrease enteric methane production in dairy cows. *J. Dairy Sci.* **2009**, *92*, 2809–2821. [CrossRef]
53. Ayoub, M.M.; Ahmed, H.A.; Sadek, K.M.; Alagawany, M.; El-Hack, M.E.A.; Othman, S.I.; Allam, A.A.; Abdel-Latif, M.A. Effects of liquid yucca supplementation on nitrogen excretion, intestinal bacteria, biochemical and performance parameters in broilers. *Animals* **2019**, *9*, 1097. [CrossRef]
54. Alghirani, M.M.; Chung, E.; Sabri, D.; Tahir, M.; Kassim, N.A.; Kamalludin, M.H.; Nayan, N.; Jesse, F.; Sazili, A.Q.; Loh, T.C. Can *Yucca schidigera* be used to enhance the growth performance, nutrient digestibility, gut histomorphology, cecal microflora, carcass characteristic, and meat quality of commercial broilers raised under tropical conditions? *Animals* **2021**, *11*, 2276. [CrossRef]
55. Bafundo, K.W.; Duerr, I.; McNaughton, J.L.; Johnson, A.B. The effects of a quillaja/yucca saponin combination on performance, *Clostridium perfringens* counts and percentage of Salmonella positive broiler chickens. *EC Vet. Sci.* **2021**, *6*, 40–45.
56. Onbasilar, E.E.; Erdem, E.; Unal, N.; Kocakaya, A.; Torlak, E. Effect of *Yucca schidigera* additions to different litter materials on broiler performance, footpad dermatitis and litter characteristics. *Eur. Poult. Sci.* **2014**, *78*, 13. [CrossRef]
57. Wang, J.P.; Kim, I.H. Effect of caprylic acid and *Yucca schidigera* extract on production performance, egg quality, blood characteristics, and excreta microflora in laying hens. *Br. Poult. Sci.* **2011**, *52*, 711–717. [CrossRef] [PubMed]

58. Begum, M.; Hossain, M.M.; Kim, I.H. Effects of caprylic acid and *Yucca schidigera* extract on growth performance, relative organ weight, breast meat quality, haematological characteristics and caecal microbial shedding in mixed sex Ross 308 broiler chickens. *Vet. Med.* **2015**, *60*, 635–643. [CrossRef]

59. Genova, J.L.; Melo, A.D.B.; Rupolo, P.E.; Carvalho, S.T.; Costa, L.B.; Carvalho, P.L.D. A summary of feed additives, intestinal health and intestinal alkaline phosphatase in piglet nutrition. *Czech. J. Anim. Sci.* **2020**, *65*, 281–294. [CrossRef]

60. Nguyet, L.T.Y.; Keeratikunakorn, K.; Kaeoket, K.; Ngamwongsatit, N. Antibiotic resistant Escherichia coli from diarrheic piglets from pig farms in Thailand that harbor colistin-resistant mcr genes. *Sci. Rep.* **2022**, *12*, 9083. [CrossRef] [PubMed]

61. Yang, Z.; Wang, Y.; He, T.; Bumbie, G.Z.; Wu, L.; Sun, Z.; Sun, W.; Tang, Z. Effects of dietary *Yucca schidigera* extract and oral *Candida utilis* on growth performance and intestinal health of weaned piglets. *Front. Nutr.* **2021**, *8*, 685540. [CrossRef] [PubMed]

62. Katsunuma, Y.; Otsuka, M.; Nakamura, Y.; Toyoda, A.; Takada, R.; Minato, H. Effects of the administration of *Yucca shidigera* saponins on pigs intestinal microbial population. *Nihon Chikusan Gakkaiho.* **2000**, *71*, 594–599. [CrossRef]

63. Adegbeye, M.J.; Elghandour, M.M.M.Y.; Monroy, J.C.; Abegunde, T.O.; Salem, A.Z.M.; Barbabosa-Pliego, A.; Faniyi, T.O. Potential influence of Yucca extract as feed additive on greenhouse gases emission for a cleaner livestock and aquaculture farming—A review. *J. Clean. Prod.* **2019**, *239*, 118074. [CrossRef]

64. Jayanegara, A.; Wina, E.; Takahashi, J. Meta-analysis on methane mitigating properties of saponin-rich sources in the rumen: Influence of addition levels and plant sources. *Asian-Australas. J. Anim. Sci.* **2014**, *27*, 1426–1435. [CrossRef]

65. Narvaez, N.; Wang, Y.; McAllister, T. Effects of extracts of *Humulus lupulus* (hops) and *Yucca schidigera* applied alone or in combination with monensin on rumen fermentation and microbial populations in vitro. *J. Sci. Food Agric.* **2013**, *93*, 2517–2522. [CrossRef]

66. Canul-Solis, J.R.; Pineiro-Vazquez, A.T.; Briceno-Poot, E.G.; Chay-Canul, A.J.; Alayon-Gamboa, J.A.; Ayala-Burgos, A.J.; Aguilar-Perez, C.F.; Solorio-Sanchez, F.J.; Castelan-Ortega, O.A.; Ku-Vera, J.C. Effect of supplementation with saponins from *Yucca schidigera* on ruminal methane production by Pelibuey sheep fed *Pennisetum purpureum* Grass. *Anim. Prod. Sci.* **2014**, *54*, 1834–1837. [CrossRef]

67. Galindo, J.; González, N.; Abdalla, A.L.; Alberto, M.; Lucas, R.C.; Dos Santos, K.C.; Santos, M.R.; Louvandini, P.; Moreira, O.; Sarduy, L. Effect of a Raw Saponin Extract on Ruminal Microbial Population and In Vitro Methane Production with Star Grass (*Cynodon nlemfuensis*) Substrate. *Cuba. J. Agric. Sci.* **2016**, *50*, 77–88. Available online: http://scielo.sld.cu/scielo.php?script=sci_abstract&pid=S2079-34802016000100010&lng=es&nrm=iso&tlng=en (accessed on 8 August 2022).

68. Wang, Y.X.; McAllister, T.A.; Yanke, L.J.; Xu, Z.J.; Cheeke, P.R.; Cheng, K.-J. In vitro effects of steroidal saponins from *Yucca schidigera* extract on rumen microbial protein synthesis and ruminal fermentation. *J. Sci. Food Agric.* **2000**, *80*, 2114–2122. [CrossRef]

69. Guyader, J.; Eugène, M.; Doreau, M.; Morgavi, D.P.; Gérard, C.; Martin, C. Tea saponin reduced methanogenesis *in vitro* but increased methane yield in lactating dairy cows. *J. Dairy Sci.* **2017**, *100*, 1845–1855. [CrossRef] [PubMed]

70. Sahoo, S.P.; Kaur, D.; Sethi, A.P.S.; Sharma, A.K.; Chandra, M. Evaluation of *Yucca schidigera* extract as feed additive on performance of broiler chicks in winter season. *Vet. World* **2015**, *8*, 556–560. [CrossRef]

71. Benamirouche, K.; Baazize-Ammi, D.; Hezil, N.; Djezzar, R.; Niar, A.; Guetarni, D. Effect of probiotics and *Yucca schidigera* extract supplementation on broiler meat quality. *Acta Sci. Anim. Sci.* **2020**, *42*, e48066. [CrossRef]

72. Rodríguez, I.; Honorio, C.; Ramirez, J.; Leon, Z.; Alarcon, W. Effect of a natural anticoccidial based on saponins of *Yucca schidigera* and *Trigonella foenum-graecum* on the control of coccidiosis in broilers. *Rev. De Investig. Vet. Del Perú* **2019**, *30*, 1196–1206. [CrossRef]

73. Bafundo, K.W.; Männer, K.; Duerr, I. The combination of quillaja and yucca saponins in broilers: Effects on performance, nutrient digestibility and ileal morphometrics. *Br. Poult. Sci.* **2021**, *62*, 589–595. [CrossRef] [PubMed]

74. Alfaro, D.M.; Silva, A.V.F.; Borges, S.A.; Maiorka, F.A.; Vargas, S.; Santin, E. Use of *Yucca schidigera* extract in broiler diets and its effects on performance results obtained with different coccidiosis control methods. *J. Appl. Poult. Res.* **2007**, *16*, 248–254. [CrossRef]

75. Fan, X.Q.; Xiao, X.J.; Chen, D.W.; Yu, B.; He, J.; Yu, J.; Luo, J.Q.; Luo, Y.H.; Wang, J.P.; Yan, H.; et al. *Yucca schidigera* extract decreases nitrogen emission via improving nutrient utilisation and gut barrier function in weaned piglets. *J. Anim. Physiol. Anim. Nutr.* **2021**. [CrossRef]

76. Gebhardt, J.T.; Woodworth, J.C.; Tokach, M.D.; DeRouchey, J.M.; Goodband, R.D.; Loughmiller, J.A.; De Souza, A.L.P.; Rincker, M.J.; Dritz, S.S. Determining the influence of chromium propionate and *Yucca schidigera* on growth performance and carcass composition of pigs housed in a commercial environment. *Transl. Anim. Sci.* **2019**, *3*, 1275–1285. [CrossRef]

Article

The Effect of Cotton Lint from Agribusiness in Diets on Intake, Digestibility, Nitrogen Balance, Blood Metabolites and Ingestive Behaviour of Rams

Anderson Zanine [1], Wanderson Castro [2], Daniele Ferreira [1], Alexandre Sousa [2], Henrique Parente [1], Michelle Parente [1], Edson Santos [3], Luiz Geron [4], Anny Graycy Lima [1], Marinaldo Ribeiro [5], Arlan Rodrigues [1], Cledson Sá [1], Renata Costa [1], Thiago Vinicius Nascimento [6], Francisco Naysson Santos [1,*] and Fagton Negrão [7]

[1] Department of Animal Science, Federal University of Maranhão, Chapadinha 65500-000, MA, Brazil
[2] Department of Animal Science, Federal University of Mato Grosso, Cuiabá 78735-901, MT, Brazil
[3] Department of Animal Science, Federal University of Paraíba, Areia 58397-000, PB, Brazil
[4] Department of Animal Science, State University of Mato Grosso, Pontes e Lacerda 78250-970, MT, Brazil
[5] Department of Animal Science, Federal University of Goiás, Goiânia 74690-900, GO, Brazil
[6] Department of Veterinary Medicine at Sertão, Federal University of Sergipe, Nossa Senhora da Glória 49680-000, SE, Brazil
[7] Department of Animal Science, Federal Institute of Education, Colorado do Oeste 76993-000, RO, Brazil
*   Correspondence: nayssonzootecnista@gmail.com

Citation: Zanine, A.; Castro, W.; Ferreira, D.; Sousa, A.; Parente, H.; Parente, M.; Santos, E.; Geron, L.; Lima, A.G.; Ribeiro, M.; et al. The Effect of Cotton Lint from Agribusiness in Diets on Intake, Digestibility, Nitrogen Balance, Blood Metabolites and Ingestive Behaviour of Rams. *Agriculture* **2022**, *12*, 1262. https://doi.org/10.3390/agriculture12081262

Academic Editor: Daniel Simeanu

Received: 7 July 2022
Accepted: 17 August 2022
Published: 19 August 2022

Publisher's Note: MDPI stays neutral with regard to jurisdictional claims in published maps and institutional affiliations.

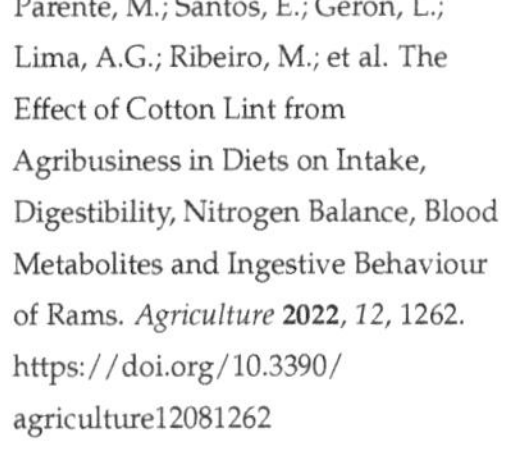

Simple Summary: The confinement system using traditional ingredients to feed animals, such as ground corn and soybean meal, has a high cost. Thus, alternative feeds that are cheaper, available in the region, and that present good nutritional quality are recommended. In this way, the residue, cottonseed lint, which is an agroindustry by-product, is an alternative feed that is cheaper than traditional feed ingredients. However, it has a high content of fiber and lignin, which is unattractive for non-ruminants' diets, but may be an interesting option in ruminant feeding.

Abstract: This study aimed to evaluate intake, digestibility, nitrogen balance, ingestive behavior, and blood parameters of sheep fed with cotton lint levels. Twenty rams weighing 30.2 ± 3.7 kg and aged 12 ± 1.3 months were distributed in a completely randomized design, with four treatments and five repetitions. Diets consisted of 50% roughage and 50% concentrate. The treatments consisted of replacing corn with cotton lint at levels of 0, 70, 140, and 210 g/kg of dry matter (DM) of the diets. The animals' feeding behavior was determined in the last three days of the experimental period. Data were subjected to regression analysis. Decreased linear effect ($p < 0.05$) was observed for the nutritional fraction's intake. However, neutral detergent fibre (NDF) intake and plasma urea-N were not affected ($p > 0.05$) by lint levels. Apparent digestibility of DM, crude protein (CP), ethereal extract (EE), and non-fibrous carbohydrate (NFC) were affected ($p < 0.05$), except for total carbohydrate. There was a decreased linear effect ($p < 0.05$) for the intake efficiency of DM and NDF in g/h. The nitrogen balance (g/day) and glucose levels (mg/dL) were reduced with the addition of lint in the diet. The addition of cotton lint up to 70 g/kg in DM can be used over a short-term period.

Keywords: alternative feed; degradability; fractions; sustainability

1. Introduction

The world population is growing, and thus there is a greater demand for food production. Thus, the productive chain is searching for alternatives to optimize production. In the food sector, the meat industry seeks to optimize its production using the feedlot, a technique for finishing animals faster that requires high technology. However, one of the primary issues is to limit the feed cost, which is about 70–80% of the final cost [1].

In view of the above, alternative products that replace traditional feeds, such as corn and soybean meal, are often a solution to reduce costs [1], and in this case, the residues from the cotton agribusiness have gained prominence in recent years [2]. According to Moreira et al. [3], the cotton culture can originate several products, from the direct ones, such as lint and oil, to the by-products, such as cottonseed, hull, and cottonseed meal.

The residual cottonseed lint is considered one of the by-products of cotton, obtained through the delinting process, where the fiber is separated from the cottonseed through chemical or physical treatments. This residue of lignocellulosic fibers is generally used in the manufacture of paper money, and the production of cotton wool and surgical fabrics. The lint content varies according to the cotton cultivar and the extraction process, ranging from 4 to 8% dry matter of the cottonseed after extraction [4]. The cottonseed lint residue has a high content of metabolizable energy, 13.29 kJ/kg DM, due to the high level of ether extract 268.7 g/kg DM [5].

According to the Companhia Nacional de Abastecimento (CONAB) [6], the 2017 harvest in the state of Mato Grosso, Brazil accounted for 56.5% of Brazilian cotton production, being the largest cotton producer in Brazil, with its by-products demonstrating a high production scale. Every year an average of 35 million hectares of cotton are planted across the planet and world demand has gradually increased since the 1950s, at an average annual growth of 2% [6]. The trend of production growth this year should consolidate Brazil as the fourth-largest producer and second-largest exporter. In the harvest planted in 2019, 3 million tons were produced. In 2017, 1.5 million tons were produced. For 2022, production is estimated between 2.6 and 2.8 million tons. There are many studies developed with cottonseed, hull and meal, whereas they are scarce or non-existent in the literature concerning lint for animal feed [3,7,8].

Thus, knowledge of the physical and chemical properties of this by-product is extremely important since it has specific characteristics in relation to other cotton by-products. Therefore, studies assessing both the chemical composition of by-products and animal tests help in the evaluation of feed, since it allows for determining the nutritional value of alternative feeds [9–11].

In view of the above, we hypothesised that by-products such as cottonseed lint can replace conventional ingredients in the diet to some level, and due to the chemical composition, lint can be an alternative ingredient to ground corn.

In this context, the aim of this study was to evaluate the addition of cotton lint in the diet of confined sheep on intake, feeding behavior, digestibility, nitrogen balance, and blood parameters.

2. Materials and Methods

2.1. Study Location

This study was conducted in the metabolism shed of the experimental area of the Animal Science Faculty at the Federal University of Mato Grosso, Rondonópolis Campus, Mato Grosso State, Brazil (16°28′ S, 50°34′ W), under the Federal University of Mato Grosso Animal Use and Care Committee (Protocol Number 23108.046399/13-4).

2.2. Animals, General Procedures, and Experimental Diets

A total of 20 mixed-breed rams with an average age of 12 ± 1.3 months and average weight of 30.2 ± 3.7 kg were used in a completely randomized design with four treatments and five replicates. The experiment was performed in April 2015 and lasted 21 days, with 15 days for adaptation to diets, environment, and management, and 6 days for sample collection. The experimental diets consisted of 50% corn silage and 50% concentrate on a DM basis, formulated to meet the nutritional needs of animals weighing 30 kg and gaining 150 g/day [12]. Soybean meal, ground corn, lint, and urea were used in the formulation of concentrates. The lint was derived from a cotton processing company located in the city of Primavera do Leste, MT, Brazil. The treatments consisted of replacing the ground corn with cottonseed lint at levels of 0, 70, 140, and 210 g/kg dry matter of the diets (Table 1).

Table 1. Chemical composition of experimental diets ingredients, maize silage (MS), ground corn (GC), soybean meal (SM), cottonseed lint (LIN), and urea.

Chemical Composition of Ingredients (g/kg DM)					
	MS	**GC**	**SM**	**LIN**	**Urea**
Dry matter [1]	327.30	910.00	896.70	900.20	990
Crude protein	58.90	81.30	480.10	29.70	2812.5
Ethereal extract	31.60	53.60	130.05	18.90	0
Ash	59.30	11.50	67.60	67.80	0
NDFap [2]	454.44	112.87	105.41	407.85	0
NDF [3]	541.80	125.25	128.74	487.80	0
ADF [4]	324.50	31.33	83.61	280.50	0
NFC [5]	395.80	740.08	216.4	475.80	0
TC [6]	850.20	853.36	322.24	883.60	0

[1] Dry matter in g/kg Natural matter; [2] NDFap: Neutral detergent fibre corrected to ash and protein; [3] NDF: Neutral detergent fibre; [4] ADF: Acid detergent fiber; [5] NFC: Non-fibrous carbohydrate; [6] TC: Total carbohydrate.

The NDF corrections for ash and protein (NDFap) were performed according to Licitra et al. [13]. The total carbohydrates (TC) were calculated according to Sniffen et al. [14] as portrayed in Equation (1):

$$TC = 100 - (\%CP + \%Ash + \%EE) \tag{1}$$

The non-fibrous carbohydrates (NFC) were calculated in the manner proposed to Hall et al. [15] according to Equation (2):

$$NFC = 100 - [\%CP + \%NDFap + \%EE + \%ash]. \tag{2}$$

Degradability data for the fractions of the ingredients were also estimated. The protein fractions (A, B1 + B2, B3, and C) were determined according to Licitra et al. [13]; carbohydrate fractions (A + B1, B2, and C) were proposed by Hall et al. [15] and Cabral et al. [16]; and estimation of the degradation of DM, CP, and NDF (Table 2) was according to Orskov and McDonald [17].

Table 2. Carbohydrate and protein fractionation, and degradation parameters of DM, CP, NDF from cottonseed lint.

	MS *	**GC** *	**SM**	**LIN**
Carbohydrate Fractionation (%TC)				
$A + B_1$	34.00	76.60	73.58	54.75
B_2	45.39	21.50	17.32	15.53
C	20.61	1.90	9.10	29.72
Protein Fractionation (% CP)				
A	32.71	11.70	18.72	33.83
$B_1 + B_2$	44.54	75.12	75.47	37.58
B_3	10.41	11.13	3.21	5.79
C	12.34	2.05	2.60	22.80
DM Degradation Parameters				
a	28.60	19.60	13.29	10.21
b	50.44	74.08	92.04	59.91
kd	0.036	0.072	0.003	0.081
CP Degradation Parameters				
a	56.34	23.27	7.65	4.03
b	33.11	55.37	90.74	74.20
kd	0.040	0.054	0.070	0.187

Table 2. *Cont.*

	MS *	GC *	SM	LIN
NDF Degradation Parameters				
a	0.89	3.93	10.33	5.65
b	67.25	39.61	52.24	55.99
kd	0.051	0.243	0.098	0.112

* CQBAL datas [18]; TC = Total carbohydrate; A + B_1 = Non-fibrous carbohydrate (%TC); B_2 = potentially digestible neutral detergent fibre (%TC); C = unavailable; A = non-protein nitrogen; B_1 + B_2 = true protein degradable in the rumen; B_2 = true protein with intermediate degradation rate in the rumen; B_3 = true protein with slow degradation rate in the rumen, C = unavailable or cell-wall bound true protein; (a) water-soluble fraction; (b) water-insoluble fraction potentially degradable in the rumen as a function of time; (kd) rate of degradation per fermentative action of b.

The rams were dewormed with Invermectin 1%-Ivomec-Merial® and were placed in metabolic cages provided with feeders, feed troughs, and drinking fountains. The animals were offered *ad libitum* water and mineral salt (Coperphós ovino®) with guaranteed levels (g/kg) as follows: zinc 30.00; calcium 17.70; manganese 12.00; phosphorus 8.00; fluorine (maximum) 8.00; iodine 6.00; copper 5.50; sulphur 2.00; sodium 0.40; selenium 0.15.

In the last three days, the ingestive behavior and the water intake of each animal were quantified by the daily difference between the supplied volume and the leftover water in graduated buckets. The buckets were washed at every use and two of them were allocated in the installation with water in order to measure evaporation losses [19].

Feed intake was measured daily by the difference in weight between the offered feed and the leftovers. The amounts of daily diets supplied per animal were adjusted in order to allow 10% of leftovers on a natural matter basis.

In the last six days of the experiment, daily samples of the supplied feed, leftovers, and feces were obtained, which were frozen and subsequently homogenized in order to obtain the respective composite samples. The diets were fed twice a day, with an interval of 8 hours between them, in a total mixed ratio (t 7.30 a.m. and 3.30 p.m.) (Table 3).

Table 3. Proportion of ingredients and chemical composition of diets with cottonseed lint levels, and average of body weight animals fed each level.

Ingredients (g/kg DM)	Cottonseed Lint Levels (g/kg)			
	0	70	140	210
Maize silage	500.00	500.00	500.00	500.00
Ground corn	424.00	352.20	280.70	209.00
Soybean meal	76.00	76.30	76.20	76.30
Lint	0.00	70.00	140.00	210.00
Urea	0.00	1.50	3.10	4.70
Chemical composition of diets (g/kg DM)				
Dry matter [1]	617.64	617.07	616.51	615.95
Organic matter	960.05	958.87	957.90	956.93
Crude protein	100.41	101.01	101.73	102.53
Ethereal extract	48.41	45.92	43.40	40.89
Ash	39.95	41.13	42.10	43.07
NDFap [2]	283.09	303.57	324.03	344.50
NDF [3]	333.79	358.98	384.16	409.34
ADF [4]	181.89	199.30	216.69	234.08
Hemicelullose	151.90	159.68	167.47	175.26
NFC [5]	528.14	508.37	488.74	469.01
TC [6]	811.41	812.09	812.90	813.60
TDN [7] estimated	747.31	733.75	720.20	706.65
Metabolisable energy [8] (Mj/kgDM)	11.25	11.05	10.84	10.63

Table 3. *Cont.*

Ingredients (g/kg DM)	Cottonseed Lint Levels (g/kg)			
	0	**70**	**140**	**210**
Animal's body weight in each group				
Body weight (kg)	29.19	28.91	32.63	29.90

[1] Dry matter in g/kg Natural matter; [2] NDFap: Neutral detergent fiber corrected to ash and protein; [3] NDF: Neutral detergent fibre; [4] ADF: Acid detergent fibre; [5] NFC: Non-fibrous carbohydrate; [6] TC: Total carbohydrate. [7] TDN: Total digestible nutrients. [8] Metabolizable energy was estimated using the equation provided by the NRC [12].

The total collection of feces was performed with the aid of individual collector bags adapted to the animals through a breastplate. The samples were collected twice a day, always before the provided diet. The samples were then weighed and stored under $-20\,°C$ for further analyses.

2.3. Chemical, Intake, Digestibility, Nitrogen Balance and Blood Analyses

The samples of silage, lint, concentrate, leftovers, and feces were stored in polyethylene containers for further analysis of dry matter (DM), organic matter (OM), crude protein (CP), ether extract (EE), mineral matter (MM), neutral detergent fiber (NDF) and acid detergent fiber (ADF), according to methodologies described by Silva and Queiroz [20].

The DM and nutrient digestibility coefficients of the diets were calculated based on the difference between the amounts ingested and excreted in the feces.

The digestibility coefficients (DC) from DM, CP, NDF, and EE were calculated through the Equation (3):

$$DC = \frac{(\text{kg of the portion ingested} - \text{kg of the portion excreted})}{\text{kg of the portion ingested}} \times 100 \qquad (3)$$

Samples of diets, feces and leftovers were stored frozen. Before performing the analysis, they were thawed, mixed, homogenized, and pre-dried in a convection oven at $65\,°C$, then ground in a Thomas Wiley mill on 1 mm sieves.

The urine of each animal was collected in the last three days and put in buckets arranged under the cages containing 50 mL of 50% hydrochloric acid in order to avoid losses by volatilization of urinary NH_3. The total urine volume of each animal was recorded and a 10% aliquot was taken, which was packed in plastic flasks and frozen for further analysis. Nitrogen balance (NB) was obtained by the difference between total nitrogen ingested and total nitrogen excreted in feces and in urine [21].

Plasmatic ureic nitrogen (PUN) was determined through blood collection, in Vacutainer tubes with ethylenediamine tetra acetic acid (EDTA), which was performed 3 h after feeding in the last day of collection, obtaining the plasma, which was analyzed by colorimetric-enzymatic method on a biochemical analyzer (Thermo Plate TP Analyser®) for urea determination, using commercial kits (Urea 500, Labtest®).

For reading the glucose concentration, a digital glucose meter (Accu-Chek®) equipped with test tape (tape reagent) was used. The measurement was performed immediately after collecting blood from the animals.

2.4. Assessment of Feeding Behaviour

The animals' feeding behavior was determined in the last three days of the experimental period by quantifying the time intervals spent on feeding, rumination, and leisure throughout 24 h [22]. To record the time of these activities, visual observation of animals was adopted every 10 min.

Trained observers were strategically positioned in the shed so as not to disturb the animals. In the same period, the number of cud chewings (MMnb, n/bolus) and the time taken to ruminate each bolus (MMtb, sec/bolus) were counted using a digital stopwatch. The minimum and maximum temperature averages were obtained every 1 h, within 24 h.

Chewing averages and rumination time were obtained through observations of rumen boluses made every 30 min, within each 24-h evaluation period [23]. The time and number of chews for each rumen bolus per animal within this period were recorded. The variables g DM/bolus and NDF/bolus were obtained by dividing the average individual intake of each fraction by the number of boluses ruminated per day (within 24 h).

The efficiencies of feeding and rumination, in g of DM/h and g of NDF/h were obtained by dividing the average daily intake of DM and NDF divided by the time spent in feeding and rumination within 24 h, respectively. These and other variables, such as the number of boluses ruminated per day (NBR), the total chewing time (TCT), and the number of cud chewings per day (CCD) were obtained according to the methodology described by Polli et al. [22] and Burguer et al. [23]. During the night data collection, the environment was maintained with artificial lighting.

The number of eating, rumination, and idleness periods were counted by the number of activity sequences observed in the worksheet. The average daily duration of these activity periods was calculated by dividing the total duration of each activity (eating, rumination, and idleness in min/day) by their respective number of periods.

2.5. Statistical Analysis

The experiment was performed in a completely randomized design with four treatments and five replicates (animals) per treatment. Data on intake, apparent digestibility, nitrogen balance, blood glucose and serum urea were submitted to analysis of variance and regression testing. The statistical model is represented by Equation (4):

$$Y_{ij} = \mu + T_i + \beta\left(X_i - \overline{X}\right) + e_{ij} \tag{4}$$

where Y_{ij} was the observed dependent variable, μ—overall average; T_i—fixed effect of treatment i; in which i = 0, 70, 140 and 210 g/kg; β—regression coefficient or functional relationship with the covariate; X_i—observed value of the covariate (body weight) applied to the experimental unit; $\overline{X}$—average of the covariate; j—random error associated with each observation and j repetitions, where j = 1, 2, 3, 4, and 5.

The data were subjected to analysis of variance using the PROC GLM command of the SAS statistical package (SAS version 9.1, 2003), and the means were subjected to regression analysis using the PROC REG command of the SAS® (9.1) statistical package (SAS University Edition). Analysis of ingestive behavior also included a *t*-test in SAS® (9.1) to verify the effect of the period (daytime and nighttime). Significance was declared when $p \leq 0.05$.

3. Results

3.1. Nutrient Intake

The addition of cotton lint promoted a decreasing linear effect ($p < 0.01$) in the intake of CP and NFC (Table 4). Considering the diet effectively consumed, despite the diets being completely supplied, it can be observed that there was selectivity of the diet by the animals (Table 4), indicating that the addition of lint at the 210 g/kg level led to the almost exclusive intake of roughage. The diets effectively consumed were modified with the addition of lint and caused a linear decreasing effect ($p < 0.01$) in the fractions CP, EE, NFC, but the lint promoted a linear increasing effect ($p < 0.01$) in the NDF and TC of (Table 4).

3.2. Nutrient Digestibility

The apparent digestibility of DM, CP, EE, and NFC presented a decreased linear effect ($p < 0.05$) with the addition of the lint in the diet (Table 5). In contrast, the NDF digestibility presented an increasing linear effect ($p < 0.0001$) and the TC digestibility was not influenced ($p = 0.5036$) by the addition of lint.

Table 4. Average intakes of dry matter (DMI), crude protein (CPI), ether extract (EEI), neutral detergent fiber (NDFI), non-fibrous carbohydrates (NFCI), total carbohydrates (TCI), total digestible nutrients (TDNI), indigestible neutral detergent fiber (iNDFI), and water (H_2OI) of rams fed with cottonseed lint inclusion levels diets.

Variable	Cottonseed Lint Levels (g/kg)				SEM [1]	Regression Equation	*p*-Value	
	0	70	140	210			L [2]	Q [3]
	Intake g/day							
DMI	986.56	796.68	722.31	517.81	23.78	$\hat{Y} = 755.84$	0.114	0.986
CPI	91.58	72.59	55.61	39.10	3.858	$\hat{Y} = 92.623 - 0.1613x$	<0.001	0.400
NDFI	210.90	195.87	217.94	183.17	1.494	$\hat{Y} = 201.97$	0.362	0.513
EEI	57.38	44.01	37.87	26.37	16.50	$\hat{Y} = 41.41$	0.405	0.136
NFCI	590.04	454.01	383.37	253.21	14.39	$\hat{Y} = 422.61 - 0.5106x$	0.005	0.382
TCI	801.20	650.05	601.45	436.47	29.93	$\hat{Y} = 622.29$	0.076	0.215
H_2OI	641.88	838.25	468.75	638.33	24.13	$\hat{Y} = 646.80$	0.067	0.462
	Relation between intake and body weight g/kg							
DMI BW	24.67	24.30	20.71	17.77	0.964	$\hat{Y} = 32.052 - 0.026x$	0.034	0.176
DMI $BW^{0.75}$	77.84	64.42	53.24	41.92	0.101	$\hat{Y} = 2.62 - 0.0043x$	<0.001	0.239
NDFI BW	7.10	6.89	6.74	6.43	0.032	$\hat{Y} = 6.79$	0.054	0.381
H_2OI BW	22.24	29.13	15.93	23.69	0.416	$\hat{Y} = 22.75$	0.498	0.072
	Diet effectively consumed g/kg							
CPI	92.17	90.89	76.50	68.39	2.438	$Y = 94.84 - 1.225x$	<0.001	0.081
NDFI	205.24	244.14	301.22	353.93	13.49	$Y = 200.66 + 7.189x$	<0.001	0.414
EEI	59.36	55.52	52.64	51.16	0.934	$Y = 58.795 - 0.393x$	0.003	0.389
NFCI	606.70	571.70	531.70	489.44	10.80	$Y = 608.65 - 5.597x$	<0.001	0.678
TCI	812.20	816.06	833.13	843.55	3.011	$Y = 809.57 + 1.587x$	<0.001	0.057

[1] SEM: Standard error mean, [2] L: linear, [3] Q:quadratic.

Table 5. Apparent digestibility of dry matter (DM), crude protein (CP), ether extract (EE), neutral detergent fiber (NDF), non-fibrous carbohydrates (NFC), and total carbohydrates (TC) of rams fed.

Variable (gDM)	Cottonseed Lint (g/kg DM)				SEM [1]	Regression Equation	*p*-Value	
	0	70	140	210			L [2]	Q [3]
DM	707.45	690.42	638.82	643.15	11.453	$Y = 706.63 - 3.493x$	0.015	0.602
CP	609.27	548.29	416.78	335.84	25.461	$Y = 620.32 - 13.597x$	<0.001	0.486
NDF	330.68	348.47	421.55	483.63	17.070	$Y = 316.30 + 7.599x$	<0.001	0.315
EE	895.75	881.49	840.24	799.19	10.556	$Y = 903.80 - 4.727x$	<0.001	0.325
NFC	883.38	864.06	816.21	783.58	11.073	$Y = 888.89 - 4.961x$	<0.001	0.642
TC	694.88	721.62	672.24	680.59	14.487	$Y = 692.33$	0.504	0.764

[1] SEM: Standard error mean, [2] L: linear, [3] Q:quadratic.

3.3. Feeding Behaviour

The ingestive behavior presented an increasing linear effect ($p < 0.0001$) with the addition of lint for the intake time. The DM intake rate indicated a decreasing linear effect ($p < 0.0001$), similar to that observed for the rumination time ($p < 0.0001$) and for the DM rumination rate ($p < 0.001$) (Table 6).

3.4. Nitrogen Fractions

The intake of total nitrogen, digested nitrogen, and nitrogen balance were reduced linearly ($p < 0.01$) with the addition of lint in the diets; however, the addition of lint did not affect ($p = 0.194$) the fecal and urinary nitrogen (Table 7). The plasmatic ureic nitrogen was not affected ($p = 0.172$) by the addition of lint; however, the blood glucose was reduced linearly ($p = 0.036$) with the addition of lint (Table 7).

Table 6. Ingestive behavior of rams fed with cottonseed lint inclusion levels diets.

Variable	Cottonseed Lint (g/kg DM)				SEM [1]	Regression Equation	*p*-Value	
	0	70	140	210			L [2]	Q [3]
Chewing number/bit	70.60	60.32	66.65	58.01	2.138	Y = 63.90	0.093	0.838
Time chewing/bit	44.93	40.27	48.71	43.00	1.315	Y = 44.23	0.809	0.830
Intake time (min)	375.37	416.44	453.70	476.30	8.901	Y = 379.25 + 0.4901x	<0.001	0.063
IRDM (gDM/h)	175.03	145.24	96.27	64.73	9.801	Y = 177.30 − 0.5427x	<0.001	0.412
IRNDF (gNDF/h)	33.95	31.23	28.97	22.89	2.112	Y = 29.26	0.073	0.689
Rumination (min)	333.23	299.91	274.08	254.06	7.618	Y = 329.82 − 0.3762x	<0.001	0.388
RRDM (gDM/h)	198.44	177.28	149.23	139.77	9.292	Y = 196.79 − 0.2915x	0.014	0.874
RRNDF (gNDF/h)	43.16	49.29	46.25	49.52	4.965	Y = 47.06	0.743	0.896
Idle (min)	731.40	723.75	710.22	709.64	8.032	Y = 718.98	0.272	0.888

[1] SEM: Sandard error mean, [2] L: linear, [3] Q:quadratic, IRDM—intake rate of dry matter; IRNDF—intake rate of neutral detergent fiber; RRMS—rumination rate of dry matter; RRNDF—rumination rate of neutral detergent fiber.

Table 7. Average values of values of the nitrogen fractions in urine, feces, and plasma for diets containing cottonseed lint levels.

Variable	Cottonseed Lint (g/kg DM)				SEM [1]	Regression Equation	*p*-Value	
	0	70	140	210			L [2]	Q [3]
[4] IN	14.65	11.61	8.90	6.24	0.950	$\hat{Y}$ = 47.917 − 0.1293x	<0.001	0.486
[5] FN	7.96	5.27	5.51	4.48	0.670	$\hat{Y}$ = 10.88	0.541	0.194
[6] UN	2.16	1.71	1.79	1.78	0.076	$\hat{Y}$ = 26.186 − 0.0631x	<0.001	0.876
[7] DIGN	6.70	6.34	3.38	1.76	0.549	$\hat{Y}$ = 7.355+ 0.0341x	<0.001	0.043
[8] NB	4.54	4.63	1.59	−0.02	0.550	$\hat{Y}$ = 5.193 − 0.0239x	<0.001	0.223
[9] PUN	20.44	25.22	28.72	19.88	2.339	$\hat{Y}$ = 23.57	0.933	0.172
[10] Glucose	79.20	76.80	62.00	61.62	3.548	$\hat{Y}$ = 80.036 − 0.0965x	0.036	0.880

[1] SEM—standard error mean; [2] L—linear; [3] Q—quadratic; [4] IN—intake nitrogen (g/day); [5] FN—fecal nitrogen (g/day); [6] UN—urinary nitrogen (g/day); [7] DIGN—digested nitrogen (g/day); [8] NB—nitrogen balance(g/day); [9] PUN—plasmatic ureic nitrogen (mg/dL); [10] glucose—blood glucose (mg/dL).

4. Discussion

For every 10 g of lint added per kg of DM in the diet, there was a decrease of 2.83 g of CP consumed. According to the NRC [12], lambs with an average weight of 30 kg require an intake of 1.03 kg/day of DM and 97 g/day of CP, but all treatments were below the recommended, where intakes of DM were 986.56 g, 796.68 g, 722.31 g, and 517.81 g, and of CP g/day were 91.58, 72.59, 55.61, and 39.10, respectively. The addition of lint reduced the DM intake and hence the various bromatological fractions, except for the NDF due to the selectivity of animals that started to consume more hay than concentrate when the lint was added. The animals' selectivity can be evidenced through the diet effectively consumed (Table 4).

This result may be a consequence of the industrial extraction process of the lint since sulphuric acid is used and may generate residue in the final product (lint) [4]. This factor can be considered highly relevant in terms of acceptability because it is known that the odor, texture, color, and taste possibly affect the feed intake by animals. Nevertheless, Mertens [24] reported that acceptability is defined as the characteristic of feed associated with taste, olfactory, and visual activities that influence the intake of animals. There was no difference for the NDF intake; however, considering the diet effectively consumed as the ratio between the amount of NDF and DM consumed, and the NDF of the diet effectively consumed was linearly affected. It increased with the addition of lint, in which corn silage had a greater amount of this fraction (Table 1). The lint is also a source of NDF and ADF. Thus, the addition of lint promoted an increase of NDF in the concentrate, i.e., the animals received a diet with a higher content of NDF, and reduced NFC and EE.

The decrease of the apparent digestibility of DM, CP, EE, and NFC with the addition of lint may be related to the quality of the diet effectively ingested (Table 4). The composition of the fractions DM, CP, and TC of the corn grain, primarily observed in fractions $A + B_1$ (Table 2), are superior to those of the lint. Therefore, with the increased addition of lint, the animals started to eat less feed and of lesser quality, thus precluding the multiplication of the ruminal microorganisms, primarily reducing the apparent digestibility of the fractions cited above. Thereby, the rumen environment had a lower pH reduction, which apparently seems to have provided a more suitable environment for the development of fibrolytic bacteria, and thus a higher digestibility rate of the NDF, as observed by Santos et al. [25]. The treatment for extraction of lint uses sulphuric acid, which weakens the ether bonds between lignin and cellulose, making cellulose more digestible in the rumen, but with a slow degradation rate [26–28]. The data related to the NDF degradation of corn grain was higher than lint; however, the addition of lint decreased the degradation of other fractions due to the higher concentration of the undegradable fraction (C) of protein and carbohydrates, in addition to lower degradation of the DM and CP (Table 2), corroborating the linear reduction of digestibility with the addition of lint (Table 5).

Diets with the addition of lint promoted an increase in unavailable fractions of CP, such as neutral detergent insoluble nitrogen (NDIN). This fraction of CP is adhered to the plant cell wall and is generally indigestible. Thus, the CP digestibility is influenced by the increase of this fraction [29,30]. This fact is evidenced by observing the fractionation of CP in which the values for the fraction C (Table 2), which is undegradable, was higher in cotton lint compared to corn grain.

The apparent digestibility of EE decreased 4.73 g/kg with the addition of each 1 g/kg DM addition of lint in the diet (Table 5). The EE digestibility followed the decreasing linear effect of DM. The NFC digestibility was linearly reduced. This was possibly due to the smaller ruminal microbial population likely a function of the lower availability of CP and lack of synchrony between the availability of CP and energy [31].

There was no effect for the digestibility of TC as lint was added to the diet. This lack of effect is due to the increasing linear effect on the digestibility of NDF and decreasing linear effect on the digestibility of NFC. As TC is the sum of these fractions, the sum of these antagonistic behavior effects promotes the lack of effect by the addition of lint.

The ingestion time of the diet was likely negatively influenced by the reduced palatability of the diet with the addition of lint. Therefore, the animals went more often to the trough to eat, trying to consume the amount of diet sufficient to satisfy their requirements. However, even increasing the ingestion time, the animals were unable to compensate such demand, since they reduced the consumption of DM with the addition of lint (Table 6). The combined effect of reduced DM intake with increased ingestion time resulted in a reduced DM intake rate; the animals demonstrated a lower efficiency in DM intake. This effect was not observed in the NDF intake, due the increasing content of NDF observed with the addition of lint (Table 6), likely because of the lower rate of DM disappearance in the rumen and lower rate of passage.

Rumination time was affected in a linear decreasing way by the addition of lint despite an increase in NDF intake, which likely led to an increasing effect of rumination time. Despite it being emphasised in the literature that only physically effective NDF would have this effect, fibers from lint do not have the physical effect due to their reduced length [30].

The NB decreased with the addition of lint in the diet (Table 7), a factor that may be associated with a lower DM intake and the decrease of CP effectively consumed in the diets when lint was added. According to the NRC [12], the NB allows assessing the nutritional quality of diets and food, since it expresses how much of the absorbed nitrogen was used for animal metabolism [32,33]. The addition of lint in 210 g/kg DM promoted a negative NB, associated with the linear reduction in blood glucose provided by the addition of lint. Thus, it can be inferred that the addition of lint will promote a greater mobilization of the animal's body reserves, culminating in weight loss.

The addition of lint reduced the intake of concentrate (Table 4), which likely promoted a smaller rumen microbial population associated with a lower production of SCFA, primarily propionic acid, which is a glucose precursor via gluconeogenesis. Generally, glucose is not variable in ruminants, but when the animals are in a negative balance due to insufficient feed intake, blood glucose is altered, indicating a negative energy balance. This biological signalling is generally mediated by several signals that stimulate the mobilization of body energy reserves [33]. However, if inclusion in values up to 70 g in the diet for a short period of time is used, we may not have deleterious effects on the physiological parameters of the animals. Considering that the performance of the animals was not evaluated, it is possible that the inclusion at this level in short-term feeding may be a possibility to improve feed conversion [34]. The presence of lint can modulate the rate of passage and disappearance of food particles in the rumen.

According to Cavalcante et al. [35], the concentration in the diet or the differentiated intake of DM may cause variations in nitrogen intake [36]. In this research, the variation was to the detriment of the decreasing effect of DM intake. Another possible explanation was the reduced concentration of CP in the diet effectively consumed when lint was added. Although they were formulated to be isoproteined, there were variations in the consumption of the diets (Table 4). The glucose concentration reduced 0.96 mg/dL for each 1 g/kg DM of lint added in the diet, indicating a lower energy supply for the animals that received lint, a factor that may be related to the lower nutrient consumption when the by-product was added. However, even with the reduction, these values were within the reference range for sheep, which is 50–80 mg/dL [37].

The PUN concentration had no significant effect as the lint was added to the diet, with an average of 23.57 mg/dL, (Table 7). Menezes et al. [38] describes that the range in plasma urea concentration considered ideal for sheep is from 20.0 to 60.0 mg/dL, which was observed in this experiment. Lira et al. [39] reported that the plasma urea concentration is related to the part of digestible nitrogen that will be directed or be excreted via urine, or will be reused via saliva, increasing the efficiency of use of dietary nitrogen.

According to Gonzáles and Scheffer [40], the liver synthesizes urea from ammonia derived from the metabolism of amino acids and the absorption of rumen ammonia from protein degradation in the rumen, being directed through the blood for renal or mammary excretion when there is excess protein in the diet or even recycled via saliva when the diet has CP levels below requirements.

The addition of lint promoted a linear reduction in almost all parameters related to the use of the diet, even modifying the animals' ingestive behavior. The bromatological composition of lint suggests that this by-product can be tested as a source of roughage in the future.

5. Conclusions

The addition of up to 70 g/kg of cotton lint can be included in the diet of rams without adversely affecting DM intake and without altering digestibility, nitrogen balance, plasma urea, and glucose concentrations. These results suggest a possible negative effect on dry matter intake when using levels above 70 g/kg of cotton lint.

Author Contributions: Conceptualization, A.Z., W.C., D.F. and E.S.; methodology, A.Z and A.S.; software, A.S and M.R.; validation, A.Z., H.P., M.P., L.G. and M.R.; formal analysis, D.F., A.G.L. and T.V.N.; investigation, A.Z. and W.C.; resources, A.Z., M.P., M.R. and F.N.; data curation, W.C. and A.S.; writing—original draft preparation, W.C., A.S., M.R. and F.N.; writing—review and editing, A.Z., D.F., E.S., A.G.L., A.R., C.S., R.C. and T.V.N.; visualization, H.P., A.G.L., A.R., C.S., R.C., T.V.N., F.N.S. and F.N.; supervision, A.Z.; project administration, A.Z., D.F. and L.G.; funding acquisition, A.Z., H.P., M.P., E.S. and F.N. All authors have read and agreed to the published version of the manuscript.

Funding: This study was supported by the Coordination for the Improvement of Higher Education Personnel (CAPES-Brazil, Finance Code 001) by the Maranhão State Research Foundation (FAPEMA-Brazil) and the Federal Institute of Education, Science and Technology of Rondônia (IFRO/DEPESP/Colorado do Oeste).

Institutional Review Board Statement: The study was conducted in accordance with the Declaration of the Federal University of Mato Grosso Animal Use and Care Committee (Protocol Number 04/2008).

Data Availability Statement: The data presented in this study are available upon request from the corresponding author.

Acknowledgments: The authors wish to thank the Foundation for Research Scientific and Technological Development of Maranhão (FAPEMA-Brazil), the Coordination for the Improvement of Higher Education Personnel (CAPES-Brazil, finance code 001), and the Federal Institute of Education, Science and Technology of Rondônia (IFRO/DEPESP/Colorado do Oeste) for their financial support.

Conflicts of Interest: The authors declare no conflict of interest.

References

1. Costa, J.B.; Oliveira, R.L.; Silva, T.M.; Ribeiro, O.L.; Ribeiro, R.D.X.; Pinto, L.F.B.; Nascimento, T.V.C. Análise econômica da terminação de cordeiros em condições de confinamento utilizando torta de licuri (Syagrus coronata (Mart.) Becc.). *Rev. Bras. Saúde Prod. Anim.* **2019**, *20*, e0252019. [CrossRef]
2. Geron, L.J.V.; Souza, A.L.; Zanine, A.M.; Pierangeli, M.A.P.; Ferreira, D.J.; Neto, E.S.; Paula, E.J.H.; Diniz, L.C.; Santos, I.S.; Zanin, S.F. Viabilidade técnica e econômica do uso de diferentes níveis de grãos secos de destilaria com solúveis (*Zea mays* L.) em borregas terminadas em confinamento. *Bol. Ind. Anim.* **2018**, *75*. [CrossRef]
3. Moreira, F.B. Subprodutos do algodão na alimentação de ruminantes. *Publicações Em Med. Veterinária E Zootec.* **2008**, *2*, 356. Available online: http://www.pubvet.com.br/texto.php?id=356 (accessed on 22 February 2021).
4. Queiroga, V.P.; Mata, M.E.R.M.C. Métodos de deslintamento apropriados para sementes de Algodão orgânicas e convencionais. *Rev. Bras. Prod. Agroindust.* **2018**, *20*, 83–101. Available online: http://www.deag.ufcg.edu.br/rbpa/rev201/rev2019.pdf (accessed on 2 July 2021).
5. Bo, Y.K.; Yang, H.J.; Wang, W.X.; Liu, H.; Wang, G.Q.; Yu, X. Metabolizable energy, in situ rumen degradation and in vitro fermentation characteristics of linted cottonseed hulls, delinted cottonseed hulls and cottonseed linter residue. *Asian-Austral. J. Anim. Sci.* **2012**, *25*, 240. [CrossRef]
6. Companhia Nacional de Abastecimento. Indicadores da Agropecuária. Disponível. Available online: http://www.conab.gov.br/OlalaCMS//uploads/arquivos/16_06_07_15_10_33_22_conjuntura_de_algodao_30_05_a_03_06_2016 (accessed on 2 July 2020).
7. Ali, A.I.; Wassie, S.E.; Korir, D.; Goopy, J.P.; Merbold, L.; Butterbach-Bahl, K.; Dickhoefer, U.; Schlecht, E. Digesta passage and nutrient digestibility in Boran steers at low feed intake levels. *J. Anim. Physiol. Anim. Nut.* **2019**, *103*, 1325–1337. [CrossRef]
8. Gallo, A.; Valsecchi, C.; Masseroni, M.; Cannas, A.; Ghilardelli, F.; Masoero, F.; Atzori, A.S. An observational study to verify the influence of different nutritional corn silage-based strategies on efficient use of dietary nutrients, fecal fermentation profile, and profitability in a cohort of intensive dairy farms. *Ital. J. Anim. Sci.* **2022**, *21*, 228–243. [CrossRef]
9. Alves, E.M.; dos Pedreira, M.S.; de Oliveira, C.A.S.; Aguiar, L.V.; Pereira, M.L.A.; Almeida, P.J.P. Comportamento ingestivo de ovinos alimentados com farelo da vagem de algaroba associado a níveis de ureia. *Acta Sci. Anim. Sci.* **2010**, *32*, 439–445. [CrossRef]
10. Castro, W.J.R.; Zanine, A.M.; Souza, A.L.; Ferreira, D.J.; Geron, L.J.V.; Leão AGNegrão, F.M.; Ferro, M.M. Inclusion of different levels of common-bean residue in sheep diets on nutrient intake and digestibility. *Semin. Ciências Agrárias* **2016**, *37*, 369–380. [CrossRef]
11. Meneghetti, C.C.; Domingues, J.L. Características nutricionais e uso de subprodutos da agroindústria na alimentação de bovinos. *Rev. Eletrônica Nutr.* **2008**, *5*, 512–536. Available online: https://www.nutritime.com.br/arquivos_internos/artigos/052V5N2P512_536_MAR2008.pdf (accessed on 2 July 2020).
12. National Research Council (NRC). *Nutrient Requirements of Small Ruminants: Sheep, Goats, Cervids, and New World Camelids*; The National Academies Press: Washington, DC, USA, 2007.
13. Licitra, G.; Hernandez, T.M.; Van Soest, P.J. Standardization of procedures for nitrogen fractionation of ruminant feeds. *Anim. Feed Sci. Technol.* **1996**, *57*, 347–358. [CrossRef]
14. Sniffen, C.J.; O'Connor, J.D.; Van Soest, P.J.; Fox, D.G.; Russell, J.B. A net carbohydrate and protein system for evaluating cattle diets: II. Carbohydrate and protein availability. *J. Anim. Sci.* **1992**, *70*, 3562–3577. [CrossRef]
15. Hall, M.B. *Calculation of Non-Structural Carbohydrate Content of Feeds That Contain Non-Protein Nitrogen*; University of Florida: Gainesville, FL, USA, 2000; p. A-25.
16. Cabral, L.S.; Valadares Filho, S.C.; Detmann, E. Rates of digestion of protein and carbohydrate fractions for silage corn and elephant grass hay, Tifton-85 and soybean meal. *Rev. Bras. Zoot.* **2004**, *33*, 1573–1580. [CrossRef]
17. Orskov, E.R.; Mcdonald, I. The estimation of protein degradability in the rumen from incubation measurements weighted according to rate of passage. *J. Agric. Sci.* **1979**, *92*, 499–503. [CrossRef]
18. Valadares Filho, S.C.; Machado, P.A.S.; Chizzotti, M.L.; Amaral, H.F.; Magalhães, K.A.; Rocha Junior, V.R. CQBAL 3.0. Tabelas Brasileiras de Composição de Alimentos para Bovinos. Available online: https://www.cqbal.com.br (accessed on 1 January 2019).

19. Souza, E.J.O.; Guim, A.; Batista, A.M.V.; Albuquerque, D.B.; Monteiro, C.C.F.; Zumba, E.R.F.; Torres, T.R. Comportamento ingestivo e ingestão de água em caprinos e ovinos alimentados com feno e silagem de maniçoba. *Rev. Bras. Saúde Prod. Anim.* **2010**, *11*, 1056–1067.
20. Silva, D.J.; Queiroz, A.C. *Análises de Alimentos (Métodos Químicos e Biológicos)*, 3rd ed.; Editora UFV: Viçosa, MG, Brazil, 2002; 235p.
21. Silva, R.R.; do Prado, I.N.; de Carvalho, G.G.P.; de Santana Junior, H.A.; Da Silva, F.F.; Dias, D.L.S. Efeito da utilização de três intervalos de observações sobre a precisão dos resultados obtidos no estudo do comportamento ingestivo de vacas leiteiras em pastejo. *Ciênc. Anim. Bras.* **2008**, *9*, 319–326. Available online: https://www.revistas.ufg.br/vet/article/view/1205 (accessed on 2 July 2020).
22. Polli, V.A.; Restle, J.; Senna, D.B.; Almeida, S.R.S. Aspectos relativos à ruminação de bovinos e bubalinos em regime de confinamento. *Rev. Bras. Saúde Prod. Anim.* **1996**, *25*, 987–993. [CrossRef]
23. Burger, P.J.; Pereira, J.C.; Queiroz, A.C.; Silva, J.F.C.; Valadares Filho, S.C.; Cecon, P.R.; Casali, A.D.P. Comportamento ingestivo em bezerros holandeses alimentados com dietas contendo diferentes níveis de concentrado. *Rev. Bras. Zoot.* **2000**, *29*, 236–242. [CrossRef]
24. Mertens, D.R. Using fiber and carbohydrate analyses to formulate dairy rations. In *Informational Conference with Dairy and Forages Industries*; US Dairy Forage Research Center: Madison, WI, USA, 1996.
25. Santos, J.W.; Cabral, L.S.; Zervoudakis, J.T.; Abreu, J.G.; Souza, A.L.; Pereira, G.A.C.; Reverdito, R. Farelo de arroz em dietas para ovinos. *Rev. Bras. Saúde Prod. Anim.* **2010**, *11*, 193–201.
26. Mertens, D.R. Regulation of forage intake. In *Forage Quality, Evaluation, and Utilization*; American Society of Agronomy: Madison, WI, USA, 1994; pp. 450–493.
27. Palmquist, D.L. Digestibility of cotton lint fiber and whole oilseeds by ruminal microorganisms. *Anim. Feed Sci. Technol.* **1995**, *56*, 231–242. [CrossRef]
28. Silva, M.L.F.; de Carvalho, G.G.P.; Silva, R.R.; da Magalhães, T.S.; Viana, P.T.; de Rufino LM, A.; Santos, A.V.; Azevedo, J.A.G.; Freitas Júnior, J.E.; de Nascimento, C.O.; et al. Effect of calcium lignosulfonate supplementation on metabolic profiles of confined lambs. *Environ. Sci. Pollut. Res.* **2018**, *25*, 19953–19961. [CrossRef] [PubMed]
29. Meinert, M.C.; Delmer, D.P. Changes in biochemical composition of the cell wall of the cotton fiber during development. *Plant Physiol.* **1977**, *59*, 1088–1097. [CrossRef] [PubMed]
30. Berchielli, T.T.; Pires, A.V.; Oliveira, S.G. *Nutrição de Ruminantes*; FUNEP: Jaboticabal, Brazil, 2011.
31. Gonçalves Neto, J.; dos Pedreira, M.S.; de Silva, H.G.O.; Alves, E.M.; dos Santos, E.J.; Da Silva, Á.C.; Perazzo, A.F.; Corrêa, Y.R. Tipos de uréia e fontes de carboidratos nas dietas de cordeiros: Síntese de proteína microbiana e balanço de nitrogênio. *REDVET Rev. Elect. Vet.* **2017**, *18*, 1–15. Available online: https://www.redalyc.org/pdf/636/63653009064.pdf (accessed on 4 March 2021).
32. Bringel, L.M.L.; Neiva, J.N.M.; Araújo, V.L.; Bomfim, M.A.D.; Restle, J.; Ferreira, A.C.H.; Lôbo, R.N.B. Consumo, digestibilidade e balanço de nitrogênio em borregos alimentados com torta de dendê em substituição à silagem de capim-elefante. *Rev. Bras. Zoot.* **2011**, *40*, 1975–1983. [CrossRef]
33. Kozloski, G.V. Bioquímica dos Ruminantes. In *Fundação de Apoio a Tecnologia e Ciência*; UFSM: Santa Maria, Brazil, 2017; 203p.
34. Dayani, O.; Dadvar, P.; Afsharmanesh, M. Effect of dietary whole cottonseed and crude protein level on blood parameters and performance of fattening lambs. *Small Rum. Res.* **2011**, *97*, 48–54. [CrossRef]
35. Cavalcante, M.A.B.; Pereira, O.G.; Valadares Filho, S.C.; Ribeiro, K.G.; Pacheco, L.B.B.; Araújo, D.; Lemos, V.M.C. Níveis de proteína bruta em dietas para bovinos de corte: Parâmetros ruminais, balanço de compostos nitrogenados e produção de proteína microbiana. *Rev. Bras. Zoot.* **2006**, *35*, 203–210. [CrossRef]
36. Oliveira Junior, R.C.; Pires, A.V.; Fernandes, J.J.R.; Susin, I.; Santos, F.A.P.; Araújo, R.C. Substituição total do farelo de soja por uréia ou amiréia, em dietas com alto teor de concentrado, sobre a amônia ruminal, os parâmetros sangüíneos e o metabolismo do nitrogênio em bovinos de corte. *Rev. Bras. Zoot.* **2004**, *33*, 738–748. [CrossRef]
37. Kaneko, J.J.; Harvey, J.W.; Bruss, M.L. *Clinical Biochemistry of Domestic Animals*; Academic Express: San Diego, CA, USA, 1997; 9321p.
38. Menezes, D.R.; Araújo, G.G.L.; Oliveira, R.L.; Bagaldo, A.R.; Silva, T.M.; Santos, A.P. Balanço de nitrogênio e medida do teor de uréia no soro e na urina como monitores metabólicos de dietas contendo resíduo de uva de vitivinícolas para ovinos. *Rev. Bras. Saúde Prod. Anim.* **2006**, *7*, 169–175. Available online: https://repositorio.ufba.br/bitstream/ri/1932/1/769-3012-2-PB.pdf (accessed on 4 March 2021).
39. De Lira, F.R.A.; De Oliveira, V.S.; De Santos, G.R.A.; Da Silva, M.A.; De Oliveira, A.G.; Gouveia, J.S.S. Monitoramento proteico em rebanhos de vacas leiteiras em Sergipe. *Semin. Ciências Agrárias* **2013**, *34*, 3043. [CrossRef]
40. Gonzáles, F.H.D.; Scheffer, J.F.S. Perfil sanguíneo: Ferramenta de análise clínica, metabólica e nutricional. In *Anais do I Simpósio de Patologia Clínica Veterinária da Região Sul do Brasil*; Gonzáles, F.H.D., Campos, R., Eds.; Gráfica da UFRGS: Porto Alegre, Brazil, 2003; pp. 73–90.

Article

Research Regarding Correlation between the Assured Health State for Laying Hens and Their Productivity

Alexandru Usturoi [1], Marius-Giorgi Usturoi [2], Bogdan-Vlad Avarvarei [2], Claudia Pânzaru [2], Cristina Simeanu [2], Mădălina-Iuliana Usturoi [3], Mihaela Spătaru [4], Răzvan-Mihail Radu-Rusu [2], Marius-Gheorghe Doliş [2,*] and Daniel Simeanu [1]

[1] Department of Control, Expertise and Services, Faculty of Food and Animal Sciences, "Ion Ionescu de la Brad" University of Life Sciences, 8 Mihail Sadoveanu Alley, 700489 Iasi, Romania

[2] Department of Animal Resources and Technology, Faculty of Food and Animal Sciences, "Ion Ionescu de la Brad" University of Life Sciences, 8 Mihail Sadoveanu Alley, 700489 Iasi, Romania

[3] Laboratory of Medical Analysis, Providenta Hospital, 115 Nicolina Avenue, 700714 Iasi, Romania

[4] Department of Public Health, Faculty of Veterinary Medicine, "Ion Ionescu de la Brad" University of Life Sciences, 8 Mihail Sadoveanu Alley, 700489 Iasi, Romania

* Correspondence: mariusdolis@uaiasi.ro

Abstract: Predictions show the possibility of banning birds' rearing in batteries. From this reason, we aimed to study the welfare conditions assured to birds accommodated in lofts in comparison with those reared in improved batteries. The research targeted ISA Brown hybrids monitored over a period of 25–55 weeks. The batches were represented by birds that were differently reared in halls provided with lofts compared to with improved batteries. The research was carried out in real production conditions. Biochemical indicators were determined, using a BA 400 analyzer produced by BioSystems, as well as quantitative ones using specific formulas based on productions, consumptions, and batch outputs. A cumulated production of 199.24 eggs/week/head was realized in the loft, versus 199.98 in the battery, at a mean laying intensity of 91.82% and 92.17%. Batch output was 4.14% (loft) and 2.98% (battery). Mean consumption registered a level of 122.20 g m.f./head/day for birds in the loft and 115.87 g for the ones from the battery, and feed conversion index was 133.09 g m.f./egg, compared to 125.69. The aviary system ensures optimal conditions to express the birds' natural behaviors, with a positive impact on the metabolic functions, resulting in a good state of health and high productive levels, comparable to those of birds exploited in batteries.

Keywords: rearing system; birds' welfare condition; biochemical analysis; productive parameters

Citation: Usturoi, A.; Usturoi, M.-G.; Avarvarei, B.-V.; Pânzaru, C.; Simeanu, C.; Usturoi, M.-I.; Spătaru, M.; Radu-Rusu, R.-M.; Doliş, M.-G.; Simeanu, D. Research Regarding Correlation between the Assured Health State for Laying Hens and Their Productivity. *Agriculture* **2023**, *13*, 86. https://doi.org/10.3390/agriculture13010086

Academic Editor: Shugeng Wu

Received: 17 December 2022
Accepted: 26 December 2022
Published: 28 December 2022

1. Introduction

Both fowl welfare and the performance of laying hybrids are affected by environmental factors, and they are still to be tuned within the context of modern industrialized aviculture.

The adaptability of hybrids to housing conditions is crucial in reaching efficiency while welfare is observed. Fowl's response to environmental factors relies on breed particularities. Stress markers are lower in less active, lymphatic genotypes (Green-legged Partridge hen) compared to the sensitive, active, and quite excitable ones (Leghorn) that are less adapted in large-scale farms [1]. Such differences also occur between purebreds and commercial hybrids, i.e., Sussex hens have the best levels of welfare markers, while ISA Browns have the lowest ones among commercial lines. However, mortality and aggressiveness seem to be more intense in pure breeds, and lower welfare occurs, compared to commercial hybrids that easily adapt to closed housing [2].

The European Union Council regulates the minimum comfort for laying hens and has also prohibited (since 2012) classic cage batteries as fowl housing [3,4]. Starting from this premise, there were designed various alternative rearing solutions to allow for the expression of flow productive potential [5,6], and to ensure wellbeing [7]. Housing in limited spaces (conventional cages) reduces resistance to diseases and decreases the fowl's comfort state markers [8].

Consequently, starting from the premise of ensuring welfare conditions, improved cage batteries (increased room) [9,10] and furniture endorsements (nesting, resting perches, sand bath) were initially designed to support high yields and superior egg quality [11–13]. The fowl that benefited from the enlarged vital space yielded more egg mass and eggs with better specific weight, Fe, Mg, and glucose level compared to those reared on narrower surfaces [14].

The second stage of the changes consisted in housing laying hens in aviaries (lofts allowing for the expression of natural instincts but also generating superior egg yields [15]. The major disadvantage of the system is the increased incidence of mechanical accidents (fractures) [16], but this was mitigated through covered edgy equipment parts with polyurethane [17]. Freedom of movement, however, raised the aggressive behavior, a problem solved by introducing males into the flocks of laying hens (egg yields also improved) [18]. In aviary housing, hens' serum immunoglobulin Y was higher and improved the specific antibody response in industrially reared fowl [19].

Serum biochemical traits, performance of production, and livability of fowl are traits depicting the fowl's wellbeing state [20,21]. The levels of plasmatic proteins, Mg, glucose, cholesterol, and P decrease as egg-laying intensity increases [22], while calcium decreases during the laying peak [23]. However, such changes are not strictly correlated with a certain housing system [24]. Various stressor factors induce fear [25] and generate changes in the concentration of basal plasma corticosterone and serotonin, with direct effects on eggs yield, weight, and flock livability [26]. The stress generated by fowl transportation affects the ratio of heterophiles/lymphocytes and plasmatic glycoprotein 1-acid [27]. Beak clipping, a procedure that limits the negative consequences of aggressive behavior and improves production performance [28], is tough, and therefore a stressor. Fasting applied to induce artificial molting induces a significant decrease in the serum glucose and an intense increase in plasmatic cholesterol and GSH-Px activity (poor plumage quality, tonic immobility reaction, etc.) and negatively affects the welfare condition [29,30].

Bird feeding has generated many studies in which, for example, direct correlations were detected between the level of essential amino acids and the incidence of feather plucking or mortality through cannibalism [31]. It has been shown that when fed in cage batteries, low dietary protein levels (13%) lead to egg yield drops and decreases in plasmatic uric acid, triglycerides, and albumin compared to fowl fed 16% crude protein diets [32,33]. Low dietary metabolizable energy and crude protein can cut feeding costs, whilst serum creatine kinase activity rises and serum triglyceride and cholesterol decrease [34]. A high proportion of cereals (wheat) in feed can induce poorer plumage quality and thermoregulation issues [35,36], while using nonconventional feedstuffs (dried olive pulp) induces improvement in the lipid profile of eggs (PUFA increases and SFA decreases, AI and TI decrease and h/H ratio increases) [37].

The welfare and performance of laying depend on many factors and the interactions between them, but especially on the applied rearing system and its particularities. Consequently, we must find these correlations as well as the way to implement the solutions in real production conditions.

2. Materials and Methods

2.1. Animals and Housing

The study took place in a private production unit in Galati County, Romania, starting in July. We used the commercial ISA Brown hybrid due to its adaptability to local conditions. The experimental groups were represented by birds from two production halls. The first batch was raised in a hall equipped with Natura Nova Twin model aviary (initial population of 35,950 heads), and the second in a hall with improved Eurovent batteries (initial population of 10,217 heads). The research started when the birds were 25 weeks old and ended when they were 55 weeks old (a total of 31 weeks).

The housing of the hens was carried out in accordance with the requirements of the technical manual of the hybrid [38] and of the equipment manufacturer [39]. In both cases, the particular density standards for integrating the farm under the welfare regime were followed.

The feeding of the birds respected the nutritional management strategy used in the work unit. An identical combined diet was given to the two groups of birds, but the nutritional values varied according to the quantity and quality of eggs laid [40]. The combination feed 21-5A (PB = 18.6%; EM = 2748.6 kcal/kg; Ca = 3.82%; P = 0.40%) was given between the ages of 25 and 40 weeks, and the combined feed 21-5B (PB = 16.43%; EM = 2786.0 kcal/kg; Ca = 3.8%; P = 0.36%) was given to animals between the ages of 41 and 55 weeks.

2.2. Data Collection and Data Processing

The technical data (feed consumption, egg production, batch outputs) necessary for elaborating the paper were taken from the farm registers and the computerized devices that served the rearing halls. The monitoring was carried out daily, with a weekly presentation of the data. Later, standardized formulas were applied to obtain the values for various indicators [41].

$$\text{Mean flock (heads/week)} = (\text{heads at the beginning of the week} - \text{heads at the end of the week})/2$$

$$\text{Laying intensity (\%)} = [\text{total egg production (pieces/week/shelter)} \times 100]/[\text{mean flock (heads)} \times 7 \text{ days}]$$

$$\text{Individual egg production (egg/week/head)} = \text{total egg production (pieces/week/shelter)}/\text{mean flock (heads)}$$

$$\text{Daily mean consumption (g/head/day)} = \text{consumed fodders (kg/period)} \times 1000/\text{mean flock (heads)}/\text{number of days}$$

$$\text{Feed conversion index (g m.f./egg)} = \text{consumed fodders (kg/period)} \times 1000/\text{egg production (pieces/period)}$$

The situation of outputs from the batch and their causes were counted by the farm manager and the veterinarian of the unit, and later reported numerically and percentage-wise to the total population (weekly situation and for the total period). Biochemistry samples were taken at 25, 35, 45, and 55 weeks by the veterinarian. They were harvested from 35 specimens from each growth hall, in accordance with the specific methods [42,43]. After harvesting, the samples were processed with the ByoSystems BA 400 analyzer.

The main experimental data were processed by calculating arithmetic mean, mean standard deviation, and variability coefficient (using an algorithm included in the Microsoft Excel software).

3. Results

3.1. Biochemical Indicators

In the case of hens from the improved batteries, the cholesterol level oscillated between a minimum of 153.72 mg/dL, the value recorded at the beginning of the research, and a maximum of 206.14 mg/dL in the 55th week. For hens reared in the aviary, the swing limits varied between 148.81 mg/d (25th week) and 195.21 mg/dL (55th week) (Table 1).

The level of blood triglycerides showed quantitative increases with the advancing age of birds, with the mention that higher values were detected in birds from improved batteries (189.56–206.19 mg/dL) and somewhat lower values were detected in those from aviaries (186.12–201.34 mg/dL).

Table 1. Biochemical indicators for the studied birds.

Specification		n	Birds' Age							
			25 Weeks		35 Weeks		45 Weeks		55 Weeks	
			$\overline{X} \pm s_{\overline{x}}$	V%	$\overline{X} \pm s_{\overline{x}}$	V%	$\overline{X} \pm s_{\overline{x}}$	V%	$\overline{X} \pm s_{\overline{x}}$	V%
Cholesterol	A	35	148.81 ± 8.36	19.36	162.35 ± 12.18	18.31	175.36 ± 15.22	23.16	195.21 ± 12.68	24.33
(mg/dL)	B	35	153.72 ± 10.14	21.14	164.14 ± 16.23	20.19	178.12 ± 19.46	25.31	206.14 ± 17.33	25.14
Triglycerides (mg/dL)	A	35	186.12 ± 5.34	16.12	191.37 ± 8.38	15.68	197.15 ± 8.55	17.22	201.34 ± 8.46	16.97
	B	35	189.56 ± 7.22	17.22	194.26 ± 9.55	17.12	203.41 ± 9.31	19.14	206.19 ± 10.15	19.45
Total protein	A	35	3.78 ± 0.72	12.20	4.12 ± 0.95	13.26	4.35 ± 0.51	10.46	4.87 ± 0.93	13.12
(g/dL)	B	35	3.54 ± 0.57	17.56	4.06 ± 0.78	18.21	4.19 ± 0.47	9.58	4.56 ± 0.77	10.68
Calcium	A	35	8.94 ± 0.30	16.32	8.75 ± 0.21	17.38	10.16 ± 0.24	15.31	12.14 ± 0.51	15.96
(mg/dL)	B	35	8.36 ± 0.25	12.44	8.16 ± 0.32	15.44	9.85 ± 0.31	12.14	11.61 ± 0.38	14.32
Phosphorous (mg/dL)	A	35	6.87 ± 0.69	20.15	6.42 ± 0.36	18.26	8.05 ± 0.93	22.17	9.02 ± 1.19	21.38
	B	35	6.31 ± 0.52	18.34	6.11 ± 0.25	15.17	7.81 ± 0.75	20.19	8.77 ± 0.92	20.71
Glucose	A	35	208.23 ± 14.96	15.19	219.14 ± 17.16	13.61	238.45 ± 20.02	18.14	252.19 ± 16.77	15.36
(mg/dL)	B	35	211.10 ± 15.23	17.32	223.08 ± 18.17	15.48	256.12 ± 19.33	19.38	270.14 ± 17.16	17.81
Uric acid	A	35	8.97 ± 0.51	12.95	7.85 ± 0.38	15.11	7.39 ± 0.26	15.19	6.89 ± 0.21	20.62
(mg/dL)	B	35	10.14 ± 0.48	13.34	9.43 ± 0.41	14.24	9.08 ± 0.21	17.26	7.24 ± 0.19	22.14
Urea	A	35	4.95 ± 0.41	15.31	5.19 ± 0.27	16.21	5.57 ± 0.55	23.15	5.72 ± 0.22	16.96
(mg/dL)	B	35	5.12 ± 0.38	19.76	5.34 ± 0.32	17.49	5.82 ± 0.71	24.21	5.95 ± 0.24	17.21
ALT	A	35	87.86 ± 12.32	13.35	89.24 ± 15.93	10.16	94.19 ± 12.38	17.31	98.83 ± 15.46	23.94
(U/l)	B	35	89.56 ± 14.16	14.26	91.31 ± 16.41	11.35	95.26 ± 14.12	17.16	101.11 ± 17.32	24.13
AST	A	35	231.22 ± 15.73	15.26	241.44 ± 15.27	15.77	267.85 ± 17.06	24.08	294.39 ± 12.08	24.31
(U/l)	B	35	236.16 ± 16.52	18.19	246.19 ± 16.11	15.27	272.11 ± 16.19	24.25	300.41 ± 19.55	25.14

A: rearing in loft. B: rearing in battery.

The amount of total protein had an average level of 3.54 g/dL at the beginning of the analyzed period (25th week) and 4.56 g/dL at its end (55th week) in the birds raised in battery. In the case of the livestock in the aviary, the total protein content increased from 3.78 g/dL, as it was in the 25th week, to 4.87 g/dL, as found at the end of the study.

Blood calcium recorded lower values at 35 weeks (8.16 g/dL for battery hens and 8.75 g/dL hens in the aviary) and higher at 55 weeks (11.61 g/dL-battery and 12.14 g/dL-aviary). Phosphorus had a similar evolution, with lower levels at 35 weeks (6.11 mg/dL for battery hens and 6.42 mg/dL in aviary birds) and higher in 55 weeks (8.77 mg/dL-battery and 9.02 mg/dL-aviary) (Table 1).

In birds aged 25 weeks, the amount of glucose was at a close level between the two groups (208.23 mg/dL-aviary and 211.10 mg/dL-improved batteries), after which progressive increases were registered towards the end of the investigations, when they were 252.19 mg/dL in aviary birds and 270.14 mg/dL in those in improved batteries.

The level of uric acid recorded higher values at the beginning of laying (week 25), was 8.97 mg/dL in birds from the aviary and 10.14 mg/dL for those from modified batteries, after which the values decreased in parallel with the age of the birds until up to 6.89 mg/dL (aviary) and at 7.24 mg/dL (improved batteries). Urea did not fluctuate significantly between batches or between reference intervals, with the beginning of the research showing values of 4.95 mg/dL (birds from the aviary) and 5.12 mg/dL (those from the batteries), while in the last period it showed values of 5.72 mg/dL and d 5.95 mg/dL, respectively (Table 1).

The determined values for alanine aminotransferase (ALT) oscillated between 89.56 U/L (25-week-old birds) and 101.11 U/L (55-week-old chickens) in the flock from the improved batteries and between 87.86 U/L (25-week-old chicken's weeks) and 98.83 U/L (55-week-old hens) in those raised in the aviary. As for aspartate aminotransferase (AST), it showed limits of variation in the range of 236.16–300.41 U/L, as determined in hens from improved batteries and in the range of 231.22–294.39 U/L in those in the aviary.

3.2. Numerical Egg Production and Laying Intensity

Average weekly egg production ranged between 6.09 eggs/head (55th week of control) and 6.53 eggs/head (in weeks 31, 32, and 33 of control) in hens reared in the aviary and between 6.17 eggs/head (55th week of bird life) and 6.57 eggs/head (in weeks 31 and 32) in those in improved batteries. For the total studied period (25–55 weeks), the individual egg production was 199.24 eggs/head for the specimens in the aviary and 199.98 eggs/head for those in the battery (Table 2).

Table 2. Numerical egg production and laying intensity.

Birds' Age (Weeks)	Rearing in Loft						Rearing in Battery					
	Mean Flock (Head)	Total Egg Production (Pieces/ Week/ Shelter)	Individual Egg Production			Laying Intensity (%)	Mean Flock (Head)	Total Egg Production (Pieces/ Week/ Shelter)	Individual Egg Production			Laying Intensity (%)
			(Egg/ Week/ Head)	(Egg/ Day/ Head)	Cumulated (Egg /Week/ Head)				(Egg/ Week/ Head)	(Egg/ Day/ Head)	Cumulated (Egg/ Week/ Head)	
25	35,927.5	229,512	6.39	0.91	6.39	91.26	10,214.5	65,445	6.41	0.91	6.41	91.53
26	35,884.5	229,639	6.40	0.91	12.79	91.42	10,209	65,717	6.44	0.91	12.85	91.96
27	35,838	230,671	6.44	0.91	19.23	91.95	10,202.5	66,261	6.49	0.92	19.34	92.78
28	35,787	230,767	6.45	0.92	25.68	92.12	10,198	66,367	6.51	0.92	25.85	92.97
29	35,742.5	231,032	6.46	0.92	32.14	92.34	10,194.5	66,409	6.51	0.93	32.36	93.06
30	35,705	231,815	6.49	0.92	38.63	92.75	10,189	66,573	6.53	0.93	38.89	93.34
31	35,670.5	232,564	6.52	0.93	45.15	93.14	10,182	66,940	6.57	0.93	45.46	93.92
32	35,635	232,657	6.53	0.93	51.68	93.27	10,174	66,838	6.57	0.93	52.03	93.85
33	35,600.5	232,482	6.53	0.93	58.21	93.29	10,164	66,573	6.55	0.93	58.58	93.57
34	35,567.5	232,142	6.53	0.93	64.74	93.24	10,154.5	66,404	6.54	0.93	65.12	93.42
35	35,535.5	231,958	6.53	0.93	71.27	93.25	10,146.5	66,274	6.53	0.93	71.65	93.31
36	35,498	231,340	6.52	0.93	77.79	93.10	10,137.5	66,151	6.52	0.93	78.17	93.22
37	35,455.5	230,766	6.51	0.92	84.30	92.98	10,127.5	65,973	6.51	0.93	84.68	93.06
38	35,415.5	230,158	6.50	0.92	90.80	92.84	10,117	65,826	6.51	0.92	91.19	92.95
39	35,371	229,696	6.49	0.92	97.29	92.77	10,108.5	65,686	6.50	0.92	97.69	92.83
40	35,323.5	229,214	6.49	0.92	103.78	92.70	10,099	65,723	6.49	0.92	104.18	92.79
41	35,275.5	228,829	6.49	0.92	110.27	92.67	10,087.5	65,394	6.48	0.92	110.66	92.61
42	35,232	228,497	6.48	0.92	116.75	92.65	10,077.5	65,308	6.48	0.92	117.14	92.58
43	35,188	228,187	6.48	0.92	123.23	92.64	10,069	65,218	6.48	0.92	123.62	92.53
44	35,137	227,709	6.48	0.92	129.71	92.58	10,061	65,159	6.48	0.92	130.10	92.52
45	35,084	227,120	6.47	0.92	136.18	92.48	10,053.5	65,096	6.47	0.92	136.57	92.50
46	35,028.5	225,902	6.45	0.92	142.63	92.13	10,045.5	64,939	6.46	0.92	143.09	92.35
47	34,970.5	224,794	6.43	0.91	149.06	91.83	10,035	64,653	6.44	0.92	149.47	92.04
48	34,916	224,175	6.42	0.91	155.48	91.72	10,023.5	64,460	6.43	0.91	155.90	91.87
49	34,863.5	223,179	6.40	0.91	161.88	91.45	10,011.5	63,997	6.39	0.91	162.29	91.32
50	34,806.5	222,107	6.38	0.91	168.26	91.16	9997.5	63,754	6.38	0.91	168.67	91.10
51	34,753	219,747	6.32	0.90	174.58	90.33	9983	63,403	6.35	0.90	175.02	90.73
52	34,703.5	216,494	6.24	0.89	180.82	89.12	9970	62,909	6.31	0.90	181.33	90.14
53	34,651	214,930	6.20	0.88	187.02	88.61	9956.5	62,454	6.27	0.89	187.60	89.61
54	34,597.5	212,007	6.13	0.87	193.15	87.54	9943.5	61,753	6.21	0.88	193.81	88.72
55	34,545	210,427	6.09	0.87	199.24	87.02	9929.5	61,277	6.17	0.88	199.98	88.16

At the beginning of the studied period (week 25), the laying rate recorded good levels, with 91.26% in hens reared in the aviary and 91.53% in those in modified batteries. The highest laying intensity (laying peak) was reached in the 33rd week of life by the birds raised in the aviary (93.29%) and in the 31st week by those in improved batteries (93.92%), after which it decreased progressively, so that at the end of the investigations (week 55), the levels achieved were 87.02% for hens in aviaries and 88.16% for those maintained in modified batteries (Table 2).

3.3. Analysis of Batch Outputs

The rate of exits from the batch, including the generative causes, represented another criterion based on which the welfare condition ensured to the laying hens operating in the two rearing systems tested (aviary vs. battery) was assessed (Table 3).

It was found that in the birds raised in the aviary, the number of specimens that left the batch in the first 5 weeks was 227 heads, corresponding to a mortality of 0.62%/week, while in the hens raised in improved batteries, only 25 cases were recorded, resulting in a mortality of only 0.25%/week. Next, the rate of exits from the herd showed fluctuating values, being 0.1% (36 heads) in the 30th week, 0.14% (49 heads) in the 40th week, 0.14% (48 heads) in the 50th week, and 0.15% (52 heads) in the 55th week for hens in the house equipped with an aviary and 0.06% (6 heads), 0.12% (12 heads), 0.15% (15 heads), and 0.17% (17 heads) for specimens from the hall equipped with batteries (Table 3).

Table 3. Situation of exits from flock at the studied hens.

Birds' Age (Weeks)	Rearing in Loft				Rearing in Battery			
	Birds' Flock		Cumulated Exits from Flock				Cumulated Exits from Flock	
	at the Beginning of the Week (Heads)	at the End of the Week (Heads)	Heads	%	at the Beginning of the Week (Heads)	at the End of the Week (Heads)	Heads	%
25	35,950	35,905	45	0.12	10,217	10,212	5	0.05
26	35,905	35,864	86	0.23	10,212	10,206	11	0.11
27	35,864	35,812	138	0.37	10,206	10,199	18	0.18
28	35,812	35,762	188	0.51	10,199	10,197	20	0.20
29	35,762	35,723	227	0.62	10,197	10,192	25	0.25
30	35,723	35,687	263	0.72	10,192	10,186	31	0.31
31	35,687	35,654	296	0.81	10,186	10,178	39	0.39
32	35,654	35,616	334	0.92	10,178	10,170	47	0.49
33	35,616	35,585	365	1.01	10,170	10,158	59	0.61
34	35,585	35,550	400	1.11	10,158	10,151	66	0.68
35	35,550	35,521	429	1.19	10,151	10,142	75	0.77
36	35,521	35,475	475	1.32	10,142	10,133	84	0.86
37	35,475	35,436	514	1.43	10,133	10,122	95	0.97
38	35,436	35,395	555	1.54	10,122	10,112	105	1.07
39	35,395	35,347	603	1.67	10,112	10,105	112	1.14
40	35,347	35,300	650	1.80	10,105	10,093	124	1.26
41	35,300	35,251	699	1.94	10,093	10,082	135	1.37
42	35,251	35,213	737	2.05	10,082	10,073	144	1.46
43	35,213	35,163	787	2.19	10,073	10,065	152	1.54
44	35,163	35,111	839	2.34	10,065	10,057	160	1.62
45	35,111	35,057	893	2.49	10,057	10,050	167	1.69
46	35,057	35,000	950	2.65	10,050	10,041	176	1.78
47	35,000	34,941	1009	2.82	10,041	10,029	188	1.90
48	34,941	34,891	1059	2.96	10,029	10,018	199	2.01
49	34,891	34,836	1114	3.12	10,018	10,005	212	2.14
50	34,836	34,777	1173	3.29	10,005	9990	227	2.29
51	34,777	34,729	1221	3.43	9990	9976	241	2.43
52	34,729	34,678	1272	3.58	9976	9964	253	2.55
53	34,678	34,624	1326	3.73	9964	9949	268	2.70
54	34,624	34,571	1379	3.99	9949	9938	279	2.81
55	34,571	34,519	1431	4.14	9938	9921	296	2.98

During the entire study period (25–55 weeks), the total exits from the flock of hens raised in the aviary were at a level of 4.14% (1431 heads from an initial flock of 35,950 heads), while for those bred in battery this figure was only 2.98% (296 heads from an initial herd of 10,217 heads).

3.4. Causes of Batch Outputs

The main cause of batch exits was represented by mechanical accidents, but with large differences between the two growth systems used. Thus, for hens in the aviary, the mortality due to accidents was 49.27% (705 heads) with limits between 42.59% (45th week) and 55.55% (30th week) of total deaths (1431 heads), while in hens raised in improved batteries, batch losses due to accidents were only 39.53% (117 heads) with limits between 20.0% (38th week) and 100% (28th week) out of a total of 296 specimens withdrawn from the population (Table 4).

Another cause of exits from the batch referred to obstetric diseases, which, in the batch of birds raised in the aviary, represented 35.64% of the total losses (510 heads out of the total exits), with a minimum of 18.0% recorded in week 28 and a maximum of 50.0% in the 42nd week. In the batch grown in improved batteries, the average value of the output from the batch due to obstetrical diseases was 33.11% (98 heads of total losses), oscillating between 0.0% in the 28th week and 60.0% as they were in the 25th week.

The exits from the batch due to internal diseases were at much lower levels compared to the other causes, being only 15.09% in the case of specimens raised in the aviary (216 heads out of total losses) and 27.36% in those raised in improved batteries (81 chapters of total losses); in both cases, the analyzed parameter had an asymptotic evolution, with limits of 2.08–36.00%/week in hens from the aviary and 0.0–55.56%/week in those from improved batteries (Table 4).

Table 4. Causes of batch outputs.

Birds' Age (Weeks)	Rearing in Loft						Rearing in Battery					
	Total Exits (Heads/Week)	Causes					Total Exits (Heads/Week)	Causes				
		Accidents		Obstetrical Diseases		Internal Diseases		Accidents		Obstetrical Diseases		Internal Diseases
		Head	%	Head	%	Head	%		Head	%	Head	%	Head	%

Birds' Age (Weeks)	Total Exits (Heads/Week)	Accidents Head	Accidents %	Obstetrical Diseases Head	Obstetrical Diseases %	Internal Diseases Head	Internal Diseases %	Total Exits (Heads/Week)	Accidents Head	Accidents %	Obstetrical Diseases Head	Obstetrical Diseases %	Internal Diseases Head	Internal Diseases %
25	45	21	46.67	11	24.44	13	28.89	5	2	40.00	3	60.00	-	-
26	41	19	46.34	13	31.71	9	21.95	6	3	50.00	2	33.33	1	16.67
27	52	26	50.00	12	23.08	14	26.92	7	3	42.86	3	42.86	1	14.28
28	50	23	46.00	9	18.00	18	36.00	2	2	100	-	-	-	-
29	39	19	48.72	16	41.03	4	10.26	5	3	60.00	2	40.00	-	-
30	36	20	55.55	14	38.89	2	5.55	6	2	33.33	3	50.00	1	16.67
31	33	16	48.48	13	39.39	4	12.12	8	2	25.00	3	37.50	3	37.50
32	38	18	47.37	15	39.47	5	13.16	8	3	37.50	2	25.00	3	37.50
33	31	17	54.83	11	35.48	3	9.68	12	5	41.67	4	33.33	3	25.00
34	35	18	51.43	14	40.00	3	8.57	7	4	57.14	2	28.57	1	14.29
35	29	16	55.17	11	37.93	2	6.90	9	4	44.44	3	33.33	2	22.22
36	46	24	52.19	19	41.30	3	6.52	9	3	33.33	3	33.33	3	33.33
37	39	19	48.72	17	43.59	3	7.69	11	3	27.27	4	36.36	4	36.33
38	41	20	48.78	15	48.79	6	14.63	10	2	20.00	4	40.00	4	40.00
39	48	25	52.08	22	45.83	1	2.08	7	3	42.86	2	28.57	2	28.57
40	47	22	46.81	19	40.43	6	12.76	12	4	33.33	3	25.00	5	41.67
41	49	23	46.94	14	28.57	12	24.48	11	4	36.36	4	36.36	3	27.27
42	38	17	44.74	19	50.00	2	5.26	9	2	22.22	2	22.22	5	55.56
43	50	27	54.00	17	34.00	6	12.00	8	3	37.5	4	50.00	1	12.50
44	52	25	48.08	18	34.61	9	17.30	8	2	25.00	3	37.50	3	37.50
45	54	23	42.59	19	35.18	12	22.22	7	2	28.57	2	28.57	3	42.86
46	57	26	45.61	19	33.33	12	21.05	9	2	22.22	5	55.56	2	22.22
47	59	31	52.54	23	38.98	5	8.47	12	3	25.00	5	41.67	4	33.33
48	50	24	48.00	21	42.00	5	10.00	11	4	36.36	4	36.36	3	27.28
49	55	28	50.91	19	34.54	8	14.54	13	4	30.77	6	46.15	3	23.08
50	59	29	49.15	22	37.29	8	13.56	15	6	40.00	4	26.67	5	33.33
51	48	26	54.17	19	39.58	3	6.25	14	7	50.00	3	21.43	4	28.57
52	51	24	47.06	16	31.37	11	21.57	12	8	66.66	2	16.67	2	16.67
53	54	29	53.70	15	27.78	10	18.52	15	8	53.33	4	26.67	3	20.00
54	53	23	43.39	17	32.08	13	24.53	11	7	63.64	2	18.18	2	18.18
55	52	27	51.92	21	40.38	4	7.69	17	7	41.18	5	29.41	5	29.41
Total	1431	705	49.27	510	35.64	216	15.09	296	117	39.53	98	33.11	81	27.36

3.5. Consumption of Mixed Fodders

In hens in the hall equipped with an aviary, the average daily consumption of mixed feeders recorded values of 120.45 g/head/day in the first studied period (25–40 weeks) and 124.10 g/head/day in the second period (41–55 weeks), while the feed conversion index was at levels of 130.0 g m.f./egg in the period of 25–40 weeks and 136.47 g m.f./egg in the interval of 46–55 weeks.

In the case of birds operating in improved batteries, the consumptions were better, both in the age period of 25–40 weeks (average daily consumption = 113.10 g/head/day; feed conversion index = 121.55 g m.f./egg), as well as during the period of 41–55 weeks (average daily consumption = 118.87 g/head/day; feed conversion index = 130.25 g m.f./egg).

Over the total period studied (25–55 weeks), the most convenient consumption of combined feed was achieved by the birds in the house equipped with improved batteries, which had an average daily consumption of 115.87 g/head/day (average effective = 10.092 head; feed consumed = 253.763 kg) and a feed conversion index of 125.69 g n.c./egg (egg production = 2.018.934 egg). Comparatively, hens reared in the aviary registered an average consumption of 122.20 g/head/day (average effective = 35.281 heads; feed consumed = 935.551 kg) and a feed conversion index of 133.07 g m.f./egg (egg production = 7.030.517 egg) (Table 5).

Table 5. Consumption of mixed fodders for studied hens.

Age Period (Weeks)	Specification	Exploitation System	
		in Loft	in Battery
25–40 16 weeks (112 days)	Mean flock (heads)	35.622	10.164
	Consumed fodders (kg/period)	480.553	128.747
	Daily mean consumption (g/head/day)	120.45	113.10
	Egg production (pieces/period)	3.696.413	1.059.160
	Feed conversion index (g m.f./egg)	130	121.55
41–55 15 weeks (105 days)	Mean flock (heads)	34.917	10.016
	Consumed fodders (kg/period)	454.998	125.016
	Daily mean consumption (g/head/day)	124.10	118.87
	Egg production (pieces/period)	3.334.104	959.774
	Feed conversion index (g m.f./egg)	136.47	130.25
25–55 31 weeks (217 days)	Mean flock (heads)	35.281	10.092
	Consumed fodders (kg/period)	935.551	253.763
	Daily mean consumption (g/head/day)	122.20	115.87
	Egg production (pieces/period)	7.030.517	2.018.934
	Feed conversion index (g m.f./egg)	133.07	125.69

4. Discussion

4.1. Biochemical Indicators

The determination of the main biochemical indicators was aimed at evaluating the comfort state provided to the studied hens, in correlation with the adopted breeding method (aviary vs. battery).

From the general analysis of the data regarding blood markers, it emerged that in most cases, their level was higher in hens raised in the aviary, regardless of age, a finding that reveals the fact that this production system ensures suitable wellbeing, as well as a good balance between neurohormonal function and metabolic activity.

The second finding was that the level of biochemical indicators was inversely related to the egg-laying intensity of the birds, in the sense that their level in the blood increased towards the end of egg laying, given the decrease in the number of nutrients necessary for egg formation (the rate of egg laying was increasingly reduced towards the end of the spawning period).

The high and even very high values of the coefficients of variation for each of the biochemical characters studied can be attributed to the physiological state in which the birds were at the time of sampling (with the egg in different stages of formation, with the egg prepared for oviposition, or with the egg already expelled).

For example, the level of cholesterol and triglycerides (components involved in the formation of lipids in the yolk) showed significant increases between the birds at the beginning of laying (week 25) and those at the end of laying (week 55), these being 31.18–34.10% for cholesterol and 8.17–8.77% for triglycerides, respectively. A similar situation was recorded for blood proteins (role in the formation of egg white), which increased by 28.84% in hens from the aviary and by 20.63% in those from the improved batteries.

In clinically healthy hens examined every five weeks of laying, Suchy et al. [22] found the following limits for biochemical parameters: protein = 47.43–60.45 g/L; glucose = 13.36–14.97 mmol/L; cholesterol = 2.73–6.18 mmol/L; calcium = 5.26–7.19 mmol/L; phosphorus = 1.25–1.90 mmol/L; magnesium = 1.08–1.42 mmol/L; sodium = 141.03–148.30 mmol/L; potassium = 3.40–5.00 mmol/L. The authors stated that the plasma levels of some biochemical parameters were directly influenced by the variations in laying capacity.

Triglyceride concentration determined in free-range Hy-Line hens was significantly lower than in those raised in battery cages ($p < 0.05$) [9].

The use of diets with reduced levels of crude protein in the Lohmann LSL hybrid significantly reduced the content of uric acid and albumin in the blood (variant with

13% CP) compared to the control group (16% CP); in contrast, blood triglyceride levels were higher ($p < 0.05$) in hens fed 14% and 13% CP than in those fed the control diet [32]. On the other hand, it was found that diets with low levels of metabolizable energy and crude protein reduced serum triglyceride and cholesterol concentrations in Gushi chickens, regardless of the density provided in the rearing cages (20–50 head/m^2) [34].

Glucose used in energy metabolism was determined in smaller amounts in young birds (they were more active), and higher in birds aged 55 weeks (they were much quieter), with the differences between the two periods being 21.11% in the case of specimens from the aviary and 27.97% for those from the battery.

In an experiment that looked at the influence of fasting on blood parameters in different genotypes (Bovans, ISA Brown, and Ross-508), it was found that the lack of food caused a significant decrease in glucose levels and a significant increase in plasma cholesterol, which suggests a reduction in the welfare of the hens [29].

It is known that for the formation of the eggshell, a bird's blood must transport 100–150 mg Ca/h [44], and if calcium is not absorbed quickly at the intestinal level, calcemia can be established within only 10–12 min [37]. In the studied birds, the blood level of calcium was influenced only by the rate of egg formation (the lowest values were in hens in full production—week 35, and the highest were in hens at the end of laying—week 55). The amount of phosphorus in the blood evolved similarly, which had lower values at high laying intensities (35th week) and higher values at low laying intensities (55th week), with the mention that the oscillations recorded were generated by the different status of each specimen from which the sample was collected (phosphorus in the blood increases significantly during the formation of the egg shell in the uterus).

Comparing the blood plasma mineral profile (Ca, P, K, Mg, Zn, Cu, and Se) in ISA Brown hens from three different batteries (traditional batteries, improved batteries, and deep litter system) did not reveal a significant effect of the system of growth on the monitored indicators, with the obtained values falling within the physiological range [24].

In Leghorn hens, a direct correlation was found between the surface provided in the rearing battery and the level of some blood indicators; for example, in the variant with 500 cm^2 cage/head, significantly lower levels ($p < 0.05$) were found for plasma calcium and plasma uric acid, but significantly higher levels ($p < 0.05$) were found for iron, magnesium, and plasma glucose compared to specimens that benefited from 2000 cm^2/head [14].

The uric acid level recorded higher values at the first determination performed (week 25), but it was significantly reduced at the last one, by 23.19% in the hens in the aviary and by 28.60% in those in the battery, while the amount of urea increased by 15.55% (aviary) and by 16.21% (battery) between the first harvest and the last harvest, respectively.

In Hisex Brown hens reared in different systems (battery and permanent litter) and with different diets (reduced levels of crude protein), a significant interaction was observed between the experimental factors and the level of serum protein, uric acid, and total blood cholesterol ($p < 0.05$). Diets with low CP levels decreased serum protein and cholesterol in litter-reared birds, as well as uric acid in battery birds, but without exceeding physiological limits [33].

The determined values for blood enzymes indicated fluctuations concerning liver metabolism, generated by the different laying rhythms of the birds, but in both cases, higher values were recorded in young birds (aged 25 weeks) and lower values were recorded in old ones (week 55). Thus, the values for alanine aminotransferase (ALT) increased by 12.48% in hens from the aviary and by 12.89% in those from batteries, while aspartate aminotransferase (AST) showed quantitative increases of 27.32% (aviary) and 27.21% (battery), respectively, between the two mentioned age periods.

4.2. Numerical Egg Production

Between the two rearing systems tested (battery vs. aviary) there is a clear difference regarding the freedom of movement of the birds and what corresponds to the level of ensured wellbeing [45], but additional energy consumption is also implied, to the detriment of the productivity of the birds [46].

Contrary to expectations, the experimental factor applied in our research (the rearing system) did not generate significant differences in terms of individual egg production, these being only 0.35% in favor of those in the battery (199.24 eggs/head vs. 199.98 eggs/head); compared to the theoretical potential of the ISA Brown hybrid, the egg production of hens in the aviary was lower by 1.85%, and of the battery hens by only 1.49%.

This indicates that the functionality of the reproductive apparatus of the hens in the aviary was stimulated by the superior welfare conditions, materialized in a high and constant numerical egg production, similar to that of the battery-reared birds [47].

This aspect is also confirmed by the fact that both groups of birds benefited from the same technological parameters (light program, microclimate, and food), the only difference being that the hens in the aviary had the opportunity to manifest their instincts much more freely, with positive effects on the neurohormonal system.

4.3. Laying Intensity

At the beginning of the investigations (the 25th week of the birds' life), the egg-laying intensity recorded values lower than the theoretical potential of the hybrid used (lower by 4.74% in the hens in the aviary and by 4.47% in the ones in the batteries), the probable cause being not achieving optimal body weights at the time of spawning (the information comes from the data provided by the unit where the research was carried out).

This aspect also had an impact on the age at which the maximum egg-laying intensity was reached (the egg-laying peak), this being achieved in the 31st week of life for the specimens raised in the battery and in the 33rd week for the birds in the aviary; the delay in the laying peak in caged hens was due to both lower body weights and disruptions to the normal ovulatory cycle, amid the constant agitation of birds that benefited from large areas of movement.

During the 31 weeks of investigations, the studied hens achieved a good average egg-laying intensity (91.81% in the specimens raised in the aviary and 92.17% in those operated in improved batteries), very close to the potential of the hybrid used (lower by 0.82% and 0.06%, respectively) and similar to those presented in the specialized literature.

Thus, for ISA Brown hens, an average egg-laying intensity of 91.42% was reported for specimens raised in improved batteries and only 75.94% for those in classic batteries, in the age period of 37–45 weeks [21].

In an experiment that aimed at the effect of the presence of males on the productivity of ISA Brown hens raised on permanent litter (6.6 head/m^2), an average egg-laying intensity of 76.21% was obtained in the version without males and of 84.4% in the version with introduced 1 male for 10 females; the control period was 18–31 weeks [18].

Egg production also depends on the technique of rearing replacement juveniles. Thus, in the case of adult ISA Brown hens raised in an aviary, an average egg-laying intensity of 85% was obtained when they came from chicks raised in a youth aviary and only 84% in specimens from chicks raised in a classic battery [15].

4.4. Batch Outputs Rate

The improved batteries limit the movement space of each bird [30], while the aviary is an equipment that respects the welfare condition, because it gives the birds the possibility to move on much larger surfaces, both vertically (between the floors of the same section) and horizontally (between two neighboring sections) [48].

Compared to the theoretical mortality rate specific to the ISA Brown hybrid for the age period of 25–55 weeks (3.1%), the birds studied by us had a 1.04% higher mortality rate in the specimens raised in the aviary (4.14%) and a 0.12% lower rate in those grown in the improved battery (2.98%).

Similar results were reported by Huneau-Salauen et al. [11], who found a mortality rate of 4.2% in ISA Brown hens at a density of 40 heads/cage in improved Zucami batteries (768 cm^2/bird) and a mortality rate of only 2.4% in the version with 20 heads/cage.

From the direct observations of the two groups of birds, it turned out that the hens in the aviary had a much more active behavior, constantly moving between the areas of interest (feeders, waterers, and nests) and fully manifesting their instincts (perching, snorting, etc.). This freedom of movement generated a greater number of mechanical accidents (in contact with the hard parts of the equipment), but also of internal diseases, the temporal distribution of which was somewhat uniform throughout the entire period studied.

In an experiment that looked at the effect of genotype and the content of essential amino acids (methionine + cystine) on the feather chipping phenomenon, ISA Brown hens had a mortality rate of 17.5% (not attributed to the provided diet), compared to 2.4% in New Hampshire hens and 0% in White Leghorns [31].

4.5. Causes for Batch Outputs

For the exits from the batch studied by us, three main causes were identified (accidents, obstetric diseases, and internal diseases) whose weight was directly influenced by the breeding system used.

From this point of view, it was found that batch losses due to mechanical accidents (especially wing and leg fractures) were much more frequent in hens from the aviary, higher by 9.74% than those from the battery, the incriminating factor in this case being the movement of birds on much larger surfaces than in the battery; another factor generating accidental mortality is the effect of fights between birds to establish the social hierarchy, at least for the beginning of the investigations. The highest rate of mechanical accidents was recorded in the age period of 46–55 weeks (37.87% for hens in the aviary and 47.86% for those in the battery compared to total accidents), due to the increase in bone fragility (higher consumption of Ca for the shell eggs), increasing body weights, and quite possibly, the decrease in the agility of the birds as they get older (especially in the aviary).

In the growth conditions ensured by the aviary, ISA Brown hens recorded a 1.37% higher proportion of flock exits (due to fractures and injuries at the level of the sternal carina) compared to Dekalb White hens [17].

As for the rate of exits from the batch due to obstetrical diseases, it was higher by 2.53% in hens in the aviary, due to the fact that they moved much more in order to satisfy their natural instincts, thus affecting the regularity of the ovulatory cycle. It should be noted that the highest levels for obstetrical diseases were in the period of 25–45 weeks (62.35% for hens in the aviary and 59.18% for those in the battery out of total exits caused by obstetrical diseases), when an increased incidence of "peritonitis" was found "vitelline" with various forms of ovaritis due to the lack of correlation between the intense rhythm of ovulation and the insufficient amount of hormones stimulating the maturation of ovarian follicles.

There were also cases of birds with abdominal peritonitis (the oviduct can no longer capture the mature follicles and they fall into the abdominal cavity), and towards the end of the studied period, the number of birds with uterine prolapse, caused by the increased volume of the eggs, also increased quite a lot, and with both categories requiring the withdrawal of birds from the batch.

The exits from the batch caused by the manifestation of internal diseases registered a higher rate in the hens in batteries compared to those in the aviary, the difference between the batches being 12.27%. Among the internal diseases, a high incidence of mortality was due to the "fatty liver" syndrome detected only in the specimens in the battery (massive deposits of fat on the abdomen and mesentery due to lack of movement), especially after the age of 40 weeks. Another cause was the manifestation of necrotic enteritis, whose etiological agent is *Clostridium perfringens*, a bacterial species that colonizes the intestines of birds (the replacement chicks were probably not treated for this disease), but which can contaminate the administered chicks, the air admitted to the sheds, or even the bedding used [49].

4.6. Consumption of Mixed Fodders

This productive indicator was correlated with the egg-laying intensity of the birds but influenced by the freedom of movement granted to them by the state of ensured well-being.

For example, the numerical egg production from 25–40 weeks was 104.21 eggs/head (battery) compared to only 103.77 eggs/head (aviary), hence a lower feed conversion index by 6.95% in hens raised in improved batteries; in the age period of 41–55 weeks, the egg production of the birds was reduced (95.49 pcs./head in the aviary and 95.82 pcs./head in the batteries), which led to higher values of the feed conversion index, with the mention that hens reared in improved batteries still recorded more favorable consumption levels (4.77% lower than aviary hens).

The obtained data showed that hens in the aviary had a higher average daily consumption than those in the battery (by 6.50% in the period of 25–40 weeks and by 4.40% in the period of 41–55 weeks), due exclusively to the additional energy expenses caused by permanent movement in the much more generous space offered by the aviary.

The consumption of combined feed calculated for the entire period studied was at levels similar to those reported in different experiments, both for hens in the aviary (daily consumption = 122.20 g n.c./head/day; feed conversion index = 133.07 g n.c./egg) and especially for those in the shed with improved batteries (average consumption = 115.87 g n.c./head/day; conversion index = 125.69 g n.c./egg), which proves that both types of equipment tested by us ensure conditions of exteriorization of the productive potential of hybrids specialized in egg production.

When reared in the improved battery (768 cm^2/bird) for 53 weeks, ISA Brown hens achieved an average daily consumption of 111.4–113.2 g d.c./head/day and a conversion index of 2.04–2.06 kg d.c./kg eggs, depending on the number of birds in the cage (20 vs. 40 heads/cage) [11].

ISA Brown hens weaned and in battery cages with different interior layouts (classic nest vs. nest lined with artificial turf or plastic mesh) had average daily intakes of 110.6–111.1 g n.c./head/day [28].

5. Conclusions

Our investigations concluded that the exploitation of laying hens in the aviary system ensures the most optimal conditions for the externalization of the birds' natural behaviors, with positive repercussions on the metabolic functions, resulting in a good state of health and high productive levels, comparable to those of birds raised in the battery. However, we believe that it is necessary to continue research in this direction by evaluating molecular, biochemical, or hormonal markers that accurately reflect the different types of stress to improve the productivity and quality of life of birds exploited for egg production.

Author Contributions: Conceptualization, A.U., M.-G.D., C.P., M.-G.U., R.-M.R.-R., M.-I.U. and D.S.; methodology, A.U., M.-I.U., M.S., C.P., B.-V.A. and D.S.; software, M.-G.D. and C.P.; validation, A.U., C.P. and M.-G.D.; formal analysis, M.-G.U., C.S. and R.-M.R.-R.; investigation, A.U., M.-G.D., C.P., M.-I.U. and M.S.; data curation, A.U. and M.-G.D.; writing—original draft preparation, A.U., B.-V.A., C.P., M.-I.U. and M.-G.U.; writing—review and editing, A.U., M.-G.D., M.S. and C.S.; supervision, A.U. and M.-G.D. All authors have read and agreed to the published version of the manuscript.

Funding: This research received no external funding.

Institutional Review Board Statement: Not applicable.

Data Availability Statement: Not applicable.

Acknowledgments: We thank to the SC Condor Matca SRL, Galati County, Romania.

Conflicts of Interest: The authors declare no conflict of interest.

References

1.   Rozempolska-Rucinska, I.; Czech, A.; Kasperek, K.; Zieba, G.; Ziemianska, A. Behaviour and stress in three breeds of laying hens kept in the same environment. *S. Afr. J. Anim. Sci.* **2020**, *50*, 272–280. [CrossRef]
2.   Sosnowka-Czajka, E.; Herbut, E.; Skomorucha, I.; Muchacka, R. Welfare levels in heritage breed vs. Commercial laying hens in the litter system. *Ann. Anim. Sci.* **2011**, *11*, 585–595. [CrossRef]
3.   Craig, J.V.; Swanson, J.C. Welfare perspectives on hens kept for egg-production. *Poult. Sci.* **1994**, *73*, 921–938. [CrossRef] [PubMed]
4.   Savory, C.J.; Jack, M.C.; Sandilands, V. Behavioural responses to different floor space allowances in small groups of laying hens. *Br. Poult. Sci.* **2006**, *47*, 120–124. [CrossRef]
5.   Zotte, A.D.; Gottardo, F.; Ravarotto, L.; Bertuzzi, S. The welfare evaluation of laying hens reared with alternative housing systems. *Ital. J. Anim. Sci.* **2003**, *2*, 462–464.
6.   Appleby, M.C.; Walker, A.W.; Nicol, C.J. Development of furnished cages for laying hens. *Br. Poult. Sci.* **2002**, *43*, 489–500. [CrossRef]
7.   Roberts, J. Welfare standards for laying hens. Achieving sustainable production of eggs. *Anim. Welf. Sustain.* **2017**, *17*, 85–97.
8.   Bozkurt, Z.; Bayram, I.; Bulbul, A.; Aktepe, O.C. Effects of Strain, Cage Density and Position on Immune Response to Vaccines and Blood Parameters in Layer Pullets. *Kafkas Univ. Vet. Fak. Derg.* **2008**, *14*, 191–204.
9.   Yang, H.M.; Yang, Z.; Wang, W.; Wang, Z.Y.; Sun, H.N.; Ju, X.J.; Qi, X.M. Effects of different housing systems on visceral organs, serum biochemical proportions, immune performance and egg quality of laying hens. *Eur. Poult. Sci.* **2014**, *78*, 48.
10.   Blokhus, H.J.; Fiks van Niekerk, T.; Bessei, W. The LayWel projects: Welfare implications of changes in productions systems for laying hens. *Worlds Poult. Sci. J.* **2007**, *63*, 101–114. [CrossRef]
11.   Huneau-Salauen, A.; Guinebretiere, M.; Taktak, A.; Huonnic, D.; Michel, V. Furnished cages for laying hens: Study of the effects of group size and litter provision on laying location, zootechnical performance and egg quality. *Animals* **2011**, *5*, 911–917. [CrossRef] [PubMed]
12.   Ozenturk, U.; Yildiz, A. Assessment of Egg Quality in Native and Foreign Laying Hybrids Reared in Different Cage Densities. *Braz. J. Poult. Sci.* **2021**, *22*, 85–97. [CrossRef]
13.   Guinebretiere, M.; Huneau-Salauen, A.; Huonnic, D.; Michel, V. Plumage condition, body weight, mortality, and zootechnical performances: The effects of linings and litter provision in furnished cages for laying hens. *Poult. Sci.* **2013**, *92*, 51–59. [CrossRef] [PubMed]
14.   Saki, A.A.; Zamani, P.; Rahmati, M.; Mahmoudi, H. The effect of cage density on laying hen performance, egg quality, and excreta minerals. *J. Appl. Poult. Res.* **2012**, *21*, 467–475. [CrossRef]
15.   Colson, S.; Michel, V.; Arnould, C. Welfare of laying hens housed in cages and in aviaries: What about fearfulness? *Arch. Geflugelkd.* **2006**, *70*, 261–269.
16.   Fossum, O.; Jansson, D.S.; Etterlin, P.E.; Vagsholm, I. Causes of mortality in laying hens in different housing systems in 2001 to 2004. *Acta Vet. Scand.* **2009**, *51*, 3–11. [CrossRef]
17.   Stratmann, A.; Froehlich, E.K.F.; Harlander-Matauschek, A.; Schrader, L.; Toscano, M.J.; Wuerbel, H.; Gebhardt-Henrich, S.G. Soft Perches in an Aviary System Reduce Incidence of Keel Bone Damage in Laying Hens. *PLoS ONE* **2015**, *10*, e0122568. [CrossRef]
18.   De Oliveira Pereira, D.C.; Da Silva Miranda, K.O.; Dematte Filho, L.C.; Pereira, G.V.; De Stefano Piedade, S.M.; Berno, P.R. Presence of roosters in an alternative egg production system aiming at animal welfare. *Rev. Bras. De Zootec.-Braz. J. Anim. Sci.* **2017**, *46*, 175–184. [CrossRef]
19.   Marcq, C.; Marlier, D.; Beckers, Y. Improving adjuvant systems for polyclonal egg yolk antibody (IgY) production in laying hens in terms of productivity and animal welfare. *Vet. Immunol. Immunopathol.* **2015**, *165*, 54–63. [CrossRef]
20.   Sossidou, E.N.; Elson, H.A. Hens' welfare to egg quality: An European perspective. *Worlds Poult. Sci. J.* **2009**, *65*, 709–718. [CrossRef]
21.   Corona, J.; Trompiz, J.; Jerez, N.; Gomez, A.; Rincon, H. Effect of warehouse type on productive variables and egg quality in laying hens Isa Brown. *Rev. Fac. Agron. Univ. Zulia* **2016**, *32*, 345–360.
22.   Suchy, P.; Strakova, E.; Vecerek, V.; Sterc, P. Biochemical studies of blood in hens during the laying period. *Czech J. Anim. Sci.* **2001**, *46*, 383–387.
23.   Pavlik, A.; Pokludova, M.; Zapletal, D.; Jelinek, P. Effects of housing systems on biochemical indicators of blood plasma in laying hens. *Acta Vet. Brno* **2007**, *76*, 339–347. [CrossRef]
24.   Pavlik, A.; Lichovnikova, M.; Jelinek, P. Blood Plasma Mineral Profile and Qualitative Indicators of the Eggshell in Laying Hens in Different Housing Systems. *Acta Vet. Brno* **2009**, *78*, 419–429. [CrossRef]
25.   Ito, S.; Tanka, T.; Yoshimoto, T. Effects of an 'enrichment feeder' on behavior and feather conditions of caged laying hens. *Anim. Sci. J.* **2002**, *73*, 149–153. [CrossRef]
26.   De Haas, E.N.; Kemp, B.; Bolhuis, J.E.; Groothuis, T.; Rodenburg, T.B. Fear, stress, and feather pecking in commercial white and brown laying hen parent-stock flocks and their relationships with production parameters. *Poult. Sci.* **2013**, *92*, 2259–2269. [CrossRef]
27.   De Marco, M.; Miro, S.M.; Tarantola, M.; Bergagna, S.; Mellia, E.; Gennero, M.S.; Schiavone, A. Effect of genotype and transport on tonic immobility and heterophil/lymphocyte ratio in two local Italian breeds and Isa Brown hens kept under free-range conditions. *Ital. J. Anim. Sci.* **2014**, *12*, e78. [CrossRef]

28. Guinebretiere, M.; Mika, A.; Michel, V.; Balaine, L.; Thomas, R.; Keita, A.; Pol, F. Effects of Management Strategies on Non-Beak-Trimmed Laying Hens in Furnished Cages that Were Reared in a Non-Cage System. *Animals* **2020**, *10*, 399. [CrossRef]
29. Altan, O.; Oguz, I.; Erbayraktar, Z.; Yimaz, O.; Bayraktarl, H.; Sis, B. The effect of prolonged fasting and refeeding on GSH-Px activity. plasma glucose and cholesterol levels and welfare of older hens from different genotypes. *Arch. Geflugelkd.* **2005**, *69*, 185–191.
30. Lay, D.C.; Fluton, R.M.; Hester, P.Y. Hen welfare in different housing systems. *Poult. Sci.* **2011**, *90*, 278–294. [CrossRef]
31. Kjaer, J.B.; Sorensen, P. Feather pecking and cannibalism in free-range laying hens as affected by genotype. dietary level of methionine plus cysteine, light intensity during rearing and age at first access to the range area. *Appl. Anim. Behav. Sci.* **2002**, *76*, 21–39. [CrossRef]
32. Torki, M.; Nasiroleslami, M.; Ghasemi, H.A. The effects of different protein levels in laying hens under hot summer conditions. *Anim. Prod. Sci.* **2017**, *57*, 927–934. [CrossRef]
33. Viana, E.F.; De Carvalho Mello, H.H.; Carvalho, F.B.; Café, M.B.; Mogyca Leandro, N.S.; Arnhold, E.; Stringhini, J.H. Blood biochemical parameters and organ development of brown layers fed reduced dietary protein levels in two rearing systems. *Anim. Biosci.* **2022**, *35*, 444–452. [CrossRef] [PubMed]
34. Wan, Y.I.; Ma, R.; Li, Y.; Liu, W.; Li, J.Y.; Zhan, K. Effect of a Large-sized Cage with a Low Metabolizable Energy and Low Crude Protein Diet on Growth Performance. Feed Cost and Blood Parameters of Growing Layers. *J. Poult. Sci.* **2021**, *58*, 70–77. [CrossRef]
35. Abrahamsson, P.; Tauson, R.; Elwinger, K. Effects on production, health and egg quality of varying proportions of wheat and barley in diets for two hybrids of laying hens kept in different housing systems. *Acta Agric. Scand. Sect. A-Anim. Sci.* **1996**, *46*, 173–182. [CrossRef]
36. Rodriguez-Aurrekoetxea, A.; Estevez, I. Use of space and its impact on the welfare of laying hens in a commercial free-range system. *Poult. Sci.* **2016**, *95*, 2503–2513.
37. Dedousi, A.; Kritsa, M.Z.; Stojcic, M.D.; Sfetsas, T.; Sentas, A.; Sossidou, E. Production Performance, Egg Quality Characteristics, Fatty Acid Profile and Health Lipid Indices of Produced Eggs, Blood Biochemical Parameters and Welfare Indicators of Laying Hens Fed Dried Olive Pulp. *Sustainability* **2022**, *14*, 3157. [CrossRef]
38. ISA Brown—Hendrix ISA. Available online: https://www.hendrix-isa.com/en/find-out-more-on-our-breeds-dekalb_white-shaver_white-bovans_brown-isa_brown/isa-brown/ (accessed on 11 July 2021).
39. Natura Nova—Big Dutchman. Available online: https://www.bigdutchman.de/de/legehennenhaltung/produkte/detail/natura-nova/ (accessed on 11 June 2021).
40. Simeanu, D. *Nutriția și Alimentația Animalelor*; Ion Ionescu de la Brad: Iași, Romania, 2018; ISBN 978-973-147-303-1.
41. Usturoi, M.G. *Creşterea Păsărilor*; Ion Ionescu de la Brad: Iaşi, Romania, 2008; ISBN 978-973-147-012-2.
42. Burtis, C.A.; Ashwood, E.R.; Bruns, D.E. *Tielz Textbook of Clinical Chemistry and Molecular Diagnostics*, 4th ed.; WB Saunders Co.: Philadelphia, PA, USA, 2005.
43. Young, D.S.; Friedman, R.B. *Effects of Disease on Clinical Laboratory Tests*, 4th ed.; AACC Press: Washington, DC, USA, 2001.
44. Burtis, C.A.; Ashwood, E.R.; Bruns, D.E. *Tietz Textbook of Clinical Chemistry and Molecular Diagnostics*, 5th ed.; WB Saunders Co.: Philadelphia, PA, USA, 2012.
45. Tactacan, G.B.; Guenter, W.; Lewis, N.J.; Rodriquez-Lecompte, J.C.; House, J.D. Performance and welfare of laying hens in conventional and enriched cages. *Poult. Sci.* **2009**, *88*, 698–707. [CrossRef]
46. Singh, R.; Cheng, K.; Silversides, F. Production performance and egg quality of four strains of laying hens kept in conventional cages and floor pens. *Poult. Sci.* **2009**, *88*, 256–264. [CrossRef]
47. Petričević, V.; Škrbić, Z.; Lukić, M.; Petričević, M.; Dosković, V.; Rakonjac, S.; Marinković, M. Effect of genotype and age of laying hens on the quality of eggs and egg shells. *Sci. Papers Ser. D Anim. Sci.* **2017**, *60*, 166–170.
48. Widowski, T.; Hemsworth, P.J.; Barnett, J.; Rault, L. Laying hen welfare I. Social environment and space. *Worlds Poult. Sci. J.* **2016**, *72*, 333–342. [CrossRef]
49. Srinivasan, P.; Balasubramaniam, G.A.; Murthy, T.R.G.K.; Balachandran, P. Bacteriological and pathological studies of egg peritonitis in commercial layer chicken in Namakkal area. *Asian Pac. J. Trop. Biomed.* **2013**, *3*, 988–994. [CrossRef] [PubMed]

Article

Determination of the Optimal Dietary Amino Acid Ratio Based on Egg Quality for Japanese Quail Breeder

Lizia C. Carvalho [1], Dimitri Malheiros [2], Michele B. Lima [1], Tatyany S. A. Mani [1], Jaqueline A. Pavanini [1], Ramon D. Malheiros [2] and Edney P. Silva [1,*]

[1]   Department of Animal Sciences, UNESP—Universidade Estadual Paulista, São Paulo 14883900, Brazil
[2]   Prestage Department of Poultry Science, North Carolina State University, Raleigh, NC 27607, USA
*   Correspondence: edney.silva@unesp.br

Abstract: The objective of this study was to determine the ideal amino acid ratio for Japanese quail based on egg quality. In total, 120 Japanese quail were used. A completely randomized design was used with 12 treatments and 10 replicates per treatment. The treatments consisted of a balanced protein (BP) and the subsequent 11 diets were obtained by the 40% deletion of the BP a specific test for Lys, Met + Cys, Thr, Trp, Arg, Gly + Ser, Val, Ile, Leu, His, and Phr + Tyr. The trial lasted for 25 days. At the end of the trial, egg weight (EW), albumen height, albumen diameter, albumen index, yolk height, yolk diameter, yolk index, Haugh unit, eggshell weight (ESW), and eggshell percentage were measured. The ideal ratio was calculated when a statistical difference was detected using Dunnett's test. Only the EW and ESW variables differed from those of BP. The ideal amino acid ratios considering Lys as 100 for EW and ESW were Met + Cys 82 and 83, Thr 60 and 68, Trp 18 and 21, Arg 109 and 112, Gly + Ser 99 and 102, Val 77 and 87, Ile 61 and 67, Leu 155 and 141, His 34 and 37, Phe + Try 134 and 133, respectively.

Keywords: egg weight; shell weight; fractional reduction; deletion method; reproduction

Citation: Carvalho, L.C.; Malheiros, D.; Lima, M.B.; Mani, T.S.A.; Pavanini, J.A.; Malheiros, R.D.; Silva, E.P. Determination of the Optimal Dietary Amino Acid Ratio Based on Egg Quality for Japanese Quail Breeder. *Agriculture* **2023**, *13*, 173. https://doi.org/10.3390/agriculture13010173

Academic Editor: Daniel Simeanu

Received: 7 November 2022
Revised: 30 December 2022
Accepted: 6 January 2023
Published: 10 January 2023

1. Introduction

Egg formation depends on maternal nutrition [1,2]. Some studies have shown that maternal nutrition can modify the composition and characteristics of eggs and consequently affect embryonic development [1,3,4]. Among nutrients, amino acids are essential for egg formation.

The ideal profile of dietary amino acids must support the protein synthesis of target tissues, which in this research consists of proteins that make up eggs, such as low- and high-density lipoproteins, phosvitain and livetin, which make up the yolk, ovoalbumin, ovotransferin, ovomucoid, ovoglubin and ovomucin that form the albumen [5,6]. Glucosamine, glycosaminoglycans, elastin, collagen (I, V, X), osteopontin, and clusterin are present in the eggshell membrane [5,7].

These proteins are synthesized in the liver, magnum, and uterus (or eggshell gland) to form the yolk, albumen, and eggshell, respectively [8], and are related to the dietary supply of amino acids [9,10]. The essential amino acids arginine (Arg), histidine (His), isoleucine (Ile), leucine (Leu), lysine (Lys), methionine (Met), phenylalanine (Phe), threonine (Thr), tryptophan (Trp), and valine (Val) are vital for the constitution of embryonic tissues [11]. The main effects of amino acids are related to production aspects [12], such as the effect of Lys on egg production [12]. Research has linked the effects of Ile and Met on bird fertility [13,14]. Kim et al. [15] and Ullah et al. [16] described the outcomes of imbalances involving Leu, Ile, and Val on egg quality, especially the albumen and eggshell, which are able to regulate protein synthesis [17], and alter the integrity of egg membranes. These findings were corroborated by other studies [18–20], which evaluated the effects of Arg, Trp, and Thr.

Several methods are available to determine the daily intake of each essential amino acid based on egg quality assessment to ensure proper embryo formation [14,20–23]. The ideal amino acid profile is used to establish essential amino acid requirements in proportion to Lys requirements. Nitrogen balance has been established as a criterion commonly used for growing animals [24–26], commercial layers [27], and broiler breeders [28]. The deletion technique has been preferred to establish ideal amino acid profiles due to the possibility of studying all essential amino acids concurrently. The challenge in this research was to apply the method to variables related to egg quality, such as egg weight (EW), format index, and yolk and albumen content. All of these variables are sensitive in detecting the effects of limiting a specific amino acid in the diet [2,4,29]. However, no research has been conducted using this information to establish an ideal amino acid profile. The present methodology allows for the establishment of an optimal ratio of all essential amino acids simultaneously, with the same group of animals and employing the same control diet, which attenuates environmental effects [28,30]. Therefore, this study aimed to establish the ideal ratio of essential amino acids (Lys, Met + Cys, Thr, Trp, Arg, Gly + Ser, Val, Ile, Leu, His, and Phe + Tyr) for Japanese quail breeding based on egg quality using the deletion method.

2. Materials and Methods

Location and ethics approval. This study was conducted in the Poultry Sector of the Animal Science Department of the Universidade Estadual Paulista (UNESP/FCAV) in accordance with ethical standards and approved by the Ethics Committee for the Use of Animals under protocol 012203/17.

2.1. Housing, Animals and Experimental Design

Experiments were conducted in a climatic chamber composed of refrigerators and exhausters that maintained the temperature at 24 °C. The birds were housed individually in galvanized wire cages measuring 0.26 m × 0.37 m × 0.36 m, equipped with a linear feeder and nipple drinkers throughout the experimental period. The light program maintained throughout the experimental period consisted of 16 h of light and 8 h of darkness. A total of 120 Japanese quail breeding at 16 weeks of age, during the peak laying period, were used. The birds were standardized by weight and egg production and distributed by experimental units. A completely randomized design was used with 12 treatments and 10 replicates per treatments.

2.2. Experimental Diets

In this study, a control diet was formulated to form a balanced protein (BP) with all of the nutritional requirements for Japanese quail as estimated by Rostagno et al. [31] for commercial Japanese quail since it does not provide nutritional requirements for breeders. Nitrogen and essential amino acids were provided by corn, soybean meal, corn gluten meal, and crystalline amino acids (Table 1).

The other experimental diets, total of 11 diets with different limiting amino acids, were obtained by deletion BP using corn starch (Tables 2 and 3). The total amino acid contents of the ingredients used in the formulation were analyzed by Evonik Industries AG, São Paulo, Brazil using a near-infrared spectrometer (NIRs) before formulating diets. The values were converted into digestible basis using digestibility coefficients from Rostagno et al. [31]. The deletion was 40% of the amino acid requirement to be evaluated in each treatment, and the other nutrients and energy were recomposed to meet the same BP level, except for the test amino acid, which was depleted by 40%, according to the methodologies described by Dorigam et al. [28]. The nutritional levels of amino acids in experimental diets were described in Table 3.

Table 1. Composition of the control diet (balanced protein).

Item	Content, g/kg
Corn	647.6
Soyabean meal (47%)	120.8
Corn Gluten (60%)	52.1
Dicalcium phosphate	11.5
Limestone	70.6
Sodium chloride	3.4
Potassium chloride	3.4
L-lysine (55%)	3.8
DL-methionine (99%)	9.5
L-threonine (98%)	2.7
L-tryptophan	1.0
L-arginine	4.8
L-glycine	1.3
L-valine	1.3
L-histidine	1.5
L-phenylalanine	0.8
L-glutamate	10.0
Choline chloride (60%)	1.6
Premix—Vitaminic [1]	0.2
Premix—Mineral [1]	0.2

[1] Content per kg of the diet—vit A, 6.668 IU; vit D3, 1.668 IU; vit E, 8 IU; vit K3, 2 mg; vit B1, 1 mg, vit B2, 3.34 mg; vit B6, 2 mg; vit B12, 9 mcg/kg; niacin, 21 mg; chlorine, 0.13 g; pantothenate acid, 8 mg; folic acid, 0.46 mg/kg, biotin, 0.05 mg/kg; 0.46; copper, 8 mg/kg; iron, 6.25 mg/kg; manganese, 70 g; zinc, 25 g; iodine, 6.25 mg; selenium 1.25 mg.

Table 2. Composition of the diet for all tested amino acids.

Item	Diets, g/kg										
	Lys	Met + Cys	Thr	Trp	Arg	Gly + Ser	Val	Ile	Leu	His	Phe + Tyr
Balanced protein	600.0	600.0	596.9	598.6	596.4	590.0	600.0	596.6	597.6	600.0	600.0
Soy oil	11.5	17.7	11.6	11.5	11.6	15.1	11.5	11.6	11.5	11.5	20.6
Dicalcium phosphate	6.0	6.0	6.0	6.0	6.0	6.2	6.0	6.0	6.0	6.0	6.0
Limestone	27.9	28.0	28.2	28.0	28.2	28.6	27.9	28.2	28.1	27.9	27.9
Sodium chloride	1.5	1.5	1.5	1.5	1.5	1.5	1.5	1.5	1.5	1.5	1.5
Potassium chloride	4.8	4.8	4.8	4.8	4.8	4.9	4.8	4.8	4.8	4.8	4.8
DL-methionine (99%)	3.6	0.0	3.7	3.6	3.7	3.7	3.6	3.7	3.6	3.6	3.6
L-lysine (55%)	0.0	8.0	8.0	8.0	8.0	8.2	8.0	8.0	8.0	8.0	8.0
L-threonine (98%)	2.7	2.7	0.0	2.7	2.7	2.7	2.6	2.7	2.7	2.7	2.7
L-tryptophan	0.9	0.9	0.9	0.0	0.9	1.0	0.9	0.9	0.9	0.9	0.9
L-arginine	5.2	5.2	5.2	5.2	0.0	5.2	5.2	5.2	5.2	5.1	5.1
L-valine	3.4	3.3	3.4	3.3	3.4	3.4	0.0	3.4	3.3	3.3	3.3
L-isoleucine	2.9	2.9	2.9	2.9	2.9	3.0	2.9	0.0	2.9	2.9	2.9
L-leucine	6.7	6.7	6.7	6.7	6.7	6.8	6.7	6.7	0.0	6.7	6.7
L-glycine	5.1	5.1	5.1	5.1	5.2	0.0	5.1	5.1	5.1	5.1	5.1
L-phenylalanine	6.0	6.1	6.1	6.1	6.1	6.2	6.0	6.1	6.1	6.0	0.0
L-histidine	1.9	1.9	1.9	1.9	1.9	1.9	1.9	1.9	1.9	0.0	1.9
L-Glutamate	54.1	47.8	54.0	45.3	61.7	55.1	54.0	47.4	51.8	48.5	49.1
Choline chloride (60%)	1.4	1.4	1.4	1.4	1.4	1.4	1.4	1.4	1.4	1.4	1.4
Corn starch	100.0	51.3	100.0	100.0	100.0	54.4	99.1	100.0	100.0	99.1	48.0
Sugar	65.1	100.0	65.0	68.3	58.2	100.0	62.6	70.7	72.7	65.7	100.0
Inert (cellulose)	88.8	100.0	86.2	88.7	88.2	100.0	89.7	87.6	84.3	88.8	100.0
Premix—Vitaminic [2]	0.2	0.2	0.2	0.2	0.2	0.2	0.2	0.2	0.2	0.2	0.2
Premix—Mineral [2]	0.2	0.2	0.2	0.2	0.2	0.2	0.2	0.2	0.2	0.2	0.2

[2] Content per kg of the diet—vit A, 6.668 IU; vit D3, 1.668 IU; vit E, 8 IU; vit K3, 2 mg; vit B1, 1 mg, vit B2, 3.34 mg; vit B6, 2 mg; vit B12, 9 mcg/kg; niacin, 21 mg; chlorine, 0.13 g; pantothenate acid, 8 mg; folic acid, 0.46 mg/kg, biotin, 0.05 mg/kg; 0.46; copper, 8 mg/kg; iron, 6.25 mg/kg; manganese, 70 g; zinc, 25 g; iodine, 6.25 mg; selenium 1.25 mg.

Table 3. Nutritional levels of amino acids in experimental diets.

Items	Diets											
	BP	Lys	Met + Cys	Thr	Trp	Arg	Gly + Ser	Val	Ile	Leu	His	Phe + Tyr
Metabolizable energy (MJ/kg)	11.7	11.7	11.7	11.7	11.7	11.7	11.7	11.7	11.7	11.7	11.7	11.7
Calcium (g/kg)	30.0	30.0	30.0	30.0	30.0	30.0	30.0	30.0	30.0	30.0	30.0	30.0
Avaliable phosphorus (g/kg)	2.8	2.8	2.8	2.8	2.8	2.8	2.8	2.8	2.8	2.8	2.8	2.8
Crude protein (g/kg)	180.1	180.1	180.1	180.1	180.1	180.1	180.1	180.1	180.1	180.1	180.1	180.1
Crude fiber (g/kg)	18.8	11.3	11.3	11.2	11.3	11.2	11.3	11.1	11.2	11.2	11.3	11.3
Starch (g/kg)	423.3	341.7	298.3	340.4	341.1	340.2	340.9	299.3	340.3	340.7	340.9	296.1
Crude fat (g/kg)	33.9	31.8	37.9	31.7	31.7	31.7	31.8	37.2	31.7	31.7	31.8	40.8
NFE (g/kg)	663.1	675.0	668.8	675.1	675.1	675.2	675.0	679.7	675.1	675.1	675.0	665.9
Lysine (g/kg)	10.9	6.6	10.9	10.9	10.9	10.9	10.9	10.9	10.9	10.9	10.9	10.9
Metionine + Cystine (g/kg)	9.0	9.0	5.4	9.0	9.0	9.0	9.0	9.0	9.0	9.0	9.0	9.0
Threonine (g/kg)	6.6	6.6	6.6	3.9	6.6	6.6	6.6	6.6	6.6	6.6	6.6	6.6
Tryptophan (g/kg)	2.3	2.3	2.3	2.3	1.4	2.3	2.3	2.3	2.3	2.3	2.3	2.3
Arginine (g/kg)	12.7	12.7	12.7	12.7	12.7	7.6	12.7	12.7	12.7	12.7	12.7	12.7
Glycine + serine (g/kg)	12.5	12.5	12.5	12.5	12.5	12.5	7.5	12.5	12.5	12.5	12.5	12.5
Valine (g/kg)	8.2	8.2	8.2	8.2	8.2	8.2	8.2	4.9	8.2	8.2	8.2	8.2
Isoleucine (g/kg)	7.1	7.1	7.1	7.1	7.1	7.1	7.1	7.1	4.2	7.1	7.1	7.1
Leucine (g/kg)	16.5	16.5	16.5	16.5	16.5	16.5	16.5	16.5	16.5	9.8	16.5	16.5
Histidine (g/kg)	4.6	4.6	4.6	4.6	4.6	4.6	4.6	4.6	4.6	4.6	2.7	4.6
Phenylalanine + Tyrosine (g/kg)	14.8	14.8	14.8	14.8	14.8	14.8	14.8	14.8	14.8	14.8	14.8	8.9

3. Data Collection

The trial lasted 25 days and was divided into adaptation and egg collection, with 20 days of adaptation and 5 days of egg collection or up to a total of 15 eggs per treatment. The daily amount of feed provided was 24 g per bird and water was ad libitum. Birds were weighed at the beginning and end of experiments to determine their body weights. Feed leftovers were measured at the end of the trial to determine consumption. Egg production was measured daily. The variables evaluated were feed intake (FI. g/bird/day), crude protein intake (CPIntake. g/ave/day), body weight (g), EP (%/bird/day), egg weight (EW. g), egg mass (g/bird/day), and feed conversion by egg mass (FCR. Feed intake g/Egg weight g and Protein intake g/Protein deposition in egg g). During the data collection period and tested were immediately, eggs were collected, weighed, and broken to measure the height, diameter, albumen, and yolk index and Haugh unit. Additionally, the eggshells were evaluated for weight and percentage.

3.1. Egg Quality Analysis

At the end of the trial, egg weight (EW), albumen height, albumen diameter, albumen index, yolk height, yolk diameter, yolk index, Haugh unit, eggshell weight (ESW), and eggshell percentage were measured. Heights of albumen and yolk were measured using a digital micrometer coupled to a tripod base, and diameters were measured using digital calipers. The HU values were calculated from the logarithmic relationship between the height of the dense albumen and the EW. This formula [32] was applied to each egg collected, as described in Equation (1)

$$\text{HU} = 100 \log (\text{H} + 7.57 - 1.7 \text{ W0.37}) \tag{1}$$

where: H = albumen height in mm and W = EW. g

The index yolk (YI) and albumen (AI) were determined by considering the relationship between the height (H) and diameter (D) of the respective components (as described by Funk) [33]. Eggshells were washed with water and dried by forced air circulation at 55 °C for 72 h. After drying. the shell was weighed using a digital scale accurate to 0.01 g. Shell percentage was obtained considering the relation between shell weight (ESW) and EW.

3.2. Statistical Analysis

Data were subjected to homoscedasticity of variance and error normality tests. Upon satisfying the premises, analysis of variance, declared as significant at 0.05, was performed. and when treatment effects were detected. Dunnett's test was applied for all egg quality variables. All data were analyzed using SAS software (v.9.4; SAS Institute Inc., Cary, CA, USA, 2014).

3.3. Determination of Ideal Amino Acid: Lys Ratios

The ideal proportion of amino acids were determined following the principles of Green and Hardy [34], modified in this study to use egg quality variables.

The calculation consists of four steps: Step 1: Calculate the proportion of reduction (Yr) of the analyzed variables of each experimental unit in relation to BP as follows:

$$\text{Yr} = 100 \times (1 - \text{Yi}/\bar{\text{Y}}_{\text{BP}})$$

where Yr is the percentage of reduction in the response of the analyzed variable of each treatment; Yi is the response of each treatment and $\bar{\text{Y}}_{\text{BP}}$ is the mean value of the response of the control treatment, which received the BP.

Step 2: The Yr values were standardized considering the percentage of deletion applied in the treatments (40%) as follows: PYr = Yr/40. where PYr is the standardized Yr value for the applied deletion.

Step 3: The treatment's actual deletion ratio (RDP) was calculated as follows:

$$RDP = 40 \times [1 - (PYri/PYrmax)]$$

where 40% is the initial deletion value. PYrmax is the maximum PYr of the analyzed variable and PYri is the PYr associated with a deficient diet resulting from the second step.

Step 4: The optimal in-feed amino acid (AAI) was calculated as follows:

$$AAI = AABP - [AA\ BP \times (RDP/100)]$$

where AABP represents the concentration of the amino acid in the BP (g/kg) and RDP is the actual deletion ratio resulting from the third step.

Step 5: The ideal ratio of amino acids to Lys (IAAR) is calculated as follows:

$$IAAR = [AAI/Lys] \times 100$$

where AAI is the value found for each amino acid (Met + Cys. Thr. Trp. Arg. Gly + Ser. Val. Ile. Leu. His. and Phe + Tyr) and Lys is the value found from AAI to Lys.

4. Results

The responses obtained for the productive performance responses were significantly affected by limiting dietary and treatment BP ($p < 0.05$; Table 4). Based on the results of the Dunnett test for limitation in Lys, Thr, Try, Arg, and Val FI was affected. However, only for Val was there a difference in the intake relative to BP. The lower IF contributed to the lower ME in the Lys, Thr, Try, and Val limiting treatments. Lys limitation still affected FCR and CPCR, by 30% and 34%, respectively, when compared to BP.

Table 4. Average responses to dietary limited in amino acids.

Amino Acid	Feed Intake (g/Bird Day^{-1})	Protein Intake (g/Bird Day^{-1})	Egg Mass (g Day^{-1})	FCR [1] (g/g)	FCR [2] (g/g)
Lysine	18.62 b	2.46 a	4.15 b	4.90 a	1.87 b
Met + Cys	21.14 a	2.58 a	6.94 a	3.13 a	1.05 a
Threonine	18.83 b	2.40 a	5.09 b	4.35 a	1.62 a
Tryptophan	19.15 b	2.56 a	4.78 b	4.41 a	1.78 b
Arginine	19.05 b	2.42 a	5.79 a	3.58 a	1.17 a
Gly + Ser	22.43 a	2.84 a	6.71 a	3.86 a	1.28 a
Valine	14.78 b	1.91 b	3.81 b	3.92 a	2.08 b
Isoleucine	20.18 a	2.53 a	5.29 a	3.83 a	1.30 a
Leucine	21.39 a	2.71 a	6.66 a	3.54 a	1.31 a
Histidine	22.55 a	2.77 a	7.18 a	3.36 a	1.12 a
Phe + Try	20.39 a	2.51 a	6.63 a	3.16 a	1.24 a
BP	21.27 a	2.61 a	7.14 a	3.09 a	1.09 a
Mean ± SE	19.98 ± 1.82	2.53 ± 0.25	5.83 ± 1.79	3.77 ± 1.30	1.39 ± 0.42
p value	<0.0001	<0.0001	<0.0001	<0.0530	<0.0001

FCR [1], feed conversion ratio; FCR [2], crude protein conversion ratio; BW, body weight; BP, balanced protein. a, b, mean values with b within the line were significantly different ($p < 0.05$) compared with balanced protein, by the Dunnett test.

The limiting dietary treatments for the respective amino acid and BP treatments are presented in Table 5. According to the ANOVA results (Table 5). Only YH was not affected by the treatments ($p > 0.05$); for the other variables, it was possible to detect the effect of the treatments ($p < 0.05$).

Table 5. Effects of the dietary amino acid limitation on the egg quality of Japanese quail breeders.

Amino Acid	Variables									
	EW	ESW	ESP	AH	AD	AI	HU	YH	YD	YI
Lysine	8.36 b	0.64 b	8.08 a	3.53 a	8.27 a	0.42 a	86.57 a	7.40 a	20.61 b	0.36 a
Met + Cys	8.43 b	0.64 b	7.60 a	4.39 a	7.28 a	0.59 a	91.53 a	7.97 a	22.20 b	0.36 a
Threonine	8.37 b	0.55 b	7.27 a	4.49 a	6.99 a	0.62 a	92.03 a	7.89 a	21.13 b	0.37 a
Tryptophan	9.43 b	0.65 b	6.83 a	4.17 a	6.67 a	0.54 a	89.19 a	7.38 a	22.49 a	0.33 a
Arginine	8.71 b	0.65 b	7.54 a	4.00 a	7.60 a	0.53 a	90.56 a	7.71 a	21.58 b	0.35 a
Gly + Ser	9.22 b	0.71 b	7.76 a	4.37 a	7.59 a	0.62 a	90.52 a	7.64 a	22.39 a	0.33 a
Valine	8.22 b	0.54 b	6.64 a	3.81 a	5.53 b	0.70 a	88.32 a	7.76 a	21.64 b	0.35 a
Isoleucine	8.74 b	0.62 b	6.99 a	4.10 a	7.92 a	0.53 a	89.36 a	7.85 a	23.16 a	0.34 a
Leucine	8.16 b	0.68 b	8.38 b	4.50 a	5.68 a	0.79 b	92.23 a	8.20 a	22.73 a	0.36 a
Histidine	9.59 b	0.73 a	7.00 a	4.05 a	7.44 a	0.49 a	88.36 a	7.44 a	23.08 a	0.32 a
Phe + Try	8.46 b	0.66 b	7.84 a	4.29 a	6.48 a	0.61 a	90.78 a	7.98 a	22.81 a	0.36 a
Balanced protein	10.76 a	0.84 a	7.79 a	4.07 a	7.87 a	0.52 a	87.69 a	8.22 a	24.24 a	0.34 a
Mean ± SE	8.86 ± 1.05	0.67 ± 0.10	7.53 ± 1.12	4.15 ± 0.69	7.09 ± 1.67	0.58 ± 0.19	89.62 ± 3.94	7.79 ± 0.62	22.34 ± 1.60	0.35 ± 0.03
p value	<0.0001	<0.0001	0.0144	0.0530	<0.0001	0.0009	0.0148	0.5218	<0.0001	0.0176

EW—Egg weight (g); ESW—Egg shell weight (g); ESP—Eggshell proportion (%); AH—Albumen height (mm); AD—Albumen diameter (mm); AI—Albumen index (%); HU—Haugh unit; YH—Yolk height (mm); YD—Yolk Diameter (mm); YI—Yolk index (%). a, b. mean values with b within the line were significantly different ($p < 0.05$) compared with balanced protein. by the Dunnett test.

Limiting dietary treatments of His, Phe + Tyr, and Leu affected EW and ESW. When the means were compared using Dunnett's test, considering BP as a reference, only the EW and ESW variables were found to have significant effects on the dietary treatments ($p < 0.05$) except for the difference between BP and the His-limited dietary treatment (Table 5) which was not significant for ESW ($p > 0.05$). Therefore, only EW and ESW variables were used to establish the ideal amino acid ratio.

The EW for BP was 10.76 g, while the EW of the other dietary treatments ranged from 9.59 to 8.16 g for the limiting dietary treatments involving His and Leu, respectively. Thus, the minimum reduction was approximately 11% for His and 24% for Leu. For ESW, the mean value for BP was 0.84 g, ranging from 0.73 and 0.54 g for the His and Val limited dietary treatments, respectively, which were equivalent to a reduction of 13% and 36%. when compared to BP.

The His exhibited the least limitation in terms of the two response variables EW and ESW, compared to BP. On the other hand, Leu and Val were amino acids that presented the greatest limitation, with a visible deterioration in egg quality, especially EW and ESW ($p < 0.05$).

These results were used to calculate the optimal concentration of the respective amino acids using Leu and Val as control standards to calculate the actual deletion. Table 6 presents AAI and IAAR values of the evaluated amino acids. AAI and IAAR values differed between EW and ESW variables. The distance quantified by the standard deviation was 8% for AAI between EW and ESW and 7% for IAAR between EW and ESW.

Table 6. Summarized results of the individual amino acid deletions for egg weight (EW) and eggshell weight (ESW) of Japanese quail's breeders.

Variables	EW				ESW			
	Yr	RDP	IAA	IAAR	Yr	RDP	IAA	IAAR
Lys	22.30	3.15	1.06	100	23.32	13.82	0.94	100
Met + Cys	21.69	4.16	0.86	82	23.45	13.68	0.78	83
Thr	22.22	3.28	0.63	60	33.45	2.45	0.63	68
Trp	12.36	19.57	0.18	18	22.53	14.72	0.2	21
Arg	18.60	9.26	1.15	109	20.14	17.39	1.05	112
Gly + Ser	14.39	16.22	1.05	99	14.90	23.28	0.96	102
Val	23.61	0.98	0.81	77	35.64	0	0.82	87
Ile	18.76	9	0.65	61	25.51	11.37	0.63	67
Leu	24.21	0	1.64	155	18.59	19.13	1.33	141
His	10.84	22.09	0.36	34	14.48	23.75	0.35	37
Phe + Tyr	21.37	4.68	1.41	134	21.57	15.79	1.25	133

Yr = per cent reduction in EW and ESW (%); RDP = real deleted proportion; IAA = amino acid requirement; IAAR = optimal in-feed amino acid ratio (%).

5. Discussion

Maternal amino acid nutrition is essential for egg formation. which later supports embryonic development [3]. This study aimed to apply the deletion method to establish an ideal amino acid profile using egg quality variables. The results obtained in the present study indicated that the applied deletion of 40% of the studied amino acids (Lys, Met + Cys, Thr, Trp, Arg, Gly + Ser, Val, Ile, Leu, His, and Phe + Tyr) limited the responses of the female breeders of the Japanese quail, verifying the worsening of performance and egg quality variables (Tables 4 and 5), especially for EW and ESW. These results support the objective proposed in this research, which assumed the existence of a dose-response relationship (Table 5). The only exception was for the amino acid His; although the dose affected the EW response, the reduction found for ESW was not significantly different according to the BP Dunnett's test, which had no dietary limitation (Table 5). Previously published results [29] validated the limitation of the test amino acid Arg based on its response to EW. According to these authors, EW is the most sensitive response in Japanese quails. The convention proposed by Morris and Gous [35] has prevailed for commercial laying hens, and dietary amino acid limitations primarily affect bird egg production, with little change in EW. This understanding was corroborated by other studies carried out with commercial laying hens [36,37] and broiler breeders [19,38,39]. However, recent Japanese quail results support the fact that these birds are more sensitive to reduced levels of amino acids in their diet, and that a significant reduction in EW occurs [29,40–42].

The results of this study indicated that nutritional limitation involving the tested amino acids was able to modify EW. This effect is related to nutritional deficiency imposed on the breeder birds, which decreases embryo birth weights [1]. One hypothesis to explain the reduction in EW is that the weight of the embryo is related to maternal protein loss. causing a decrease in protein synthesis under dietary conditions deficient in essential amino acids [1,4]. The 40% limitation imposed on the diets decreased the supply of amino acids to meet the physiological processes related to the maintenance of body weight, where a small part would be available for processes related to protein deposition in the egg. Therefore. to compensate for the loss of amino acids and to maintain plasma levels [43,44], only muscle protein mobilization remains. In addition to EW, ESW was also significantly affected given the greater thickness of the eggshell membrane protein constitution. which was considered in the ESW computation.

The eggshell is a structure whose function is to protect the interior of the egg from physical and microbial agents, regulate gas, water, and metabolite exchanges, and provide mineralized components for embryo development [45]. The effects of dietary amino acid limitation on ESW have been reported, in particular regarding the protein that composes the protein matrix and influences eggshell texture [45,46]. Mann and Mann [47] identified two proteins from the ovocleidin-17 family as the main components of eggshell protein matrix. In quails, the eggshell membrane is thicker than that found in chicken eggs. This difference lies in the number of protein families in the membrane constitution; in quails, there are two families, while in eggs from chickens, there is only one family Mann and Mann [47].

The YD verified in the limited dietary treatments involving Lys, Met + Cys, Thr, Arg, and Val was significantly lower than the value obtained for BP. El-Tarabany [48] reported that yolk diameter is positively correlated with EW, corroborating the results obtained in this study.

His was the only amino acid that exhibited a difference in EW. His is an essential amino acid [49,50] which does not present immediate signs of deficiency. but a function of protein metabolism that compensates for such deficiency through hemoglobin and carnosine catabolism [51]. Robbins et al. [52] demonstrated that the growth rate of broiler chicks fed His-deficient diets was recovered by intravenous administration of L-carnosine. However, there was no increase in the plasma concentration of His, supporting the hypothesis that muscle carnosine is rapidly metabolized to His and used in priority demands such as the synthesis of regulatory proteins.

The results obtained for AAI revealed a difference of 8% when calculated based on EW and ESW. For IAAR, the difference between EW and ESW variables was approximately 8.4%. The amino acid Leu showed a greater degree of limitation. and one hypothesis may be related to its concentration in egg protein composition [1,53]. In addition, there exists potential antagonism involving Val and Ile [50,54,55]. The amino acid Leu has been considered the most efficient in protein synthesis among BCAAs (Lynch et al. 2006), since Leu induces the activation of the mTOR complex, which stimulates protein synthesis [56]. Our results corroborate the suggestion of Macelline et al. [6], who indicate that farmers should pay attention to dietary Leu levels.

The amino acid profile of the target tissue is usually used when there is no information regarding the ideal amino acid ratio. The ideal relationship obtained considering the composition of the egg presented by Bayomy et al. [57] was as follows: Met + Cys 118%, Thr 52%, Arg 35%, Gly + Ser 75%, Val 69%, Ile 57%, Leu 108%, His 35%, and Phe + Tyr 97%. The ideal ratios obtained based on the composition of meat presented by Bayomy et al. [57], was (Lys 100%) Met + Cys 37%, Thr 29%, Arg 31%, Gly + Ser 33%, Val 48%, Ile 60%, Leu

55%, His 38%, and Phe + Tyr 84%. These values are different when compared to the results obtained in this study. Suggesting that the composition of the target tissue alone may not represent the best option for establishing ideal ratios of amino acids [57,58], since the proportion of dietary amino acids is modified during digestion and absorption processes, which precede deposition in the target tissue [55].

The IAA obtained for EW and ESW estimated by the deletion method (Table 6) showed considerable variation from 0.86 to 19.13% for the same amino acid. The greatest variations were observed for Leu, Phe + Tyr, Lys, Met + Cys, Arg, Gly + Ser, Trp, Ile, His, Val, and Thr, with values of 19.13, 11.65, 11.02, 9.93, 8.97, 8.42, 6.04, 2.60, 2.13, 0.86, and 0.99%. respectively. In addition, only the IAAs of Thr and Val were higher in terms of eggshell weight; however. with a difference of 11.65% in the IAA for Lys between the variables (EW:1.06; ESW:0.94), it was possible to reduce the proportional distance between Lys and the other amino acids. Thus, the amino acid requirements increased for all when the eggshell weight was met, except for Leu and Phe + Thr.

The NRC [59] and Rostagno et al. [60], based their recommendations on compilations of studies. Due to the lack of research for breeders the requirements for breeders and commercial quails were not differentiated. However, it is expected that due to the genetic differences between these birds a modification in IAA and consequently IAAR is likely. The results of Rostagno et al. [60], indicate that IAA for all amino acids studied are higher for commercial quails, when an average was made between the IAA of EW and ESW, with difference ranging from 5.40 (Val) to 35.37% (His). When evaluations between the AAIs are specific for each variable the highest variation was for AAI for ESW, with a difference of up to 37% for the amino acids Gly + Ser and His and the lowest difference was for Val (5%). The observed for IAA of EW with highest difference for His (34%) and lowest for Leu (5%). To compare the results with NRC [59], the IAAs of NRC [59] were transformed considering the digestibility of lysine with an average of 89% [31]. When compared. Less variation was observed with proximity of the estimated results for Arg. Gly + Ser and Val (1.12, 1.04, and 0.82% in feed, respectively) for IAA from EW and Thr. Val and Phe + Try (0.66, 0.82, and 1.25% in feed, respectively) for IAA from ESW. However, for the other amino acids IAA are higher for NRC [59] results, being only for ESW results for Arg. Gly + Ser. Ile and His (1.12, 1.42, 0.80, 0.37% in feed, respectively) IAA were higher.

However, the AAI was estimated based on egg quality to determine the AAI related to Lys. The intake of amino acids by the proportion between them can be modified in relation to the AAI. When comparing the AARI for commercial quails based on the recommendations of Rostagno et al. [60]. NRC [59] and Silva and Costa [61] for Lys at 100%: 82, 70, and 74; 61, 74, and 71; 21, 19, and 19; 115, 126, and 133; 115 and 117; 75, 92, and 92; 65, 90, and 92; 150, 142, and 151; 42, 42, and 44; 135, 140, and 149 for Met + Cys, Thr, Trp, Arg, Gly + Ser, Val, Ile, Leu, His and Phr + Tyr, respectively, except for Gly + Ser from Silva and Costa [61] which was not determined. Lower ratios were observed for the recommendations of NRC [59] and Silva and Costa [61]. Rostagno et al. [60] recommended higher ratios between amino acids and Lys, approaching the IAARs determined in this study. However. the slight increase did not provide ratios close to those found for EW in Trp. Arg. Gly + Ser. Ile and His. with an increase in 19.84, 5.35, 15.93, 6.21, and 23.71% difference in the AA:Lys ratio, respectively. Corroborating with that found for ESW, where differences of 9.63, 12.64, 14.09, 6.24, and 10.69% differed for Thr. Gly + Ser. Val. Leu and His in AA:Lys ratios, respectively, His and Val being the amino acids that differed most when compared with the reference used today for formulation of rations.

The ratios for Met + Cys:Lys (82:100) and Phe + Try:Lys (135:100), did not differ with that recommended by Rostagno et al. [60] for the two IAAR determined in that study (EW and ESW = 82 and 83 for Met + Cys and 134 and 133 for Phe + Try, in percentage, relative to Lys), minimal differences were found for Thr:Lys (61:100) and Val:Lys (75:100) in EW and between Trp:Lys (21:100), Arg:Lys (115:Lys) and Ile:Lys (65:100). However, differences in ratios were found for most amino acids. Previous recommendations used for broilers may have overestimated the IAAR for all amino acids except for Val for birds production. The current IAAR recommendations for broilers in contrast to all previously used techniques apply a target tissue to ensure quality of protein synthesis and minimize lighter EW and ESW during production, thus modifications were noticed when a target trait is selected as a basis.

Understanding how the limitation of essential amino acids influences egg quality is essential for the animal category studied in this work, as the physical quality of eggs can be used as a selection trait directly and indirectly in genetic improvement [62]. Among the characteristics studied (EW and ESW), determining the optimal relationship between essential amino acids presents high heritability [63,64].

In a study by Hegab and Hanafy [65] with Japanese quail breeders, it was observed that EW influenced the hatchability of eggs and increased the weight of chicks at hatch. Regarding the characteristics of ESW, the study also found that larger ESWs have a higher hatchability rate, resulting in greater eggshell volume, pore count, and surface area [65]. Therefore, the contribution of nutrition to maintaining the desired quality of these parameters is extremely important for genetic companies and nutritionally enables animals to express their genetic potential.

6. Conclusions

In conclusions, the application of the deletion method with a limitation of 40% of dietary amino acid deletion, allowed the EW and ESW variables to be sensitive for all tested amino acids. Thus, it was possible to establish the ideal profile of essential amino acids in the diet simultaneously, focusing on a target tissue, (which in this research was EW and ESW).

Author Contributions: Conceptualization, L.C.C. and E.P.S.; Methodology, L.C.C. and E.P.S.; Software, L.C.C. and E.P.S.; Validation, L.C.C., R.D.M. and E.P.S.; Formal Analysis, L.C.C., T.S.A.M. and E.P.S.; Investigation, L.C.C. and E.P.S.; Resources, L.C.C., R.D.M. and E.P.S.; Data Curation, T.S.A.M., L.C.C. and J.A.P.; Writing—Original Draft Preparation, L.C.C., D.M. and E.P.S.; Writing—Review & Editing, L.C.C. and D.M.; Supervision, E.P.S.; Project Administration, M.B.L. All authors have read and agreed to the published version of the manuscript.

Funding: This research received no external funding.

Institutional Review Board Statement: The procedures used in this research by the Committen on Animal Use Ethics. under protocol 012203/17.

Data Availability Statement: The data can be requested to the corresponding author.

Acknowledgments: The first author acknowledges the scholarship by the CAPES Foundation and the National Council for Scientific and Technological Development (CNPq) by financial support (grant No. 432588/2016-7). This study was financed in part by the Coordenação de Aperfeiçoamento de Pessoal de Nível Superior Brasil (CAPES) Finance Code 001.

Conflicts of Interest: The authors declare no conflict of interest.

References

1. Li, F.; Shan, M.X.; Gao, X.; Yang, Y.; Yang, X.; Zhang, Y.Y.; Hu, J.W.; Shan, A.S.; Cheng, B.J. Effects of Nutrition Restriction of Fat- and Lean-Line Broiler Breeder Hens during the Laying Period on Offspring Performance. Blood Biochemical Parameters. and Hormone Levels. *Domest. Anim. Endocrinol.* **2019**, *68*, 73–82. [CrossRef] [PubMed]
2. Li, F.; Yang, Y.; Yang, X.; Shan, M.; Gao, X.; Zhang, Y.; Hu, J.; Shan, A. Dietary Intake of Broiler Breeder Hens during the Laying Period Affects Amino Acid and Fatty Acid Profiles in Eggs. *Rev. Bras. Zootec.* **2019**, *48*, 1–12. [CrossRef]
3. Güçlü, B.K.; Uyanik, F.; Işcan, K.M. Effects of Dietary Oil Sources on Egg Quality. Fatty Acid Composition of Eggs and Blood Lipids in Laying Quail. *S. Afr. J. Anim. Sci.* **2008**, *38*, 91–100.
4. Li, F.; Xu, L.M.; Shan, A.S.; Hu, J.W.; Zhang, Y.Y.; Li, Y.H. Effect of Daily Feed Intake in Laying Period on Laying Performance. Egg Quality and Egg Composition of Genetically Fat and Lean Lines of Chickens. *Br. Poult. Sci.* **2011**, *52*, 163–168. [CrossRef]
5. Guha, S.; Majumder, K.; Mine, Y. Egg Proteins. In *Handbook of Food Proteins*; Woodhead Publishing: Sawston, UK, 2011; pp. 74–84. [CrossRef]
6. Macelline, S.P.; Toghyani, M.; Chrystal, P.V.; Selle, P.H.; Liu, S.Y. Amino Acid Requirements for Laying Hens: A Comprehensive Review. *Poult. Sci.* **2021**, *100*, 101036. [CrossRef]

7. Baláž, M.; Boldyreva, E.V.; Rybin, D.; Pavlović, S.; Rodríguez-Padrón, D.; Mudrinić, T.; Luque, R. State-of-the-Art of Eggshell Waste in Materials Science: Recent Advances in Catalysis. Pharmaceutical Applications. and Mechanochemistry. *Front. Bioeng. Biotechnol.* **2021**, *8*, 612567. [CrossRef]

8. Edwards, N.A.; Verna, L.; Nir, I. The secretion and synthesis of albumen by the magnum of the domestic fowl (*Gallus domesticus*). *Comp. Biochem. Physiol.* **1976**, *53*, 183–186. [CrossRef]

9. Hurwitz, S.; Bornstein, S. The Protein and Amino Acid Requirements of Laying Hens: Suggested Models for Calculation. *Poult. Sci.* **1973**, *52*, 1124–1134. [CrossRef]

10. Bornstein, S.; Hurwitz, S.; Lev, Y. The Amino Acid and Energy Requirements of Broiler Breeder Hens. *Poult. Sci.* **1979**, *58*, 104–116. [CrossRef]

11. Fisher, H.; Johnson, D. The Amino Acid Requirement of Laying Hens. *Poult. Sci.* **1965**, *44*, 198–205. [CrossRef]

12. Kim, E.; Wickramasuriya, S.S.; Shin, T.K.; Cho, H.M.; Kim, H.B.; Heo, J.M. Estimating Total Lysine Requirement for Optimised Egg Production of Broiler Breeder Hens during the Early-Laying Period. *J. Anim. Sci. Technol.* **2020**, *62*, 521–532. [CrossRef] [PubMed]

13. Ekmay, R.D.; Salas, C.; England, J.; Cerrate, S.; Coon, C.N. Lysine Partitioning in Broiler Breeders Is Not Affected by Energy or Protein Intake When Fed at Current Industry Levels. *Poult. Sci.* **2014**, *93*, 1737–1744. [CrossRef] [PubMed]

14. Reda, F.M.; Swelum, A.A.; Hussein, E.O.S.; Elnesr, S.S. On Productive and Reproductive Performance. Egg Quality and Blood Biochemical Parameters. *Animals* **2020**, *10*, 1839. [CrossRef] [PubMed]

15. Kim, W.K.; Singh, A.K.; Wang, J.; Applegate, T. Functional Role of Branched Chain Amino Acids in Poultry: A Review. *Poult. Sci.* **2022**, *101*, 101715. [CrossRef]

16. Ullah, S.; Ditta, Y.A.; King, A.J.; Pasha, T.N.; Mahmud, A.; Majeed, K.A. Varying Isoleucine Level to Determine Effects on Performance. Egg Quality. Serum Biochemistry. and Ileal Protein Digestibility in Diets of Young Laying Hens. *PLoS ONE* **2022**, *17*, e0261159. [CrossRef]

17. Bai, J.; Greene, E.; Li, W.; Kidd, M.T.; Dridi, S. Branched-Chain Amino Acids Modulate the Expression of Hepatic Fatty Acid Metabolism-Related Genes in Female Broiler Chickens. *Mol. Nutr. Food Res.* **2015**, *59*, 1171–1181. [CrossRef]

18. Ramalho de Lima, M.; Costa, F.G.P.; Guerra, R.R.; da Silva, J.H.V.; Rabello, C.B.V.; Miglino, M.A.; Lobato, G.B.V.; Netto, S.B.S.; Dantas, L.D.S. Threonine: Lysine Ratio for Japanese Quail Hen Diets. *J. Appl. Poult. Res.* **2013**, *22*, 260–268. [CrossRef]

19. Lima, M.B.; Sakomura, N.K.; Silva, E.P.; Leme, B.B.; Malheiros, E.B.; Peruzzi, N.J.; Fernandes, J.B.K. Arginine Requirements for Maintenance and Egg Production for Broiler Breeder Hens. *Anim. Feed Sci. Technol.* **2020**, *264*, 114466. [CrossRef]

20. Kalvandi, O.; Sadeghi, A.; Karimi, A. Arginine Supplementation Improves Reproductive Performance. Antioxidant Status. Immunity and Maternal Antibody Transmission in Breeder Japanese Quail under Heat Stress Conditions. *Ital. J. Anim. Sci.* **2022**, *21*, 8–17. [CrossRef]

21. Kalvandi, O.; Sadeghi, A.; Karimi, A. Methionine Supplementation Improves Reproductive Performance. Antioxidant Status. Immunity and Maternal Antibody Transmission in Breeder Japanese Quail under Heat Stress Conditions. *Arch. Anim. Breed.* **2019**, *62*, 275–286. [CrossRef]

22. Hanafy, A.M.; Attia, F.A.M. Productive and Reproductive Responses of Breeder Japanese Quails to Different Dietary Crude Protein and L-Valine Levels. *Egypt. Poult. Sci. J.* **2018**, *38*, 735–753. [CrossRef]

23. Jiang, S.; El-Senousey, H.A.K.; Fan, Q.; Lin, X.; Gou, Z.; Li, L.; Wang, Y.; Fouad, A.M.; Jiang, Z. Effects of Dietary Threonine Supplementation on Productivity and Expression of Genes Related to Protein Deposition and Amino Acid Transportation in Breeder Hens of Yellow-Feathered Chicken and Their Offspring. *Poult. Sci.* **2019**, *98*, 6826–6836. [CrossRef] [PubMed]

24. Samadi; Liebert, F. Modelling the Optimal Lysine to Threonine Ratio in Growing Chickens Depending on Age and Efficiency of Dietary Amino Acid Utilisation. *Br. Poult. Sci.* **2008**, *49*, 45–54. [CrossRef]

25. Pastor, A.; Wecke, C.; Liebert, F. Assessing the Age-Dependent Optimal Dietary Branched-Chain Amino Acid Ratio in Growing Chicken by Application of a Nonlinear Modeling Procedure. *Poult. Sci.* **2013**, *92*, 3184–3195. [CrossRef] [PubMed]

26. Wecke, C.; Liebert, F. Improving the Reliability of Optimal In-Feed Amino Acid Ratios Based on Individual Amino Acid Efficiency Data from N Balance Studies in Growing Chicken. *Animals* **2013**, *3*, 558. [CrossRef]

27. Soares, L.; Sakomura, N.K.; Dorigam, J.C.D.P.; Liebert, F.; Sunder, A.; Nascimento, M.Q.d.; Leme, B.B. Optimal In-Feed Amino Acid Ratio for Laying Hens Based on Deletion Method. *J. Anim. Physiol. Anim. Nutr.* **2019**, *103*, 170–181. [CrossRef] [PubMed]

28. Dorigam, J.C.P.; Sakomura, N.K.; Sarcinelli, M.F.; Gonçalves, C.A.; de Lima, M.B.; Peruzzi, N.J. Optimal In-Feed Amino Acid Ratio for Broiler Breeder Hens Based on Deletion Studies. *J. Anim. Physiol. Anim. Nutr.* **2017**, *101*, 1194–1204. [CrossRef]

29. De Lima, M.B.; de Sousa, M.G.B.L.; Minussi, A.R.T.; de Carvalho, L.C.; Veras, A.G.; Malheiros, E.B.; da Silva, E.P. Arginine Requirement for Japanese Quails. *Poult. Sci.* **2022**, *101*, 101841. [CrossRef]

30. Rollin, X.; Mambrini, M.; Abboudi, T.; Larondelle, Y.; Kaushik, S.J. The Optimum Dietary Indispensable Amino Acid Pattern for Growing Atlantic Salmon (*Salmo Salar* L.) Fry. *Br. J. Nutr.* **2003**, *90*, 865–876. [CrossRef]

31. Rostagno, H.; Albino, L.F.; Donzele, J.; Gomes, P.; de Oliveira, R.; Lopes, D.; Ferreira, A.; Barreto, S.L.; Euclides, R. *Tabelas Brasileiras Para Suínos e Aves: Composição de Alimentos e Exigências Nutricionais*; Animal Science Department UFV: Viçosa, MG, Brazil, 2011; p. 252.

32. Haugh, R.R. The Haugh unit for measuring egg quality. *U. S. Egg Poult. Mag.* **1937**, *43*, 552–555.

33. Funk, E.M. The relation of the Yolk Index Determined in Natural Position to the Yolk Index as Determined after Separating the Yolk from the Albumen. *Poult. Sci.* **1948**, *27*, 367. [CrossRef]

34. Green, J.A.; Hardy, R.W.; Brannon, E.L. The Optimum Dietary Essential: Nonessential Amino Acid Ratio for Rainbow Trout (*Oncorhynchus mykiss*). Which Maximizes Nitrogen Retention and Minimizes Nitrogen Excretion. *Fish Physiol. Biochem.* **2002**, *27*, 109–115. [CrossRef]
35. Morris, T.R.; Gous, R.M. Partitioning of The Response to Protein Between Egg Number And Egg Weight. *Br. Poult. Sci.* **1988**, *29*, 93–99. [CrossRef] [PubMed]
36. Bendezu, H.C.P.; Sakomura, N.K.; Hauschild, L.; da Silva, E.P.; Dorigam, J.C.D.P.; Malheiros, E.B.; Fernandes, J.B.K. Response of Laying Hens to Methionine + Cystine Intake by Dilution Technique. *Rev. Bras. Zootec.* **2015**, *44*, 15–21. [CrossRef]
37. Silva, E.P.; Malheiros, E.B.; Sakomura, N.K.; Venturini, K.S.; Hauschild, L.; Dorigam, J.C.P.; Fernandes, J.B.K. Lysine Requirements of Laying Hens. *Livest. Sci.* **2015**, *173*, 69–77. [CrossRef]
38. Silva, E.P.; Sakomura, N.K.; Oliveira, C.F.S.; Costa, F.G.P.; Dorigam, J.C.P.; Malheiros, E.B. The Optimal Lysine and Threonine Intake for Cobb Broiler Breeder Hens Using Reading Model. *Livest. Sci.* **2015**, *174*, 59–65. [CrossRef]
39. Lima, M.B.; Sakomura, N.K.; Silva, E.P.; Dorigam, J.C.P.; Ferreira, N.T.; Malheiros, E.B.; Fernandes, J.B.K. The Optimal Digestible Valine. Isoleucine and Tryptophan Intakes of Broiler Breeder Hens for Rate of Lay. *Anim. Feed Sci. Technol.* **2018**, *238*, 29–38. [CrossRef]
40. Sarcinelli, M.F.; Sakomura, N.K.; Dorigam, J.C.P.; Silva, E.P.; Venturini, K.S.; Lima, M.B.; Gonçalves, C.A. Modelling Japanese Quail Responses to Methionine + cystine. Threonine and Tryptophan Intake. *Anim. Feed Sci. Technol.* **2020**, *263*, 114486. [CrossRef]
41. Silva, E.P.; Lima, M.B.; Sakomura, N.K.; Moraes, L.E.; Peruzzi, N.J. Weight Gain Responses of Laying-Type Pullets to Methionine plus Cystine Intake. *Animal* **2020**, *14*, s294–s302. [CrossRef]
42. Silva, E.P.D.; Sakomura, N.K.; Sarcinelli, M.F.; Dorigam, J.C.D.P.; Venturini, K.S.; Lima, M.B.D. Modeling the Response of Japanese Quail Hens to Lysine Intake. *Livest. Sci.* **2019**, *224*, 69–74. [CrossRef]
43. Fuller, M.F. *Amino Acid Requirements for Maintenance. Body Protein Accretion and Reproduction in Pigs*; CABI: Wallingford, UK, 1994.
44. Cadirci, S.; Smith, W.K. The Effect of Body Weight of Laying Hens on the Intake of Methionine-Deficient Diet. In Proceedings of the 15th European Symposium on Poultry Nutrition, Balatonfüred, Hungary, 25–29 September 2005.
45. Hincke, M.T.; Nys, Y.; Gautron, J. The Role of Matrix Proteins in Eggshell Formation. *J. Poult. Sci.* **2010**, *47*, 208–219. [CrossRef]
46. Gautron, J.; Hincke, M.T.; Nys, Y. Precursor Matrix Proteins in the Uterine Fluid Change with Stages of Eggshell Formation in Hens. *Connect. Tissue Res.* **1997**, *36*, 195–210. [CrossRef] [PubMed]
47. Mann, K.; Mann, M. Proteomic Analysis of Quail Calcified Eggshell Matrix: A Comparison to Chicken and Turkey Eggshell Proteomes. *Proteome Sci.* **2015**, *13*, 22. [CrossRef]
48. El-Tarabany, M.S. Impact of Cage Stocking Density on Egg Laying Characteristics and Related Stress and Immunity Parameters of Japanese Quails in Subtropics. *J. Anim. Physiol. Anim. Nutr.* **2016**, *100*, 893–901. [CrossRef] [PubMed]
49. Holeček, M.; Vodeničarovová, M. Effects of Histidine Supplementation on Amino Acid Metabolism in Rats. *Physiol. Res.* **2020**, *69*, 99–111. [CrossRef]
50. Kidd, M.T.; Loar, R.E. A Synopsis of Recent Work on the Amino Acid Nutrition of Layers. *J. Appl. Poult. Res.* **2021**, *30*, 100108. [CrossRef]
51. Fisher, H.; Konlande, J.; Strumeyer, D. Levels of Histidine and Histidine Derivatives in Breast Muscle of Protein-Depleted and Repleted Adult Cockerels. *Ann. Nutr. Metab.* **1975**, *18*, 120–126. [CrossRef] [PubMed]
52. Robbins, K.R.; Baker, D.H.; Norton, H.W. Histidine Status in the Chick as Measured by Growth Rate. Plasma Free Histidine and Breast Muscle Carnosine. *J. Nutr.* **1977**, *107*, 2055–2061. [CrossRef] [PubMed]
53. Kudre, T.G.; Bejjanki, S.K.; Kanwate, B.W.; Sakhare, P.Z. Comparative Study on Physicochemical and Functional Properties of Egg Powders from Japanese Quail and White Leghorn Chicken. *Int. J. Food Prop.* **2018**, *21*, 956–971. [CrossRef]
54. D'mello, J.P.F.; Lewis, D. Amino Acid Interactions in Chick Nutrition. *Br. Poult. Sci.* **1971**, *12*, 345–358. [CrossRef]
55. Smith, T.K.; Austic, R.E. The Branched-Chain Amino Acid Antagonism in Chicks. *J. Nutr.* **1978**, *108*, 1180–1191. [CrossRef] [PubMed]
56. Wu, G. Functional Amino Acids in Growth. *Adv. Nutr.* **2010**, *1*, 31–37. [CrossRef] [PubMed]
57. Bayomy, H.M.; Rozan, M.A.; Mohammed, G.M. Nutritional Composition of Quail Meatballs and Quail Pickled Eggs. *J. Nutr. Food Sci.* **2017**, *7*, 1000584. [CrossRef]
58. Friedman, M. Analysis. Nutrition. and Health Benefits of Tryptophan. *Int. J. Tryptophan Res.* **2018**, *11*, 1178646918802282. [CrossRef]
59. NRC1994. *Nutrient Requirements of Poultry*, 9th ed.; Washington: National Academy Press: Washington, DC, USA, 1994.
60. Rostagno, H.S.; Albino, L.F.T.; Donzele, J.L.; Gomes, P.C.; de Oliveira, R.F.; Lopes, D.C.; Ferreira, A.S.; Barreto, S.L.T.; Euclides, R. *Brazilian Tables for Poultry and Swine: Composition of Feedstuffs and Nutritional Requirements*; Animal Science Department UFV: Viçosa, Brazil, 2017; ISBN 9788560249725.
61. Da Silva, J.H.V.; Costa, F.G.P. *Tabela Para Codornas Japonesas e Européias*; Funep: Jaboticabal, Brazil, 2009.
62. Minvielle, F.; Oguz, Y. Effects of Genetics and Breeding on Egg Quality of Japanese Quail. *Worlds. Poult. Sci. J.* **2002**, *58*, 291–295. [CrossRef]
63. Minvielle, F.; Monvoisin, J.L.; Costa, J.; Frenot, A. Quail Lines Selected on Egg Number Either on Pureline or on Crossbred Performance. In Proceedings of the Twelth Symposium of Current Problems in Avian Genetics (Aviagen), Prague, Czech Republic, 3–5 September 1997.

64. Stino, F.K.R.; Kicka, M.A.; Kamar, G.A.; Altakreti, B.T.O. Egg Quality Traits of the Japanese Quail and Their Heritability in the Subtropics. *Arch. Geflügelkd.* **1985**, *3*, 104–108.
65. Hegab, I.M.; Hanafy, A.M. Effect of Egg Weight on External and Internal Qualities. Physiological and Hatching Success of Japanese Quail Eggs (*Coturnix japonica*). *Rev. Bras. Cienc. Avic.* **2019**, *21*, 1–8. [CrossRef]

 *agriculture*

Article

Microbiological Assessment of Broiler Compound Feed Production as Part of the Food Chain—A Case Study in a Romanian Feed Mill

Dragoș Mihai Lăpușneanu, Daniel Simeanu, Cristina-Gabriela Radu-Rusu, Roxana Zaharia and Ioan Mircea Pop *

Department of Control, Expertise and Services, Faculty of Food and Animal Sciences, "Ion Ionescu de la Brad" University of Life Sciences, 700489 Iasi, Romania
* Correspondence: popim@uaiasi.ro

Abstract: Compound feed and the raw materials used in their production are potential vectors of microbiological contamination in the food chain. The purpose of this study was to microbiologically asses raw materials (maize, wheat, soybean meal, and sunflower meal), and broiler compound feed (starter, grower, and finisher) from a representative feed mill in Romania; the microbiological contaminants that were analyzed were yeasts and molds, *Salmonella* spp., *Escherichia coli*, and *Clostridium perfringens*. Our study occured during the years 2019 and 2020; in 2019, 191 samples of raw materials and 360 samples of compound feed were analyzed and in 2020, 143 samples of raw materials and 241 samples of compound feed were analzyed. Among the tested samples of raw materials, the mean values of the yeasts and molds for maize, wheat, soybean, and sunflower meal were 1.3×10^3, 9.5×10^2, 6.4×10^2, and 7.4×10^2 cfu/g in 2019 and 1.5×10^3, 1.0×10^3, 5.2×10^2, and 7.1×10^2 cfu/g in 2020. In the analyzed compound feed samples, the mean amounts for the starter, grower, and finisher were 5.9×10^2, 4.2×10^2, and 4.2×10^2 cfu/g in 2019 and 5.3×10^2, 6.5×10^2, and 5.8×10^2 cfu/g in 2020. Potentially toxigenic fungi from *Aspergillus*, *Penicillium*, and *Fusarium* genera have been identified as the most common in all of the samples. In the raw materials, in both years the highest numbers of *Aspergillus*-positive samples were recorded: 66.6% in 2019 and 100% in 2020 for the maize samples, 50% in 2019 and 75% in 2020 for the wheat samples, 76% in 2019 and 87.5% in 2020 for the soybean meal samples and 71.4% in 2019 and 100% in 2020 for the sunflower meal. In the starter compound feed, the *Aspergillus* genera was prevailing in 2019 (46.6%), while in 2020, the species of the *Penicillium* and *Cladosporium* genera were identified in the majority of the samples (50%); for the grower and finisher compound feed, the *Aspergillus* genera was predominantly identified in 2019 (60% and 72.2% of the samples, respectively) and 2020 (61.5% and 46.6%, respectively). All of the results of the bacteriological analysis for determining the contamination with *Salmonella* spp., *E. coli*, and *Clostridium perfringens* were negative. Based on the results obtained in this study, monitoring and analysis of microbiological hazards in a feed mill will help to control and prevent contamination and have a direct impact on food safety.

Keywords: food and feed safety; yeasts and molds; *Salmonella* spp.; *Escherichia coli*; *Clostridium perfringens*

Citation: Lăpușneanu, D.M.; Simeanu, D.; Radu-Rusu, C.-G.; Zaharia, R.; Pop, I.M. Microbiological Assessment of Broiler Compound Feed Production as Part of the Food Chain—A Case Study in a Romanian Feed Mill. *Agriculture* **2023**, *13*, 107. https://doi.org/10.3390/agriculture13010107

Academic Editor: Lin Zhang

Received: 18 November 2022
Revised: 16 December 2022
Accepted: 27 December 2022
Published: 30 December 2022

1. Introduction

Compound feed is vulnerable to the introduction of bacteria throughout the production chain. The aim of the control of pathogens in feed is to ensure that they are under a critical threshold to minimize the risk to human and animal health [1]. The unnecessary or unintentional presence of pathogenic microorganisms is called microbiological contamination [2].

Potentially toxigenic fungi of field crops belong to the *Alternaria*, *Aspergillus*, *Cladosporium*, *Helminthosporium*, and *Fusarium* genera [3,4]. When cereal grains and feed are colonized by molds, there is a significant risk of contamination with their secondary metabolites [5], such as mycotoxins [6–8]. In addition to impairing the nutritional value and processing, the sanitary quality of the corn kernels can chemically alter the composition of the feed through the presence of substrates produced by microorganisms [9]. A high incidence of fungi of the *Aspergillus* genus (aflatoxigenic) was identified in maize grain stored under conditions of humidity between 13–18% [4]. Zearalenone mycotoxin in high concentrations can contaminate the carcasses of broilers, implying an anabolic effect in humans [10,11].

The genus *Salmonella* corresponds to an enteric Gram-negative, facultative anaerobe and non-spore-forming bacillus with cell diameters ranging from 0.7 to 1.5 μm and lengths from 2 to 5 μm, that belongs to the *Enterobacteriaceae* family [12]. Members of the *Salmonella* genus grow under temperatures from 7 to 48 °C, tolerating growth at water activity levels up to 0.995 and pH values between 6.5 to 7.5 [13]. *Salmonella* infection in poultry has long been categorized as a zoonotic disease of economic importance in public health [14–18]. Poultry products have been considered as the major reservoir of *Salmonella*, with approximately 200 serovars isolated from them [18–22]. Broilers and broiler products are an important source of *Salmonella* [23,24] and contaminated compound feed is considered to be one of the main sources of infection with *Salmonella* [25,26]. After contaminated feed has been ingested by broilers, *Salmonella* can multiply in their gastrointestinal tract and be eliminated in feces leading to increased multiplication on the farm [27,28]. Thermal inactivation of *Salmonella* enteridis and *Escherichia coli* O157:H7 was determined by temperatures between 54.3–64.5 °C, pH values between 4.2–9.6 with HCl or NaOH, and a NaCl concentration between 0.5–8.5% [29].

Escherichia coli (*E. coli*) is a Gram-negative bacterium of the family *Enterobacteriaceae* [30]. Regularly, it is found in the small intestine of endothermic organisms. *E. coli* is not always confined to the gut, and its ability to survive for short periods outside the body makes it possible to test samples for fecal contamination [31]. Avian colibacillosis is one of the major bacterial diseases in the poultry industry that has gained substantial attention worldwide [32]. Moreover, it is important to consider the potential for its zoonotic transmission through poultry reservoirs [33].

Clostridium perfringens (*C. perfringens*) is an anaerobic spore-forming, Gram-positive bacterium capable of producing various toxins and enzymes responsible for the lesions and associated symptoms [34,35]. The colonisation of poultry with *C. perfringens* is an early incident [36,37]. As a consequence of its ubiquity, *C. perfringens* is also recovered from broiler carcasses after refrigeration [36,37]. In broilers, Clostridia can cause necrotizing enteritis, an infection of the intestinal wall that is mainly caused by *C. perfringens* type A toxin [38]; it also constitutes a risk of transmission to humans through the food chain [39,40].

The European Union's food safety policy was reformulated at the beginning of the 2000s, in accordance with the approach of an integrative concept "from farm to fork", thus guaranteeing a high level of safety for food products in all stages of the production chain (including feed production) [41]. Regulation (EC) number 2160 of 2003 [42] ensures that appropriate and effective measures are implemented for their detection and control at all relevant stages of production, processing, and distribution, including animal feed, to reduce their prevalence and the risk to public health. According to Article 5, paragraph 3 of Regulation (EC) number 183 of 2005 [43] on feed hygiene, feed manufacturers must comply with specific microbiological criteria.

Considering the inclusion of compound feed production in the food chain, in the present work, raw materials and broiler compound feed from a feed mill in Romania were microbiologically assessed; the microbiological contaminants that were analyzed were yeasts and molds, *Salmonella* spp., *E. coli*, and *Clostridium perfringens*. Our study took place during the years 2019 and 2020, and provides new data on the microbiological evaluation of compound feed production as a part of the food chain.

2. Materials and Methods

2.1. Feed Samples

Compound feed is a balanced mixture of raw feed materials in order to satisfy the energy and nutrient requirements of a species/category of animals. In Romania, the main source of energy for compound feed is cereal seeds (maize and wheat), and soybean and sunflower meal are used as protein sources [44].

The samples of raw materials used in the production of compound feed (maize grains, wheat grains, soybean meal, and sunflower meal) and compound feed for broilers in different growth phases (starter, grower, and finisher) originated from a representative feed mill in Romania in terms of production (120,000 t/year). In 2019, 191 raw material samples were analyzed (37 maize, 15 wheat, 107 soybean meal, and 32 sunflower meal) and in 2020, 143 raw material samples were analyzed (13 maize, 9 wheat, 109 soybean meal, and 12 sunflower meal). In 2019, 360 samples of compound feed were analyzed (103 starter, 90 grower, and 167 finisher) and in 2020, 241 samples were analyzed (44 starter, 84 grower, and 113 finisher). The samples of raw materials were collected from the storage silos and the samples of compound feed were collected from the bunkers for storing the finished products of the studied unit, and they transferred to the specialized laboratories.

For the raw materials analysis, elementary samples from 25 points were collected with a manual probe inserted perpendicular to the base of the silo; for the compound feed analysis, seven elementary samples from batches of 24 tons were sampled manually with a trowel. All of the elementary samples taken constituted the global sample which was divided and homogenized with the centrifugal mechanical divider and resulted in the laboratory sample.

2.2. Yeasts and Molds Analysis

Yeasts and molds were detected in according with standard SR ISO 21527-2:2009 which describes a horizontal method for the enumeration of viable osmophilic yeasts and xerophilic molds in products intended for human consumption or the feeding of animals by means of the colony count technique at $25 \pm 1\ °C$. In order to create surface-inoculated plates, a specific selective culture medium was used. A certain amount of the initial suspension, or decimal dilutions of the suspension, was employed, depending on the anticipated number of colonies. The plates were then aerobically incubated at $25 \pm 1\ °C$ for 5 to 7 days, then the colonies were counted. The number of colonies discovered on the plates selected at dilution levels producing countable colonies was used to compute the quantity of yeasts and molds per gram of sample [45].

2.3. Salmonella spp. Analysis

Salmonella spp. was detected in according with standard SR EN ISO 6579-1:2017 which specifies a horizontal method for the detection of *Salmonella* and is applicable to the products intended for human consumption and the feeding of animals. Even when *Salmonella* is present, other *Enterobacteriaceae* or bacteria from other families are frequently present in far higher numbers. Pre-enrichment is used to enable the detection of *Salmonella* with low numbers or that have been harmed. The test portion was inoculated into buffered peptone water, and the mixture was then incubated between $34\ °C$ and $38\ °C$ for 18 h. The resulting culture was inoculated into Rappaport–Vassiliadis medium with soy (RVS broth) and a Muller–Kauffmann tetrathionate–novobiocin broth (MKTTn broth). The MKTTn broth was incubated at $37\ °C$ for 24 h whereas the RVS broth was incubated at $41.5\ °C$. From the cultures obtained, a selective solid media was inoculated with xylose lysine deoxycholate agar (XLD agar); the XLD agar was incubated at $37\ °C$ and examined after 24 h. *Salmonella* subcultures were used to subculture colonies, and the identities of the colonies were verified using the proper biochemical and serological tests [46].

2.4. Escherichia coli Analysis

Escherichia coli was detected in according with standard SR ISO 7251:2009 which gives general guidelines for the detection and enumeration of presumptive *Escherichia coli* by means of the liquid-medium culture technique and calculation of the most probable number (MPN) after incubation at 37 °C, then at 44 °C. The initial suspension was inoculated into three tubes of double-strength liquid selective enrichment medium in a predetermined amount. A predetermined amount of the initial suspension was inoculated into each of the three tubes of the single-strength liquid enrichment medium. A specific number of decimal dilutions of the initial suspension was then added to three more tubes of the single-strength medium under the identical conditions. After 24 and 48 h of incubation at 37 °C, the double- and single-strength medium tubes were checked to see if any gas was being produced. Each tube contained a liquid selective medium and was subcultured to a tube containing a double-strength medium that had produced opacity, cloudiness, or gaseous emissions, as well as each tube containing a single-strength medium that had produced gaseous emissions (EC broth). The obtained tubes were incubated at 44 °C for up to 48 h, and after 24 h and 48 h, they were checked for gas production. Each tube of media that had produced gaseous emissions was subcultured with a tube of peptone water free of indole. The obtained tubes were incubated at 44 °C for 48 h before being checked for indole formation as a result of tryptophan being broken down in the peptone component. The MPN table from Standard, which takes into account the number of tubes of single- and double-strength medium whose subcultures have produced gas in the EC broth and indole in the peptone water at 44 °C, was used to calculate the most likely number of presumptive *Escherichia coli* [47].

2.5. Clostridium perfringens Analysis

Clostridium perfringens was detected in according with standard SR EN ISO 7937:2005 which is applicable to products intended for human consumption and the feeding of animals. A predetermined amount of the original suspension was inoculated into Petri dishes as an inoculant. Under the same circumstances, additional Petri dishes were inoculated using decimal dilutions of the test sample or the initial suspension. An overlay of the same media was applied after adding a selective medium (poured-plate technique). The plates were anaerobically incubated at 37 °C for 20 h and two hours, and the distinctive colonies were enumerated. The quantity of distinctive colonies was verified, and it was computed as to how many *C. perfringens* bacteria there were per gram of sample [48].

2.6. Statistical Analysis

The data obtained from the analyses were processed and interpreted statistically. The minimum and maximum values were established and the position and variation estimators were calculated, respectively, the arithmetic mean ($\bar{x}$) and the standard deviation (s), for the samples that recorded positive results. The means and standard deviation were calculated using Microsoft Excel 2016 [49].

3. Results

3.1. Raw Materials

The results of the microbiological assessment (yeasts and molds, *Salmonella* spp., *E. coli*, and *Clostridium perfringens*) of the raw materials (maize, wheat, soybean meal, and sunflower meal) are presented in Table 1.

Table 1. Results of microbiological assessment of raw materials.

Specification		n	Positive (%)	$\bar{x}$	s	Min.	Max.	n	Positive (%)	$\bar{x}$	s	Min.	Max.
				2019 Year						**2020 Year**			
Yeasts and molds (cfu/g)	Maize	17	88.2	1.3×10^3	12.4×10^2	6.0×10^2	4.3×10^3	13	84.6	1.5×10^3	1.6×10^3	4×10^2	4.9×10^3
	Wheat	4	100	9.5×10^2	9.7×10^2	4.0×10^2	2.4×10^3	4	100	1.0×10^3	1.1×10^3	4×10^2	2.8×10^3
	Soybean meal	27	92.5	6.4×10^2	2.5×10^2	4.0×10^2	1.0×10^3	28	85.7	5.2×10^2	1.9×10^2	4×10^2	1.0×10^3
	Sunflower meal	7	100	7.4×10^2	2.5×10^2	4.0×10^2	1.0×10^3	3	100	7.1×10^2	3.0×10^2	4×10^2	1.0×10^3
Salmonella spp. /25 g	Maize	8	0	0	0	absent/25 g		0	-	-	-	-	-
	Wheat	4	0	0	0	absent/25 g		2	0	0	0	absent/25 g	
	Soybean meal	32	0	0	0	absent/25 g		39	0	0	0	absent/25 g	
	Sunflower meal	10	0	0	0	absent/25 g		4	0	0	0	absent/25 g	
E.coli (cfu/g)	Maize	7	0	0	0	0	0	0	-	-	-	-	-
	Wheat	3	0	0	0	0	0	1	0	0	0	0	0
	Soybean meal	31	0	0	0	0	0	39	0	0	0	0	0
	Sunflower meal	10	0	0	0	0	0	4	0	0	0	0	0
Clostridium perfringens (cfu/g)	Maize	5	0	0	0	0	0	0	-	-	-	-	-
	Wheat	4	0	0	0	0	0	2	0	0	0	0	0
	Soybean meal	17	0	0	0	0	0	3	0	0	0	0	0
	Sunflower meal	5	0	0	0	0	0	1	0	0	0	0	0

n—number of samples analyzed. $\bar{x}$—mean. s—standard deviation. Min.—minimum value identified. Max.—maximum value identified.

The mean values established for the analysis to determine the yeast and mold contamination of the maize were 1.3×10^3 cfu/g in 2019 for 88.2% of the positive samples, and a maximum value of 4.3×10^3 cfu/g and of 1.5×10^3 cfu/g in 2020 for 84.6% of the positive samples with a maximum value of 4.9×10^3 cfu/g. Compared to the number of analyses carried out to determine the content of yeasts and molds in soybean meal, a proportion of 92.5% of positive samples was identified in 2019 and 86.7% in 2020, and also the average values of 6.4×10^2 cfu/g in 2019 5.2×10^2 cfu/g in were identified in 2020.

Regarding the isolation proportion of potentially toxigenic fungi genera from the maize and wheat grains analyzed, the graphic representation (Figure 1) highlights the five genera that were identified (*Aspergillus*, *Penicillium*, *Fusarium*, *Cladosporium*, and *Alternaria*).

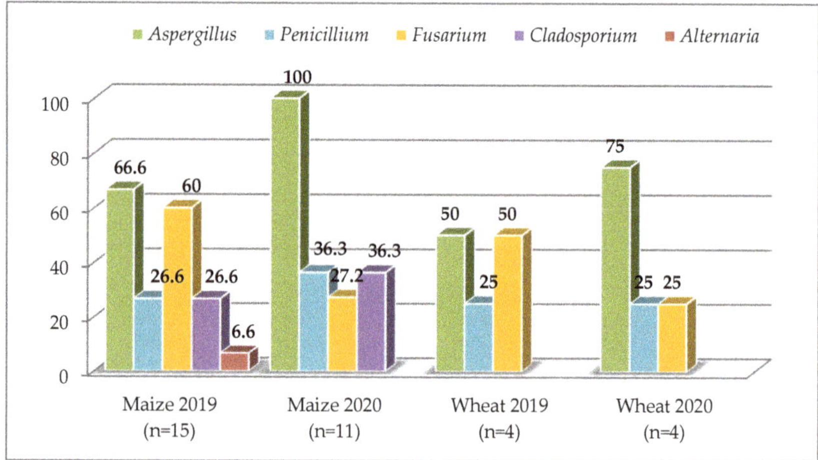

Figure 1. Proportion (%) of potentially toxigenic fungi genera in positive maize and wheat samples (n—number of positive samples).

Regarding the proportion of potentially toxigenic fungi genera identified in soybean and sunflower meal (Figure 2), five genera were identified (*Aspergillus*, *Penicillium*, *Fusarium*, *Cladosporium*, and *Mucor*).

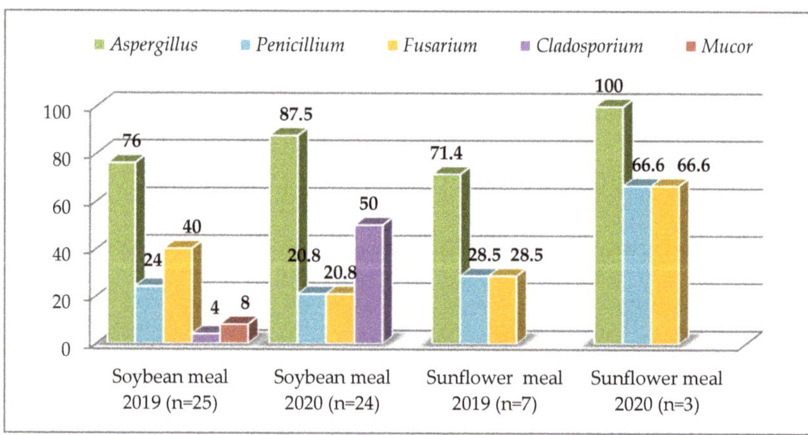

Figure 2. Proportion (%) of potentially toxigenic fungal genera in positive soybean and sunflower meal samples (n—number of positive samples).

3.2. Compound Feed

The results of microbiological assessment (yeasts and molds, *Salmonella* spp., and *E. coli*, and *Clostridium perfringens*) of broiler compound feed (starter, grower, and finisher) are presented in Table 2.

The mean values established for the analysis to determine the yeast and mold contamination of the starter compound feed were 5.9×10^2 cfu/g in 2019 for 50% of the positive samples and a maximum value of 2.9×10^3 cfu/g, and 5.3×10^2 cfu/g in 2020 for 42.8% of the positive samples with a maxim value of 9.0×10^2 cfu/g. In the case of grower compound feed, a proportion of 38.4% of positive samples was identified in 2019 and 44.8% in 2020, and also the average value of 4.2×10^2 cfu/g in 2019 respectively 6.5×10^2 cfu/g in 2020. Regarding finisher compound feed, positive samples had a percentage of 37.5% in 2019 with a mean value of 4.2×10^2 cfu/g, and 40.5% in 2020 with a mean of 5.8×10^2 cfu/g. The proportions of potentially toxigenic fungi genera identified in the starter (Figure 3), grower (Figure 4), and finisher (Figure 5) compound feed highlight that the *Aspergillus* and *Penicillium* genera were predominantly isolated.

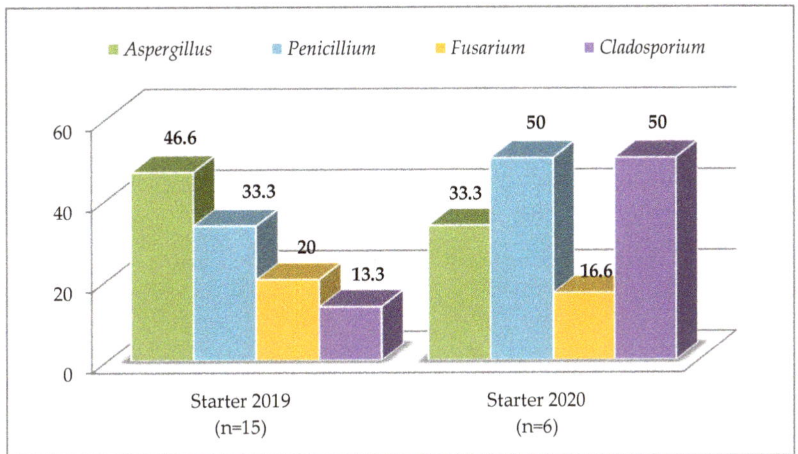

Figure 3. Proportion (%) of potentially toxigenic fungal genera in positive starter compound feed samples (n—number of positive samples).

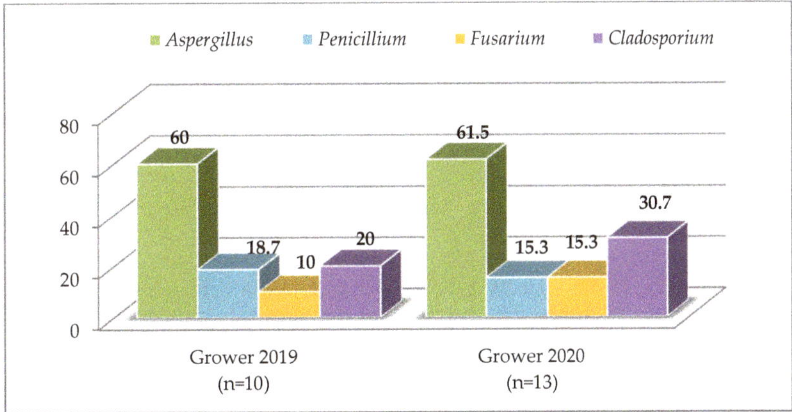

Figure 4. Proportion (%) of potentially toxigenic fungal genera in positive grower compound feed samples (n—number of positive samples).

Table 2. Results of microbiological assessment of compound feed.

Specification		n	Positive (%)	$\bar{x}$	s	Min.	Max.	n	Positive (%)	$\bar{x}$	s	Min.	Max.
				2019 Year						**2020 Year**			
Yeasts and molds (cfu/g)	Starter	30	50.0	5.9×10^2	6.4×10^2	4.0×10^2	2.9×10^3	14	42.8	5.3×10^2	1.9×10^2	4.0×10^2	9.0×10^2
	Grower	26	38.4	4.2×10^2	6.3×10^1	4.0×10^2	6.0×10^2	29	44.8	6.5×10^2	6.4×10^2	4.0×10^2	26.5×10^2
	Finisher	48	37.5	4.2×10^2	3.3×10^1	4.0×10^2	7.0×10^2	37	40.5	5.8×10^2	5.1×10^2	4.0×10^2	23.5×10^2
Salmonella spp. /25 g	Starter	31	0	0	0	absent/25 g		14	0	0	0	absent/25 g	
	Grower	26	0	0	0	absent/25 g		27	0	0	0	absent/25 g	
	Finisher	48	0	0	0	absent/25 g		37	0	0	0	absent/25 g	
E.coli (cfu/g)	Starter	29	0	0	0	0	0	14	0	0	0	0	0
	Grower	25	0	0	0	0	0	27	0	0	0	0	0
	Finisher	47	0	0	0	0	0	37	0	0	0	0	0
Clostridium perfringens (cfu/g)	Starter	13	0	0	0	0	0	2	0	0	0	0	0
	Grower	13	0	0	0	0	0	1	0	0	0	0	0
	Finisher	24	0	0	0	0	0	2	0	0	0	0	0

n—number of samples analyzed. $\bar{x}$—mean. s—standard deviation. Min.—minimum value identified. Max.—maximum value identified.

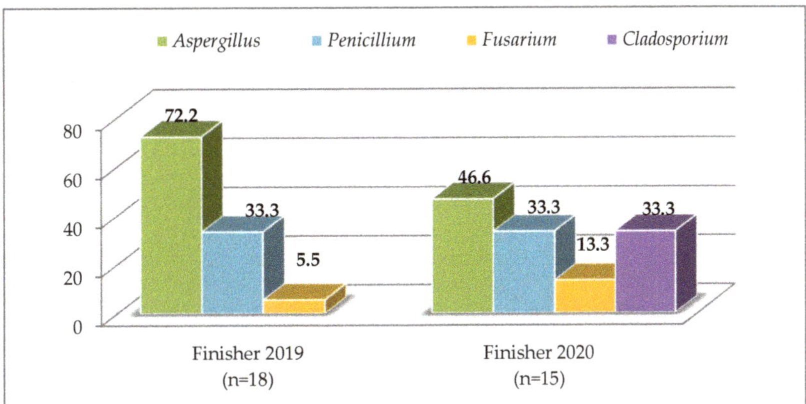

Figure 5. Proportion (%) of potentially toxigenic fungal genera in positive finisher compound feed samples (n—number of positive samples).

4. Discussion

4.1. Raw Materials

The main sources of fungal microflora in compound feed originate from raw materials of plant origin, firstly cereals [50]. Potentially toxigenic fungi are associated with oilseeds and cereals and mostly belong to the *Fusarium*, *Aspergillus*, and *Penicillium* genera [51]. It is distinguished in Figure 1 that the *Aspergillus* genus was identified in a high proportion of cases, over 50% in both years of research in contrast to the research carried out by Krnjaja et al. [52] which identified the most frequent genera as the *Fusarium* genera (92.22%). Krnjaja et al. [52] analyzed 127 maize samples for determinate fungal contamination; the total fungal count ranged from 1.0×10^1–3.0×10^6 cfu/g. In our study, the total fungal count ranged from 4.0×10^2 to 4.9×10^3 cfu/g. To determine the content of the yeasts and molds in the wheat samples, four analyses were carried out in both years of the study, with all of the results being positive. *Aspergillus* genera was isolated in more than 70% for the two types of meals analyzed in both years of the study (Figure 2). Studies have specified that fungi of the *Aspergillus* genera are the main producers of aflatoxins [53] and ochratoxin A, which in terms of their effect are the second most investigated mycotoxin after aflatoxins [54]. For the 60 most common mycotoxins occurring in feed, 48% have been shown to be produced by the *Fusarium* genus, 13% by the *Aspergillus* genus, 8% by the *Penicillium* genus, and 12% by the *Alternaria* genus [55]. In our study, high fungal colony counts for maize and wheat in both of the research years can be attributed to favorable climate conditions that were ideal for the growth of toxic mold during their growing and harvesting phases. The right temperature and humidity levels are essential for the growth of toxic fungus in grain both before and after harvest. The grain's moisture content is also one of the most crucial preconditions for grain infection during storage. A moisture level of 15% or less in maize is acceptable for safe storage.

Raw materials used in compound feed production represent an important source of *Salmonella* contamination [39]. Jones and Richardson [56] isolated *Salmonella* from maize, cottonseed meal, soybean meal, and wheat bran. In a raw materials survey, Ge B. et al. [57] identified positive samples for *Salmonella* contamination for 14 (54.2%) of 31 samples of soybean meal, 3 (60%) of 5 samples of sunflower meal, and 3 (17.6%) of 17 samples of maize. In our study, there were no positive results for *Salmonella* spp. contamination during the two years for the raw materials studied.

In addition, the results to determine raw materials contaminated with *E. coli* and *Clostridium perfringens* were negative. In their study, Da Costa et al. [58] analyzed 66 samples of raw materials (corn, wheat, barley, soybean meal, and sunflower meal) for *E. coli* contamination and identified that 89.2% of the samples were positive. Casagrande et al. [59]

demonstrated that out of the 80 raw materials' samples used in compound feed production, *C. perfringens* was isolated in 12.5% of the plant origin samples (soybean meal, and maize). Prió et al. [60] studied raw material microbial contamination and identified the presence of *C. perfringens* in 1.2% of maize samples (298 samples analyzed), 35.2% of wheat samples (85 samples analyzed), 5.3% of soybean meal samples (464 samples analyzed), and 13.3% of sunflower meal samples (70 samples analyzed). Another study conducted an analysis of 298 poultry feed ingredients and *C. perfringens* was detected in 33.89% of samples; the highest level of contamination with *C. perfringens* was observed in fish meal (55.26%) followed by bone meal (44.83%); the same as in our study, negative results of *C. perfringens* contamination were noticed in soybean meal and maize [61].

Regarding the correlation in the microbiological analysis results for raw materials intended for compound feed, all of them were in accordance with the limits allowed by legislation [62,63].

4.2. Compound Feed

Dalcero et al. [64] found the highest incidence of the *Aspergillus* (85%) and *Fusarium* genera (70%) in 130 samples of poultry feed. Rosa et al. [65] collected 96 samples of poultry feeds from four feed mills which were examined for total molds and for *Aspergillus* and *Penicillium* occurrence; the total mold counts were generally higher than 1.0×10^5 cfu/g; the *Aspergillus* and *Penicillium* species were isolated in the highest numbers. Shareef [66] realized a study in which 45 samples of poultry feed were analyzed; the most frequent fungi were from genus *Aspergillus* (88%) with a range of 1.0×10^4–5.3×10^6 cfu/g and a mean value of 2.6×10^6 cfu/g and the second most frequent were the from the genera *Penicillium* and *Mucor* (64%) with a range of 2.0×10^4–4.4×10^6 cfu/g and 3.0×10^4–2.6×10^5 cfu/g and a mean of 2.2×10^6 and 1.4×10^5 cfu/g, respectively. In a study realized by Cegielska-Radziejwska et al. [67], 45 samples of compound feed for broilers were analyzed, collected from four different feed mills in Poland; the samples consisted of compound feed for the different growing stage of broilers (starter, grower, and finisher). For the starter compound feed, the yeasts and molds ranged from 5.5×10^1–3.0×10^2 cfu/g and a mean value of 1.8×10^2 cfu/g; the determined grower compound feed values ranged from 6.0×10^1–4.7×10^2 cfu/g with a mean of 3.2×10^2 cfu/g; the finisher compound feed had values for yeasts and molds ranging from 8.5×10^1–7.0×10^3 cfu/g and a mean of 1.6×10^3 cfu/g. In the tested feed samples, fungi from the genera *Aspergillus*, *Fusarium*, *Mucor*, *Penicillium*, and *Rhizopus* were identified, with *Aspergillus* being the most commonly found genera, the same as in our study (Figures 3–5). A total of 49 samples of compound feed for broilers were analyzed by Greco at al. [68] for mycotoxigenic fungi and the most frequent mycotoxigenic fungi were those from the genus *Fusarium* (69.6), followed by *Eurotium* (52.2%), *Penicillium* (45.6%), and *Aspergillus* (43.5%). The fungal count for the compound feed was typically lower than the values found for raw materials. The pelleting procedure was applied to the studied compound feed, and it has been established that this procedure considerably reduced the fungal counts. As a result, pelleting may be the cause of the moderate frequency of some fungal genera.

Contamination of compound feed with *Salmonella* has been considered as an important source of contamination for broilers since the 1950s [69]. The potential transmission of *Salmonella* strains from feed to animals and subsequently to carcasses has been proven since the 1970s [70]. Over a period of 16 years, 23,963 samples for *Salmonella* culture and serotyping were collected from 22 stock feed mills in Australia. The percentage of positive samples ranged from 7.2% in 2003 to 2.8% in 2017. Of the 1069 positive samples, 976 were serotyped with 61 different *Salmonella* serotypes were isolated. The serotype most frequently isolated from raw materials was *Salmonella* Agona, (n = 108) whilst *Salmonella* Anatum was the serotype most frequently isolated from compound feed (n = 156) [71]. To determine *Salmonella* contamination in the studied broiler compound feeds, 105 analyses were carried out in 2019, and 78 analyses were carried out in 2020, with all of the results being negative, in accordance with the legislative regulations [62,63].

To determine the compound feed contamination with *E. coli*, 101 analyses were carried out in 2019, and 78 analyses were carried out in 2020, with all of the results being negative (Table 1). A survey on the prevalence of *Escherichia coli* in compound feed was carried out over a period of nine years in the Republic of Croatia; a total of 1688 feed samples were collected from feed mill and poultry farms, and analyses showed the presence of *E. coli* in 629 (37.3%) feed samples, which is different to our results [72]. In their study, Da Costa et al. [58] analyzed 23 samples of broiler compound feed for *E. coli* contamination and it was detected that 100% of the samples were positive. A study conducted an analysis on 378 animal feed samples to determine the prevalence of *E. coli* isolated from U.S. feed between 2005–2011; from the 65 samples of the poultry feed analyzed, 22 were positive (33.8%) [73].

In a survey conducted by Schocken-Iturrino et al. [74], they observed that among the 90 samples of broiler compound feed analyzed, 42% were contaminated by *C. perfringens*, and the mean count was 3.69×10^2 cfu/g. Tessari et al. [75] analyzed 121 samples of compound feed and reported *C. perfringens* isolated in 23 samples (19%). In comparison with the previous study, in our study all of the results were negative for *C. perfringens* contamination.

For the negative results of microbiological analysis (*Salmonella* spp., *E. coli*, and *C. perfringens*) both for the raw materials and compound feed, maybe it is because the feed mill that was studied has implemented a HACCP (hazard analysis and critical control points) system, which, as specified in Regulation (EC) number 183 of 2005 [43], can facilitate the achievement of a high level of feed safety; this may be a reason as to why our results are different from other studies.

Regarding the correlation in the microbiological analysis results for compound feed, the same as in case of raw materials, all of them were in accordance with limits allowed by national legislation [63,64].

5. Conclusions

Microbiological control of raw materials and broiler compound feed must be considered relevant due to the demands of consumers for food safety all over the food chain. This fact is possible through the introduction of an appropriate system for monitoring and analyzing microbiological contaminants in feed mill; this contributes to the control and prevention of contamination, with it having a direct impact on food safety and animal and human health. Based on the results obtained of the study, it can be concluded that the microbiological analysis is justified and necessary to assess the safety of broiler compound feed. It is necessary to continue the assessment with a complete approach to the hazards associated with food safety (physical, chemical, and biological), extended to other raw materials and categories of compound feed.

Author Contributions: Conceptualization, D.M.L. and I.M.P.; methodology, D.M.L. and R.Z.; software, D.M.L. and C.-G.R.-R.; validation, D.M.L. and D.S.; formal analysis, D.M.L., C.-G.R.-R., R.Z. and I.M.P.; investigation, D.M.L., C.-G.R.-R., R.Z. and I.M.P.; data curation, D.M.L. and D.S.; writing—original draft preparation, D.M.L. and C.-G.R.-R.; writing—review and editing, D.M.L., C.-G.R.-R. and I.M.P.; supervision, D.M.L. All authors have read and agreed to the published version of the manuscript.

Funding: This research received no external funding.

Institutional Review Board Statement: Not applicable.

Data Availability Statement: The data present in this study are available on request from the corresponding author.

Conflicts of Interest: The authors declare no conflict of interest.

References

1. Alali, W.Q.; Ricke, S.C. The ecology and control of bacterial pathogens in animal feed. In *Animal Feed Contamination-Effects on Livestock and Food Safety*, 1st ed.; Fink-Gremmels, J., Ed.; Woodhead Publishing: Cambridge, MA, USA, 2012; pp. 35–55. [CrossRef]
2. Chatterjee, A.; Abraham, J. Microbial contamination, prevention and early detection in food industry. In *Microbial Contamination and Food Degradation*; Grumezescu, A.T., Holban, A.M., Eds.; Academic Press: Amsterdam, The Netherlands, 2018; pp. 21–47. [CrossRef]
3. Wielogórska, E.; Mac Donald, S.; Elliott, C.T. A review of the efficacy of mycotoxin detoxifying agents used in feed in light of changing global environment and legislation. *World Mycotoxin J.* **2016**, *9*, 419–433. [CrossRef]
4. Pinotti, L.; Ottoboni, M.; Giromini, C.; Dell'Orto, V.; Cheli, F. Mycotoxin contamination in the EU feed supply chain: A focus on cereal byproducts. *J. Toxins* **2016**, *8*, 45. [CrossRef]
5. Atungulu, G.G.; Zhong, S.; Thote, A.; Okeyo, A.; Couch, A. Microbial Prevalence on Freshly-Harvested Long-Grain Pureline, Hybrid, and Medium-Grain Rice Cultivars. *Appl. Eng. Agric.* **2015**, *31*, 949–956. [CrossRef]
6. Bento, L.F.; Caneppele, M.A.B.; Albuquerque, M.C.F.; Kobayasti, L.; Caneppele, C.; Andrade, P.J. Occurrence of fungi and aflatoxins in corn kernels. *Rev. Do Inst. Adolfo Lutz* **2012**, *71*, 44–49.
7. Ruiz, M.J.; Macáková, P.; Juan-García, A.; Font, G. Cytotoxic effects of mycotoxin combinations in mammalian kidney cells. *Food Chem. Toxicol.* **2011**, *49*, 2718–2724. [CrossRef] [PubMed]
8. Savi, G.D.; Piacentini, K.C.; Marchi, D.; Scussel, V.M. Fumonisins B1 and B2 in the corn-milling process and corn-based products, and evaluation of estimated daily intake. *Food Addit. Contam. A* **2016**, *33*, 339–345. [CrossRef]
9. Abdollahi, M.R.; Ravindran, V.; Wester, T.J.; Ravindran, G.; Thomas, D.V. Influence of conditioning temperature on the performance, nutrient utilisation and digestive tract development of broilers fed on maize-and wheat-based diets. *Brit. Poult. Sci.* **2010**, *51*, 648–657. [CrossRef] [PubMed]
10. Briyones-Reyes, D.; Gómez-Martinez, L.; Cueva-Rolon, R. Zearalenone contamination in corn for human consumption in the state of Tlaxcala, Mexico. *Food Chem.* **2007**, *100*, 693–698. [CrossRef]
11. Liu, M.T.; Ram, B.P.; Hart, L.P.; Pestka, J.J. Indirect Enzyme-Linked Immunosorbent Assay for the Mycotoxin Zearalenone. *Appl. Environ. Microb.* **1985**, *50*, 332–336. [CrossRef]
12. Bhunia, A.K. Salmonella Enterica. In *Foodborne Microbial Pathogens*; Springer: New York, NY, USA, 2008; pp. 201–216.
13. Pui, C.F.; Wong, W.C.; Chai, L.C.; Robin, T.; Ponniah, J.; Hidayah, M.S.; Anyi, U.; Mohamad Ghazali, F.; Cheah, Y.K.; Son, R. Review article *Salmonella*: A foodborne pathogen. *Int. Food Res. J.* **2011**, *18*, 465–473.
14. Antunes, P.; Mourão, J.; Campos, J.; Peixe, L. Salmonellosis: The role of poultry meat. *Clin. Microbiol. Infect.* **2016**, *22*, 110–121. [CrossRef] [PubMed]
15. Gast, R.K. Serotype-specific and serotype-independent strategies for preharvest control of food-borne *Salmonella* in poultry. *Avian Dis.* **2007**, *51*, 817–828. [CrossRef] [PubMed]
16. Ferrari, R.G.; Rosario, D.K.A.; Cunha-Neto, A.; Mano, S.B.; Figueiredo, E.E.S.; Conte-Junior, C.A. Worldwide Epidemiology of *Salmonella* serovars in animal-based foods: A meta-analysis. *Appl. Environ. Microbiol.* **2019**, *85*, e00591-19. [CrossRef] [PubMed]
17. Popa, G.L.; Popa, M.I. *Salmonella* spp. Infection—A continuous threat worldwide. *Germs* **2021**, *11*, 88–96. [CrossRef] [PubMed]
18. Ruvalcaba-Gómez, J.M.; Villagrán, Z.; Valdez-Alarcón, J.J.; Martínez-Núñez, M.; Gomez- Godínez, L.J.; Ruesga-Gutiérrez, E.; Anaya-Esparza, L.M.; Arteaga- Garibay, R.I.; Villarruel-López, A. Non-Antibiotics Strategies to Control *Salmonella* Infection in Poultry. *Animals* **2022**, *12*, 102. [CrossRef]
19. Desin, T.S.; Köster, W.; Potter, A.A. *Salmonella* vaccines in poultry: Past, present and future. *Expert Rev. Vaccines* **2013**, *12*, 87–96. [CrossRef]
20. Shah, D.H.; Paul, N.C.; Sischo, W.C.; Crespo, R.; Guard, J. Population dynamics and antimicrobial resistance of the most prevalent poultry-associated *Salmonella* serotypes. *Poult. Sci.* **2017**, *96*, 687–702. [CrossRef]
21. Crouch, C.F.; Pugh, C.; Patel, A.; Brink, H.; Wharmby, C.; Watts, A.; van Hulten, M.C.W.; de Vries, S.P.W. Reduction in intestinal colonization and invasion of internal organs after challenge by homologous and heterologous serovars of *Salmonella* Enterica following vaccination of chickens with a novel trivalent inactivated *Salmonella* vaccine. *Avian Pathol.* **2020**, *49*, 666–677. [CrossRef]
22. Cadirci, O.; Gucukoglu, A.; Gulel, G.T.; Gunaydin, E.; Uyanik, T.; Kanat, S. Determination and antibiotic resistance profiles of *Salmonella* serotypes isolated from poultry meat. *Fresenius Environ. Bull.* **2021**, *30*, 4251–4261.
23. Tauxe, R.V. *Salmonella*: A Postmodern Pathogen. *J. Food Prot.* **1991**, *54*, 563–568. [CrossRef]
24. Bryan, F.L.; Doyle, M.P. Health Risks and Consequences of *Salmonella* and *Campylobacter jejuni* in Raw Poultry. *J. Food Prot.* **1995**, *58*, 326–344. [CrossRef]
25. Jones, F.T.; Axtell, R.C.; Rives, D.V.; Scheideler, S.E.; Tarver, F.R.; Walker, R.L.; Wineland, M.J. A Survey of *Salmonella* Contamination in Modern Broiler Production. *J. Food Prot.* **1991**, *54*, 502–507. [CrossRef]
26. Maciorowski, K.G.; Jones, F.T.; Pillai, S.D.; Ricke, S.C. Incidence, sources, and control of food-borne *Salmonella* spp. in poultry feeds. *World Poult. Sci. J.* **2004**, *60*, 446–457. [CrossRef]
27. Cason, J.A.; Cox, N.A.; Bailey, J.S. Transmission of *Salmonella* typhimurium during Hatching of Broiler Chicks. *Avian Dis.* **1994**, *38*, 583–588. [CrossRef]
28. Petkar, A.; Alali, W.Q.; Harrison, M.A.; Beuchart, L.R. Survival of *Salmonella* in organic and conventional broiler feeds as affected by temperature and water activity. *Agric. Food Anal. Bacteriol.* **2011**, *1*, 175–185.

29. Blackburn, C.d.W.; Curtis, L.M.; Humpheson, L.; Billon, C.; McClure, P.J. Development of thermal inactivation models for *Salmonella enteridis* and *Escherichia coli* O157:H7 with temperature, pH and NaCl as controlling factors. *Int. J. Food Microbiol.* **1997**, *38*, 31–44. [CrossRef]

30. Tenaillon, O.; Skurnik, D.; Picard, B.; Denamur, E. The population genetics of commensal *Escherichia coli*. *Nat. Rev. Microbiol.* **2010**, *8*, 207–217. [CrossRef]

31. Ali, M.A.; Eldin, T.A.S.; El Moghazy, G.M.; Tork, I.M.; Omara, I.I. Detection of *E. coli* O157:H7 in feed samples using gold nanoparticles sensor. *Int. J. Curr. Microbiol. App. Sci.* **2014**, *3*, 697–708.

32. Kabir, S.M.L. Avian colibacillosis and salmonellosis: A closer look at epidemiology, pathogenesis, diagnosis, control and public health concerns. *Int. J. Environ. Res. Public Health* **2010**, *7*, 89–114. [CrossRef]

33. Markland, S.M.; LeStrange, K.J.; Sharma, M.; Knie, K.E. Old Friends in New Places: Exploring the Role of xtraintestinal *E. coli* in Intestinal Disease and Foodborne Illness. *Zoonoses Public Health* **2015**, *62*, 491–496. [CrossRef]

34. Songer, J.G.; Meer, R.R. Genotyping of *Clostridium perfringens* by Polymerase Chain Reaction is a Useful Adjunct to Diagnosis of Clostridial Enteric Disease in Animals. *Anaerobe* **1996**, *2*, 197–203. [CrossRef]

35. Petit, L.; Gibert, M.; Popoff, M.R. *Clostridium perfringens*: Toxinotype and genotype. *Trends Microbiol.* **1999**, *7*, 104–110. [CrossRef] [PubMed]

36. Craven, S.E.; Stern, N.J.; Bailey, J.S.; Cox, N.A. Incidence of *Clostridium perfringens* in broiler chickens and their environment during production and processing. *Avian Dis.* **2001**, *45*, 887–896. [CrossRef] [PubMed]

37. Stanley, D.; Bajagai, Y.S. Feed Safety and the Development of Poultry Intestinal Microbiota. *Animals* **2022**, *12*, 2890. [CrossRef]

38. Schuring, M.; van Gils, B. *Clostridium perfringens*-a nutritional challenge? *Feed Mix* **2001**, *9*, 26–28.

39. Van Immersel, F.; De Zutter, L.; Houf, K.; Pasmans, F.; Haesebrouck, F.; Ducatelle, R. Strategies to control *Salmonella* in the broiler production chain. *World Poult. Sci. J.* **2009**, *65*, 367–392. [CrossRef]

40. Buzby, J.; Roberts, T. Economic costs and trade impacts of microbial foodborne ilness. *World Health Statist. Quart.* **1997**, *50*, 57–66.

41. Pop, I.M. Food quality and safety management. In *Managementul Calității și Siguranței Alimentelor*; Tipo Moldova: Iași, Romania, 2022.

42. Regulation (EC) No 2160/2003 of the European Parliament and of the Council of 17 November 2003 on the Control of Salmonella and Other Specified Food-Borne Zoonotic Agents. Available online: https://eur-lex.europa.eu/legal-content/en/ALL/?uri=CELEX:32003R2160 (accessed on 26 December 2022).

43. Regulation (EC) No 183/2005 of the European Parliament and of the Council of 12 January 2005 Laying down Requirements for Feed Hygiene. Available online: https://leap.unep.org/countries/eu/national-legislation/regulation-ec-no-1832005-european-parliament-and-council-laying#:~{}:text=(EC)%20No.-,183%2F2005%20of%20the%20European%20Parliament%20and%20of%20the%20Council,down%20requirements%20for%20feed%20hygiene.&text=The%20present%20Regulation%20lays%20down,registration%20and%20approval%20of%20establishments (accessed on 26 December 2022).

44. Pop, I.M.; Halga, P.; Avarvarei, T. Animal nutrition and feeding. In *Nutriția și Alimentația Animalelor*; Tipo Moldova: Iași, Romania, 2006.

45. *SR ISO 21527-2:2009*; Microbiology of Food and Animal Feeding Stuffs—Horizontal Method for the Enumeration of Yeasts and Moulds—Part 2: Colony Count Technique in Products with Water Activity Less than or Equal to 0.95. Romanian Standards Association: Bucharest, Romania, 2009.

46. *SR EN ISO 6579-1:2017*; Microbiology of the Food Chain—Horizontal Method for the Detection, Enumeration and Serotyping of Salmonella—Part 1: Detection of Salmonella spp. Romanian Standards Association: Bucharest, Romania, 2017.

47. *SR ISO 7251:2009*; Microbiology of Food and Animal Feeding Stuffs—Horizontal Method for the Detection and Enumeration of Presumptive Escherichia coli—Most Probable Number Technique. Romanian Standards Association: Bucharest, Romania, 2009.

48. *SR EN ISO 7937:2005*; Microbiology of Food and Animal Feeding Stuffs—Horizontal Method for the Enumeration of Clostridium perfringens—Colony-Count Technique. Romanian Standards Association: Bucharest, Romania, 2005.

49. Rayat, C.S. Applications of Microsoft Excel in Statistical Methods. In *Statistical Methods in Medical Research*, 1st ed.; Rayat, C.S., Ed.; Springer: Singapore, 2018; pp. 139–146. [CrossRef]

50. Creppy, E.E. Update of survey, regulation and toxic effects of mycotoxins in Europe. *Toxicol. Lett.* **2002**, *127*, 19–28. [CrossRef]

51. Pacin, A.M.; González, H.H.L.; Etcheverry, M.; Resnik, S.L.; Vivas, L.; Espin, S. Fungi associated with food and feed commodities from Ecuador. *Mycopathologia* **2002**, *156*, 87–92. [CrossRef]

52. Krnjaja, V.; Stanojković, A.; Stanković, S.Ž.; Lukić, M.; Bijelić, Z.; Mandić, V.; Mićić, N. Fungal contamination of maize grain samples with a special focus on toxigenic genera. *Biotechnol. Anim. Husb.* **2017**, *33*, 233–241. [CrossRef]

53. Marin, S.; Ramos, A.J.; Cano-Sancho, G.; Sanchis, V. Mycotoxins: Occurrence, toxicology, and exposure assessment. *Food Chem. Toxicol.* **2013**, *60*, 218–237. [CrossRef]

54. Völkel, I.; Schröer-Merker, E.; Czerny, C.P. The carry-over of mycotoxins in products of animal origin with special regards to its implications for the european food safety legislation. *Food Nutr. Sci.* **2011**, *2*, 852–867. [CrossRef]

55. Khoshal, A.K.; Novak, B.; Martin, P.G.P.; Jenkins, T.; Neves, M.; Schatzmayr, G.; Oswald, I.P.; Pinton, P. Co-Occurrence of DON and emerging mycotoxins in worldwide finished pig feed and their combined toxicity in intestinal cells. *Toxins* **2019**, *11*, 727. [CrossRef] [PubMed]

56. Jones, F.T.; Richardson, K.E. *Salmonella* in commercially manufactured feeds. *Poult. Sci.* **2004**, *83*, 384–391. [CrossRef] [PubMed]

57. Ge, B.; LaFon, P.C.; Carter, P.J.; McDermott, S.D.; Abbott, J.; Glenn, A.; Ayers, S.L.; Friedman, S.L.; Paige, J.C.; Wagner, D.D.; et al. Retrospective Analysis of *Salmonella, Campylobacter, Escherichia coli,* and *Enterococcus* in Animal Feed Ingredients. *Foodborne Pathog. Dis.* **2013**, *10*, 684–691. [CrossRef]

58. Da Costa, P.M.; Oliveira, M.; Bica, A.; Vaz-Pires, P.; Bernardo, F. Antimicrobial resistance in *Enterococcus* spp. and *Escherichia coli* isolated from poultry feed and feed ingredients. *Vet. Microbiol.* **2007**, *120*, 122–131. [CrossRef]

59. Casagrande, M.F.; Cardozo, M.V.; Beraldo-Massoli, M.C.; Boarini, L.; Longo, F.A.; Paulilo, A.C.; Schocken-Iturrino, R.P. *Clostridium perfringens* in ingredients of poultry feed and control of contamination by chemical treatments. *J. Appl. Poultry Res.* **2013**, *22*, 771–777. [CrossRef]

60. Prió, P.; Gasol, R.; Soriano, R.C.; Pérez-Rigau, A. Effect of raw material microbial contamination over microbiological profile of ground and pelleted feeds. *Cah. Opt. Méditerran.* **2001**, *54*, 197–199.

61. Udhayavel, S.; Thippichettypalayam Ramasamy, G.; Gowthaman, V.; Malmarugan, S.; Senthilvel, K. Occurrence of *Clostridium perfringens* contamination in poultry feed ingredients: Isolation, identification and its antibiotic sensitivity pattern. *Anim. Nutr.* **2017**, *3*, 309–312. [CrossRef]

62. *Order no. 249 of March 31, 2003 for the Approval of the Norms Regarding the Quality and Sanitation Parameters for the Production, Import, Quality Control, Marketing and Use of Simple, Compound Feeds, Feed Additives, Premixtures, Energy Substances, Mineral Substances and Special Feeds;* Ministry of Agriculture, Food and Forest of Romania: București, Romania, 2003.

63. *Order no. 110 of June 16, 2020 for the Approval of the Veterinary Sanitary Norm Regarding the Maximum Limits Provided for Microorganisms in Certain Categories of Feed;* National Veterinary Sanitary and Food Safety Authority: București, Romania, 2020.

64. Dalcero, A.; Magnoli, C.; Luna, M.; Ancasi, G.; Reynoso, M.M.; Chiacchiera, S.; Miazzo, R.; Palacio, G. Mycoflora and naturally occurring mycotoxins in poultry feeds in Argentina. *Mycopathologia* **1998**, *141*, 37–43. [CrossRef]

65. Rosa, C.A.R.; Ribeiro, J.M.M.; Fraga, M.J.; Gatti, M.; Cavaglieri, L.R.; Magnoli, C.E.; Dalcero, A.M.; Lopes, C.W.G. Mycoflora of poultry feeds and ochratoxin-producing ability of isolated *Aspergillus* and *Penicillium* species. *Vet. Microbiol.* **2006**, *113*, 89–96. [CrossRef]

66. Shareef, A.M. Molds and mycotoxins in poultry feeds from farms of potential mycotoxicosis. *Iraqui J. Vet. Sci.* **2010**, *24*, 17–25. [CrossRef]

67. Cegielska-Radziejewska, R.; Stuper, K.; Szablewski, T. Microflora and mycotoxin contamination in poultry feed mixtures from western Poland. *Ann. Agric. Environ. Med.* **2013**, *20*, 30–35.

68. Greco, M.V.; Franchi, M.L.; Rico Golba, S.L.; Pardo, A.G.; Pose, G.N. Mycotoxins and Mycotoxigenic Fungi in Poultry Feed for Food-Producing Animals. *Sci. World J.* **2014**, *2014*, 968215. [CrossRef] [PubMed]

69. Williams, J.E. *Salmonellas* in poultry feeds—A worldwide review. *Poult. Sci. J.* **1981**, *37*, 97–105. [CrossRef]

70. MacKenzie, M.A.; Bains, B.S. Dissemination of *Salmonella* serotypes from raw feed ingredients to chicken carcases. *Poult. Sci.* **1976**, *55*, 957–960. [CrossRef] [PubMed]

71. Parker, E.M.; Edwards, L.J.; Mollenkopf, D.F.; Ballash, G.A.; Wittum, T.E.; Parker, A.J. *Salmonella* monitoring programs in Australian feed mills: A retrospective analysis. *Aust. Vet. J.* **2019**, *97*, 336–342. [CrossRef]

72. Sokolovic, M.; Šimpraga, B.; Amšel-Zelenika, T.; Berendika, M.; Krstulovi'c, F. Prevalence and Characterization of Shiga Toxin Producing *Escherichia coli* Isolated from Animal Feed in Croatia. *Microorganisms* **2022**, *10*, 1839. [CrossRef]

73. Ge, B.; Domesle, K.J.; Gaines, S.A.; Lam, C.; Bodeis Jones, S.M.; Yang, Q.; Ayers, S.L.; McDermott, P.F. Prevalence and Antimicrobial Susceptibility of Indicator Organisms *Escherichia coli* and *Enterococcus* spp. Isolated from U.S. Animal Food, 2005–2011. *Microorganisms* **2020**, *8*, 1048. [CrossRef]

74. Schocken-Iturrino, R.P.; Vittori, J.; Beraldo-Massoli, M.C.; Delphino, T.P.C.; Damasceno, P.R. *Clostridium perfringens* search in water and ration used in the raising of broiler in sheds of São Paulo State–Brazil. *Ciência Rural* **2010**, *40*, 197–199. [CrossRef]

75. Tessari, E.N.C.; Cardoso, A.L.S.P.; Kanashiro, A.M.I.; Stoppa, G.F.Z.; Luciano, R.L.; de Castro, A.G.M. Analysis of the presence of *Clostridium perfringens* in feed and raw material used in poultry production. *Food Nutr. Sci.* **2014**, *5*, 614–617. [CrossRef]

 *agriculture*

Article

Comparative Evaluation of the Dynamics of Animal Husbandry Air Pollutant Emissions Using an IoT Platform for Farms

Razvan Alexandru Popa [1,2], Dana Catalina Popa [1,*], Elena Narcisa Pogurschi [1], Livia Vidu [1], Monica Paula Marin [1], Minodora Tudorache [1], George Suciu [3], Mihaela Bălănescu [3], Sabina Burlacu [2,4], Radu Budulacu [2] and Alexandru Vulpe [2,4,*]

1   Formative Sciences in Animal Breeding and Food Industry Department, University of Agricultural Sciences and Veterinary Medicine, 59 Marasti Blv., 011464 Bucharest, Romania
2   R&D Department, Beam Innovation SRL, 041386 Bucharest, Romania
3   R&D Department, Beia Consult International, 041386 Bucharest, Romania
4   Telecommunications Department, University Politehnica of Bucharest, 061071 Bucharest, Romania
*   Correspondence: dana-catalina.popa@igpa.usamv.ro (D.C.P.); alex.vulpe@beaminnovation.ro (A.V.); Tel.: +40-720-210-178 (D.C.P.); +40-753-389-174 (A.V.)

Abstract: One of the major challenges of animal husbandry, in addition to those related to the economic situation and the current energy crisis, is the major contribution of this sector to atmospheric pollution. Awareness of pollution sources and their permanent monitoring in order to ensure efficient management of the farm, with the aim of reducing emissions, is a mandatory issue, both at the macro level of the economic sector and at the micro level, specifically at the level of each individual farm. In this context, the acquisition of consistent environmental data from the level of each farm will constitute a beneficial action both for the decision-making system of the farm and for the elaboration or adjustment of strategies at the national level. The current paper proposes a case study of air pollutants in a cattle farm for different seasons (winter and summer) and the correlation between their variation and microclimate parameters. A further comparison is made between values estimated using the EMEP (European Monitoring and Evaluation Programme, 2019) methodology for air pollutant emission and values measured by sensors in a hybrid decision support platform for farms. Results show that interactions between microclimate and pollutant emissions exist and they can provide a model for the farm's activities that the farmer can manage according to the results of the measurements.

Keywords: AP monitoring; IoT; AP estimation; decision support; livestock farming

Citation: Popa, R.A.; Popa, D.C.; Pogurschi, E.N.; Vidu, L.; Marin, M.P.; Tudorache, M.; Suciu, G.; Bălănescu, M.; Burlacu, S.; Budulacu, R.; et al. Comparative Evaluation of the Dynamics of Animal Husbandry Air Pollutant Emissions Using an IoT Platform for Farms. *Agriculture* **2023**, *13*, 25. https://doi.org/10.3390/agriculture13010025

Academic Editor: Daniel Simeanu

Received: 18 November 2022
Revised: 17 December 2022
Accepted: 18 December 2022
Published: 22 December 2022

1. Introduction

Currently, the exploitation of cows for milk production represents a major challenge, both from an economic point of view (the price of milk at the farm gate in relation to the current financial challenges) and in terms of its significant impact on the environment, i.e., the ecological implications at a global scale, noticed for several decades by the scientific community but widely and acutely felt in recent years.

The exploitation of animals for food production undoubtedly has a major ecological impact on the environment, represented in deforestation, crises in the provision of drinking water through the pollution of surface and underground waters, the loss of biodiversity at all its levels, emissions of greenhouse gases (approximately 30% of total emissions), pollutant emissions, etc. [1–4].

Ensuring food security is a priority policy of any modern state, but this goal must not be fulfilled under any circumstances. The concept of food security must also include environmental security in such a way that future generations are not affected.

Animal exploitation affects the environment on several levels. A first problem that arises is related to the pollution of water by increased levels of nutrients such as nitrogen and phosphorus, which causes the phenomenon of water eutrophication, with all its consequences for the biodiversity of aquatic ecosystems. For this reason, the member states of the United Nations have included this phenomenon in the list of objectives "Sustainable Development Goals (SDGs) 14, 15, 17" (https://sdgs.un.org/goals, accessed on 18 October 2022).

Loss of specific biodiversity is another consequence of animal agriculture. The growing demand for animal products has led to an increase in cultivated areas. Under the conditions of the permanent expansion of urban areas, the provision of areas for the cultivation of fodder is achieved by deforestation. The FAO (Food an Agriculture Organization) has estimated that more than 7% of land area is used to feed dairy animals [5]. Deforestation leads, in addition to massive biodiversity loss [6], to the intensification of the greenhouse effect by releasing stored carbon in various forms.

A large number of gases that cause climate change are also known air pollutants that affect our health and the environment. In many ways, improving air quality can also boost climate change mitigation efforts and vice versa, but not always. The challenge is to ensure that climate change and air policies focus on win-win scenarios.

In its 2007 assessments, the Intergovernmental Panel on Climate Change (IPCC) predicted a decline in air quality in the future due to climate change, but we do not currently have a complete understanding of how climate change might affect air quality.

It is interesting to note that many climate-related processes are controlled not by the main components of our atmosphere, but by some gases that are found only in very small quantities. The most common of these so-called waste gases, carbon dioxide, makes up only 0.0391% of the air. Any variation in these very small amounts has the ability to affect and modify our climate.

The World Health Organization specifies that almost all the world's population (99%) breathes air with levels of pollutants that significantly exceed the maximum limits of admissibility and which, obviously, affect human health, generating different categories of pathologies (https://www.who.int/health-topics/air-pollution#tab=tab_1, accessed on 18 October 2022). These pollutants are represented by ammonia, volatile organic compounds, nitrogen oxides and microscopic particles in suspension. Animal production generates such pollutants; it is estimated that approximately 8% (PM_{10}) and 4% ($PM_{2.5}$) of the total microscopic particles in suspension come from this economic sector [7].

In this context and taking into account the fact that the demand for milk and dairy products is constantly increasing, one of the major challenges of milk production is represented by reducing its impact on the environment and minimizing pollutant emissions. Obviously, in this action we are limited by the physiology of the animal. As a result, the tools at our disposal are related to nutrition and farm management (type of feed, shelters, microclimate, feed administration, manure management).

The objectives of this study are:

- To present the current knowledge on the relationship between air pollutants and animal welfare and the relationship between air pollutant variation and farm management.
- To propose a case study in a real environment where an IoT infrastructure is used for monitoring key parameters of the stable environment: gas sensors (NH_3) and PM sensors ($PM_{2.5}$, PM_1, PM_{10}).
- To estimate the air pollutants' concentrations (in two seasons—winter and summer) based on European Monitoring and Evaluation Programme (EMEP) methodology and to compare the estimated values with the monitored concentrations.
- To study the behavior of air pollutants in correlation with micro-climate parameters.

The paper is organized as follows. Section 2 reviews the state of the art in assessing the relationship between air pollutants, animal welfare and farm management. Section 3 describes the architecture of the platform for AP monitoring and outlines the case study conducted using the platform. Sections 4 and 5 present the results obtained and the conclusions drawn, respectively.

2. State of the Art

Pollutant emissions from animal husbandry must benefit from a holistic treatment, as they influence each other. In order to understand the phenomenon, it is necessary that the physiology of cows be introduced in the context of trophic relations in a biocenosis. Thus, it must be accepted that the Eltonian pyramid is a simplistic representation of trophic relationships, and the very strict labeling of a trophic link is a gross error. The stability of a biocenosis by optimizing energy and nutrient flows would not be possible without some plasticity in the component species. Thus, cows can be considered the most inefficient herbivores, trophically speaking, and, to use a figure of speech, "the most carnivorous of herbivores." For this reason, cows are the most inefficient organisms regarding nitrogen utilization. It has been found that between 50 and 80% of consumed nitrogen is excreted as urea and other nitrogen compounds through feces and urine [8] (a consequence of physiology and trophic position), which represent important sources of ammonia emissions [9]. Most of the nitrogen is present in the urine and a smaller part in the feces. Regardless of the operating conditions, in the stable or on the pasture, at the time of excretion and the mixture of urine and feces, the nitrogen in the form of urea in the urine is converted into an unstable mixture of ammonia (NH_3) and ammonium (NH_4) under the action of urease from feces, resulting ammonia volatilization [10].

Numerous research papers have highlighted the direct relationship between the nitrogen content (in various forms) of cow manure and their nutrition. The crude protein content of the ration influences ammonia emissions through the manure, even establishing a linear relationship between them [11]. Decreasing the crude protein content of the ration is proving to be an effective means of reducing ammonia emissions from dairy cow manure [12], with research showing a 45% reduction in ammonia emissions as a result of lowering from 17% to 13.5% of crude protein content in cow diets [13]. Swensson [14] finds that ammonia emissions will be three times higher when the crude protein content of the ration increases from 13 to 19%. Similar results, namely the highlighting of the direct relationship between crude protein intake and ammonia emissions, are reported by other authors [15–18].

It should be emphasized that the efficiency of nitrogen use in the metabolism of dairy cows is also influenced by other nutritional factors apart from crude protein intake. For example, in cows, the nitrogen content of milk, under the conditions of a diet with similar levels of protein, depends on the content and composition of carbohydrates in the ration [19]. The nitrogen content of milk will decrease (so we expect higher values in urine and feces) on a high-fiber diet compared to a high-starch diet.

However, it should be kept in mind that ammonia emissions from manure (urine) also depend on other factors. Research has shown that the type of soil, atmospheric humidity, temperature, wind speed or air currents in the shelter influence this aspect and cause large variations in ammonia losses from urine: between 25–50% [9] or between 4–52% [20]. In this regard, significant results were achieved [21] stating that ammonia emissions increase with temperature, and that this is directly related to the type of floor in the shelter and the management of manure. Significant differences occur depending on the cow maintenance system. In open systems the manure will be deposited directly on the soil and there will thus be a rapid conversion of urea to ammonia (i.e., high emissions), while in closed systems, with the maintenance of cows in the shelter, the emissions are lower as a result of the regular removal of manure from the shelter [22,23]. Inside the stable, ammonia emissions differ significantly depending on the type of flooring. Solid floors generate higher emissions than those in the form of a grid [24], because the former facilitate the mixing of feces and urine. Numerous research papers have shown that the temperature inside the stable is also an important factor in ammonia emission, its high values causing higher ammonia emissions (through its relation to urease activity), which is also generating seasonal differences [25,26], especially in temperate climates.

Microscopic particles are particles suspended in the air and produced in various types of industries and agriculture. A high concentration of them affects human and animal health. In animal husbandry in general, and in cow farms in particular, the sources of air pollution with microscopic particles could be feed administration [27] and feed management (wet vs. dry food, feed distribution system, feed storage), waste burning [28], stable cleaning, manure management, animal movement, animal maintenance (on bedding or not), ventilation rate or the microclimate in the shelter, while an indirect origin is oxidation of ammonia or other precursor gases [29–31].

Obviously, in agriculture, as in other types of industries, the degree of pollution with microscopic particles both at the global level and at the point level (stable, farm) depends on a number of factors: climatic zone, season, geographical peculiarities, humidity etc. [32]. The limitation of air pollution with microscopic particles belongs exclusively to the management of the farm, given the fact that the other influencing factors cannot be controlled, but only possibly influenced to a small extent.

One possible way of limiting particle emissions is by carrying out certain farm operations (particle generators) during the night. This is based on the fact that during the day the concentration of particles can increase 10–15 times [33] as a result of the simultaneous action of four factors: moisture at the soil surface, air humidity, the angle of the sun's rays and temperature.

Manure management can result in a decrease in farm-level particulate emissions. Removing litter and storing it in non-concentrated forms can be a useful action, along with maintaining a layer of 2–3 cm of concentrated manure mixed with soil [34].

Other research has highlighted that reducing particle pollution in cow farms can be possible through the use of a sprinkler system [35], but this solution is debatable under the current water crisis.

The reduction of atmospheric pollutant emissions must represent a priority of each economic sector, and the lack or inefficiency of technical and/or legal mechanisms for their permanent monitoring attracts behaviors that are not related to ensuring sustainability. Obviously, these mechanisms must be subject to a general legal framework, but the awareness of each farmer of his contribution to the total amount of emissions at the national level is a matter of common sense and absolutely mandatory. In this sense, the implementation of technical solutions for monitoring emissions at point level (at the farm level) and for alerting in case of exceeding them would allow farmers to achieve efficient management and create a particularly useful national system that would enable strategy correction in real time.

The growing urbanization trend has led to a great deal of research conducted on the modification of air pollutant emissions due to climate modification. For example, a study has shown [36] that air quality deterioration may be caused by anthropogenic activities and land use changes. Air pollutants such as VOC, O_3, PMs or NOx are found to be correlated with density and population, but also with mean summer temperature and precipitation.

In addition, a considerable number of studies show that micro-climatic features have an impact on air quality and thermal comfort. For example, in Sri Lanka [37] indoor concentrations of CO_2, NO_2, $PM_{2.5}$, CO, VOC, temperature, relative humidity and wind speeds were measured. The findings of the study recommended the introduction of a vegetative cover around buildings in suburban areas to overcome this problem since vegetation has a favorable impact on temperature and the concentrations of several air pollutants.

3. Measurement Platform for AP Concentration Monitoring and Case Study Description

3.1. Platform Architecture

AP concentrations in a case study are obtained through two methods: (i) estimation based on EMEP/EEA methodologies and (ii) monitoring using various sensors (IoT devices). Based on the EMEP/EEA equations, an IoT-based hardware–software platform was designed, tested and implemented. The information gathered by this system (concerning stable environment, AP concentration and animal state) and the way that information is processed, stored and presented allow it to be used with ease by farmers.

The platform has been presented in detail in a previous paper [38]; Figure 1 illustrates the platform architecture.

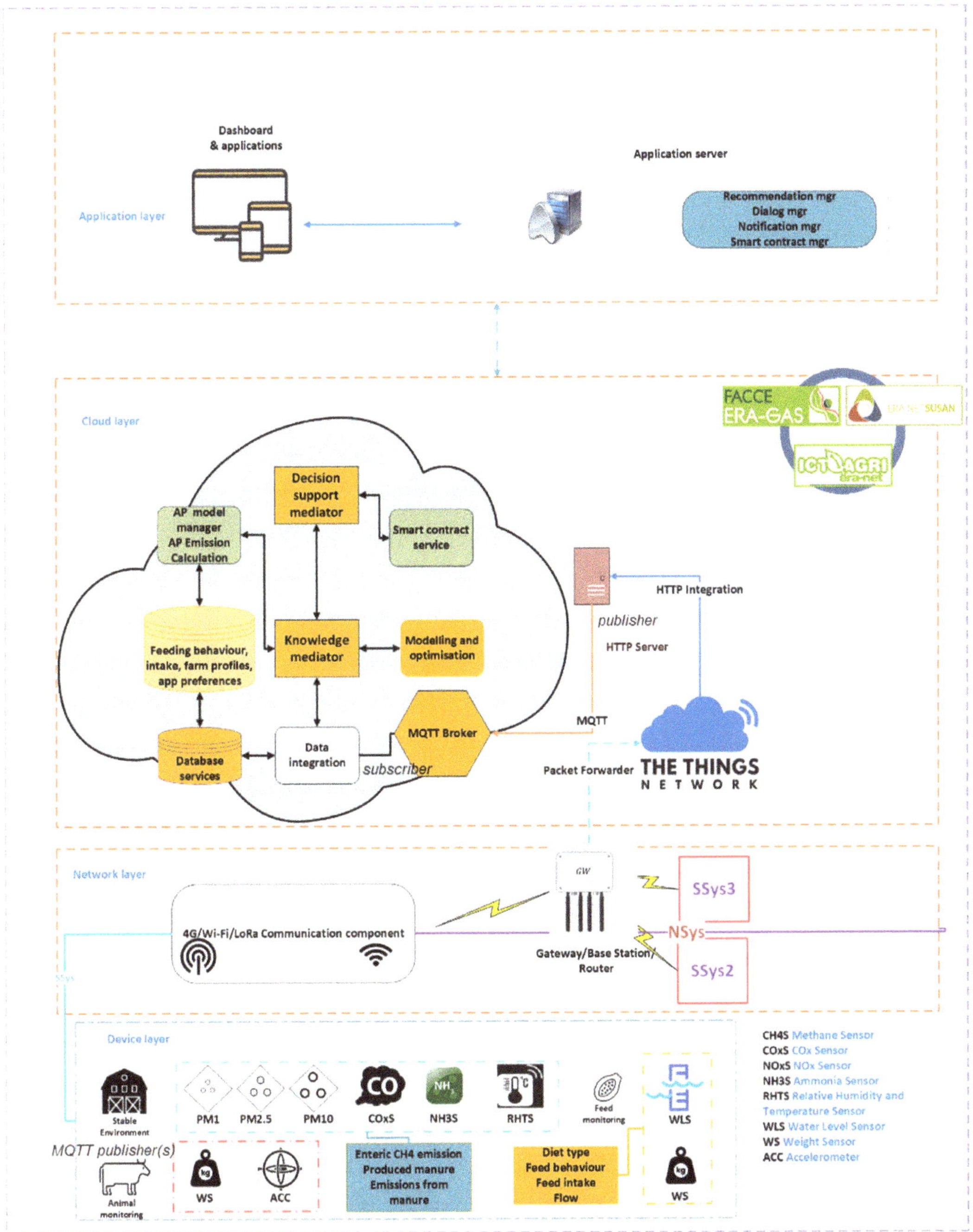

Figure 1. IoT-based AP Monitoring platform architecture.

The high-level architecture of the proposed IoT platform is divided into four layers:

- Device Layer: includes sensors (i.e., for measurement of the concentrations of APs such as NH_3, CO, CO and PMx), devices and client agents (to collect and transmit data to the IoT platform).
- Network Layer: includes the communication component (which uses low-power radio transmission technologies such as LoRa and cellular IoT) and the gateway (which sends the data packets to the next Layer).
- Cloud Layer: has the role of transforming data into knowledge. In this way, intelligence is added as a higher level of services. This layer receives the data and integrates and transforms them into knowledge. The data are received through the use of The Things Network and MQTT protocol.
- Application Layer: uses the knowledge generated in the previous layer to provide an overview of the farm performance (based on specific KPIs such as productivity, AP concentrations etc.) and their visual representation (various graphic representations).

Data is collected via diverse sensors which are equipped with different wireless interfaces: either 4G, Wi-Fi or LoRa. They transmit data packets via a device that allows connection to the Internet, called Base Station (for 4G), Router (for Wi-Fi) or Gateway (for LoRa). The data packets are transmitted via MQTT protocol either directly (for 4G and Wi-Fi) or via the Things Network (for LoRa). Finally an MQTT broker receives the data which is at this point being used by the Cloud Layer.

The platform presented uses open-source software and offers the possibility of long-term operation (>10 years) for IoT devices within the platform without battery replacement. We argue that the platform is sustainable (i) from an environmental point of view and (ii) in terms of the human resources needed for the operation and future extension of the platform.

3.2. Case Study

The input data for the estimation of atmospheric pollutant emissions are provided from a dairy cow farm, which has in operation (intensive system, no pasture or paddock) 120 dairy cows, 40 heifers and primiparous and 40 youth heads (3–9 months). The animals are from the Montbeliarde and Friza breeds. The stable is cleaned twice a day, and the feeding is done with the technological trailer used for feed distribution. The manure is separated into a solid fraction (deposited on the solid storage platform) and a liquid fraction (in the sealed lagoon).

The monitoring of the farm environment is done using the IoT infrastructure listed on Table 1 which transmits data wirelessly according to the architecture in Figure 1. These on-farm wireless sensors transmit information needed to determine the source of AP emissions and the costs associated with these emissions. Figure 2 shows that the IoT devices are carefully positioned in locations where they do not interfere with the daily activities of the animals.

Table 1. Sensors used in the IoT infrastructure.

Sensor Name	Parameter	Measurement Unit	Minimum Measured Value	Maximum Measured Value
BME280 [1]	Temperature	°C	0	65
BME280	Humidity	% RH	0	100
BME280	Pressure	kPa	30	110
MICS-6814 [2]	CO	ppm	30	1000
MICS-6814	NO_2	ppm	0.05	5
CCS_811 [3]	CO_2	ppm	350	10,000
SEN0237-A [4]	O_2	%	0	30
CCS811	VOC	ppm	30	400
OPC-N2 [5]	PM1	$\mu g/m^3$	0	1
OPC-N2	PM2.5	$\mu g/m^3$	0	2.5
OPC-N2	PM10	$\mu g/m^3$	0	10

[1] https://www.bosch-sensortec.com/products/environmental-sensors/humidity-sensors-bme280/, accessed on 18 October 2022; [2] https://www.sgxsensortech.com/content/uploads/2015/02/1143_Datasheet-MiCS-6814-rev-8.pdf, accessed on 18 October 2022. [3] https://learn.adafruit.com/adafruit-ccs811-air-quality-sensor, accessed on 18 October 2022. [4] https://ro.farnell.com/dfrobot/sen0237-a/analog-dissolved-oxy-sensor-kit/dp/3517931, accessed on 18 October 2022. [5] http://www.aqmd.gov/aq-spec/product/alphasense, accessed on 18 October 2022.

Figure 2. IoT devices installed in the farm and their configuration.

4. Comparison between Estimated and Monitored AP Concentrations

The usefulness of a support platform for monitoring environmental factors and pollutant emissions lies in two aspects: it is necessary on the one hand from the perspective of environmental protection (for monitoring emissions), and on the other hand from the perspective of farm management (as a support system in the decision-making process through emission monitoring and alert systems).

4.1. AP Concentration Estimated Using EMEP Methodology

This section will present the estimation of air pollutant (AP) concentrations using the EMEP methodology.

The EMEP 2019 guideline (the co-operative programme for monitoring and evaluation of the long-range transmission of air pollutants in Europe—unofficially European Monitoring and Evaluation Programme) proposes different methods of estimating the emissions of pollutants from animal husbandry, depending on how many parameters are known in each case. For example, Tier 1 is the simplest method for estimating the pollutant emission as it consists of the multiplication of the default emission factor by the number of animals specific to each category (this method was used for PM emissions estimation).

The next level, if more parameters are known, can use the tier 2 method, which involves specific parameters. In our study, we used tier 2 for estimating NH_3 emissions (EMEP/EEA guideline, 2019) and we combined this value with tier 2 for the calculation of excreted nitrogen from the IPCC (Intergovernmental Panel on Climate Change) 2019 guideline [39].

Methane emissions were estimated using the manure management N-flow tool (https://www.eea.europa.eu/publications/emep-eea-guidebook-2019/part-b-sectoral-guidance-chapters/4-agriculture/manure-management-n-flow-tool, accessed on 18 October 2022).

There are no specific equations in the IPCC or EMEP guidelines for CO and in our case study the data refers exclusively to the sensors' measurements. Table 2 presents the parameters and equations used for AP emissions.

Table 2. Parameters used for the estimation of AP emissions and methodology.

Crt. No.	Parameter	Guideline	Equation/Table Number in IPCC, 2019 and EMEP, 2019
		Calculated parameters	
1	N_{ex}	IPCC, 2019	10.31 A
2	N_{intake}	IPCC, 2019	10.32
3	$N_{retention}$	IPCC, 2019	10.33
4	NE_g	IPCC, 2019	10.6
5	m_{hous_N}	EMEP, 2019	5
6	m_{hous_TAN}	EMEP, 2019	10
7	$m_{hous\_solid\_N}$	EMEP, 2019	14
8	E_{hous_solid}	EMEP, 2019	16
9	$E_{storage\_solid}$	EMEP, 2019	34
10	E_{MMS_NH3}	EMEP, 2019	46
		Default values	
1	X_{TAN}	EMEP, 2019	Table 3.9
2	$EF_{housing}$	EMEP, 2019	Table 3.9
3	$EF_{PM2.5}, EF_{PM10}$	EMEP, 2019	Table 3.5

The primary data needed to calculate excreted nitrogen (N_{ex}) (equations 10.32 and 10.33—see Table 2) are provided by the farm. The rations administered to all three categories fluctuate during the year, but the differences between the summer and winter seasons are insignificant. Depending on the animal's age and the category of exploitation the rations were calculated using the structure and chemical composition of the feed in each case. Feeding dairy cattle is very important for ensuring constant and qualitative production throughout the year. Romanian farmers with an important herd of dairy cows choose the option of establishing a unique forage recipe, so that the animals on the farm benefit from the same food all year round. The uniformity of the feed contributes to ensuring a uniform amount of milk and does not allow farmers to change the qualitative parameters of the milk.

In our case study, during the summer season dairy cows, primiparous and heifers receive the green mass (5 kg maximum, introduced gradually, because the rumen is not accustomed to this type of feed after the winter season), and in the cold season for these categories the green mass is replaced with beer mug, without important changes in the value of the ingested energy. Between the summer and winter seasons there exist differences regarding N_{ex} values (dairy cattle, primiparous and heifers). For the young (3–9 months), a single ration is provided throughout the year, and for primiparous and heifers there are no differences between seasonal N_{ex} values.

Table 3 presents the feed structure of rations for both seasons and for all animal categories.

Table 3. Fodder type used in all cattle categories and chemical structure (per kilo).

Category	PB (g/kg)	GB (g/kg)	CelB (g/kg)	SEN (g/kg)	GE (Kcal/kg)
Barley straw	32.00	14.00	390.00	381.00	3772.91
Alfalfa hay	120.00	30.00	330.00	340.00	3969.90
Corn silage	22.00	8.00	85.00	127.00	1138.58
Corn kernels	90.00	40.00	22.00	710.00	3960.88
Barley kernels	100.00	20.00	56.00	678.00	3857.50
Rape seed meal	350.00	25.00	130.00	312.00	4163.24
Wheat bran	150.00	40.00	105.00	530.00	3951.05
Soybean meal	443.00	14.00	63.00	311.00	4265.60
Beer wort	50.40	15.20	38.10	70.00	907.09
Green mass	31.00	4.80	60.00	70.00	802.22

where, PB = crude proteine; GB = crude fat; CelB = crude cellulose; SEN = non-nitrogenous extractable substances; GE = gross energy intake.

The feed categories presented in Table 3 are cut to the required dimensions, mixed and homogenized, and are presented in the form of a unique mixture, balanced according to the physiological and production requirements of the animals.

For all categories, the unique mixture is administered throughout the year with the technological trailer. In summer green mass is added, and in winter beer wort is incorporated into the mixture. Table 4 presents the composition of that mixture for dairy cattle, primiparous and heifers during the year (summer period and winter period).

Table 4. Composition of unique mixture.

Category	Animal Category					
	Dairy Cattle		Primiparous and Heifers		Youth (3–9 Months)	
	Summer	Winter	Summer	Winter	Summer	Winter
	kg/Head/Day	kg/Head/Day	kg/Head/Day	kg/Head/Day	kg/Head/Day	kg/Head/Day
Barley straw	0.5	0.5	1.5	1.5	1.0	1.0
Alfalfa hay	2.0	2.0	3.5	3.5	2.0	2.0
Corn silage	20.0	20.0	15.0	17.0	9.0	9.0
Corn kernels	3.5	3.8	3.5	4.0	2.5	2.5
Barley kernels	1.5	1.5	1.5	1.0	0.3	0.3
Rape seed meal	2.5	2.5	2.0	2.0	0.3	0.3
Wheat bran	2.0	2.0	-	-	-	-
Soybean meal	2.0	2.0	2.0	2.0	-	-
Beer wort	-	8.0	-	3.0	-	-
Green mass	8.0	-	5.0	-	-	-
Total	42.0	42.3	34.0	34	15.1	15.1

A total mixed ration (TMR) was administered consisting of alfalfa hay, barley straw, corn silage, corn and barley grain colza meal, soya meal, the proportion of feeds varying according to the need for nutrients in the categories of tested cattle.

During the summer the TMR was supplemented with green mass represented by alfalfa, which is a food rich in the nutrients necessary for production. During the winter the TMR was supplemented with by-products represented by beer wort.

The rations are balanced from the point of view of both macro- and micro-nutrients, which allows farmers to obtain increased milk production, a development of pregnancy in optimal conditions and an average daily gain corresponding to the young cattle.

For the formulation and optimization of the rations administered to the Holstein taurine categories, a list of fodder was established that included fibrous, coarse and concentrated fodder. This feed structure of the rations has been optimized so as to cover the entire nutrient requirement for dairy cows, pregnant cows and heifers and young cattle.

In the composition and optimization of the rations, fodder and by-products of plant origin were introduced in order to efficiently use the local fodder resources and with an optimal cost price.

To calculate the caloricity of the gross energy intake of each recipe or ration, the following equivalences were considered [35] (p. 114):

$$1 \text{ g crude protein} = 5.72 \text{ kcal};$$

$$1 \text{ g crude fat} = 9.5 \text{ kcal};$$

$$1 \text{ g crude fibers} = 4.79 \text{ kcal};$$

$$1 \text{ g SEN (non-nitrate extractable substances)} = 4.17 \text{ kcal}.$$

The GE calculation formula [35] is (p. 131):

$$GE \text{ (kcal/kg)} = 5.72 \cdot GP + 9.5 \cdot GB + 4.79 \cdot CelB + 4.17 \cdot SEN$$

where:

GE = gross energy intake
GP = crude protein
GB = crude fat
CelB = crude fibers
SEN = non-nitrate extractable substances

The rations were calculated according to this equation, and the values of crude protein, crude fat, crude fibers and non-nitrate extractable substances were taken from the tables with the feed chemical composition [35] (pp. 513–517).

In accordance with the requirements of the IPCC 2019 that the energy be expressed in MJ/kg in the calculation of ratios, we multiplied the values by 10 to express the caloricity for 1 kg (tables give the value of these nutrients expressed as a percentage, for 100 g).

The total value of the ration, expressed in kcal, was divided by 239 in order to obtain the equivalence in MJ (Mega Joules).

The equivalence relations are as follows [35] (p. 114):

$$1 \text{ MJ} = 239 \text{ kcal}$$

where MJ = megajoule and Kcal = kilocalory

For each feed category, the values of crude protein, crude fat, crude fibers and non-nitrate extractable substances are included in a table [35] (pp. 513–517); these table values are multiplied by the caloricity specific to each nutrient (5.72 kcal for 1 g of crude protein, etc.), followed by the adding of the caloricity of each nutrient and the achievement of the respective forage caloricity. This value is multiplied by the number of feed kilograms specified in the ration.

Table 5 presents total gross energy (GE) expressed in MJ per head and per day, for each category of animals and for both seasons.

Table 5. Total gross energy intake (MJ/head/day).

Category / Season	Dairy Cows	Heifers and Primiparous	Youth (3–9 Months)
Winter	330.91	270.33	143.40
Summer	335.90	280.10	143.40

Table 6 presents the specific parameters used for calculation of ammonia emissions.

Table 6. Parameters' values used for calculated excreted nitrogen (N_{ex}).

Category / Parameter	Dairy Cows		Heifers and Primiparous		Youth (3–9 Months)
Days of life	365		365		180
Heads number	120		40		40
AAP	365		365		19.73
Season	Summer	Winter	Summer	Winter	All year
GE (MJ/head/day)	330.91	335.90	270.33	280.10	143.40
CP% (%)	0.143	0.160	0.161	0.170	0.205
Milk (kg/head/day)	30	28	-	-	-
Milk% (%)	1.92	1.92	-	-	-
WG (kg/day)	0.2	0.2	0.4	0.4	0.9
NE_g (MJ/head/day)	1.96	1.96	3.93	8.31	6.05
N_{ex} (kg/head/year)	119.83	139.78	135.06	147.55	81.40

where: AAP = average annual population = number of animals produced annually × days of live. GE = gross energy intake (MJ/head/day). CP% = percent crude protein in dry matter (%). Milk = milk production (kg/head/day). Milk% = percent of protein in milk, calculated as [1.9 + 0.4 × %Fat], where %Fat was determined by milk analyzer (Farm Eco 25) = 4% (%). WG = weight gain (kg/day). NE_g = net energy for growth, calculated in livestock characterization, based on current weight, mature weight, rate of weight gain, and IPCC constants (MJ/head/day). N_{intake} = daily N consumed per animal of each category (kg N/head/day). $N_{retention}$ = amount of daily N intake by head of animal (kg N/head/day). N_{ex} = annual N excretion rates (kg N/head/year).

The percentage of crude protein in dry matter (CP%) was calculated based on the chemical composition of each fodder and then multiplied by the proportion of feed in the total ratio.

Table 7 presents the estimated NH_3 emissions.

Table 7. Estimated NH_3 emissions during the monitoring period.

Animal Category	NH_3 (t/Year)		
	Season		
	Summer (185 Days)	Winter (180 Days)	All Year (365 Days)
Dairy cattle	1.21	1.40	2.61
Primiparous and heifers	0.48	0.5	0.98
Young (3–9 months)		0.92	0.92
Total		**4.51**	

Table 8 presents the estimated microscopic particles emissions.

Table 8. Microscopic particles emissions ($PM_{2.5}$, PM_{10}) (kg PM/year).

Category	Heads No	Life Days	AAP	EF		Emissions (kg/Year)	
				PM_{10}	$PM_{2.5}$	PM_{10}	$PM_{2.5}$
Dairy cows	120	365	110	0.63	0.41	75.6	49.2
Heifers and primiparous	40	365	50	0.63	0.41	25.2	16.4
Young	40	180	19.73	0.27	0.18	10.8	7.2
Total						**111.6**	**72.8**

4.2. AP Concentration Monitored Using Sensors

The validation of an interactive platform model that can be used for farm management from the perspective of microclimate parameters and pollutant emissions must be based on the study of their behavior, alone and in relation to one another. The complexity of the relationships between them, of a physical-chemical nature, makes taking a parameter out of context and studying its behavior, abstracting it from its interaction with others, a dead end.

In this sense, using the data obtained from two studied seasons (summer and winter) regarding the microclimate parameter values (humidity and temperature) and the studied pollutant emission values, respectively **711** records in the winter season and **597** records in the summer season, the correlations between them and their meaning and significance were determined. Tables 9 and 10 present the correlation values between the studied parameters and their significance over two measurement seasons.

Table 9. Correlation between microclimate and pollutant concentrations during the winter season (711 recordings).

Specification	HUM	NH_3	PM_1	PM_{10}	$PM_{2.5}$	T^0C
HUM	-	0.56[HS] t = 17.99	0.43[HS] t = 12.68	0.02[NS] t = 0.53	0.24[HS] t = 6.58	-0.67[HS] t = 24.03
NH_3		-	0.41[HS] t = 11.97	0.12[S] t = 3.21	0.33[HS] t = 9.31	-0.02[NS] t = 0.53
PM_1			-	0.13[S] t = 3.49	0.71[HS] t = 26.84	-0.23[HS] t = 6.29
PM_{10}				-	0.42[HS] t = 12.32	0.06[NS] t = 1.60
$PM_{2.5}$					-	-0.02[NS] t = 0.53

HUM = humidity (%). t = correlation significance. S = significant, $p < 0.05$. S = significant, $p < 0.01$. HS = high significant, $p < 0.001$. NS = nonsignificant, $p > 0.05$. Critical $t_{0.05} = 1.96$. Critical $t_{0.01} = 2.58$. Critical $t_{0.001} = 3.29$.

Table 10. Correlation between microclimate and pollutant concentrations during the summer season (597 recordings).

Specification	HUM	NH_3	PM_1	PM_{10}	$PM_{2.5}$	T^0C
HUM	-	0.25^{HS} t = 6.30	0.18^{HS} t = 4.46	-0.31^{HS} t = 7.95	-0.18^{HS} t = 4.46	-0.89^{HS} t = 47.61
NH_3		-	0.11^{S} t = 2.70	-0.03^{NS} t = 0.73	0.002^{NS} t = 0.048	-0.25^{HS} t = 6.29
PM_1			-	0.39^{HS} t = 10.33	0.65^{HS} t = 20.86	-0.01^{NS} t = 0.24
PM_{10}				-	0.80^{HS} t = 32.52	0.32^{HS} t = 8.23
$PM_{2.5}$					-	0.25^{HS} t = 6.30

HUM = humidity (%). t = correlation significance. S = significant, $p < 0.05$. S = significant, $p < 0.01$. HS = high significant, $p < 0.001$. NS = nonsignificant, $p > 0.05$ Critical $t_{0.05}$ = 1.96. Critical $t_{0.01}$ = 2.58. Critical $t_{0.001}$ = 3.29.

From the analysis of the results presented in Tables 9 and 10 we can observe the existence of correlations with varying degrees of significance between the values of the microclimate parameters and the pollutant emissions from the stable, different between the two seasons. This cumulative behavior raises certain problems related to the management of the farm. Thus, in the winter season, we can observe that the humidity in the animal stable correlates intensely negatively with the temperature, the increase of one implicitly leading to the decrease of the other. Furthermore, the increase in humidity is not related to PM_{10} emissions, the latter being exclusively related to farm management, especially feed administration, stable cleaning actions and other possible occurrences. As a result, the increase in PM_{10} emissions cannot be attributed to the existence of vapor supersaturation of the air, and the eventual switching on of fans to reduce humidity would do nothing more than circulate microscopic particles through the air. Conversely, increased humidity will maintain in the air a high concentration of the other types of particulate matter (PM_1 and $PM_{2.5}$) as well as ammonia. In the case of ammonia, when the temperature increases, the proportion of stable emissions becomes lower. Regarding particulate matter, it is noted that its concentration in the air is related exclusively to the concentration of ammonia, and not at all to the temperature.

In the warm season, increased humidity will lead to a decrease in the concentration of PM_{10} and $PM_{2.5}$. In the summer, it seems that the concentrations of particulate matter in the air (PM_{10} and $PM_{2.5}$) do not correlate with those of ammonia, and consequently the reduction of particle emissions must be attributed to activities related to food administration, cleaning, the evacuation of solid waste, and the reduction of ammonia by good ventilation. But the significantly positive correlation between PM_{10} and temperature indicates that, during the warm season, the reduction of PM_{10} emissions is done by lowering the temperature, that is, by ventilation and possibly by air humidification systems (sprinklers).

A clearer picture of the interactions between microclimate and pollutant emissions is presented in Figures 3 and 4 for the two seasons, showing the correlations between the observations by trendline using zero-intercept linear regression.

Analysis of the scatterplots presented in Figures 3 and 4 graphically reveals the links between the analyzed variables. Thus, the shape of the cloud of points, the direction of the regression line and the value of R^2 emphasize the strength of the links. Strongly correlated values are highlighted by the existence of a cloud of points aligned along the regression line. The interruption of the cloud of points (Figure 3m) suggests the existence, with a preponderance, of extreme values of PM_{10} and $PM_{2.5}$ in the winter season. Points that lie outside the cloud suggest either outliers (possible registration errors) or drastic increases in a certain parameter. Obviously, the direction of the regression line suggests the algebraic sign of the correlation.

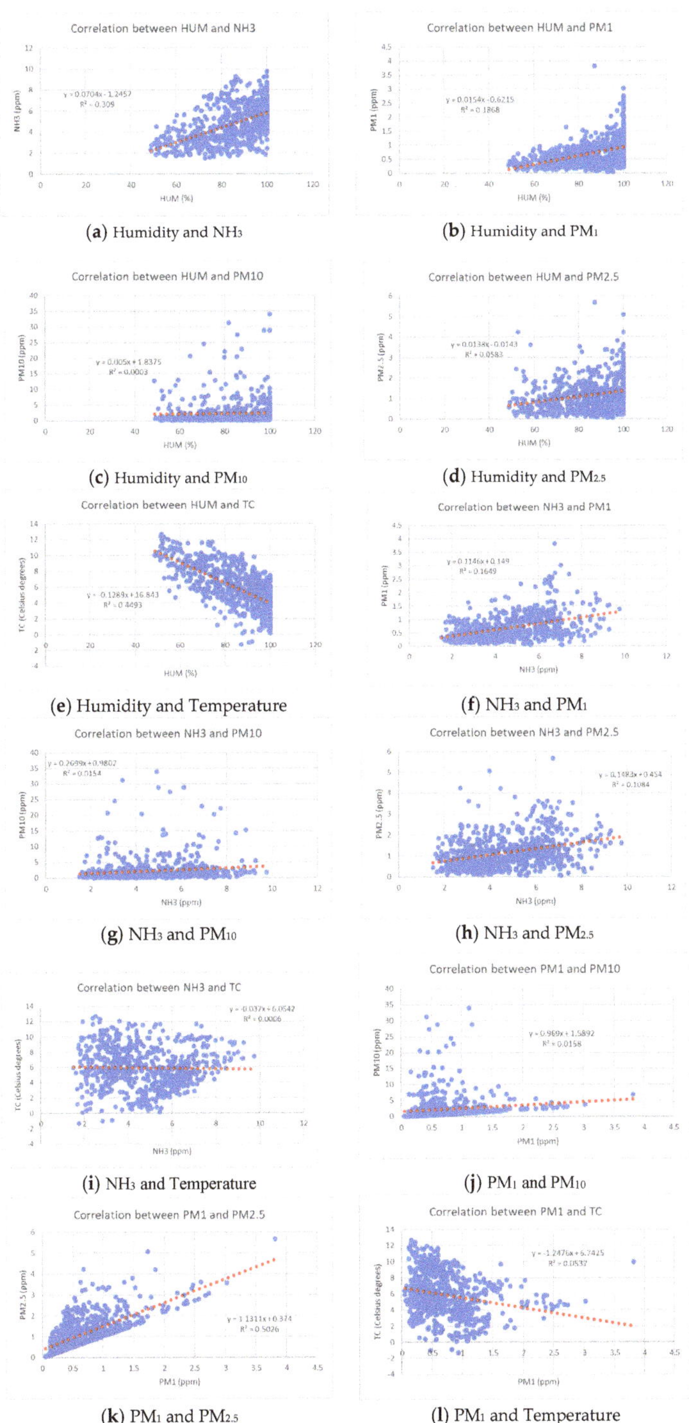

(**a**) Humidity and NH₃   (**b**) Humidity and PM₁

(**c**) Humidity and PM₁₀   (**d**) Humidity and PM₂.₅

(**e**) Humidity and Temperature   (**f**) NH₃ and PM₁

(**g**) NH₃ and PM₁₀   (**h**) NH₃ and PM₂.₅

(**i**) NH₃ and Temperature   (**j**) PM₁ and PM₁₀

(**k**) PM₁ and PM₂.₅   (**l**) PM₁ and Temperature

Figure 3. *Cont.*

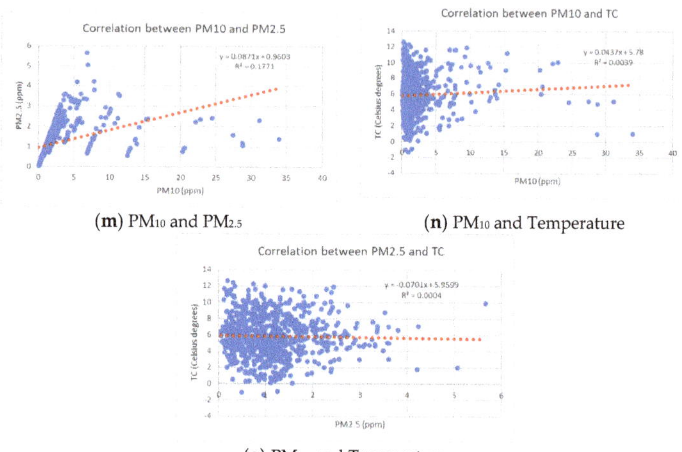

(**m**) PM₁₀ and PM₂.₅   (**n**) PM₁₀ and Temperature

(**o**) PM₂.₅ and Temperature

Figure 3. Relationship between different daily measured microclimate values and pollutants. The red line represents the 0−intercept line regression, during the winter season.

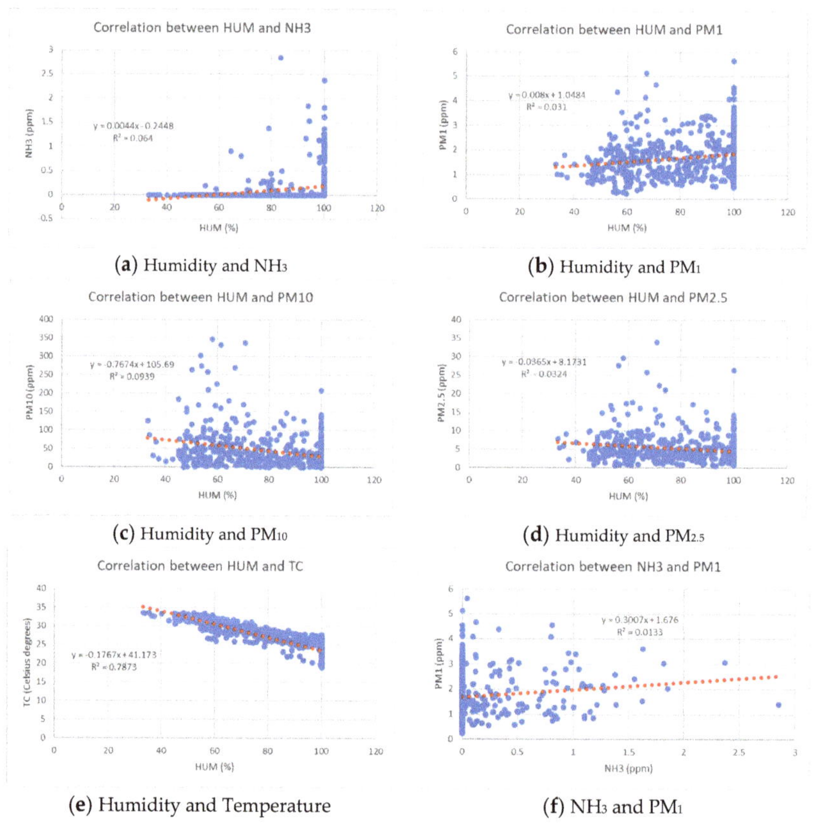

(**a**) Humidity and NH₃   (**b**) Humidity and PM₁

(**c**) Humidity and PM₁₀   (**d**) Humidity and PM₂.₅

(**e**) Humidity and Temperature   (**f**) NH₃ and PM₁

Figure 4. *Cont.*

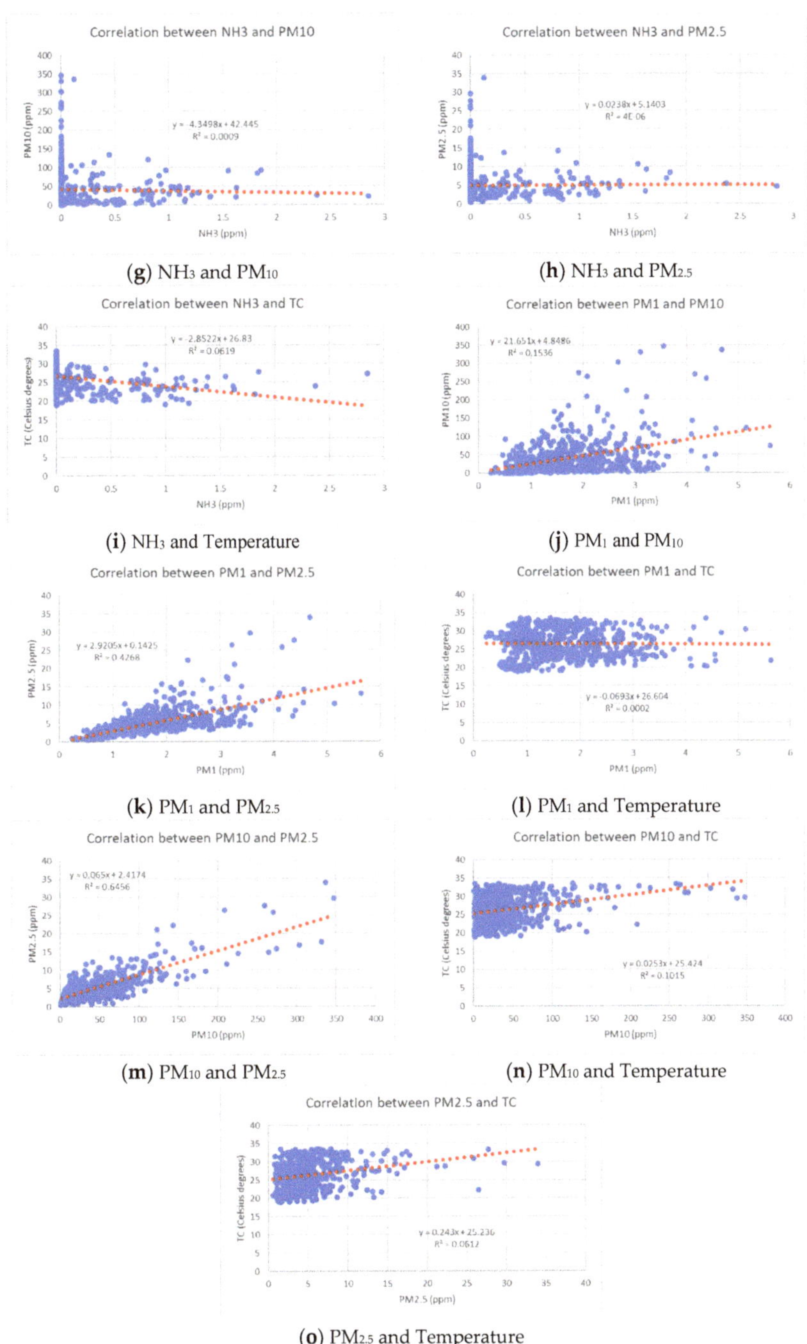

(**g**) NH₃ and PM₁₀

(**h**) NH₃ and PM₂.₅

(**i**) NH₃ and Temperature

(**j**) PM₁ and PM₁₀

(**k**) PM₁ and PM₂.₅

(**l**) PM₁ and Temperature

(**m**) PM₁₀ and PM₂.₅

(**n**) PM₁₀ and Temperature

(**o**) PM₂.₅ and Temperature

Figure 4. Relationship between different daily measured microclimate values and pollutants. The red line represents the 0-intercept line regression, during the summer season.

All these observations, which reveal a complexity in the behavior of microclimate parameters and pollutant emissions, complicate farm management activities, and the existence of a platform that allows permanent monitoring of air quality through sensors, as well as an alert system, becomes useful in interpreting the causality of concrete situations that go beyond the limits of admissibility and in the optimization of the decision-making system regarding the welfare of animals, workers and the environment.

Figure 5a–f presents the microclimate values and pollutant concentrations in the stable, throughout one day, in the two analyzed seasons. Measurements were made by placing sensors approximately 1 m above the animals' heads and approximately 2.5 m from the floor.

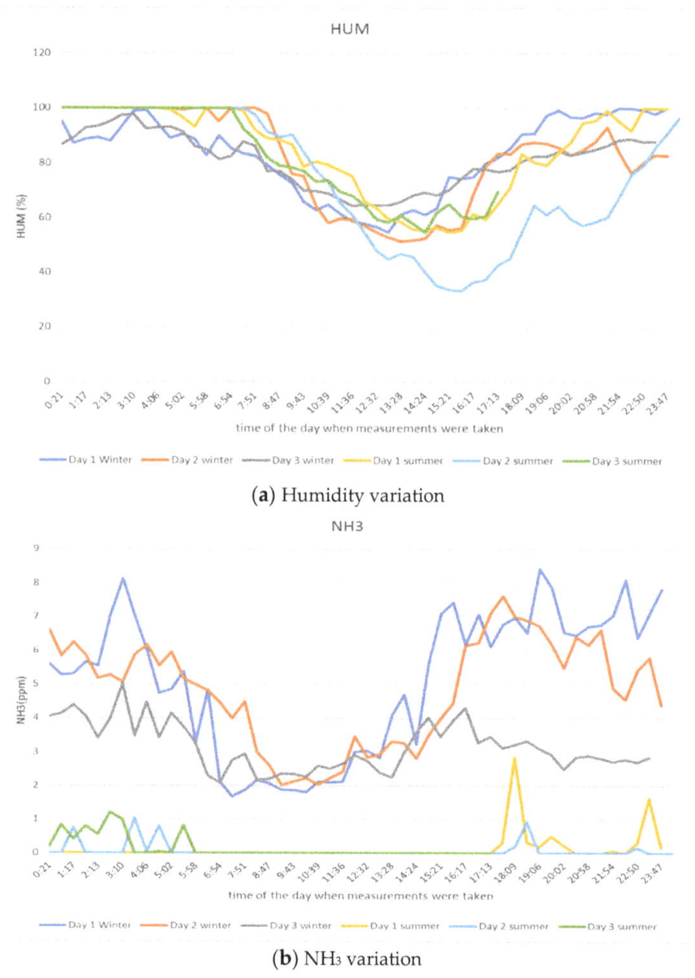

(**a**) Humidity variation

(**b**) NH₃ variation

Figure 5. *Cont.*

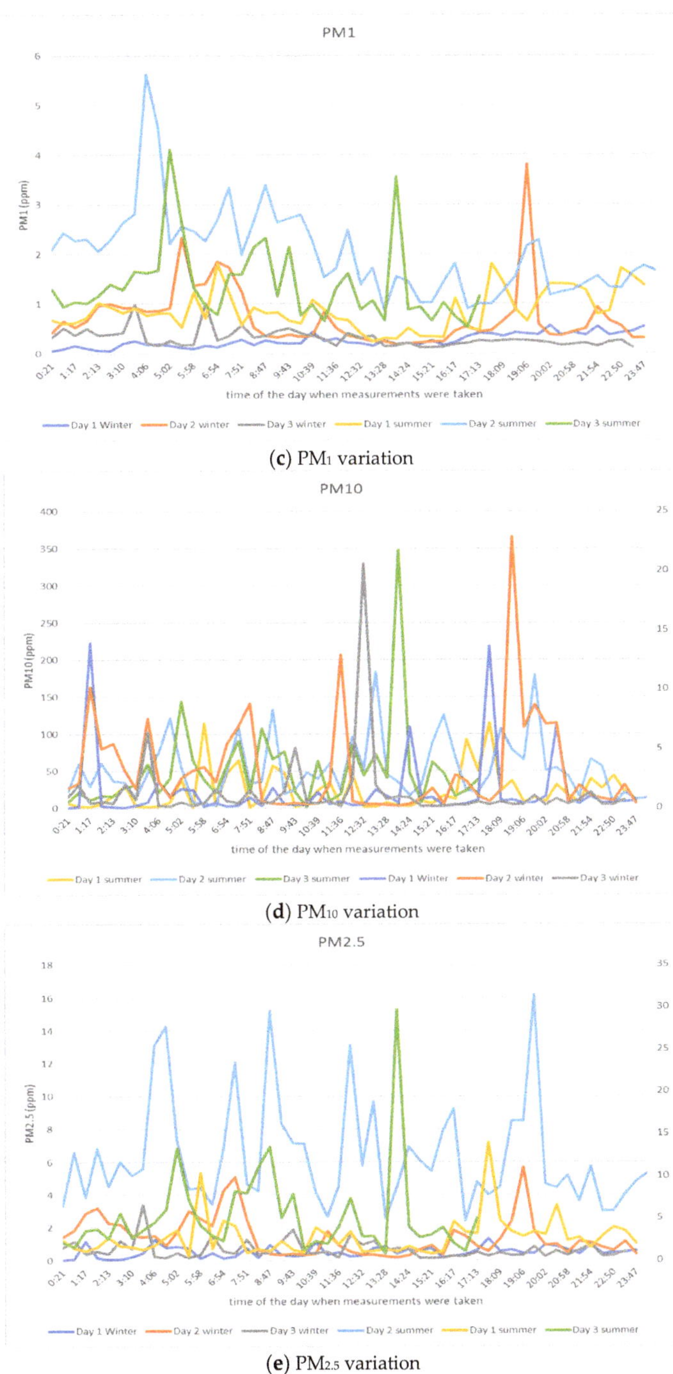

(**c**) PM₁ variation

(**d**) PM₁₀ variation

(**e**) PM₂.₅ variation

Figure 5. *Cont.*

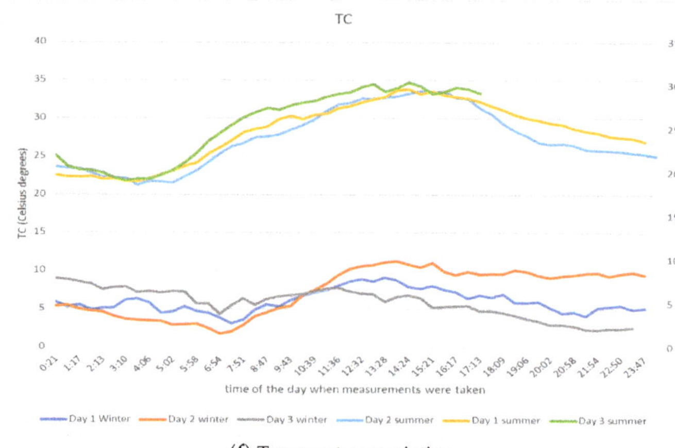

(**f**) Temperature variation

Figure 5. Microclimate values and pollutant concentrations at 1 m above the animals' heads in the barn, during one day, in both seasons: winter (January) and summer (August).

We can observe that, during the day, the concentration of ammonia increases when the stables are closed (during the night and afternoon to evening), a consequence of the lack of ventilation and the increase in humidity. Although the increase in ammonia concentration is apparently not related to temperature, it favors the increase in ammonia emissions by maintaining metabolic processes and intensifying urease activity in manure. Zero values for ammonia concentration during the warm season (no records) can be explained by its negative correlation with temperature, when the sensor was not able to record any emission (which does not mean that it did not exist, however, but simply that the values were below the sensitivity of the sensor).

The variation in PM_{10} concentration captures the times of the day when food is administered and stable cleaning is performed. During both seasons, the concentration of PM_{10} is also influenced by the increase in ammonia.

$PM_{2.5}$ concentrations throughout the day vary in the warm season following a similar pattern to PM_{10}, capturing the same activities, but in the cold season, throughout the day, $PM_{2.5}$ concentration variations most likely capture certain activities that are not are related to animal husbandry (possibly smoking in the barn by the farm hands, sheltered from the cold temperatures outside—this would also explain the PM_1 variations).

Regarding the observed difference between the values estimated by the EMEP/EEA guideline equations (2019) and the values recorded by the sensors, this phenomenon is because the estimates do not consider the natural and artificial ventilation of the stable, being used instead for an overall assessment of atmospheric air quality. Estimates and measurements may be similar only under experimental conditions, or if the stable were permanently closed. For this reason, for point sources of air pollutants from livestock (farms) it is recommended that farmers use sensors and create alert systems that adjust pollutant concentrations in real time (support for farm management decisions) to ensure animal welfare.

5. Conclusions

This article presents a case study of air pollutant emissions and their correlation with animal welfare and farm management in two different seasons.

Our use-case from the Milanovici farm showed that the estimated air pollutant concentration exhibits complex behavior that correlates with the micro-climate parameters. Therefore, for efficient farm management it is important to treat them as a whole and not individually.

We conclude that estimates using the EMEP methodology do not take into account the natural and artificial ventilation of the stable since they are used for an overall assessment of atmospheric air quality, and it is recommended that sensors are used and alert systems that adjust pollutant concentrations in real time (support for farm management decisions) are created to ensure animal welfare for point sources of air pollutants from livestock farms.

Author Contributions: Conceptualization, R.A.P., A.V., D.C.P. and L.V.; methodology, R.A.P., D.C.P., A.V., G.S. and M.P.M.; software, M.B., A.V., R.B. and S.B.; validation, R.A.P., D.C.P., A.V., M.T. and E.N.P.; writing—original draft preparation, R.A.P., D.C.P., M.B. and S.B.; writing—review and editing, A.V., D.C.P. and M.B.; project administration, A.V., D.C.P. and G.S.; funding acquisition, R.A.P. All authors have read and agreed to the published version of the manuscript.

Funding: This research was partially supported by a grant of the Romanian National Authority for Scientific Research and Innovation, CCCDI–UEFISCDI, projects no. ERANET-ERAGAS-ICTAGRI3-FarmSusteinaBl-1 and ERANET-ERAGAS-ICTAGRI3-FarmSusteinaBl-2 (FarmSustainaBl) within PNCDI III, and funded in part by a grant of Ministry of Research, Innovation and Digitization, CNCS/CCCDI-UEFISCDI, project number ERANET-ICT-AGRI-FOOD-Solution4Farming, within PNCD III. The authors acknowledge the financial support through the partners of the Joint Call of the Cofund ERA-Nets SusCrop (Grant N° 771134), FACCE ERA-GAS (Grant N° 696356), ICT-AGRI-FOOD (Grant N° 862665) and SusAn (Grant N° 696231) for the project Solution4Farming. The APC was funded by the University of Agricultural Sciences and Veterinary Medicine.

Institutional Review Board Statement: Not applicable.

Data Availability Statement: The data presented in this study are available on request from the corresponding authors.

Acknowledgments: The authors are grateful for the support of the Milanovici farm for making available their facilities for sensor installation and data collection and to Beia Cercetare SRL for providing the unified messaging and cloud data storage services.

Conflicts of Interest: The authors declare no conflict of interest. The funders had no role in the design of the study; in the collection, analyses, or interpretation of data; in the writing of the manuscript, or in the decision to publish the results.

Abbreviations

In the manuscript, we used the following abbreviations and chemical symbols:

AAP	Average Annual Population
ANFIS-GP	Adaptive Neuro-Fuzzy Inference Systems with Grid Partitioning
ANFIS-SC	Adaptive Neuro-Fuzzy Inference Systems with Subtractive Clustering
AP	Air Pollutant
AQI	Air Quality Index
AQM	Air Quality Monitoring
CFC	Cloud Farm Controller
CoAP	Constrained Application Protocol
CP	Crude Protein
EF	Emission Factor
EMEP	European Monitoring and Evaluation Programme
EPA	United States Environmental Protection Agency
EX-ACT	EX-Ante Carbon-balance Tool
FEM	Farm Emissions Model
HTTP	Hypertext Transfer Protocol
IoT	Internet of Things
IPCC	Intergovernmental Panel on Climate Change
KF	Kalman Filter
KPI	Key Performance Indicator

LEACH	Low Energy Adaptive Clustering Hierarchy Aggregation
LFC	Local Farm Controller
LMC	Litter Moisture Content
MLP	Multilayer Perceptron
MLR	Multiple Linear Regression
MQTT	Message Queuing Telemetry Transport
NPM	National Practices Model
pH	Potential of Hydrogen
PM	Microscopic Particles
PMx	Microscopic Particles less than x microns in diameter, where x {1, 2.5, 10}
REST	Representational State Transfer
WSN	Wireless Sensor Network
CO	carbon monoxide
CO_2	carbon dioxide
CH_4	methane
N_2O	nitrous oxide
NO_2	nitrogen dioxide
NH_3	ammonia
O_3	ozone
SO_2	sulfur dioxide

References

1. Vermeulen, S.J.; Campbell, B.M.; Ingram, J.S.I. Climate change and food systems. *Annu. Rev. Environ. Resour.* **2012**, *37*, 195–222. [CrossRef]
2. Tilman, D.; Clark, M.; Williams, D.R.; Kimmel, K.; Polasky, S.; Packer, C. Future threats to biodiversity and pathways to their prevention. *Nature* **2017**, *546*, 73–81. [CrossRef] [PubMed]
3. Clark, M.; Hill, J.; Tilman, D. The diet, health, and environment trilemma. *Annu. Rev. Environ. Resour.* **2018**, *43*, 109–134. [CrossRef]
4. Crippa, M.; Solazzo, E.; Guizzardi, D.; Monforti-Ferrario, F.; Tubiello, F.N.; Leip, A. Food systems are responsible for a third of global anthropogenic GHG emissions. *Nat. Food* **2021**, *2*, 198–209. [CrossRef]
5. FAO. *The Global Dairy Sector: Fact*; Food and Agricultural Organization of the United Nations: Rome, Italy, 2019.
6. Diaz, S.; Settele, E.; Brondizio, E.; Ngo, H.; Gueze, M.; Agard, J.; Arneth, A.; Balvanera, P. Summary for Policymakers of the Global Assessment Report on Biodiversity and Ecosystem Services of the Intergovernmental Science-Policy Platform on Biodiversity and Ecosystem Services. *IPBS* **2019**, *45*, 680–681.
7. Cambra-López, M.; Aarnink, A.J.A.; Zhao, Y.; Calvet, S.; Torres, A.G. Airborne particulate matter from livestock production systems: A review of an air pollution problem. *Environ. Pollut.* **2010**, *158*, 1–17. [CrossRef]
8. Moore, K.D.; Young, E.; Wojcik, M.D.; Martin, R.S.; Gurell, C.; Bingham, G.E.; Pfeiffer, R.L.; Prueger, J.H.; Hatfield, J.L. Ammonia measurements and emissions from a California dairy using point and remote sensors. *Trans. ASABE* **2014**, *57*, 181–198.
9. Hristov, A.N.; Hanigan, M.; Cole, A.; Todd, R.; McAllister, T.A.; Ndegwa, P.M.; Rotz, A. Review: Ammonia emissions from dairy farms and beef feedlots. *Can. J. Anim. Sci.* **2011**, *91*, 1–35. [CrossRef]
10. Powell, J.M.; Broderick, G.A.; Misselbrook, T.H. Seasonal diet affects ammonia emissions from tie-stall dairy barns. *J. Dairy Sci.* **2008**, *91*, 857–869. [CrossRef]
11. James, T.; Meyer, D.; Esparza, E.; Depeters, E.J.; Perez-Monti, H. Effects of dietary nitrogen manipulation on ammonia volatilization from manure from holstein heifers. *J. Dairy Sci.* **1999**, *82*, 2430–2439. [CrossRef]
12. Frank, B.; Persson, M.; Gustafsson, G. Feeding dairy cows for de- creased ammonia emissions. *Livest. Prod. Sci.* **2002**, *76*, 171–179. [CrossRef]
13. Frank, B.; Swensson, C. Relationship between content of crude pro- tein in rations for dairy cows and milk yield, concentration of urea in milk, and ammonia emissions. *J. Dairy Sci.* **2002**, *85*, 1829–1838. [CrossRef] [PubMed]
14. Swensson, C. Relationship between content of crude protein in rations for dairy cows, N in urine and ammonia release. *Livest. Prod. Sci.* **2003**, *84*, 125–133. [CrossRef]
15. Broderick, G.A. Effects of varying dietary protein and energy levels on the production of lactating dairy cows. *J. Dairy Sci.* **2003**, *86*, 1370–1381. [CrossRef]
16. Colmenero, J.J.O.; Broderick, G.A. Effect of dietary crude protein concentration on milk production and nitrogen utilization in lactating dairy cows. *J. Dairy Sci.* **2006**, *89*, 1704–1712. [CrossRef] [PubMed]
17. Aguerre, J.M.; Wattiaux, M.A.; Hunt, T.; Larget, B.R. Effect of dietary crude protein on ammonia-N emission measured by herd nitrogen mass balance in a freestall dairy barn managed under farm-like conditions. *Animal* **2010**, *4*, 1390–1400. [CrossRef]
18. Powell, M.J.; Wattiaux, M.A.; Broderick, G.A. Evaluation of milk urea nitrogen as a management tool to reduce ammonia emissions from dairy farms. *J. Dairy Sci.* **2011**, *94*, 4690–4694. [CrossRef]
19. Cantalapiedra-Hijar, G.; Peyraud, J.L.; Lemosquet, S.; Molina-Alcaide, E.; Boudra, H.; Nozière, P.; Marty-Ortigues, I. Dietary carbohydrate composition modifies the milk N efficiency in late lactation cows fed low crude protein diets. *Animal* **2014**, *8*, 275–285. [CrossRef] [PubMed]

20. Oenema, O.; Bannink, A.; Sommer, S.G.; van Groenigen, J.W.; Velthof, G.L. Gaseous Nitrogen Emissions from Livestock Farming Systems. In *Nitrogen in the Environment: Sources, Problems, and Management*, 2nd ed.; Hatfield, J.L., Follett, R.F., Eds.; Academic Press: San Diego, CA, USA, 2008.

21. Zhang, G.; Strøm, J.S.; Li, B.; Rom, H.B.; Morsing, S.; Dahl, P.; Wang, C. Emission of ammonia and other contaminant gases from naturally ven- tilated dairy cattle buildings. *Biosyst. Eng.* **2005**, *92*, 355–364. [CrossRef]

22. Montes, F.; Meinen, R.; Dell, C.; Rot, A.; Hristov, A.N.; Oh, G.; Waghorn, G.; Gerber, P.J.; Henderson, B.; Makkar, H.P.S.; et al. Mitigation of methane and nitrous oxide emissions from animal operations: II. A review of manure management mitigation options. *J. Anim. Sci.* **2013**, *91*, 5070–5094. [CrossRef]

23. Leytem, B.A.; Dungan, R.S.; Bjornberg, D.L.; Koehn, A.C. Greenhouse gas and ammonia emissions from an open-freestall dairy in southern Idaho. *J. Environ. Qual.* **2013**, *42*, 10–20. [CrossRef] [PubMed]

24. Pereira, J.; Misselbrook, T.H.; Chadwick, D.R.; Coutinhoc, J.; Trindade, H. Ammonia emissions from naturally ventilated dairy cattle buildings and outdoor concrete yards in Portugal. *Atmos. Environ.* **2010**, *44*, 3413–3421. [CrossRef]

25. Van der Stelt, B.; Temminghoff, E.J.M.; Van Vliet, P.C.J.; Van Riemsdijk, W.H. Volatilization of ammonia from manure as affected by manure additives, temperature and mixing. *Bioresour. Technol.* **2007**, *98*, 3449–3455. [CrossRef] [PubMed]

26. Saha, C.K.; Ammon, C.; Berg, W.; Fiedler, M.; Loebsin, C.; Sanftleben, P.; Brunsch, R.; Amon, T. Seasonal and diel variations of ammonia and methane emissions from a naturally ventilated dairy building and the associated factors in- fluencing emissions. *Sci. Total Environ.* **2014**, *468*, 53–62. [CrossRef]

27. Oenema, O.; Velthof, G.; Klimont, Z.; Winiwarter, W. *Emissions from Agriculture and Their Control Potentials*; TSAP Report #3, Version 2.1, Service Contract on Monitoring and Assessment of Sectorial Implementation Actions, (ENV.C.3/SER/2011/0009): 27; International Institute for Applied Systems Analysis: Laxenburg, Austria, 2012.

28. Sankey, J.B.; Eitel, J.U.H.; Glenn, N.F.; Germino, M.J.; Vierling, L.A. Quantifying relationships of burning, roughness, and potential dust emission with laser altimetry of soil surfaces at submeter scales. *Geomorphology* **2011**, *135*, 181–190. [CrossRef]

29. Place, S.E.; Mitloehner, F.M. Invited review: Contemporary environmental issues: A review of the dairy industry's role in climate change and air quality and the potential of mitigation through improved production efficiency. *J. Dairy Sci.* **2010**, *93*, 3407–3416. [CrossRef]

30. Sutton, M.; Howard, C.; Erisman, J. *The European Nitrogen Assessment: Sources, Effects and Policy Perspectives*; Cambridge University Press: Cambridge, UK, 2011.

31. Lelieveld, J.; Evans, J.S.; Fnais, M.; Giannadaki, D.; Pozzer, A. The contribution of outdoor air pollution sources to premature mortality on a global scale. *Nature* **2015**, *525*, 367–371. [CrossRef]

32. D'Amato, G.; Cecchi, L. Effects of climate change on environmental factors in respiratory allergic diseases. *Clin. Exp. Allergy* **2008**, *38*, 1264–1274. [CrossRef]

33. Auvermann, B.A. Texas/New Mexico Open-Lot Research. In Proceedings of the Western Dairy Air Quality Symposium, Sacramento, CA, USA, 20 April 2011.

34. Auvermann, B.A.; Romanillos, A. Effect of Increased Stocking Density on Fugitive Dust Emissions of PM10 from Cattle feedyards. In Proceedings of the International Meeting of the Air and Waste Management Association, Salt Lake City, UT, USA, 18–22 June 2000.

35. Bonifacio, H.F.; Maghirang, R.G.; Razote, E.B.; Auvermann, B.W.; Harner, J.P.; Murphy, J.P.; Guo, L.; Sweeten, J.M.; Hargrove, W.L. Particulate control efficiency of a water sprinkler system at a beef cattle feedlot in Kansas. *Trans. ASABE* **2011**, *54*, 295–304. [CrossRef]

36. Fan, C.; Tian, L.; Zhou, L.; Hou, D.; Song, Y.; Qiao, X.; Li, J. Examining the Impacts of Urban Form on Air Pollutant Emissions: Evidence from China. *J. Environ. Manag.* **2018**, *212*, 405–414. [CrossRef]

37. Premachandra, D.S.P.R.D.; Nayantha, K.H.D.; Jayasinghe, C. Investigation on Micro-Climatic Features Affecting the Indoor Air Quality in Suburbs, University of Muratuwa. In Proceedings of the 5th International Conference on Sustainable Built Environment, Kandy, Sri Lanka, 12–15 December 2014.

38. Popa, R.A.; Popa, D.C.; Gheorghe, E.M.; George, S.; Bălănescu, M.; Paştea, D.; Vulpe, A.; Marius, V.; Drăgulinescu, A.M. Hybrid platform for assessing air pollutants released from animal husbandry activities for sustainable livestock agriculture. *Sustainability* **2021**, *13*, 9633. [CrossRef]

39. IPCC. *2019 Refinement to the 2006 IPCC Guidelines for National Greenhouse Gas Inventories*; Calvo Buendia, E., Tanabe, K., Kranjc, A., Baasansuren, J., Fukuda, M., Ngarize, S., Osako, A., Pyrozhenko, Y., Shermanau, P., Federici, S., Eds.; IPCC: Geneva, Switzerland, 2019.

 *agriculture*

Article

Phenolic and Total Flavonoid Contents and Physicochemical Traits of Romanian Monofloral Honeys

Aida Albu [1], Răzvan-Mihail Radu-Rusu [2,*], Daniel Simeanu [1], Cristina-Gabriela Radu-Rusu [1] and Ioan Mircea Pop [1]

[1] Department of Control, Expertise and Services, Faculty of Food and Animal Sciences, "Ion Ionescu de la Brad" University of Life Sciences, 8 Mihail Sadoveanu Alley, 700489 Iasi, Romania

[2] Department of Animal Resources and Technologies, Faculty of Food and Animal Sciences, "Ion Ionescu de la Brad" University of Life Sciences, 8 Mihail Sadoveanu Alley, 700489 Iasi, Romania

* Correspondence: radurazvan@uaiasi.ro

Abstract: Since ancient times, honey has been appreciated not only for its sensorial traits, but also for the observed effects in rejuvenation and treatment against several bad health conditions, when used externally or internally, along with other beehive products, such as pollen, propolis and royal jelly. Today, it is known that such effects are generated by compounds bearing antimicrobial, anti-inflammatory, and antioxidative features (enzymes, polyphenolic molecules). The purpose of this study was to assess the total phenolic and flavonoid contents of 28 samples of Romanian raw monofloral honey (acacia; linden; rapeseed, sunflower and mint), and to establish their correlations with several qualitative parameters. Pearson's test revealed a strong positive correlation between total phenolic content and total flavonoids ($r = 0.76$) and color intensity ($r = 0.72$). For total flavonoid content, correlations were strongly positive with color intensity ($r = 0.81$), ash content ($r = 0.76$) and electrical conductivity ($r = 0.73$). The relevant levels of polyphenols and flavonoids identified in the analyzed honey types demonstrate its antioxidant potential, with essential nutritional and sanogenic features in human nutrition.

Keywords: honey; quality; phenolic content; flavonoid content; Pearson's correlation

Citation: Albu, A.; Radu-Rusu, R.-M.; Simeanu, D.; Radu-Rusu, C.-G.; Pop, I.M. Phenolic and Total Flavonoid Contents and Physicochemical Traits of Romanian Monofloral Honeys. *Agriculture* **2022**, *12*, 1378. https://doi.org/10.3390/agriculture12091378

Academic Editor: Muraleedharan G. Nair

Received: 24 July 2022
Accepted: 25 August 2022
Published: 2 September 2022

Publisher's Note: MDPI stays neutral with regard to jurisdictional claims in published maps and institutional affiliations.

1. Introduction

The Romanian beekeeping sector has developed throughout the last few years. Both conventional and organic honey production has increased because Romania has large areas of melliferous plants [1]. The variety of melliferous flora in Romania provides the possibility of producing many types of honey: monofloral, multifloral and honeydew honey [2]. The chemical composition of honey consists mainly of sugars, about 80%, mainly glucose and fructose, 15–17% water, 0.1–0.4% protein and 0.2% ash, while other components contained in small quantities provide some special properties [3]. The quality of honey is closely related to its properties (sensorial, physical, chemical, nutritional and sanogenic traits) that, in turn, are largely influenced by the type of melliferous flora, geographical and environmental factors of the region, by the final operations (processing, packaging, storage place and time) and manipulations [4–6]. The price and purchase frequency of honey depends on some traits that consumers seek in perceiving the quality of honey, such as: color, texture, clarity and flavor. Maintaining health through an appropriate diet led, in the last few years, to a change in preferences in honey consumption, namely, consumers are interested in the nutritional properties of this unique natural food [7]. The antimicrobial, anti-inflammatory and antioxidative activities of honey are properties that have been recognized for their beneficial effects on the human body. Many studies have shown that the composition and antioxidant activity of honey depends on several factors that can directly or indirectly affect its quality [8,9]. Honey's antioxidant capacity has been correlated with the levels of certain molecules: enzymes, polyphenolic compounds (phenolic acids, phenolic acid derivates, flavonoids), proteins, amino acids and other compounds. Flavonoids belong to a larger group of vegetal phenolic compounds. These bioactive molecules, originating in plants, are brought together with the collected nectar, and they have been used in other studies as floral markers to identify the geographical and botanical origin of honey [6,10,11]. In the literature, there are reports of correlations between certain physicochemical parameters: electrical conductivity and total ash content [12–14], pH and moisture, pH and acidity, acidity and ash [15,16]. Sant'ana et al., observed that darker honey has a higher content of phenolic compounds (flavonoids) and minerals [16]; other correlations, such as total phenolic content and total flavonoid content with color have also been studied [17–19]. In the present work, the therapeutic qualities of honey issued from different geographic regions, and, in particular, their antioxidant features, were studied (especially the polyphenolic and flavonoids compounds). Within this conjuncture, our study brings forth new data on the total phenolic and total flavonoid contents, as well as their correlations with several qualitative parameters of raw honey collected from eastern parts of Romania throughout 2019.

2. Materials and Methods

2.1. Honey Samples

In 2019, the following 28 samples of raw monofloral honey types were collected directly from beekeepers: acacia (A), eight samples; linden (L), seven samples; rapeseed (R), five samples; sunflower (SF), five samples; and mint (M), three samples. The honey bee samples were issued from the Eastern and Southeastern Romanian sites of Iasi County (A1, A2, A3, A4, L1, L2, R1, R2, SF1, SF2), Vaslui County (A5, A6, L3, L4, L5, R3, R4, SF3), Botosani County (A7, A8, L6, L7, R5, SF4, SF5) and Tulcea County (M1, M2, M3) (Figure 1). Each analytical sample was taken from fully filled and sealed individual jars of 400 mL capacity (honey content of 0.5 kg/jar). The botanical origin was declared by the beekeepers based on the naturally occurring floral species or on cultivated crops that the bees had harvested during the flowering period of the year. Thereafter, the company that purchased the honey from the beekeepers applied melissopalynological methods in order to certify the floral origin, prior to its processing and marketing. The raw honey samples were kept at 20 ± 3 °C in the dark. The preparation stage of the raw honey for analysis consisted in liquefying the crystallized samples at 40 °C in a water bath (manufacturer: Memmert GMBH—Schwabach, Germany). They were subsequently homogenized and filtered through a gauze.

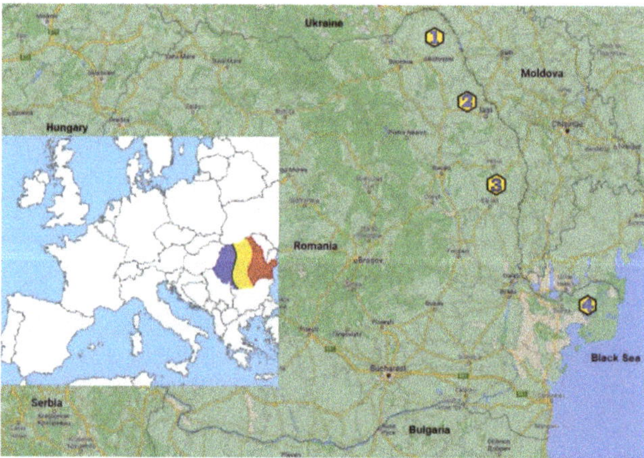

Figure 1. Geographic regions of collected honey samples (1—Botosani County; 2—Iasi County; 3—Vaslui County; 4—Tulcea County).

2.2. Color

The color of the honey samples was determined using the Shimadzu UV-1700 Pharma Spec instrument (manufacturer: Shimadzu Corporation, Analytical Instruments Division, Kyoto, Japan). Aqueous honey solutions of 50% (*w/v*) were centrifuged at 3200 rpm for 5 min in a Universal 320 Hettich centrifuge (manufacturer: Hettich GMBH—Tuttlingen, Germany) [20]. The absorbance units, measured at 635 nm wavelength, were converted into mm Pfund using the following calculation:

$$Pfund\ (mm) = -38.7 + 371.39 \times Abs \tag{1}$$

where Pfund = the honey color value on the Pfund scale (mm); Abs = the value of the absorbance read at the wavelength 635 nm.

2.3. Water-Insoluble Solids

Ten grams of honey sample weighed on the PI-214 Denver analytical balance (manufacturer: Denver Instrument GMBH—Gottingen, Germany) were dissolved in distilled water, filtered through filter paper with constant weight, and washed several times. After being dried in an ESAC 100 oven at 105 °C (manufacturer: SC Electronic April Aparatura Electronica Speciala S.R.L.—Cluj-Napoca, Romania), the content of water-insoluble solids was calculated by the difference between the filter paper with water-insoluble solids weight and filter paper weight, and the result are expressed as percentages [21].

2.4. Refractive Index, Moisture and Solid Substances

The refractive index (RI) of samples was determined using the Abbé Kruss AR 2008 refractometer (manufacturer: Kruss Scientific GMBH, Hamburg, Germany), then, moisture content (M) was taken from the table of correspondence between the water content and the refractive index at 20 °C [21]. Solid substance content (SS), expressed as a percentage, was calculated as the difference between 100 and moisture content.

2.5. Total Soluble Solids and Specific Gravity

The total soluble solids (TSS) represented by soluble sugars, expressed as Brix degrees, were obtained from the table of correspondence between the refractive index at 20 °C and the Brix degrees [22].

Specific gravity was assessed by a gravimetric method, using the pycnometer device. The results are expressed in g/cm^3 [5].

2.6. pH and Free Acidity

Using the Multi 3320 multiparameter (manufacturer: WTW GMBH, Weilheim, Germany), the pH values were measured in an aqueous honey solution (10 g of honey in 75 mL of distilled water). In the same solution, free acidity was measured by titration with 0.1 N NaOH (Chemical Company, Iasi, Romania) solution using phenolphthalein (Chimreactiv, Bucharest, Romania) as a color indicator [21,23,24].

2.7. Ash and Electrical Conductivity

Assessment of ash content was carried out using a gravimetric method after honey samples were calcinated in the Nabertherm B180 furnace (manufacturer: Nabertherm GMBH, Lilienthal, Germany); the results are expressed in g/100 g. Electrical conductivity was measured with the Multi 3320 multiparameter (manufacturer: WTW GMBH, Weilheim, Germany). The 20% solution (the weighted honey was calculated as dry matter) was formed with ultrapure water produced by the Barnstead Easy Pure II system (manufacturer: Thermo Fisher Scientific Co. Ltd., North Liberty, IA, USA); the results are expressed in $\mu S\ cm^{-1}$ [21,23].

2.8. Total Phenolic Content and Total Flavonoid Content

Total phenolic content and total flavonoid content were determined according to the Folin–Ciocalteu method with minor modification.

For total phenolic content, a 10% honey-based alcoholic solution (solution 1:1 of methanol (Merck KGaA, Darmstadt, Germany) with acidified water (deionized water at pH = 2 with HCl (Merck KGaA, Darmstadt, Germany)) was homogenized and filtered through filter paper. An aliquot of honey solution was mixed with 0.2 mL of Folin–Ciocalteu's phenol reagent (Merck KGaA, Darmstadt, Germany) for 5 min, and to it was added 75g/L Na_2CO_3 (Merck KGaA, Darmstadt, Germany) until the volume was 10 mL. The solution was kept in the dark, at room temperature, for 30 min prior to measuring it at 742 nm wavelength, against a blank sample using the Shimadzu UV-1700 Pharma Spec instrument (manufacturer: Shimadzu Corporation, Analytical Instruments Division, Kyoto, Japan). To obtain the calibration curve with five calibration points (concentration range of 2–12 mg L^{-1}; y = 0.089x + 0.1147; R^2 = 0.9972), gallic acid (Merck KGaA, Darmstadt, Germany) was used as the standard. The maximum absorption was recorded at 742 nm for a spectrum range of 700–800 nm. The results are expressed in mg of gallic acid equivalents (GAE)/100 g [24,25].

The same honey-based alcoholic solution was used to assess the total flavonoid content. The same volume of 2% $AlCl_3$ (Merck KGaA, Darmstadt, Germany) was added to this honey solution. After 10 min, absorbance was read at 430 nm. To obtain the calibration curve with six calibration points (concentration range of 0.5–5 mg L^{-1}; y = 0.1331x + 0.0112; R^2 = 0.9997), quercetin (Sigma-Aldrich, St. Louis, MO, USA) was used as the standard. The maximum absorption was recorded at 430 nm for a spectrum range of 400–500 nm. The results are expressed in mg of quercetin equivalents (QE)/100 g [17,25].

2.9. Statistical Analyses

All analyses were conducted in triplicate. The analytical data were statistically processed for main descriptors and analysis of variance—one-way ANOVA—via GraphPad Prism 9.4.0 (673) software. Correlations between physicochemical parameters were tested using Pearson's correlation coefficient (r) between total phenolic and total flavonoid content and the other investigated parameters. These r values were grouped using OriginPro 2022 software and presented as a heatmap. Graphics were built, and circular hierarchical cluster analysis (HCA) was performed, using the same software.

3. Results

3.1. Physicochemical Analyses

The results of some studied parameters (color, water-insoluble solids, refractive index, moisture, solid substance content, total soluble solids, specific gravity) are presented in Table 1.

Table 1. Parameter (color, water-insoluble solids, refractive index, moisture, solid substances, total soluble solids, specific gravity) values of honey samples.

Type	Descriptive Statistics	Color mm Pfund	WIS %	RI	M %	SS %	TSS %	SG %
Acacia 8 samples	Min–Max	0.2–7.5	0.035–0.108	1.488–1.498	15.41–19.49	80.51–84.59	79.03–83.06	1.420–1.448
	Mean ± SD	3.9 ± 2.29	0.079 ± 0.03	1.494 ± 0.00	16.98 ± 1.21	83.02 ± 1.21	81.51 ± 1.20	1.437 ± 0.01
	CV	71.20	33.81	0.21	7.15	1.46	1.47	0.58
Linden 7 samples	Min–Max	21.7–26.7	0.062–0.107	1.488–1.493	17.28–19.20	80.80–82.72	79.32–81.19	1.422–1.435
	Mean ± SD	24.5 ± 1.75	0.090 ± 0.01	1.491 ± 0.00	18.10 ± 0.70	81.90 ± 0.70	80.40 ± 0.68	1.430 ± 0.00
	CV	7.14	16.19	0.12	3.87	0.86	0.85	0.33
Rapeseed 5 samples	Min–Max	52.5–61.0	0.074–0.107	1.486–1.495	16.77–20.07	79.93–83.23	78.28–81.73	1.410–1.439
	Mean ± SD	55.6 ± 3.29	0.095 ± 0.01	1.491 ± 0.00	18.21 ± 1.35	81.79 ± 1.35	80.27 ± 1.39	1.428 ± 0.01
	CV	5.92	13.42	0.23	7.40	1.65	1.74	0.80
Sunflower 5 samples	Min–Max	36.9–82.9	0.060–0.114	1.487–1.494	16.93–19.60	80.40–83.07	78.92-81.58	1.420–1.438
	Mean ± SD	61.5 ± 18.92	0.080 ± 0.02	1.491 ± 0.00	18.27 ± 1.23	81.73 ± 1.23	80.23 ± 1.22	1.429 ± 0.01
	CV	30.77	26.87	0.21	6.76	1.51	1.53	0.59

Table 1. *Cont.*

Type	Descriptive Statistics	Color mm Pfund	WIS %	RI	M %	SS %	TSS %	SG %
Mint 3 samples	Min–Max	42.8–86.1	0.047–0.087	1.489–1.496	16.07–18.79	81.21–83.93	79.73–82.40	1.425–1.444
	Mean ± SD	68.0 ± 22.52	0.072 ± 0.02	1.493 ± 0.00	17.31 ± 1.38	82.69 ± 1.38	81.18 ± 1.35	1.435 ± 0.01
	CV	33.13	30.52	0.24	7.95	1.66	1.66	0.64
p values (ANOVA)	A vs. L	$p = 7.7 \times 10^{-9}$	$p = 0.3633$	$p = 0.0049$	$p = 0.0067$	$p = 0.0067$	$p = 0.0070$	$p = 0.0127$
	A vs. R	$p = 4.5 \times 10^{-12}$	$p = 0.1050$	$p = 0.0070$	$p = 0.0073$	$p = 0.0073$	$p = 0.0062$	$p = 0.0028$
	A vs. SF	$p = 4.3 \times 10^{-11}$	$p = 0.9999$	$p = 0.0037$	$p = 0.0040$	$p = 0.0040$	$p = 0.0045$	$p = 0.0076$
	A vs. M	$p = 3.9 \times 10^{-11}$	$p = 0.8471$	$p = 0.9247$	$p = 0.9342$	$p = 0.9342$	$p = 0.9299$	$p = 0.9477$
	L vs. R	$p = 3.7 \times 10^{-12}$	$p = 0.9303$	$p = 0.9995$	$p = 0.9985$	$p = 0.9985$	$p = 0.9966$	$p = 0.9402$
	L vs. SF	$p = 4.1 \times 10^{-12}$	$p = 0.5542$	$p = 0.9935$	$p = 0.9898$	$p = 0.9898$	$p = 0.9914$	$p = 0.9906$
	L vs. M	$p = 4.3 \times 10^{-12}$	$p = 0.1344$	$p = 0.3291$	$p = 0.3507$	$p = 0.3507$	$p = 0.3647$	$p = 0.4127$
	R vs. SF	$p = 0.4897$	$p = 0.2120$	$p = 0.9997$	$p = 0.9998$	$p = 0.9998$	$p = 0.9999$	$p = 0.9986$
	R vs. M	$p = 0.0318$	$p = 0.0398$	$p = 0.2936$	$p = 0.2843$	$p = 0.2843$	$p = 0.2705$	$p = 0.1685$
	SF vs. M	$p = 0.5265$	$p = 0.8441$	$p = 0.2203$	$p = 0.2183$	$p = 0.2183$	$p = 0.2352$	$p = 0.2672$

WIS—water-insoluble matter. RI—refractive index. M—moisture. SS—solid substances. TSS—total soluble substances. SG—specific gravity. SD—standard deviation; CV—coefficient of variation.

The results of some parameters (pH, free acidity, ash, electrical conductivity, total phenolic content, total flavonoid content) are presented in Table 2.

Table 2. Parameter (pH, free acidity, ash, electrical conductivity, total phenolic content, total flavonoid content) values of honey samples.

Type	Descriptive Statistics	pH	FA meq kg^{-1}	Ash %	EC mS cm^{-1}	TPC mg GAE/100 g	TFC mg QE/100 g
Acacia 8 samples	Min–Max	4.14–4.72	6.8–15.4	0.040–0.100	0.130–0.220	11.10–17.92	0.44–1.63
	Mean ± SD	4.36 ± 0.18	11.3 ± 2.82	0.066 ± 0.02	0.173 ± 0.03	13.88 ± 2.39	0.86 ± 0.40
	CV	4.21	24.84	32.02	17.83	17.21	46.92
Linden 7 samples	Min–Max	4.14–4.81	12.5–37.2	0.157–0.333	0.397–0.623	20.30–29.29	1.01–3.14
	Mean ± SD	4.42 ± 0.24	27.7 ± 7.93	0.246 ± 0.06	0.506 ± 0.09	24.37 ± 3.08	2.02 ± 0.78
	CV	5.51	28.65	25.99	17.03	12.65	38.70
Rapeseed 5 samples	Min–Max	3.62–4.26	19.9–44.0	0.085–0.135	0.197–0.290	19.70–24.74	1.33–3.12
	Mean ± SD	4.00 ± 0.25	29.1 ± 9.62	0.101 ± 0.02	0.224 ± 0.04	21.72 ± 1.98	2.00 ± 0.69
	CV	6.17	33.05	19.84	16.88	9.10	34.42
Sunflower 5 samples	Min–Max	3.25–5.03	21.6–47.0	0.127–0.428	0.328–0.637	20.60–28.84	1.63–3.92
	Mean ± SD	4.09 ± 0.67	28.8 ± 10.43	0.251 ± 0.11	0.428 ± 0.12	25.12 ± 3.26	2.52 ± 0.90
	CV	16.45	36.18	44.50	28.04	12.96	35.64
Mint 3 samples	Min–Max	3.80–4.20	24.3–40.0	0.134–0.238	0.220–0.551	42.06–50.82	2.04–3.97
	Mean ± SD	4.02 ± 0.21	30.6 ± 8.29	0.202 ± 0.06	0.394 ± 0.17	47.20 ± 4.58	3.05 ± 0.97
	CV	5.16	27.09	29.17	42.17	9.70	31.71
p values (ANOVA)	A vs. L	$p = 7.7 \times 10^{-8}$	$p = 5.5 \times 10^{-10}$	$p = 3.2 \times 10^{-13}$	$p = 2.1 \times 10^{-11}$	$p = 3.5 \times 10^{-10}$	$p = 1.6 \times 10^{-6}$
	A vs. R	$p = 4.5 \times 10^{-12}$	$p = 7.7 \times 10^{-10}$	$p = 0.3660$	$p = 0.3081$	$p = 2.8 \times 10^{-10}$	$p = 1.5 \times 10^{-5}$
	A vs. SF	$p = 4.3 \times 10^{-11}$	$p = 1.5 \times 10^{-9}$	$p = 3.9 \times 10^{-14}$	$p = 1.3 \times 10^{-9}$	$p = 6.1 \times 10^{-12}$	$p = 6.2 \times 10^{-10}$
	A vs. M	$p = 3.9 \times 10^{-11}$	$p = 1.7 \times 10^{-8}$	$p = 4.6 \times 10^{-7}$	$p = 5.9 \times 10^{-8}$	$p = 3.9 \times 10^{-11}$	$p = 4.7 \times 10^{-12}$
	L vs. R	$p = 3.7 \times 10^{-12}$	$p = 0.9776$	$p = 8.8 \times 10^{-10}$	$p = 4.6 \times 10^{-10}$	$p = 0.0420$	$p = 0.9999$
	L vs. SF	$p = 4.1 \times 10^{-12}$	$p = 0.9897$	$p = 0.9986$	$p = 0.0415$	$p = 0.9252$	$p = 0.1906$
	L vs. M	$p = 4.3 \times 10^{-12}$	$p = 0.8508$	$p = 0.3167$	$p = 0.0068$	$p = 7.3 \times 10^{-12}$	$p = 0.0021$
	R vs. SF	$p = 0.4897$	$p = 0.9999$	$p = 3.4 \times 10^{-9}$	$p = 8.7 \times 10^{-9}$	$p = 0.0091$	$p = 0.2251$
	R vs. M	$p = 0.0318$	$p = 0.9883$	$p = 0.0007$	$p = 2.9 \times 10^{-5}$	$p = 4.8 \times 10^{-8}$	$p = 0.0034$
	SF vs. M	$p = 0.5265$	$p = 0.9783$	$p = 0.2585$	$p = 0.8516$	$p = 5.5 \times 10^{-6}$	$p = 0.3357$

pH—pH value. FA—Free Acidity. Ash—Crude ash, total minerals content. EC—electrical con-ductivity. TPC—Total Phenolic Content. TFC—Total Flavonoid Content.

Figures 2 and 3 present the mean content, standard error and mean line values of selected parameters of five floral honey samples.

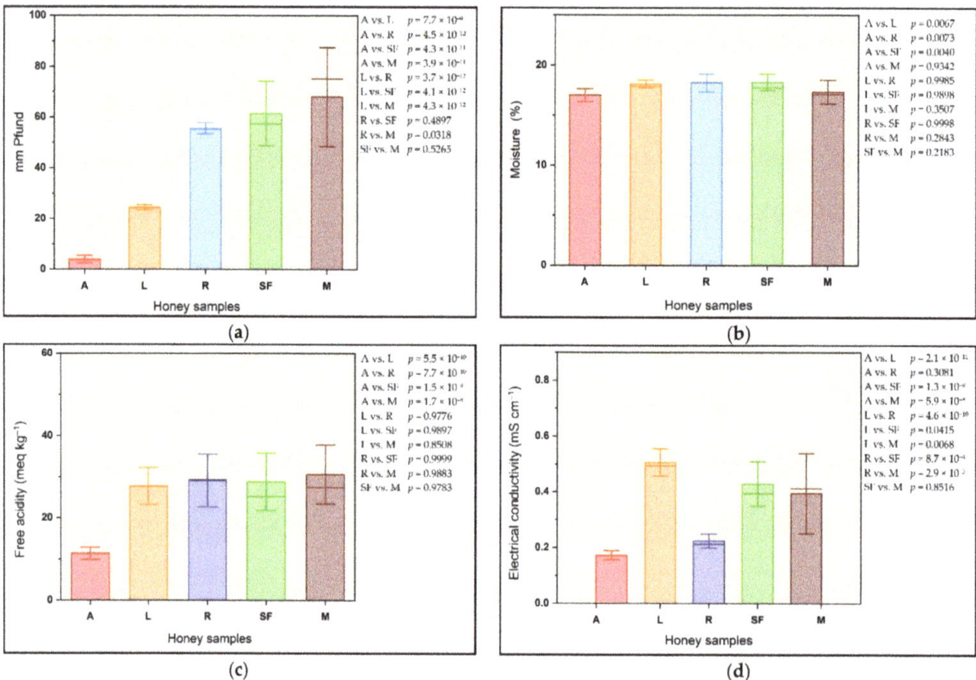

Figure 2. Mean content, standard error and mean line values of selected honey sample parameters: (**a**) mm Pfund; (**b**) moisture; (**c**) free acidity; (**d**) electrical conductivity.

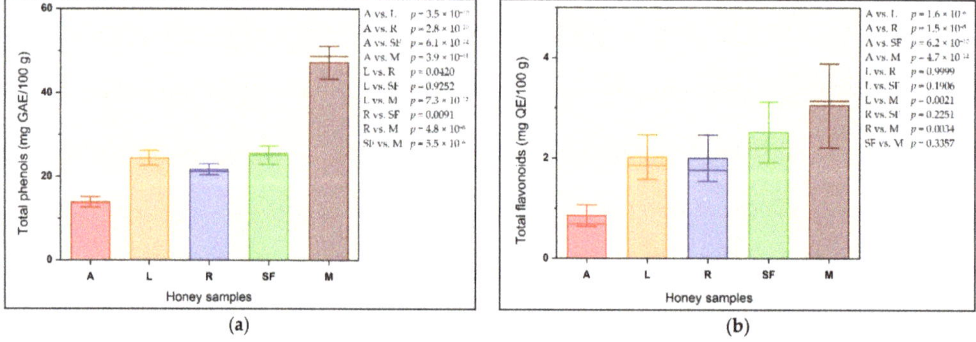

Figure 3. Mean content, standard error and mean line values of some honey sample parameters: (**a**) total phenols; (**b**) total flavonoids.

3.2. Correlations between Physicochemical Parameters

The correlations between all studied parameters are presented on a heatmap; these results were determined using Pearson's correlation coefficient (Figure 4).

The correlation between ash and electrical conductivity is presented in Figure 5.

The relationship between the flavonoid content and honey color is shown in a visual graph; data are expressed as % of total studied honey samples (Figure 6).

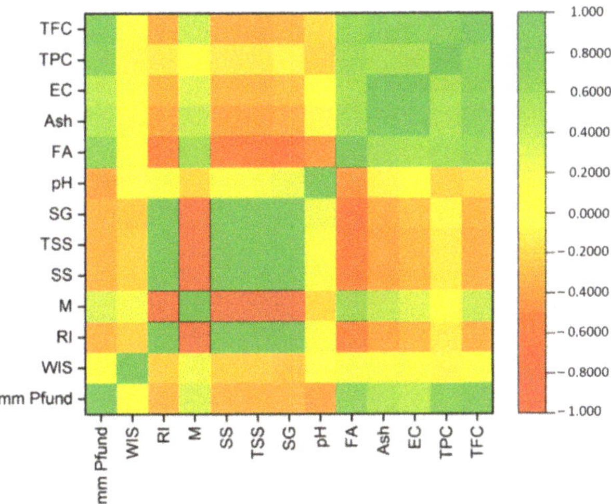

Figure 4. Pearson's correlation coefficient results presented on a heatmap.

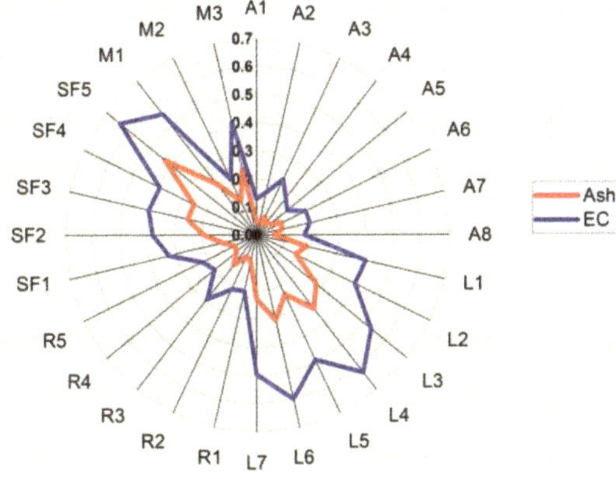

Figure 5. Radial graph of correlation between two parameters: ash content (%) and electrical conductivity (EC) (mS cm^{-1}).

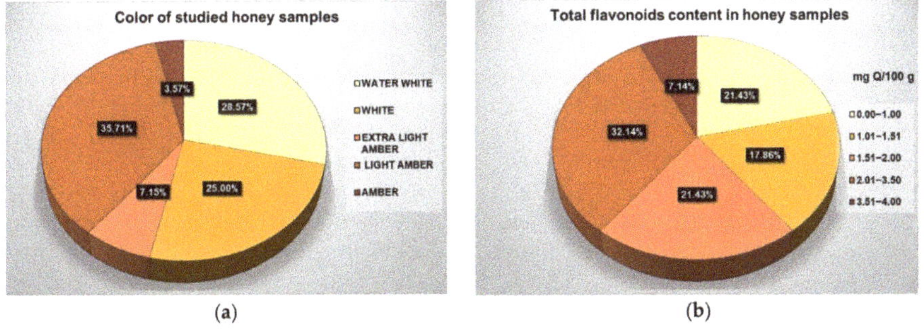

(**a**)   (**b**)

Figure 6. Visual presentation of correlation between (**a**) color and (**b**) total flavonoid content of honey samples.

In Figure 7, five colored clusters are distinguished, highlighting the connections between the parameters of the honey samples.

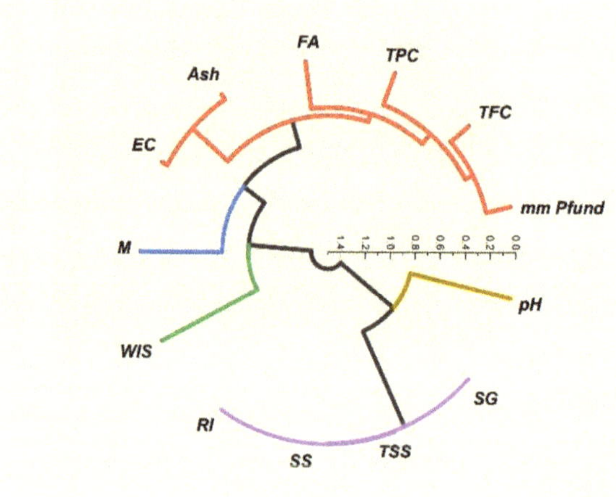

Figure 7. Circular hierarchical cluster analysis (HCA) of studied honey sample parameters.

4. Discussion

4.1. Water-Insoluble Solids (WIS)

In honey, materials such as wax particles, pollen, bee parts, plant material pieces and other foreign elements that contaminate the honey, represent water-insoluble solids (WIS) [26]. An increased amount of WIS causes not only a lower quality of honey, but also an increased opportunity for the development of yeasts, which can create favorable conditions for the fermentation process, leading to further decreasing of honey quality. Beekeepers perform centrifugal and filtering operations on their bees' honey, but most of them do not sell it directly on the market; therefore, raw honey may have a higher content of WIS. In our studied honey samples, water-insoluble solids varied from 0.035% in acacia honey, to 0.114% in sunflower honey. Seven samples of raw monofloral honey (two acacia, two linden, two rapeseed, one sunflower) had WIS contents between 0.101% and 0.114%, which are above the threshold level (0.1%) established by the legislation [27].

4.2. Color

The color of honey has an important visual impact on the consumers. In different parts of the world, honey is consumed primarily because of its color hue: some consumers prefer light-colored honey, others dark-colored. This honey color palette depends on certain factors, such as the botanical origin, the composition of the nectar, the amount of pigments, the extraction operations, temperature, storage conditions and time [3,28]. The color of the studied honey samples varied from water-white (0.2 mm Pfund) in acacia honey samples to amber (86.1 mm Pfund) in mint honey samples. The average color intensity increased as follows: acacia honey < linden honey < rapeseed honey < sunflower honey < mint honey (Figure 2a). The color of the studied acacia honey samples was similar to the color of acacia honey from Croatia [29]; various studies show that for the same botanical origin, honey samples from different regions exhibit various colors; for example, sunflower honey color intensity varied from 33.66 mm Pfund to 114.00 mm Pfund (Table 3).

Table 3. Color intensity of different types of floral honey in our study and from literature sources.

	Literature source	Country	mm Pfund				
			Acacia	Linden	Rapeseed	Sunflower	Mint
	Our study	Romania	0.20–7.50	21.70–26.70	52.50–61.00	36.90–82.90	42.80–86.10
[30]	Chiş and Purcărea (2017)	Romania	-	-	-	61.3; 70.5; 88.7	-
[4]	Al et al. (2009)	Romania	11.00–45.00	36.00–54.00	-	79.00–83.00	-
[2]	Pauliuc and Oroian (2020)	Romania	-	-	-	32.87–47.52	-
[3]	Pauliuc et al. (2020)	Romania	-	-	29.40	37.60	74.30
[31]	Pauliuc et al. (2022)	Romania	12.87	35.64	36.14	33.66	63.86
[32]	Aazza et al. (2013)	Portugal	-	-	-	97.60	-
[33]	Chirsanova et al. (2021)	R. Moldova	-	22.00–38.00	-	39.00–41.00	-
[29]	Flanjak et al. (2016)	Croatia	1.00–8.00	10.00–29.00	-	-	-
[20]	Raţiu et al. (2020)	Poland	-	68.01; 76.80	34.34–114.07	62.07; 114.00	-
[34]	Smetanska et al. (2021)	Germany	26.51	-	-	-	-
[35]	Živković et al. (2019)	Serbia	20.00	70.00	-	-	-

Previous studies of honey parameters revealed correlations between the color intensity and the quantity of minerals (lowest content found in light-colored honey samples) [36] with phenolic or flavonoid content (color intensity increased as phenolic content grew higher) [16]. Chiş and Purcărea, showed that the colors of three Romanian sunflower honey samples became darker, i.e., the mm Pfund increased, over time (2 years) from 70.5 mm Pfund, 88.7 mm Pfund and 61.3 mm Pfund to 73.4 mm Pfund, 101.4 mm Pfund and to 68.5 mm Pfund, respectively [30].

4.3. Refractive Index, Moisture and Solid Substance Content, Total Soluble Solids and Specific Gravity

The refractive index (RI), as read from the refractometer, ranged from 1.486 in rapeseed honey samples to 1.498 in acacia honey samples. The range in moisture content (M) was between 15.41% (acacia) and 20.07% (rapeseed), while the highest average moisture content (18.27%) was registered in sunflower honey (Figure 2b). Solid substance contents (SS) were between 79.93% (rapeseed) and 84.59% (acacia) (Table 1). Of the 28 honey samples, only one had a moisture content above the 20% (water content) value, recommended for honey, according to Romanian standards and international regulations [21,27]. The moisture content of honey samples studied in other research publications [2,29,37–39] ranged between 14.45% and 22.8%; there were some honey samples reported with moisture contents exceeding the value of 20% established by the legislation (Table 4). A low content of moisture provides quality assurance for a long time period; high moisture content is a favorable condition for the occurrence of the fermentation phenomenon, thus leading to changes in both sensory and physicochemical properties, and thus, to the decrease in the quality of the bee product [12,34].

Table 4. Some physicochemical properties of the five different types of floral honey in our study and from literature sources.

	Literature Source	Country	Moisture (%)	pH	Free Acidity (meq kg^{-1})	Ash (%)	EC (mS cm^{-1})	TPC mgGAE/100 g	TFC mgQE/100 g
					Acacia				
	Our study	Romania	15.41–19.49	4.14–4.72	6.8–15.4	0.04–0.10	0.130–0.220	11.10–17.92	0.44–1.63
[4]	Al et al. (2009)	Romania	16.60–19.80	-	-	0.03–0.28	-	2.00–39.00	0.91–2.42
[31]	Pauliuc et al. (2022)	Romania	15.96	4.31	3.86	-	0.12	-	-
[37]	Stihi et al. (2016)	Romania	16.7–22.8	3.65–4.63	-	-	0.097–0.268	-	-
[40]	Atanassova et al. (2012)	Bulgaria	16.9	3.23	-	-	0.159	-	-
[29]	Flanjak et al. (2016)	Croatia	14.6–19.9	-	-	-	0.1–0.161	2.82–5.20	-
[41]	Rostislav et al. (2016)	Czech Republic	17	3.82	9.6	-	0.18	23.84	0.87
[42]	Alzahrani et al. (2012)	Germany	17	5.4	-	-	-	62.75	-
[11]	Smetanska et al. (2021)	Germany	18.83	4.10	-	-	-	21.457	-
[43]	Attanzio et al. (2016)	Italy	-	-	-	-	-	18.2	7.6
[44]	Di Marco et al. (2018)	Italy	-	-	-	-	-	10.72	3.31
[45]	Gośliński et al. (2021)	Poland	-	-	-	-	-	76.3	-

Table 4. *Cont.*

	Literature Source	Country	Moisture (%)	pH	Free Acidity (meq kg^{-1})	Ash (%)	EC (mS cm^{-1})	TPC mgGAE/100 g	TFC mgQE/100 g
[39]	Milek et al. (2021)	Poland	17.17	3.77	20.85	-	0.31	14.081	-
[46]	Tomczyk et al. (2019)	Poland	17.73	3.79	25.6	-	0.42	47	0.32
[47]	Milosavljević et al. (2021)	Serbia	14.5–18.5	-	6.6–15.5	0.04–0.15	0.083–0.174	58.17–142.61	-
[48]	Sakač et al. (2022)	Serbia	16.4; 17.3	3.90; 4.51	13.8; 16.3	-	0.114; 0.136	13.5; 14.4	-
[35]	Živković et al. (2019)	Serbia	-	-	-	-	-	37.93	-
[46]	Tomczyk et al. (2019)	Slovakia	17.86	3.71	16.1	-	0.20	20	0.14
[38]	Akgün et al. (2021)	Turkey	14.45–21.62	-	12–21	-	0.14–0.27	1–3	-
					Linden				
	Our study	Romania	17.28–19.20	4.14–4.81	12.5–37.2	0.157–0.333	0.397–0.623	20.30–29.29	1.01–3.14
[4]	Al et al. (2009)	Romania	16.70–19.10	-	-	0.19–0.30	-	16–38	4.7–6.98
[31]	Pauliuc et al. (2022)	Romania	16.75	4.05	14.55	-	0.33	-	-
[37]	Stihi et al. (2016)	Romania	17.2–18.8	3.84–4.35	-	-	0.202–0.346	-	-
[40]	Atanassova et al. (2012)	Bulgaria	17.1	4.04	-	-	0.689	-	-
[29]	Flanjak et al. (2016)	Croatia	15.9–20.00	-	-	-	0.497–0.628	6.62–12.10	-
[41]	Rostislav et al. (2016)	Czech Republic	16	4.06	14.9	-	0.39	45.04	1.88
[44]	Di Marco et al. (2018)	Italy	-	-	-	-	-	26	5.5
[49]	Dżugan et al. (2018)	Poland	-	-	-	-	-	30.27–54.95	-
[45]	Gośliński et al. (2021)	Poland	-	-	-	-	-	91.3	-
[39]	Milek et al. (2021)	Poland	20.30	14.13	25.50	-	0.640	43.69	-
[46]	Tomczyk et al. (2019)	Poland	17.76	3.81	34.2	-	0.53	38	0.5
[48]	Sakač et al. (2022)	Serbia	15.8; 17.1	4.62; 4.72	14.5; 16.1	-	0.488; 0.608	53.7; 67.3	-
[35]	Živković et al. (2019)	Serbia	-	-	-	-	-	71.49	-
[46]	Tomczyk et al. (2019)	Slovakia	18.35	3.90	21.6	-	0.23	35	0.26
					Rapeseed				
	Our study	Romania	16.77–20.07	3.62–4.26	19.9–44.0	0.085–0.135	0.197–0.290	19.70–24.74	1.33–3.12
[3]	Pauliuc et al. (2020)	Romania	18.4	4.22	16	-	0.162	19.9	20.2
[31]	Pauliuc et al. (2022)	Romania	17.31	4.11	17.33	-	0.15	-	-
[40]	Atanassova et al. (2012)	Bulgaria	19.7	3.33	-	-	0.181	-	-
[49]	Dżugan et al. (2018)	Poland	-	-	-	-	-	20.54–31.08	-
[45]	Gośliński et al. (2021)	Poland	-	-	-	-	-	101.6	-
[46]	Tomczyk et al. (2019)	Poland	17.86	3.88	18.6	-	0.23	25	0.32
[48]	Sakač et al. (2022)	Serbia	18.4; 19.4	4.01; 4.10	16.3; 21.3	-	0.191; 0.224	11.5; 11.9	-
[46]	Tomczyk et al. (2019)	Slovakia	17.45	3.61	13.6	-	0.16	21	0.14
					Sunflower				
	Our study	Romania	16.93–19.60	3.25–5.03	21.6–47.0	0.127–0.428	0.328–0.637	20.60–28.84	1.63–3.92
[50]	Chiş and Purcărea (2015)	Romania	18.7	3.656	22.36	0.112	0.301	-	-
[30]	Chiş and Purcărea (2017)	Romania	-	-	-	-	-	48.6–132.5	-
[4]	Al et al. (2009)	Romania	17.80–19.70	-	-	0.35–0.40	-	20.00–45.00	11.53–15.33
[2]	Pauliuc and Oroian (2020)	Romania	16.23–20.39	3.65–4.34	15.94–47.32	-	0.315–0.441	-	-
[3]	Pauliuc et al. (2020)	Romania	18.4	3.94	31.6	-	0.362	21.1	22.8
[31]	Pauliuc et al. (2022)	Romania	16.95	4.04	18.32	-	0.31	-	-
[37]	Stihi et al. (2016)	Romania	17	3.67	-	-	0.188	-	-
[45]	Gośliński et al. (2021)	Poland	-	-	-	-	-	82.4	-
[32]	Aazza et al. (2013)	Portugal	19.2	3.84	25.50	0.15	0.235	36.69	1.93
[33]	Chirsanova, A. et al., 2021	R. Moldova	16.05–17.52	3.68–4.05	-	0.31–0.49	-	-	-
[47]	Milosavljević et al. (2021)	Serbia	17.4–19.8	-	18.5–39.4	0.12–0.30	0.189–0.359	25.45–61.09	-
[48]	Sakač et al. (2022)	Serbia	17.0	3.38	28.9	-	0.366	27.5	-
[51]	Živkov-Baloš et al. (2021)	Serbia	14.6–18.6	-	20.40–36.4	0.05–0.30	0.22–0.54	-	-
					Mint				
	Our study	Romania	16.07–18.79	3.80–4.20	24.3–40.0	0.134–0.238	0.220–0.551	42.06–50.82	-
[3]	Pauliuc et al. (2020)	Romania	17.7	4.20	26.9	-	0.474	23.7	25.7
[31]	Pauliuc et al. (2022)	Romania	16.24	4.52	33.17	-	0.60	-	-
[52]	Boussaid et al. (2018)	Tunisia	19.80	-	-	0.13	0.43	119.42	-

Minimum/maximum values of total soluble solids of 78.28 °Brix/83.06 °Brix were found in rapeseed/acacia honey samples. Minimum/maximum values of specific gravity (1.410 g/cm^3/1.448 g/cm^3) were also found in rapeseed/acacia samples.

4.4. pH and Free Acidity

The values of pH were found to be within a 3.25–5.03 range, while the free acidity ranged between 6.8 meq kg^{-1} in acacia honey, and 47.0 meq kg^{-1} in sunflower honey; all free acidity values were situated below the threshold (50 meq kg^{-1}) value established by the legislation [27]. The lowest average free acidity value was found in acacia samples, 11.3 meq kg^{-1} (Figure 2c). Honey pH and acidity are correlated with the internal amount of microorganisms, enzymatic activity and the presence of organic acids. A low pH indicates a good environment that inhibits microorganism growth, while the free acidity level is a biochemical marker for honey samples freshness [34,51]. Similar values of free acidity were found among Serbian acacia honey samples [47]; these were the lowest values when compared to other types of honey (Table 4). The data on the free acidity of sunflower honey varied greatly, between 21.6 meq kg^{-1} and 47.0 meq kg^{-1}. In their study, Pauliuc and Oroian, found even wider variation limits of free acidity for other Romanian sunflower honey types, from 15.94 meq kg^{-1} to 47.32 meq kg^{-1} [2]. The values of pH and free acidity, obtained in other studies from different countries are presented in Table 4.

4.5. Ash and Electrical Conductivity

The highest average ash content was found in sunflower samples (0.251%), while the lowest (0.066%) was measured in acacia samples. The electrical conductivity ranged between 0.130 mS cm^{-1} in acacia samples and 0.637 mS cm^{-1} in sunflower samples. The lowest average electrical conductivity value was registered in acacia honey samples (0.173 mS cm^{-1}) (Figure 2d). All values were measured below the maximal threshold of 0.8 mS cm^{-1}, specified by Directive 2001/110/EC of the European Union [3,27] for blossom honey. Electrical conductivity has become an increasingly common analysis; its value is related to the quantity of minerals, organic acids and proteins. Additionally, the assessment of electrical conductivity is useful in certifying honey authenticity/adulteration, as it facilitates discriminating between blossom honey types and honeydew [3,20,51]. Highly similar results were obtained from investigations carried out on the same floral type of honey, with values also below 0.8 mS cm^{-1} (Table 4).

4.6. Total Phenolic Content and Total Flavonoid Content

Throughout the year, from spring to late autumn, bees mix the nectars of flowers or secretions of other plants with their fluids to produce blossom honey or honeydew. The composition of honey, the amount of nutrients, especially the antioxidant compounds (phenols, flavonoids, flavones), are influenced by the botanical family, the richness of flora diversity, and/or by the geographical area [28,45]. Honey can be considered a healing food due to the antioxidant role conferred mainly by phenolic compounds, such as phenolic acids and flavonoids. The amount of antioxidant compounds depends mainly on the floral source and on the geographical origin [34,35]. The highest amplitude of variation for the phenolic content was found in the linden honey (8.99 mg GAE/100 g of phenolic content between the analytical limits). The minimal individual sample value of phenols (11.10 mg GAE/100 g) was found in acacia honey, while the maximal value (50.82 mg GAE/100 g) was measured in mint honey. The lowest average value of phenolic content was calculated in acacia honey samples (13.88 mg GAE/100 g), followed ascendingly by rapeseed (21.72 mg GAE/100 g), linden (24.37 mg GAE/100 g) and sunflower (25.12 mg GAE/100 g); thus, the richest content was found in mint honey samples (47.20 mg GAE/100 g) (Figure 3a). The values of total phenol content obtained by Milosavljević et al. in Serbian acacia honey were much higher in comparison to our findings (58.17 to 142.61 mg GAE/100 g) [47]. The total phenol content assessed in honey bee samples from several European countries have been reported, ranging from 1 mg GAE/100 g in Turkish acacia honey [38] to 142.61 mg GAE/100 g in Serbian acacia honey [47] (Table 4). Honey is considered to be a food that, when kept under optimal conditions, can maintain its nutritional qualities for a long time. However, the antioxidant content is influenced negatively as storage time extends. In their study, Chiş and Purcărea found that the total phenolic content of three sunflower honey samples decreased over time (2 years) from 68.3 mg GAE/100 g to 60.3 mg GAE/100 g, from 132.5 mg GAE/100 g to 98.2 mg GAE/100 g, and from 48.6 mg GAE/100 g to 40.5 mg GAE/100 g, respectively [30]. In our findings, total flavonoid content ranged between the limits of 0.44 mg QE/100 g in acacia honey and 3.97 mg QE/100 g in mint honey (Table 2).

The highest average content was recorded in mint samples (3.05 mg QE/100 g) followed, in descending order, by sunflower honey (2.52 mg QE/100 g), and linden and rapeseed honey (2.02–2.00 mg QE/100 g); thus, the lowest average flavonoid content was calculated in acacia honey (0.86 mg QE/100 g) (Figure 3b). Table 4 shows the results obtained in other studies related to flavonoid content in the same floral types of honey. As can be seen, the data vary considerably from one type of honey to another, and from one country to another country, within the same honey variety. The presence of phenols and flavonoids reaffirms the antioxidant properties of honey.

4.7. Correlations between Honey Parameters

Correlations between the physicochemical parameters of the studied honey samples were revealed using the Pearson correlation coefficient r value; these results are arranged and presented as a heatmap in Figure 4. Moisture (M) content had a strong negative correlation with four parameters: refractive index (RI) ($r = -1.0$), solid substances (SS) ($r = -1.0$), total soluble substances (TSS) ($r = -1.0$) and specific gravity (SG) ($r = -0.99$). There were some strong positive linear correlations ($r = +1$) between three parameters: refractive index (RI), solid substances (SS) ($r = +1$) and total soluble solids (TSS); strong positive linear correlations ($r = +0.99$) were also noticed between specific gravity (SG) and refractive index (RI), solid substances (SS) and total soluble solids (TSS). Another strong positive correlation ($r = +0.95$) stood out between ash (Ash) and electrical conductivity (EC) (Figure 5). Phenolic compounds exhibit a close positive relationship with color, thus explaining the Pearson coefficient of +0.72 determined between total phenol content (TPC) and color (mm Pfund), and +0.81 between total flavonoid content (TFC) and color (mm Pfund) (Figure 6). There was also a close positive relationship between these two compounds and antioxidant properties, the Pearson coefficient having a high value of +0.76. Total phenol content (TPC) exhibited moderate positive correlations with free acidity (FA) ($r = +0.57$), electrical conductivity (EC) ($r = +0.53$) and with ash (Ash) ($r = +0.51$). Total flavonoid content (TFC) had strong positive correlations with ash (Ash) ($r = +0.76$) and electrical conductivity (EC) ($r = +0.73$), and a moderate positive correlation with free acidity (FA) ($r = +0.65$). Moderate positive correlations were also found between free acidity (FA) and color (mm Pfund) ($r = +0.68$) and moisture (M) ($r = +0.59$). There are many other studies exploring the correlations between honey parameters. Pontis et al. [17] reported strong correlations between total phenolic content and color ($r = +0.967$), total flavonoid content and color ($r = +0.924$) of Brazilian honey. Al Farsi et al. found strong correlations for color vs. flavonoids (+0.999) and color vs. phenols (+0.974) in honey samples of *Apis mellifera* honeybees from regions within the Sultanate of Oman [18], with higher Pearson coefficient values compared to those obtained in our study. Total flavonoid content was significantly correlated with color ($r = +0.82$) in Brazilian honey samples, as reported by De Almeida et al. [53]. Cimpoiu et al. obtained a strong linear correlation of +0.86 between total phenolic content and the color intensity in Romanian honey [19]. It is known that pigments provide the color hue of honey, specifically those pigments with antioxidant properties, such as carotenoids. Many studies have shown strong correlations between color intensity and antioxidant compounds: Kavanagh et al. in Irish honey (color/TPC, $r = +0.6$) [54]; Ciappini et al., in Argentinian honey (flavonoid content and color, $r = +0.93$) [55]; Moniruzzaman et al. [56] in Bangladesh honey samples (color with phenolic acids, r = +0.943 and flavonoids $r = 0.926$); Aazza et al. [32] in Portuguese honey (color/TPC, $r = +0.685$; color/TFC, $r = 0.843$); Živković et al. [35] in Serbian honey (color/TPC, $r = +0.815$; color/TFC, $r = 0.771$); Flanjak et al. [29] in Croatian honey (color/TPC, $r = +0.925$). The correlations between the total phenol (TPC) and total flavonoid (TFC) contents were also confirmed in a study on monofloral honey from Sicily (Italy) ($r = +0.919$), a higher value than that found in our research [43]. Al Farsi et al. found strong positive correlations between flavonoids and phenolics ($r = +0.977$) [18], as did Živković et al., who observed a correlation of +0.967 between TPC and TFC in Serbian honey samples [35]. Aazza et al., on commercial Portuguese honey, found strong correlation between TPC and TFC ($r = +0.861$) [32], while in Romanian honey, correlation reached a level of 0.7512 [8]. Flanjak et al., found a strong statistically significant positive correlation between honey EC and TPC (r = +0.837) [29], while in another study, this correlation had a lower value (r = +0.55) [54]. In sunflower honey samples from Serbia, Živkov Baloš et al. found a moderate correlation between electrical conductivity and ash mass fraction (r = +0.611) [51]. To compare similarities between studied parameters, we used a color circular hierarchical cluster analysis (HCA). Figure 7 shows the formation of five clusters, with the following parameters in the first cluster: refractive index (RI), solid substances (SS), total soluble substances (TSS) and specific gravity (SG). These parameters exhibited strong positive linear correlations. The second cluster also consisted of parameters with strong relationships: total flavonoid content (TFC), color (mm Pfund), free acidity (FA), electrical conductivity (EC) and ash (Ash). The parameters forming other clusters were moisture content (M), water-insoluble solids (WIS) and pH.

5. Conclusions

The results of this study show many strong positive correlations, especially between antioxidant compound levels and color. Dark-colored honey samples had higher phenolic and flavonoid contents, in comparison with light-color honey samples.

The relevant levels of polyphenols and flavonoids identified in the analyzed honey demonstrate its antioxidant potential, as well as its essential nutritional and sanogenic features in human nutrition.

Author Contributions: Conceptualization, A.A. and D.S.; methodology, A.A. and R.-M.R.-R.; software, A.A. and R.-M.R.-R.; validation, A.A. and D.S.; formal analysis, A.A., R.-M.R.-R., C.-G.R.-R. and I.M.P.; investigation, A.A., R.-M.R.-R., C.-G.R.-R. and I.M.P.; data curation, A.A. and D.S.; writing—original draft preparation, A.A., R.-M.R.-R.; writing—review and editing, A.A., R.-M.R.-R. and D.S.; supervision, A.A. All authors have read and agreed to the published version of the manuscript.

Funding: This research received no external funding.

Institutional Review Board Statement: Not applicable.

Informed Consent Statement: Not applicable.

Data Availability Statement: The data present in this study are available on request from the corresponding author.

Conflicts of Interest: The authors declare no conflict of interest.

References

1. Pocol, C.B.; Šedík, P.; Brumă, I.S.; Amuza, A.; Chirsanova, A. Organic Beekeeping Practices in Romania: Status and Perspectives towards a Sustainable Development. *Agriculture* **2021**, *11*, 281. [CrossRef]
2. Pauliuc, D.; Oroian, M. Organic Acids and Physico-Chemical Parameters of Romanian Sunflower Honey. *Food Environ. Saf. J.* **2020**, *19*, 148–155.
3. Pauliuc, D.; Dranca, F.; Oroian, M. Antioxidant Activity, Total Phenolic Content, Individual Phenolics and Physicochemical Parameters Suitability for Romanian Honey Authentication. *Foods* **2020**, *9*, 306. [CrossRef]
4. Al, M.L.; Daniel, D.; Moise, A.; Bobis, O.; Laslo, L.; Bogdanov, S. Physico-chemical and bioactive properties of different floral origin honeys from Romania. *Food Chem.* **2009**, *112*, 863–867. [CrossRef]
5. Popescu, N.; Meica, S. Bee products and their chemical analysis. In *Produsele Apicole si Analiza lor Chimica*; Diacon Coresi: Bucuresti, RO, USA, 1997.
6. Da Silva, P.M.; Gauche, C.; Gonzaga, L.V.; Costa, A.C.O.; Fett, R. Honey: Chemical composition, stability and authenticity. *Food Chem.* **2016**, *196*, 309–323. [CrossRef]
7. Testa, R.; Asciuto, A.; Schifani, G.; Schimmenti, E.; Migliore, G. Quality Determinants and Effect of Therapeutic Properties in Honey Consumption. An Exploratory Study on Italian Consumers. *Agriculture* **2019**, *9*, 174. [CrossRef]
8. Ciucure, C.T.; Geană, E.-I. Phenolic compounds profile and biochemical properties of honeys in relationship to the honey floral sources. *Phytochem. Anal.* **2019**, *30*, 481–492. [CrossRef] [PubMed]
9. Geana, E.I.; Ciucure, C.T. Establishing authenticity of honey via comprehensive Romanian honey analysis. *Food Chem.* **2020**, *306*, 125595. [CrossRef]

10. Ciulu, M.; Spano, N.; Pilo, M.I.; Sanna, G. Recent Advances in the Analysis of Phenolic Compounds in Unifloral Honeys. *Molecules* **2016**, *21*, 451. [CrossRef] [PubMed]

11. De-Melo, A.A.M.; de Almeida-Muradian, L.B.; Sancho, M.T.; Pascual-Maté, A. Composition and properties of *Apis mellifera* honey: A review. *J. Apic. Res.* **2017**, *57*, 5–37. [CrossRef]

12. Ahmida, N.H.; Elagori, M.; Agha, A.; Elwerfali, S.; Ahmida, M.H. Physicochemical, heavy metals and phenolic compounds analysis of Libyan honey samples collected from Benghazi during 2009–2010. *Food Nutr. Sci.* **2013**, *4*, 33–40. [CrossRef]

13. El Sohaimy, S.A.; Masry, S.H.D.; Shehata, M.G. Physicochemical characteristics of honey from different origins. *Ann. Agric. Sci.* **2015**, *60*, 279–287. [CrossRef]

14. Majewska, E.; Druzynska, B.; Wołosiak, R. Determination of the botanical origin of honeybee honeys based on the analysis of their selected physicochemical parameters coupled with chemometric assays. *Food Sci. Biotechnol.* **2019**, *28*, 1307–1314. [CrossRef]

15. Krishnasree, V.; Ukkuru, P.M. Quality Analysis of Bee Honeys. *Int. J. Curr. Microbiol. Appl. Sci.* **2017**, *6*, 626–636. [CrossRef]

16. Sant'Ana, L.D.; Ferreira, A.B.; Lorenzon, M.C.A.; Berbara, R.L.L.; Castro, R.N. Correlation of total phenolic and flavonoid contents of Brazilian honeys with color and antioxidant capacity. *Int. J. Food Prop.* **2014**, *17*, 65–76.

17. Pontis, J.A.; Costa, L.A.M.A.D.; Silva, S.J.R.D.; Flach, A. Color, phenolic and flavonoid content, and antioxidant activity of honey from Roraima, Brazil. *Food Sci. Technol.* **2014**, *34*, 69–73. [CrossRef]

18. Al-Farsi, M.; Al-Amri, A.; Al-Hadhrami, A.; Al-Belushi, S. Color, flavonoids, phenolics and antioxidants of Omani honey. *Heliyon* **2018**, *4*, e00874. [CrossRef] [PubMed]

19. Cimpoiu, C.; Hosu, A.; Miclaus, V.; Puscas, A. Determination of the floral origin of some Romanian honeys on the basis of physical and biochemical properties. *Spectrochim. Acta Part A Mol. Biomol. Spectrosc.* **2013**, *100*, 149–154. [CrossRef] [PubMed]

20. Ratiu, I.A.; Al-Suod, H.; Bukowska, M.; Ligor, M.; Buszewski, B. Correlation study of honey regarding their physicochemical properties and sugars and cyclitols content. *Molecules* **2020**, *25*, 34. [CrossRef] [PubMed]

21. Romanian Standards Association. SR (Romanian Standard) 784-3:2009: Honey Bee. Part 3: Analytical Methods. Available online: https://e-standard.eu/en/standard/174480 (accessed on 16 April 2018).

22. USDA. *Extracted Honey Grading Manual, United States Department of Agriculture. Standards for Honey Grading*; USDA: Washington DC, USA, 1985. Available online: https://www.ams.usda.gov/sites/default/files/media/Extracted_Honey_Inspection_Instructions%5B1%5D.pdf (accessed on 10 December 2018).

23. Bogdanov, S. Harmonized Methods of the International Honey Commission. 2009. Available online: https://www.ihc-platform.net/ihcmethods2009.pdf (accessed on 30 May 2018).

24. Sereia, M.J.; Março, P.H.; Perdoncini, M.R.G.; Parpinelli, R.S.; de Lima, E.G.; Anjo, F.A. Chapter 9: Techniques for the Evaluation of Physicochemical Quality and Bioactive Compounds in Honey. In *Honey Analysis*; INTECH: London, UK, 2017; pp. 194–214.

25. Bobis, O.; Marghitas, L.; Rindt, I.K.; Niculae, M.; Dezmirean, D. Honeydew honey: Correlations between chemical composition, antioxidant capacity and antibacterial effect. *Sci. Pap. Anim. Sci. Biotechnol.* **2008**, *41*, 271–277.

26. Pourtallier, J.; Taliercio, Y. Honey Control Analyses. *Apiacta* **1972**, *1*, 2–5.

27. European Commission. Council Directive 2001/110/CE concerning honey. *Off. J. Eu. Communities* **2002**, *10*, 47–52.

28. Becerril-Sánchez, A.L.; Quintero-Salazar, B.; Dublán-García, O.; Escalona-Buendía, H.B. Phenolic Compounds in Honey and Their Relationship with Antioxidant Activity, Botanical Origin, and Color. *Antioxidants* **2021**, *10*, 1700. [CrossRef] [PubMed]

29. Flanjak, I.; Kenjerić, D.; Bubalo, D.; Primorac, L. Characterisation of Selected Croatian Honey Types Based on the Combination of Antioxidant Capacity, Quality Parameters, and Chemometrics. *Eur. Food Res. Technol.* **2016**, *242*, 467–475. [CrossRef]

30. Chiş, A.M.; Purcărea, C. Storage effect on antioxidant capacities of some monofloral honey. *Studia Univ. Vasile Goldiş Life Sci. Ser.* **2017**, *27*, 91–97.

31. Pauliuc, D.; Dranca, F.; Ropciuc, S.; Oroian, M. Advanced Characterization of Monofloral Honeys from Romania. *Agriculture* **2022**, *12*, 526. [CrossRef]

32. Aazza, S.; Lyoussi, B.; Antunes, D.; Miguel, M.G. Physicochemical Characterization and Antioxidant Activity of Commercial Portuguese Honeys. *J. Food Sci.* **2013**, *78*, 1159–1165. [CrossRef]

33. Chirsanova, A.; Capcanari, T.; Boistean, A. Quality Assessment of Honey in Three Different Geographical Areas from Republic of Moldova. *Food Nutr. Sci.* **2021**, *12*, 962–977. [CrossRef]

34. Smetanska, I.; Alharthi, S.S.; Selim, K.A. Physicochemical, antioxidant capacity and color analysis of six honeys from different origin. *J. King Saud Univ. Sci.* **2021**, *33*, 101447. [CrossRef]

35. Živković, J.; Sunarić, S.; Stanković, N.; Mihajilov-Krstev, T.; Spasić, A. Total phenolic and flavonoid contents, antioxidant and antibacterial activities of selected honeys against human pathogenic bacteria. *Acta Pol. Pharm. Drug Res.* **2019**, *76*, 671–681. [CrossRef]

36. Kędzierska-Matysek, M.; Teter, A.; Stryjecka, M.; Skałecki, P.; Domaradzki, P.; Rudaś, M.; Florek, M. Relationships Linking the Colour and Elemental Concentrations of Blossom Honeys with Their Antioxidant Activity: A Chemometric Approach. *Agriculture* **2021**, *11*, 702. [CrossRef]

37. Stihi, C.; Chelarescu, E.D.; Duliu, O.G.; Toma, L.G. Characterization of Romanian honey using physico-chemical parameters and the elemental content determined by analytical techniques. *Rom. Rep. Phys.* **2016**, *68*, 362–369.

38. Akgün, N.; Çelik, Ö.F.; Kelebekli, L. Physicochemical properties, total phenolic content, and antioxidant activity of chestnut, rhododendron, acacia and multifloral honey. *J. Food Meas. Charact.* **2021**, *15*, 3501–3508. [CrossRef]

39. Miłek, M.; Bocian, A.; Kleczyńska, E.; Sowa, P.; Dżugan, M. The Comparison of Physicochemical Parameters, Antioxidant Activity and Proteins for the Raw Local Polish Honeys and Imported Honey Blends. *Molecules* **2021**, *26*, 2423. [CrossRef] [PubMed]
40. Atanassova, J.; Lazarova, Y.; Lazarova, M. Pollen and inorganic characteristics of Bulgarian unifloral honeys. *Czech J. Food Sci.* **2012**, *30*, 520–526. [CrossRef]
41. Rostislav, H.; Petr, T.; Ćavar, Z.S. Characterisation of phenolics and other quality parameters of different types of honey. *Czech J. Food Sci.* **2016**, *34*, 244–253. [CrossRef]
42. Alzahrani, H.A.; Alsabehi, R.; Boukraâ, L.; Abdellah, F.; Bellik, Y.; Bakhotmah, B.A. Antibacterial and Antioxidant Potency of Floral Honeys from Different Botanical and Geographical Origins. *Molecules* **2012**, *17*, 10540–10549. [CrossRef]
43. Attanzio, A.; Tesoriere, L.; Allegra, M.; Livrea, M.A. Monofloral honeys by Sicilian black honeybee (*Apis mellifera* ssp. sicula) havehigh reducing power and antioxidant capacity. *Heliyon* **2016**, *2*, e00193. [CrossRef]
44. Di Marco, G.; Gismondi, A.; Panzanella, L.; Canuti, L.; Impei, S.; Leonardi, D.; Canini, A. Botanical Influence on Phenolic Profile and Antioxidant Level of Italian Honeys. *J. Food Sci. Technol.* **2018**, *55*, 4042–4050. [CrossRef]
45. Gośliński, M.; Nowak, D.; Szwengiel, A. Multidimensional Comparative Analysis of Bioactive Phenolic Compounds of Honeys of Various Origin. *Antioxidants* **2021**, *10*, 530. [CrossRef]
46. Tomczyk, M.; Tarapatskyy, M.; Dżugan, M. The influence of geographical origin on honey composition studied by Polish and Slovak honeys. *Czech J. Food Sci.* **2019**, *37*, 232–238. [CrossRef]
47. Milosavljević, S.; Jadranin, M.; Mladenović, M.; Tešević, V.; Menković, N.; Mutavdžić, D.; Krstić, G. Physicochemical parameters as indicators of the authenticity of monofloral honey from the territory of the Republic of Serbia. *Maced. J. Chem. Chem. Eng.* **2021**, *40*, 57–67. [CrossRef]
48. Sakač, M.; Jovanov, P.; Marić, A.; Četojević-Simin, D.; Novaković, A.; Plavšić, D.; Škrobot, D.; Kovač, R. Antioxidative, Antibacterial and Antiproliferative Properties of Honey Types from the Western Balkans. *Antioxidants* **2022**, *11*, 1120. [CrossRef] [PubMed]
49. Dżugan, M.; Tomczyk, M.; Sowa, P.; Grabek-Lejko, D. Antioxidant Activity as Biomarker of Honey Variety. *Molecules* **2018**, *23*, 2069. [CrossRef] [PubMed]
50. Chiş, A.M.; Purcărea, C. The phisico-chemical caracterisation of sun flower honey from bihor county. *An. Univ. Din Oradea Fasc. Ecotoxicologie Zooteh. Şi Tehnol. Ind. Aliment.* **2015**, *14*, 107–114.
51. Živkov-Baloš, M.M.; Jakšić, S.M.; Popov, N.S.; Polaček, V.A. Characterization of Serbian sunflower honeys by their physicochemical characteristics. *Food Feed. Res.* **2021**, *48*, 1–8. [CrossRef]
52. Boussaid, A.; Chouaibi, M.; Rezig, L.; Hellal, R.; Donsì, F.; Ferrari, G.; Hamdi, S. Physicochemical and bioactive properties of six honey samples from various floral origins from Tunisia. *Arab. J. Chem.* **2018**, *11*, 265–274. [CrossRef]
53. De Almeida, A.M.M.; Oliveira, M.B.S.; Da Costa, J.G.; Valentim, I.B.; Goulart, M.O.F. Antioxidant Capacity, Physicochemical and Floral Characterization of Honeys from the NorthEast of Brazil. *Rev. Virtual Química* **2016**, *8*, 57–77. [CrossRef]
54. Kavanagh, S.; Gunnoo, J.; Passos, T.M.; Stout, J.C.; White, B. Physicochemical Properties and Phenolic Content of Honey from Different Floral Origins and from Rural versus Urban Landscapes. *Food Chem.* **2019**, *272*, 66–75. [CrossRef] [PubMed]
55. Ciappini, M.; Gatti, M.; Di Vito, M. El Color Como Indicador Del Contenido de Flavonoides En Miel. *Rev. Cienc. Tecnol.* **2013**, *15*, 59–63.
56. Moniruzzaman, M.; Yung, A.C.; Azlan, S.A.B.M.; Sulaiman, S.A.; Rao, P.V.; Hawlader, M.N.I.; Gan, S.H. Identification of phenolic acids and flavonoids in monofloral honey from Bangladesh by high performance liquid chromatography: Determination of antioxidant capacity. *BioMed Res. Int.* **2014**, *2014*, 737490. [CrossRef] [PubMed]

Article

Quality Profile of Several Monofloral Romanian Honeys

Ioan Mircea Pop [1,†], Daniel Simeanu [1,*], Simona-Maria Cucu-Man [2,*], Aurel Pui [2] and Aida Albu [1,†]

[1] Department of Control, Expertise and Services, Faculty of Food and Animal Sciences, "Ion Ionescu de la Brad" University of Life Sciences, 8 Mihail Sadoveanu Alley, 700489 Iasi, Romania

[2] Department of Chemistry, Faculty of Chemistry, "Alexandru Ioan Cuza" University of Iasi, 11 Carol I Blvd, 700506 Iasi, Romania

* Correspondence: dsimeanu@uaiasi.ro (D.S.); sman@uaic.ro (S.-M.C.-M.)

† These authors contributed equally to this work.

Abstract: The objective of this research was to evaluate some quality-defining physicochemical parameters (moisture, specific gravity, pH, free acidity, ash, electrical conductivity, total phenols, and total flavonoids content, K, Ca, Mg, Na, and P) of seven Romanian monofloral honeys (linden, acacia, rapeseed, sunflower, mint, raspberry, and chestnut) collected in 2017. The investigated quality parameters are mainly within the recommended limits set by standards for honey. Sample analyses indicate the presence of antioxidants, such as TPC (17.9–73.2 mg GAE/100 g) and TFC (0.84–4.81 mg QE/100 g), and high amounts of K (101–1462 mg kg^{-1}), Ca (58.3–167.5 mg kg^{-1}), Mg (24.8–330.6 mg kg^{-1}), Na (94.5–233.3 mg kg^{-1}), and P (34.1–137.2 mg kg^{-1}). The Pearson's correlations between some parameters (such as color/TFC, color/Mg, color/P, EC/Ash, mm Pfund/TFC, TPC/TFC, K/Ash, P/Mg), together with PCA, HCA, and ANOVA statistics, highlight three main factors that explain the variability in the dataset and could be attributed to stability, mineral, and color/antioxidant contributions. FTIR spectra confirm the authenticity of all the monofloral honeys. The results and data processing confirm the influence of environmental elements (soil, water, air) on the honey composition and highlight the quality of honey, as a complete food and a therapeutic product.

Keywords: honey; phenols; flavonoids; minerals; FTIR; Pearson's correlation

Citation: Pop, I.M.; Simeanu, D.; Cucu-Man, S.-M.; Pui, A.; Albu, A. Quality Profile of Several Monofloral Romanian Honeys. *Agriculture* **2023**, *13*, 75. https://doi.org/10.3390/agriculture13010075

Academic Editor: Alessandra Durazzo

Received: 20 November 2022
Revised: 22 December 2022
Accepted: 23 December 2022
Published: 27 December 2022

1. Introduction

Honey is considered one of the healthiest foods because any change of the environment quality produces a negative impact on the health of bees. Bees have important contributions to pollination, not only of wild plants that have nectar, but also of cultivated ones. Along with other pollinators, bees play an important role in maintaining biodiversity [1]. One of the measures to preserve biodiversity is to keep bees safe. Healthy bees have an impact over the quality of products of the hive. Honey is one of the main products of the hive. In Romania, the diversity of melliferous plants as well as those that are cultivated ensures a diversified production of polyfloral and especially monofloral honey [2]. Monofloral honey is a type of honey that comes mostly from the nectar of a single plant species, is more valuable, and has a higher price compared to the polyfloral honey [3]. It is known that the quality properties of honey differ from one year to another, and responsible for this, first, are the climatic variations of the last years and other factors such as the pastoral regions, environmental quality (soil, water, air), and mainly the botanical origin [4–7]. The nutritional and therapeutic quality of honey has been known for a long time, and therefore its use in different fields (nutrition, medicine, cosmetics, etc.) has been continuously increasing.

Honey contains many substances (mainly sugars, water, organic acids, proteins, vitamins, phenolic compounds, minerals, pigments, etc.) that, in perfect harmony, give it its known properties [4,8–10]. Qualitatively, honey must comply with a series of legislative requirements, not more than 20% moisture content, not less than 60 g/100 g sugars' content, free acidity not more than 50 milliequivalents acid per 1000 g, etc. [11]. The moisture level is important, because with a higher content of moisture, honey can ferment and other parameters such as viscosity, taste, and odor change and contribute to the depreciation of honey [10]. All the studied European honeys were acidic, and a low pH inhibits the growth of microorganisms. Free acidity is considered a freshness indicator and the flavors of honey are due to the presence of organic acids [2,12–16]. Honeydew and blossom honey differ through composition. The value of electrical conductivity established by legislation [11] delimits the botanical origin of blossom honey from honeydew, and it could be considered a criterion for determining a possible honey adulteration. This parameter is correlated with ash, salts, concentrations of mineral elements, etc. [8,10,12,17,18]. Other constituents of honey that have a special role are polyphenols and flavonoids. The quantity and the type of these antioxidant compounds in honey come mainly from flowers, which give this beehive product its antioxidant properties, and are responsible for its therapeutic qualities [19,20]. Each mineral and trace element in honey varies depending on the geographical as well as the floral origin. Minerals in honey come from the environment, mainly from the soil, from which plants absorb them. Plants absorb both essential minerals for human health (K, Ca, Mg, Na, P, Cu, Mn, Fe) and toxic ones (Pb, Cd, Hg). The traceability of these mineral elements from soil to honey can negatively influence the quality of the hive products. The presence and quantity of minerals in honey could reflect the quality and degree of pollution, which is why honey has also been considered as a possible indicator of the environment quality [9]. Many researchers have shown some correlations between compounds of honey that could help to identify some common characteristics of different types of honey from different geographical areas [21–28].

The analyses carried out on honey involve the structural destruction of the sample, the consumption of chemical reagents, and consequently, take a long time to obtain viable results. In recent years, Fourier transform infrared (FTIR) spectroscopy has been used in honey research. This is a nondestructive analytical method used to scan, to identify, or to highlight certain substances or chemical groups present in the composition of the tested samples and the chemical properties of the samples. FTIR spectroscopy yields valuable information on the quality of honey and is important to verify the authenticity of the honey and its possible adulteration [29–31]. The technique was applied on different types of honey samples to determine the botanical or geographical origin, and to identify some chemical compounds that define the specific quality of honey, such as carbohydrates (fructose, glucose, sucrose) [32–35].

The aim of this study was to evaluate the quality of some monofloral honey types produced in Romania by using classical physicochemical methods and to show the similarities or molecular differences with a nondestructive method—FTIR spectroscopy.

2. Materials and Methods

2.1. Honey Samples

The honey samples, from *Apis mellifera* species, collected in 2017, come from two companies that collect, process, and then sell the honey (producers 1 and 2) and from four private beekeepers (producers 3, 4, 5, and 6) (Table 1). The honey samples originate from mainly the eastern part of Romania, as well as the northwestern and southeastern parts (Figure 1).

The botanical origin of raw honey samples was established by a company that purchases the honey from the beekeepers, by using melissopalynological methods. Three jars of 750 g each were collected from every type of honey. All samples were kept in the dark at laboratory temperature ($22 \pm 2\ °C$). The crystallized samples were liquefied at $40\ °C$ in a water bath (Memmert GMBH—Schwabach, Germany), homogenized, and filtered through gauze.

Table 1. Sample description.

Sample	Producer	Location	Geographical Location	Sample	Producer	Location	Geographical Location
L1	1	Barnova Iasi	47°05′32″ N 27°38′14″ E	L3	3	Popesti Iasi	47°08′42″ N 27°15′33″ E
A1	1	Lunca Prutului Iasi	47°27′0″ N 27°33′0″ E	A3	3	Baltati Iasi	47°13′42″ N 27°06′32″ E
RP1	1	Barnova Iasi	47°05′32″ N 27°38′14″ E	RP3	3	Baltati Iasi	47°13′42″ N 27°06′32″ E
SF1	1	Barnova Iasi	47°05′32″ N 27°38′14″ E	SF3	3	Popesti Iasi	47°08′42″ N 27°15′33″ E
M1	1	Danube Delta Tulcea	45°20′ N 29°30′ E	L4	4	Aroneanu Iasi	47°12′50″ N 27°36′00″ E
RB1	1	Bistrita mountains	47°12′, 25°67′ E	A4	4	Aroneanu Iasi	47°12′50″ N 27°36′00″ E
C1	1	Satu Mare county	47°47′24″ N 22°53′24″ E	L5	5	Tomesti Iasi	47°07′07″ N 27°42′48″ E
L2	2	Raducaneni Iasi	46°57′33″ N 27°58′41″ E	A5	5	Tomesti Iasi	47°07′07″ N 27°42′48″ E
RP2	2	Raducaneni Iasi	46°57′33″ N 27°58′41″ E	L6	6	Roscani Iasi	47°26′18.12″ N 27°25′42.78″ E
M2	2	Danube Delta Tulcea	45°20′ N 29°30′ E	A6	6	Roscani Iasi	47°26′18.12″ N 27°25′42.78″ E
RB2	2	Dobrovat Iasi	46°57′55″ N 27°43′18″ E				

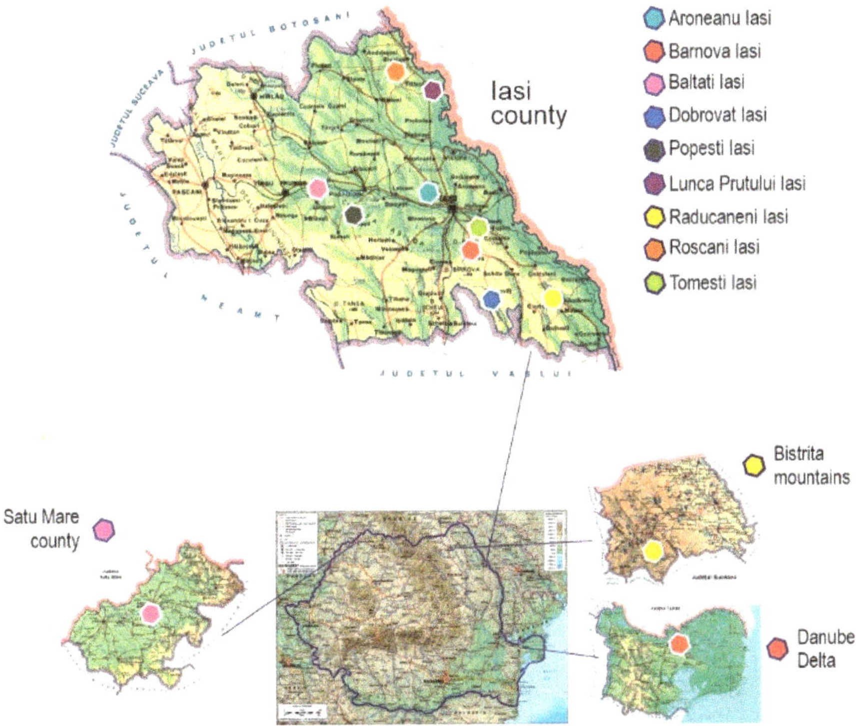

Figure 1. Map showing the locations where the honey samples were produced.

2.2. Pfund Value and Color

The Pfund value and the color of honey samples were determined using the method described by Ratiu et al. [17,26,36]. A 50% (*w/v*) honey aqueous solution was centrifuged at 3200 rpm (UNIVERSAL 320 HETTICH centrifuge, Hettich GMBH—Tuttlingen, Germany) and the absorbance was measured at 635 nm using a Shimadzu UV-1700 Pharma Spec spectrophotometer (Shimadzu Corporation, Analytical Instruments Division, Kyoto, Japan). The absorbance units were converted to mm Pfund with the relation:

$$\text{Pfund (mm)} = -38.7 + 371.39 \times \text{Abs} \tag{1}$$

where Pfund represents the honey color value in the Pfund scale (mm), and Abs is the absorbance at 635 nm.

2.3. Refractive Index, Moisture, Solid Substances, and Specific Gravity

Moisture content (M), expressed as a percentage, was taken from the table of correspondence between the water content and the refractive index at 20 °C [37] after reading the refractive index on an ABBÉ Kruss AR 2008 refractometer (Kruss Scientific GMBH, Hamburg, Germany) and applying the appropriate temperature correction. Solid substances' content (SS), expressed as a percentage, was calculated as the difference between 100 and the moisture content. Specific gravity was determined by the gravimetric method, using a pycnometer device. The results were expressed in g/cm^3 [38,39].

2.4. pH and Free Acidity

The pH values were measured in a honey solution (10 g of honey in 75 mL of distilled water) using the MULTI 3320 multiparameter (WTW GMBH, Weilheim, Germany). Free acidity was determined by titration with 0.1 N NaOH (Chemical Company, Iasi, Romania) of a honey solution (10 g of honey in 75 mL of distilled water) using phenolphthalein (Chimreactiv, Bucuresti, Romania) as a color indicator. The free acidity was expressed in g/cm^3 [10,37,39].

2.5. Ash and Electrical Conductivity

Ten grams of each honey sample were weighed in porcelain crucibles and were calcinated in a furnace (Nabertherm B180, Nabertherm GMBH, Lilienthal, Germany), and the ash content was expressed in g/100 g. The 20% solution (the mass of honey was calculated as dry matter) with ultrapure water (Barnstead EASY PURE II, Thermo Fisher Scientific Co. Ltd.; Marietta, OH, USA) was measured with the MULTI 3320 multiparameter (WTW GMBH, Weilheim, Germany). The electrical conductivity was expressed in $mS\,cm^{-1}$ [37,39].

2.6. Total Phenols Content and Total Flavonoids Content

The extraction of the total phenols and total flavonoids was carried out with an alcoholic solution (1:1 equal parts of methanol (Merck KGaA, Darmstadt, Germany) and acidified water (deionized water at pH = 2 with HCl (Merck KGaA, Darmstadt, Germany)). The 10% honey solution obtained by mixing the corresponding amount of honey with the prepared alcoholic solution was homogenized and filtered through filter paper. An aliquot of filtered honey solution was mixed with 0.2 mL of Folin–Ciocalteu's phenol reagent (Merck KGaA, Darmstadt, Germany) for 5 min and 75 g/L of Na_2CO_3 (Merck KGaA, Darmstadt, Germany) was added to a total volume of 10 mL. The solution was spectrophotometrically analyzed at 742 nm, after incubating in the dark at room temperature for 30 min (Shimadzu UV-1700 Pharma Spec, Shimadzu Corporation, Analytical Instruments Division, Kyoto, Japan). The calibration curve was linear (y = 0.0993x + 0.0741; R^2 = 0.9991) in the concentration range of 2–12 mg L^{-1} gallic acid (Merck KGaA, Darmstadt, Germany). The total phenols content was expressed as mg of gallic acid equivalents (GAE)/100 g [10,40].

Total flavonoids were determined in the same alcoholic solution of honey prepared for the determination of total polyphenols. Equal volumes of 2% $AlCl_3$ (Merck KGaA, Darmstadt, Germany) and honey solution were mixed, and after 10 min, the absorbance was measured at 430 nm. A standard solution of quercetin (Sigma-Aldrich, Saint Louis, MO, USA) was prepared and used to obtain the calibration curve (concentration range 0.5–5 mg L^{-1}; y = 0.01330x + 0.0111; R^2 = 0.9998). The total flavonoids content was expressed as mg of quercetin equivalents (QE)/100 g [26,40].

2.7. Mineral Elements (K, Ca, Mg, Na, and P)

To determine the mineral elements' content, the ash resulting from the calcination of the samples was moistened with ultrapure water, evaporated in a sand bath, and after calcination for 6 h, treatment with 6 M HCl, and heating to evaporate the acid, the residue was dissolved in 0.1 M nitric acid. The extracts were then filtered, and ultrapure water was added to a total volume of 25 mL. The presence of possible contaminants during the digestion process was controlled using blanks. Phosphorus was spectrophotometrically determined with molybdovanadate reagent at 430 nm (Shimadzu UV-1700 Pharma Spec, Shimadzu Corporation, Analytical Instruments Division, Kyoto, Japan) [41]. The calibration curve was linear in the concentration range of 5–50 mg L^{-1} (y = 0.0209x + 0.0150; R^2 = 1.000). Ca and Mg were determined by flame atomic absorption spectrometry (Ca: y = 0.0264x + 0.0140, R^2 = 0.9995, 1–10 mg L^{-1}; Mg: y = 0.3997x + 0.0253, R^2 = 0.9991, 0.1–0.5 mg L^{-1}), and Na and K were determined by flame atomic emission spectrometry (Na: y = 0.0970x + 0.0017, R^2 = 0.9966, 1–10 mg L^{-1}; K: y = 0.1010x + 0.0128, R^2 = 0.9988, 1–10 mg L^{-1}) (Analytik Jena novAA 350, Analytik Jena GmbH, Jena, Germany).

2.8. FTIR Analysis

Infrared spectra were obtained by using a Jasco FT/IR-660 Plus Fourier Transform Infrared Spectrometer (Tokyo, Japan). The honey samples were liquefied at 40 °C and homogenized. A small quantity of the honey samples was incorporated into a KBr (Sigma-Aldrich, Darmstadt, Germany) pellet. Spectral measurements were recorded in the wavenumber over a domain from 4000 to 400 cm^{-1} with 32 scans, with a resolution of 4 cm^{-1} [30].

From the FTIR analysis, the raw data of 21 honey samples were saved in a file with ".jws" and ".txt" extensions. In each obtained spectrum, the transmittance vs. wavenumber was plotted. With the OriginPro 2022 software, the spectra of the honey samples were obtained and graphically displayed.

2.9. Statistical Analyses

All analyses were performed in triplicate. Statistical analyses were performed using the software package STATISTICA 12.0 (StatSoft Inc, Tusla, OK, USA). Correlations between the investigated parameters were tested using Pearson's correlation coefficient. Principal component analysis and hierarchical cluster analysis were used to obtain an overview of physicochemical parameters' contributions.

3. Results

3.1. Physicochemical Analyses

Figure 2 shows the color differences of the 21 honey samples and their classifications according to the Pfund scale.

In Table 2, the values of mm Pfund, refractive index, moisture, solid substances' content, and specific gravity are presented.

The results for pH, free acidity, ash, electrical conductivity, total phenols content, and total flavonoids content are presented in Table 3.

Table 4 lists the concentrations of mineral elements in the analyzed honeys.

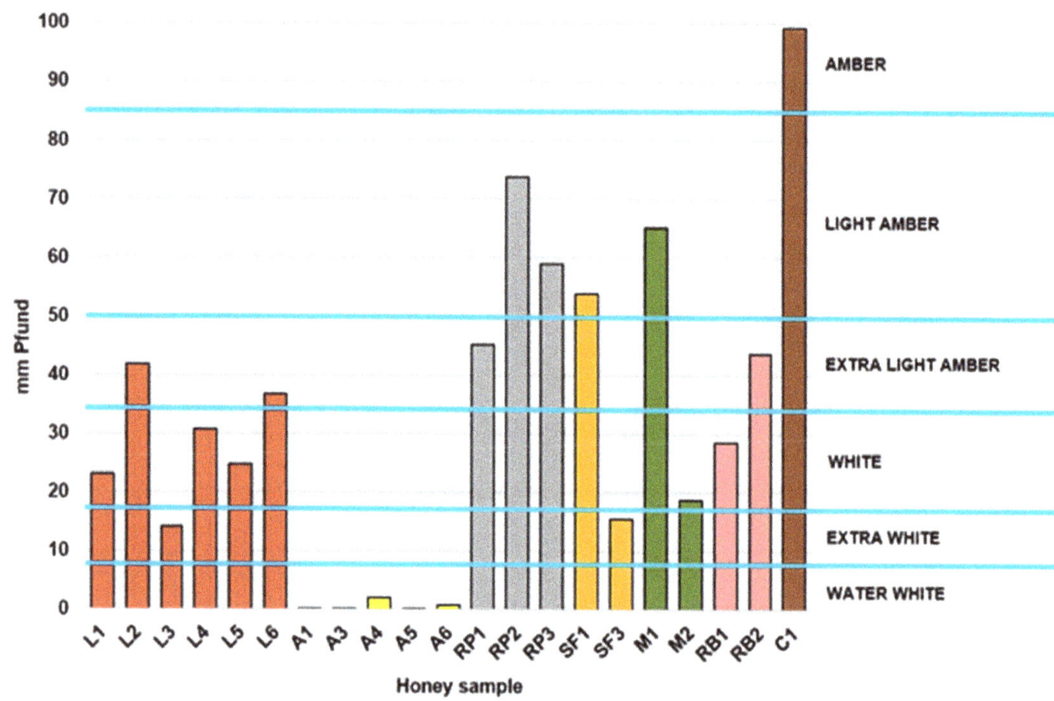

Figure 2. The color of the honey samples according to the Pfund scale.

Table 2. Parameter values (mm Pfund, refractive index, moisture, solid substances' content, and specific gravity) of seven types of honey.

Parameter	Descriptive Statistics	Type							ANOVA
		Linden 6 Samples	Acacia 5 Samples	Rapeseed 3 Samples	Sunflower 2 Samples	Mint 2 Samples	Raspberry 2 Samples	Chestnut 1 Sample	
mm Pfund	Min–Max	23.1–41.8	0.1–2.0	45.2–73.7	15.5–53.9	18.8–65.2	28.5–43.7	98.3–100.2	
	Mean ± SD	28.6 ± 9.94	0.6 ± 0.81	59.3 ± 14.30	34.7 ± 27.14	42.0 ± 32.83	36.1 ± 10.75	99.2 ± 0.59	***
	CV	34.81	132	24.12	78.27	78.17	29.77	0.59	
RI	Min–Max	1.486–1.494	1.488–1.493	1.486–1.492	1.488–1.493	1.490–1.491	1.490–1.492	1.497–1.498	
	Mean ± SD	1.489 ± 0.00	1.491 ± 0.00	1.488 ± 0.00	1.490 ± 0.00	1.490 ± 0.00	1.491 ± 0.00	1.497 ± 0.00	ns
	CV	0.20	0.15	0.24	0.21	0.05	0.09	0.01	
M %	Min–Max	17.0–20.4	17.5–19.5	17.8–20.4	17.6–19.3	18.3–18.7	17.8–18.7	15.9–16.0	
	Mean ± SD	18.9 ± 1.18	18.3 ± 0.92	19.5 ± 1.44	18.4 ± 1.24	18.5 ± 0.29	18.3 ± 0.64	15.9 ± 0.03	ns
	CV	6.24	5.04	7.36	6.75	1.55	3.49	0.21	
SS %	Min–Max	79.6–83.0	80.5–82.5	79.6–82.2	80.7–82.4	81.3–81.7	81.3–82.2	84.0–84.1	
	Mean ± SD	81.1 ± 1.18	81.7 ± 0.92	80.5 ± 1.44	81.6 ± 1.24	81.5 ± 0.29	81.8 ± 0.64	84.1 ± 0.03	ns
	CV	1.46	1.13	1.78	1.53	0.35	0.78	0.04	
SG g/cm³	Min–Max	1.415–1.437	1.421–1.434	1.415–1.432	1.422–1.434	1.426–1.429	1.426–1.432	1.444–1.445	
	Mean ± SD	1.420 ± 0.01	1.428 ± 0.01	1.421 ± 0.01	1.428 ± 0.01	1.28 ± 0.00	1.429 ± 0.00	1.445 ± 0.00	ns
	CV	0.55	0.44	0.68	0.59	0.14	0.30	0.02	

RI—refractive index, M—moisture, SS—solid substances, SG—specific gravity, SD—standard deviation, CV—coefficient of variation. Significant difference at: $p < 0.001$ (***), ns—not significant.

Table 3. Parameter values (pH, free acidity, ash, electrical conductivity, total phenols content, total flavonoids content) of seven types of honey.

Parameter	Descriptive Statistics	Type							ANOVA
		Linden 6 Samples	Acacia 5 Samples	Rapeseed 3 Samples	Sunflower 2 Samples	Mint 2 Samples	Raspberry 2 Samples	Chestnut 1 Sample	
pH	Min–Max	3.81–5.08	3.94–4.64	3.95–4.17	3.91–4.91	3.99–4.77	4.27–4.30	4.63–4.67	ns
	Mean ± SD	4.59 ± 0.44	4.17 ± 0.27	4.06 ± 0.11	4.41 ± 0.71	4.38 ± 0.55	4.29 ± 0.02	4.65 ± 0.01	
	CV	9.50	6.49	2.65	16.09	12.54	0.50	0.31	
FA meq kg^{-1}	Min–Max	23.4–38.6	12.8–25.4	24.3–46.6	35.6–50.1	38.2–45.7	25.8–42.0	49.6–49.9	**
	Mean ± SD	29.51 ± 5.20	19.8 ± 5.08	34.7 ± 11.21	42.9 ± 10.25	42.0 ± 5.33	33.9 ± 11.46	49.8 ± 0.11	
	CV	17.64	25.60	32.30	23.92	12.71	33.79	0.23	
Ash %	Min–Max	0.199–0.471	0.044–0.190	0.061–0.114	0.184–0.501	0.199–0.213	0.174–0.176	0.459–0.502	**
	Mean ± SD	0.33 ± 0.11	0.098 ± 0.06	0.088 ± 0.03	0.343 ± 0.22	0.206 ± 0.01	0.175 ± 0.00	0.483 ± 0.02	
	CV	34.75	64.12	30.56	65.45	4.81	0.81	3.30	
EC mS cm^{-1}	Min–Max	0.358–0.692	0.140–0.320	0.209–0.252	0.469–0.699	0.503–0.542	0.294–0.378	0.920–0.935	***
	Mean ± SD	0.594 ± 0.17	0.200 ± 0.08	0.2350.02±	0.584 ± 0.16	0.523 ± 0.03	0.336 ± 0.06	0.927 ± 0.01	
	CV	27.82	40.36	9.73	27.85	5.28	17.68	0.57	
TPC mg GAE/100g	Min–Max	22.0–27.5	13.7–23.1	22.4–24.1	21.10–23.90	50.3–58.7	28.2–33.5	72.5–74.1	***
	Mean ± SD	25.1 ± 2.42	17.9 ± 3.79	23.3 ± 0.87	22.50 ± 1.98	54.5 ± 5.93	30.9 ± 3.75	73.2 ± 0.51	
	CV	9.63	21.17	3.72	8.80	10.89	12.15	0.70	
TFC mg QE/100g	Min–Max	1.48–2.56	0.46–1.29	2.29–2.87	1.21–2.56	2.12–3.62	2.38–2.87	4.21–5.27	***
	Mean ± SD	1.81 ± 0.43	0.87 ± 0.32	2.43 ± 0.33	1.89 ± 0.95	2.87 ± 1.06	2.63 ± 0.35	4.81 ± 0.31	
	CV	24.0	36.18	13.35	50.64	36.96	13.20	6.41	

FA—free acidity, EC—electrical conductivity, TPC—total phenols content, TFC—total flavonoids content, SD—standard deviation, CV—coefficient of variation. Significant difference at: $p < 0.01$ (**), $p < 0.001$ (***), ns—not significant.

Table 4. Concentrations of mineral elements (K, Ca, Mg, Na, P) in seven types of honey.

Parameter	Descriptive Statistics	Type							ANOVA
		Linden 6 Samples	Acacia 5 Samples	Rapeseed 3 Samples	Sunflower 2 Samples	Mint 2 Samples	Raspberry 2 Samples	Chestnut 1 Sample	
K mg kg^{-1}	Min–Max	404–1507	75–213	46–196	455–1150	415–697	228–327	1455–1467	**
	Mean ± SD	984 ± 476	131 ± 53.96	101 ± 82.76	803 ± 491.44	556 ± 199.15	278 ± 70.00	1462 ± 1.71	
	CV	48.45	41.15	82.08	61.24	35.81	25.23	0.12	
Ca mg kg^{-1}	Min–Max	31.7–429.8	17.7–139.7	24.3–84.3	131.0–132.0	122.6–171.9	15.7–128.8	155.7–173.1	ns
	Mean ± SD	167.5 ± 143	62.2 ± 53.41	58.3 ± 30.79	131.5 ± 0.71	147.3 ± 34.86	72.3 ± 79.97	164.0 ± 5.50	
	CV	85.51	85.82	52.81	0.54	23.67	110.69	3.35	
Mg mg kg^{-1}	Min–Max	39.4–96.9	10.8–64.6	33.3–58.6	50.8–74.3	29.5–78.2	70.9–72.8	320.2–339.0	***
	Mean ± SD	59.4 ± 21.12	24.8 ± 22.52	46.2 ± 12.66	62.6 ± 16.62	53.9 ± 34.44	71.9 ± 1.34	330.6 ± 1.23	
	CV	35.54	90.80	27.38	26.57	63.95	1.87	0.37	
Na mg kg^{-1}	Min–Max	97.9–223.1	28.4–292.6	45.0–128.3	129.7–247.0	172.5–181.2	110.1–156.1	216.0–245.2	ns
	Mean ± SD	169.4 ± 49.41	94.5 ± 113.23	98.0 ± 46.03	188.4 ± 82.94	176.9 ± 6.15	133.1 ± 32.53	233.3 ± 5.79	
	CV	29.18	119.85	46.99	44.04	3.48	24.44	2.48	
P mg kg^{-1}	Min–Max	29.8–52.9	21.8–47.3	39.5–50.5	40.1–52.5	43.5–74.5	44.2–79.4	125.6–145.1	***
	Mean ± SD	42.1 ± 9.68	34.1 ± 9.69	45.4 ± 5.54	46.3 ± 8.77	59.0 ± 21.92	61.8 ± 24.89	137.2 ± 1.05	
	CV	22.98	28.41	12.21	18.94	37.15	40.28	0.76	

SD—standard deviation, CV—coefficient of variation. Significant difference at: $p < 0.01$ (**), $p < 0.001$ (***), ns—not significant.

Statistical differences based on the mean values of the determined parameters between all types of analyzed samples (one-way ANOVA) are listed in Tables 2–4. In Table 5, the statistical differences for the investigated quality parameters for pairs of different honey types are summarized.

3.2. FTIR Spectra

In Table 6, the band wavelengths of honey samples' spectra are presented. Figures 3 and 4 show the FTIR spectra of the analyzed honey samples in the 4000–400 cm^{-1} and 1700–400 cm^{-1} spectral regions.

3.3. Correlation and Multivariate Statistical Analysis

The correlations between all studied parameters are summarized in Table 7.

Table 5. Analysis of variance (ANOVA) for the determined quality parameters (pairs of honey types).

	mm Pfund	RI	M	SS	SG	pH	FA	Ash	EC	TPC	TFC	K	Ca	Mg	Na	P
L-A	***	ns	ns	ns	ns	ns	*	**	***	**	**	**	ns	*	ns	ns
L-RP	**	ns	ns	ns	ns	ns	ns	**	**	ns	ns	*	ns	ns	ns	ns
L-SF	ns	ns	ns	ns	ns	ns	*	ns	ns	ns	ns	ns	ns	ns	ns	ns
L-M	ns	ns	ns	ns	ns	ns	*	ns	ns	***	ns	ns	ns	ns	ns	ns
L-RB	ns	ns	ns	ns	ns	ns	ns	ns	ns	*	ns	ns	ns	ns	ns	ns
L-C	**	ns	ns	ns	ns	ns	*	ns	ns	***	**	ns	ns	***	ns	***
A-RP	***	ns	ns	ns	ns	ns	*	ns	ns	ns	***	ns	ns	ns	ns	ns
A-SF	*	ns	ns	ns	ns	ns	**	ns	**	ns	ns	*	ns	ns	ns	ns
A-M	*	ns	ns	ns	ns	ns	**	ns	**	***	**	**	ns	ns	ns	ns
A-RB	***	ns	ns	ns	ns	ns	ns	ns	ns	**	**	*	ns	*	ns	ns
A-C	***	ns	ns	ns	ns	ns	**	**	**	***	***	***	ns	***	ns	***
RP-SF	ns	ns	ns	ns	ns	ns	ns	ns	*	ns	ns	ns	*	ns	ns	ns
RP-M	ns	ns	ns	ns	ns	ns	ns	**	***	**	ns	*	ns	ns	ns	ns
RP-RB	ns	ns	ns	ns	ns	ns	ns	*	ns	*	ns	ns	ns	ns	ns	ns
RP-C	ns	ns	ns	ns	ns	*	ns	**	**	***	*	**	ns	**	ns	**
SF-M	ns	ns	ns	ns	ns	ns	ns	ns	ns	*	ns	ns	ns	ns	ns	ns
SF-RB	ns	ns	ns	ns	ns	ns	ns	ns	ns	ns	ns	ns	ns	ns	ns	ns
SF-C	ns	ns	ns	ns	ns	ns	ns	ns	ns	*	ns	ns	*	*	ns	ns
M-RB	ns	ns	ns	ns	ns	ns	ns	*	ns	*	ns	ns	ns	ns	ns	ns
M-C	ns	ns	ns	ns	ns	ns	ns	*	ns	ns	ns	ns	ns	ns	ns	ns
RB-C	ns	ns	ns	ns	ns	*	ns	**	ns	ns	ns	ns	*	ns	**	ns

Significant difference at: $p < 0.05$ (*), $p < 0.01$ (**), $p < 0.001$ (***), ns—not significant.

Table 6. Positions of absorption bands for the monofloral honey samples analyzed by FTIR.

Spectral Range	FT-IR Wavenumber (cm^{-1})						
	Linden	Acacia	Rapeseed	Sunflower	Mint	Raspberry	Chestnut
D1	3377–3416	3393–3415	3383–3398	3396–3408	3376–3388	3370–3416	3369
D2	2933–2934	2934–2935	2932–2933	2922–2933	2933	2929–2933	2933
D3	2117–2120	2118–2119	2115–2118	2120–2121	2116–2118	2116–2117	2119
D4	1641–1647	1636–1647	1638–1646	1637–1640	1637–1639	1640–1647	1638
D5	1452–1453	1452–1454	1453–1454	1450–1454	1452–1453	1453	1454
D6	1415–1416	1415–1418	1415–1417	1415–1418	1414–1416	1415–1416	1416
D7	1342–1350	1348–1350	1340–1342	1348	1342–1344	1339–1345	1343
D8	1256–1258	1256–1257	1256	1257	1255–1256	1255–1259	1256
D9	1144–1146	1144–1146	1144	1144–1145	1144	1144–1145	1145
D10	1055–1057	1054–1056	1056	1056	1056	1054–1057	1056
D11	919–920	919–920	918–919	919	919	919	919
D12	866–867	867	866–867	867	867–868	867	867
D13	819	819–819	818	818	818	818–819	819
D14	778–779	778–779	778	778	778	778–779	778

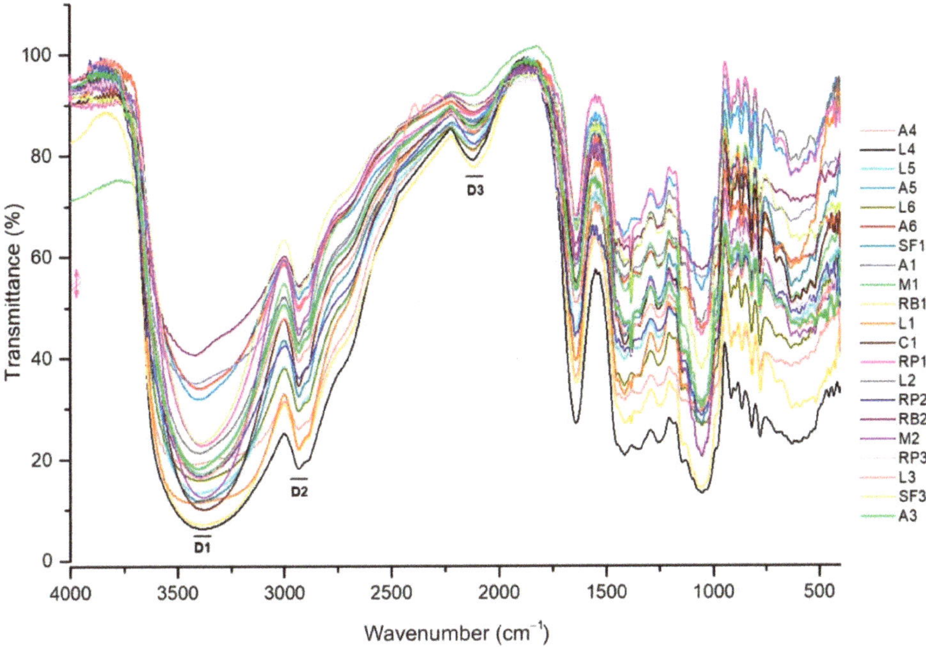

Figure 3. FTIR spectra of all studied honey samples in the spectral region from 4000 to 400 cm^{-1}.

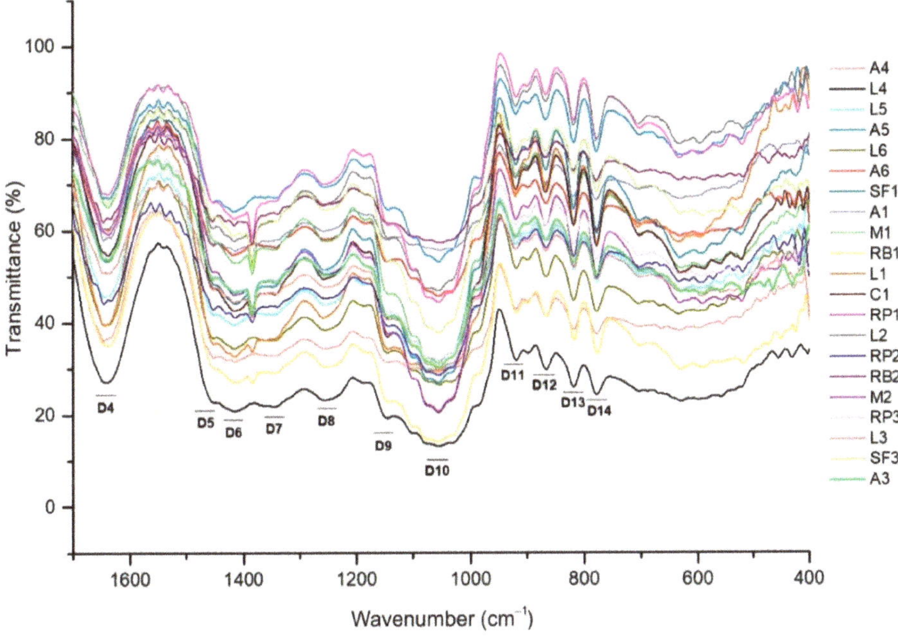

Figure 4. FTIR spectra of all studied honey samples in the spectral region from 1700 to 400 cm^{-1}.

Table 7. Pearson's correlation coefficients between the investigated honey parameters (significant at: $p < 0.05$ (*), $p < 0.01$ (**), $p < 0.001$ (***)). The values of significant correlation coefficients are marked in bold.

	RI	M	SS	Color	SG	pH	FA	Ash	EC	TPC	TFC	K	Ca	Mg	Na	P
RI	1.00															
M	**−0.99 *****	1.00														
SS	**0.99 ***	**−1.00 ***	1.00													
Color	0.61	−0.61	0.61	1.00												
SG	**0.99 ***	**−0.99 ***	**0.99 ***	0.61	1.00											
pH	0.62	−0.63	0.63	0.41	0.62	1.00										
FA	0.55	−0.55	0.55	**0.82 ***	0.55	0.52	1.00									
Ash	0.71	−0.72	0.72	0.58	0.71	**0.93 ***	0.68	1.00								
EC	0.74	−0.74	0.74	0.68	0.74	**0.92 ***	**0.76 ***	**0.97 ***	1.00							
TPC	**0.79 ***	**−0.79 ***	**0.79 ***	**0.78 ***	**0.80 ***	0.59	**0.75 ***	0.61	**0.75 ***	1.00						
TFC	0.73	−0.73	0.73	**0.95 ***	0.73	0.50	**0.83 ***	0.61	0.72	**0.90 ***	1.00					
K	0.70	−0.71	0.71	0.60	0.70	**0.95 ***	0.65	**0.98 ***	**0.98 ***	0.66	0.63	1.00				
Ca	0.45	−0.46	0.46	0.40	0.45	**0.93 ***	0.58	**0.85 ***	**0.89 ***	0.59	0.46	**0.90 ***	1.00			
Mg	**0.91 ***	**−0.91 ***	**0.91 ***	**0.84 ***	**0.91 ***	0.64	0.67	**0.77 ***	**0.81 ***	**0.82 ***	**0.88 ***	**0.78 ***	0.51	1.00		
Na	0.71	−0.72	0.72	0.63	0.72	**0.90 ***	**0.81 ***	**0.95 ***	**0.98 ***	**0.75 ***	0.70	**0.94 ***	**0.89 ***	0.75	1.00	
P	**0.91 ***	**−0.91 ***	**0.91 ***	**0.86 ***	**0.91 ***	0.59	0.72	0.71	**0.78 ***	**0.89 ***	**0.93 ***	0.72	0.47	**0.98 ***	0.73	1.00

The results of the principal component analysis are shown in Table 8 and Figure 5. In Figure 6, the hierarchical dendrogram of cluster analysis is presented.

Table 8. Loadings and explained variance (%) for the extracted principal components for the analyzed honey samples. The values of significant correlation coefficients are marked in bold.

Variable	PC 1	PC 2	PC 3
RI	**0.98**	0.03	0.15
M	**−0.99**	−0.03	−0.15
SS	**0.99**	0.03	0.15
Color	0.05	0.00	**0.93**
SG	**0.98**	0.02	0.15
pH	0.12	**0.91**	−0.23
FA	−0.06	0.03	**0.88**
Ash	0.00	**0.94**	0.19
EC	0.12	**0.90**	0.36
TPC	0.31	0.31	**0.73**
TFC	0.12	0.02	**0.98**
K	0.15	**0.88**	0.18
Ca	−0.31	0.66	0.04
Mg	0.31	0.47	0.70
Na	0.00	**0.81**	0.10
P	0.33	0.16	**0.86**
Eigenvalue	6.85	4.02	2.89
% Total variance	42.83	25.15	18.04
Cumulative %	42.83	67.97	86.01

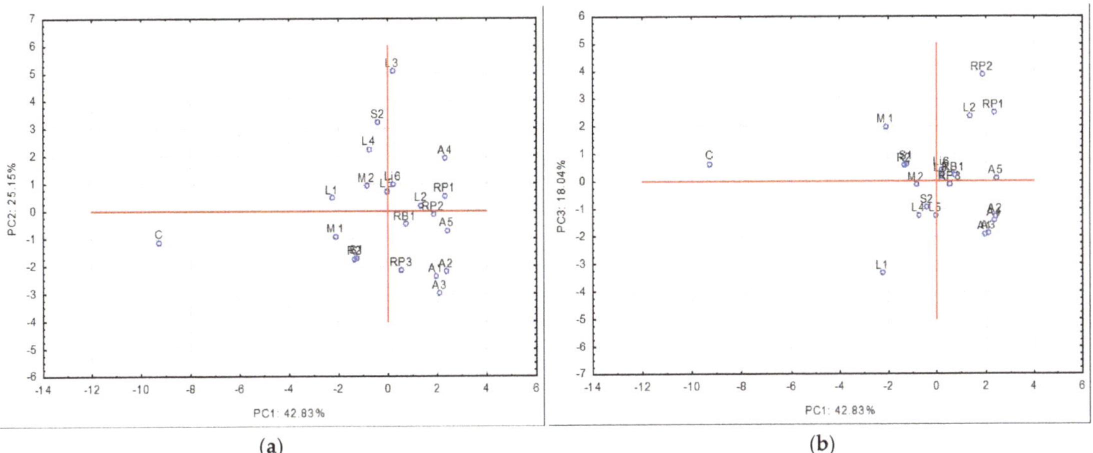

Figure 5. Graphical representation of PC for the analyzed honey samples: (**a**) scores for PC2 vs. PC1 and (**b**) scores for PC3 vs. PC1.

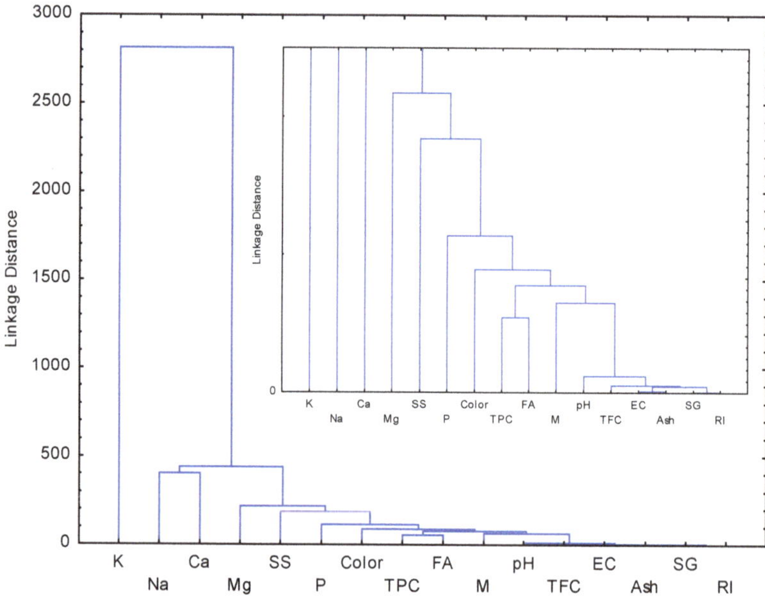

Figure 6. The hierarchical dendrogram of cluster analysis for the analyzed honey samples based on the determined quality parameters.

4. Discussion

4.1. Color

The first opinion on the quality of a food product for a consumer is the color, and subsequently the flavor and taste. Consumer preferences on honey vary, some prefer light-colored honey and others dark-colored, and these preferences can affect the price [4]. The colors of honey are grouped into seven categories: water white, extra white, white, extra light amber, light amber, amber, and dark amber, without considering the variations of shades that they can have. There are several methods to determine the color of honey, such as Pfund, Lovibond, and Jack's scale color grader [42]. The color of honey is directly influenced by various factors, such as water content and some chemical compounds (phenolic, carotenoids, minerals), pollen floral types, and geographical origin, and indirectly by technological conditions (temperature, processing/handling/storage), time, etc. [8,42–44]. In this study, the average honey color values ranged from 0.1 mm Pfund for acacia honey to 100.2 mm Pfund for chestnut honey (Table 1). The color of linden honey varied between 23.1 and 41.80 mm Pfund (extra white–extra light amber). High differences were found for mint honey of 46.4 mm Pfund and of 38.4 mm Pfund for sunflower honey. Similar high differences were found by Ratiu et al. for rapeseed and sunflower honey from Poland (Table 9) [17]. All acacia samples have low values of mm Pfund, such as those observed by Flanjak et al. for acacia honey from Croatia [45]. It is known that chestnut honey is dark-colored, as also shown in other studies when higher values were reported than those found in this study [5,42] (Table 9). The results in the present study showed the highest concentration of antioxidant compounds in the dark-colored honey. Considering the variability in the global dataset, the ANOVA indicated significant differences at the $p < 0.001$ level (Table 2), mainly based on the statistically significant differences between acacia honey and the other monofloral honey types (Table 5) and also between chestnut and linden honeys.

Table 9. Comparative results on color intensity (mm Pfund) of different floral-type honeys.

mm Pfund							Country	Literature Source
Linden	Acacia	Rapeseed	Sunflower	Mint	Raspberry	Chestnut		
23.1–41.8	0.1–2.0	45.2–73.7	15.5–53.9	18.8–65.2	28.5–43.7	98.3–100.2	Romania	This study
-	-	-	61.3; 70.5; 88.7	-	-	-	Romania	[46]
36.00–54.00	11.00–45.00	-	79.00–83.00	-	-	-	Romania	[47]
-	-	-	32.87–47.52	-	-	-	Romania	[2]
-	-	29.40	37.60	74.30	61.4	-	Romania	[8]
35.64	12.87	36.14	33.66	63.86	63.36	-	Romania	[15]
-	-	-	97.60	-	-	-	Portugal	[36]
22.00–38.00	-	-	39.00–41.00	-	-	-	R. Moldova	[48]
10.00–29.00	1.00–8.00	-	-	-	-	-	Croatia	[45]
68.01; 76.80	-	34.34–114.07	62.07; 114.00	-	87.45; 91.29	-	Poland	[17]
68.1	10.2; 21.47	85.47–114.07	-	-	101.56; 87.45	-	Czech Republic/Poland	[17]
-	26.51	-	-	-	-	-	Germany	[49]
70.00	20.00	-	-	-	-	-	Serbia	[43]
-	-	-	-	-	39–74.5	-	Romania	[50]
-	-	-	-	-	118	-	Spain	[16]
38.27–139.48	−7.26–20.92	2.08–138.56	96.3–198.91	-	-	71.3–149.42	Hungary	[42]
-	-	-	-	-	-	123–150	Spain	[5]
-	-	-	-	-	-	69.63–108.7	Portugal	[51]
33.3	12.9	26.2	52.4	-	-	87.9	Europe	[52]

4.2. Refractive Index, Moisture, and Solid Substances, and Specific Gravity

The main compounds of honey are sugars, and therefore one of its properties is to be hygroscopic. A high moisture content of honey can cause fermentation and spoilage, and these processes negatively change the physicochemical properties and undoubtedly lower the quality and price, shortening the shelf life of the product [9,10,17,21,49,52]. The moisture content of honey samples can be determined by measuring the refractive index. The refractive index of honey is directly related to the content of solid substances. The values of the refractive index (RI) were found between 1.486 and 1.498, with the range of the moisture content between 15.9% for chestnut honey and 20.4% for two samples of linden and rapeseed honey (Table 2). Three honey samples had moisture over the maximum moisture content limit of 20%, recommended for honey in international regulations and Romanian standards [11,37] (20.3% and 20.4% for RP1 and RP2, 20.4% for L3). The various values of moisture content (Table 2) can be attributed to factors such as: weather, harvesting and manipulation procedures, storage, etc. Various values, from 3.9% in acacia honey [53] to 22.8% in acacia honey [54], were also obtained in other studies conducted on the seven honey types analyzed, taken from different countries (Table 10). Honey is a viscous product, having a higher density than water. The specific gravity depends on the moisture content, and it is important to know the stored quantity of honey. The lowest specific gravity of 1.415 g/cm^3 was obtained for linden honey and the highest specific gravity of 1.445 g/cm^3 for chestnut honey. No significant differences were observed between the mean values of all four parameters, neither at the global level (by testing the variability in the dataset of seven honey types) nor between the pairs of honey types (Tables 2 and 5).

Table 10. Comparative physicochemical properties of seven types of monofloral honeys.

Country	Moisture (%)	pH	Free Acidity (meq kg^{-1})	Ash (%)	EC (mS cm^{-1})	TPC mg GAE/100 g	TFC mg QE/100 g	Literature Source
				Linden				
Romania	17.0–20.4	3.81–5.08	23.4–38.6	0.199–0.471	0.358–0.692	22.0–27.5	1.48–2.56	This study
Romania	5.4–6.0	3.6–4.7	-	-	0.410–0.730	-	-	[53]
Romania	16.70–19.10	-	-	0.19–0.30	-	16–38	4.7–6.98	[47]
Romania	16.75	4.05	14.55	-	0.33	-	-	[15]
Romania	17.2–18.8	3.84–4.35	-	-	0.202–0.346	-	-	[54]

Table 10. *Cont.*

Country	Moisture (%)	pH	Free Acidity (meq kg^{-1})	Ash (%)	EC (mS cm^{-1})	TPC mg GAE/100 g	TFC mg QE/100 g	Literature Source
				Linden				
Bulgaria	17.1	4.04	-	-	0.689	-	-	[55]
Croatia	15.9–20.0	-	-	-	0.497–0.628	6.62–12.10	-	[45]
Czech Republic	16	4.06	14.9	-	0.39	45.04	1.88	[13]
Italy	-	-	-	-	-	26	5.5	[56]
Poland	20.30	4.13	25.50	-	0.640	43.69	-	[14]
Poland	17.76	3.81	34.2	-	0.53	38	0.5	[20]
Serbia	15.8; 17.1	4.62; 4.72	14.5; 16.1	-	0.488; 0.608	53.7; 67.3	-	[57]
Serbia	-	-	-	-	-	71.49	-	[43]
Slovakia	18.35	3.90	21.6	-	0.23	35	0.26	[20]
Romania	18.54	4.83	5.88	-	0.512	-	-	[58]
				Acacia				
Romania	17.5–19.5	3.94–4.65	12.8–25.4	0.044–0.190	0.140–0.320	13.7–23.1	0.46–1.29	This study
Romania	3.9–6.2	3.7–4.3	-	-	0.110–0.270	-	-	[53]
Romania	16.60–19.80	-	-	0.03–0.28	-	2.00–39.00	0.91–2.42	[47]
Romania	15.96	4.31	3.86	-	0.12	-	-	[15]
Romania	16.7–22.8	3.65–4.63	-	-	0.097–0.268	-	-	[54]
Bulgaria	16.9	3.23	-	-	0.159	-	-	[55]
Croatia	14.6–19.9	-	-	-	0.1–0.161	2.82–5.20	-	[45]
Czech Republic	17	3.82	9.6	-	0.18	23.84	0.87	[13]
Germany	17	5.4	-	-	-	62.75	-	[59]
Germany	18.83	4.10	-	-	-	21.457	-	[49]
Italy	-	-	-	-	-	18.2	7.6	[60]
Italy	-	-	-	-	-	10.72	3.31	[56]
Poland	17.17	3.77	20.85	-	0.31	14.081	-	[14]
Poland	17.73	3.79	25.6	-	0.42	47	0.32	[20]
Serbia	14.5–18.5	-	6.6–15.5	0.04–0.15	0.083–0.174	58.17–142.61	-	[61]
Serbia	16.4; 17.3	3.90; 4.51	13.8; 16.3	-	0.114; 0.136	13.5; 14.4	-	[57]
Serbia	-	-	-	-	-	37.93	-	[43]
Slovakia	17.86	3.71	16.1	-	0.20	20	0.14	[20]
Turkey	14.45–21.62	-	12–21	-	0.14–0.27	1–3	-	[62]
Croatia	16.78–17.01	-	10.45–11.02	-	0.15–0.18	-	-	[63]
Romania	18.02	4.02	2.29	-	0.218	-	-	[58]
				Rapeseed				
Romania	17.8–20.4	3.95–4.17	24.3–46.6	0.061–0.114	0.209–0.252	22.4–24.1	2.29–2.87	This study
Romania	5.1–5.8	3.6–3.9	-	-	0.150–0.285	-	-	[53]
Romania	18.51	4.23	15.26	-	0.168	-	-	[58]
Romania	17.31	4.11	17.33	-	0.15	-	-	[15]
Bulgaria	19.7	3.33	-	-	0.181	-	-	[55]
Poland	17.86	3.88	18.6	-	0.23	25	0.32	[20]
Serbia	18.4; 19.4	4.01; 4.10	16.3; 21.3	-	0.191; 0.224	11.5; 11.9	-	[57]
Slovakia	17.45	3.61	13.6	-	0.16	21	0.14	[20]
				Sunflower				
Romania	17.6–19.3	3.91–4.91	35.6–50.1	0.184–0.501	0.469–0.699	21.10–23.90	1.21–2.56	This study
Romania	18.82	4.12	11.82	-	0.367	-	-	[58]
Romania	4.7–6.6	3.3–3.8	-	-	0.340–0.475	-	-	[53]
Romania	18.7	3.656	22.36	0.112	0.301	-	-	[64]
Romania	-	-	-	-	-	48.6–132.5	-	[46]
Romania	17.80–19.70	-	-	0.35–0.40	-	20.00–45.00	11.53–15.33	[47]
Romania	16.23–20.39	3.65–4.34	15.94–47.32	-	0.315–0.441	-	-	[2]
Romania	18.4	3.94	31.6	-	0.362	21.1	22.8	[8]
Romania	16.95	4.04	18.32	-	0.31	-	-	[15]
Romania	17	3.67	-	-	0.188	-	-	[54]
Portugal	19.2	3.84	25.50	0.15	0.235	36.69	1.93	[36]
R Moldova	16.05–17.52	3.68–4.05	-	0.31–0.49	-	-	-	[48]
Serbia	17.4–19.8	-	18.5–39.4	0.12–0.30	0.189–0.359	25.45–61.09	-	[61]
Serbia	17.0	3.38	28.9	-	0.366	27.5	-	[57]
Serbia	14.6–18.6	-	20.40–36.4	0.05–0.30	0.22–0.54	-	-	[65]
Morocco	16.9–18.5	3.52–3.8	15.3–36.7	0.12–0.20	0.43–0.52	-	-	[18]

Table 10. *Cont.*

Country	Moisture (%)	pH	Free Acidity (meq kg^{-1})	Ash (%)	EC (mS cm^{-1})	TPC mg GAE/100 g	TFC mg QE/100 g	Literature Source
				Linden				
				Mint				
Romania	18.3–18.7	3.99–4.77	38.2–45.7	0.199–0.213	0.503–0.542	50.3–58.7	2.12–3.62	This study
Romania	17.91	4.19	26.71	-	0.466	-	-	[58]
Romania	17.7	4.20	26.9	-	0.474	23.7	25.7	[8]
Romania	16.24	4.52	33.17	-	0.60	-	-	[15]
Tunisia	19.80	-	-	0.13	0.43	119.42	-	[12]
Morocco	15.6–18.3	3.53–4.07	26.6–32	0.18–0.23	0.350–0.505	-	-	[18]
				Raspberry				
Romania	17.8–18.7	4.27–4.30	25.8–42.0	0.174–0.176	0.294–0.378	28.2–33.5	2.38–2.87	This study
Romania	17.32–20.12	4.01–4.31	20.1–42.1	-	0.367–0.528	14.48–25.72	25.36–41.35	[50]
Romania	18.3	4.16	27.3	-	0.446	19.9	35.5	[8]
Romania	17.27	4.27	24.06	-	0.519	-	-	[15]
Romania	18.35	4.16	27.31	-	0.439	-	-	[58]
Poland	-	-	-	-	-	109.1	-	[19]
				Chestnut				
Romania	15.9–16.0	4.63–4.67	49.6–49.9	0.459–0.502	0.920–0.935	72.5–74.1	4.21–5.27	This study
Spain	15.83	5.31	26.25	-	1.16	35.41	-	[16]
Romania	18.2	5.10	11.2	0.63	1.21	-	-	[66]
Georgia	18.2–20.9	4.5–4.97	20.8–44.6	0.89–2.71	1.036–1.667	80.0–461.5	-	[67]
Spain	-	3.93–4.65	-	-	0.737–1.235	88.43–166.45	6.60–11.78	[5]
Spain	17.6–18.2	4.5–4.7	-	-	1.0–1.1	121.5–138.2	8.4–9.6	[68]
Portugal	15.23–16.87	4.35–4.42	19.67	-	0.98–1.14	67.88–73.39	-	[51]
Turkey	17.4–19.5	4.80–5.34	-	0.74–0.80	1.30–1.52	76.20–94.05	4.20–6.50	[69]

4.3. pH and Free Acidity

Raw honey can be collected at different maturity stages, in different seasons, and its pH ranges between 3.5 and 5.5. The pH values of the analyzed honey samples ranged between 3.81 and 5.08. The concentration of organic acids lowers the pH of honey, giving it an acidic character [10,14,65]. The more acidic the honey is, the more the fermentation process and its alteration are avoided because the medium does not favor the development and growth of microorganisms. The maximum recommended value for free acidity is 50 meq kg^{-1} [11]. In this research, only one sunflower honey sample with 50.1 meq kg^{-1} exceeded the regulated limit. As can be seen from Table 3, the lowest average value of 19.8 meq kg^{-1} for free acidity was determined in acacia honey samples and the highest average value of free acidity was found in the chestnut honey sample (49.8 meq kg^{-1}). Large intervals of free acidity were also reported in other studies, from 2.29 to 47.32 meq kg^{-1} (Table 10). While the pH of the different types of honey did not significantly differ, the free acidity showed a significant difference at the $p < 0.01$ level (Table 3), mainly due to the differentiation of acacia and linden honeys from the other types (Table 5).

4.4. Ash and Electrical Conductivity

The ash content is formed by all the minerals in the honey, an inorganic residue obtained after calcination. In general, blossom honey has an ash content lower than 0.6%, [22,70]. The lowest average mineral content of 0.088% was found in rapeseed samples and the highest average mineral content was found in chestnut honey samples (0.483%). Mărghitaş et al. found the lowest ash content of 0.03% in acacia honey, and in their study, Kharadze et al. reported a 2.1% ash content in chestnut honey (Table 10) [47,67]. Due to its strong correlation with ash, electrical conductivity was included in new quality standards. It is an important parameter, that helps to identify the honey origin, and therefore to differentiate the blossom honey from the honeydew. Electrical conductivity measures inorganic and ionizable organic substances [8,12,17,18,23,65]. The lowest average value of electrical conductivity was found in acacia honey (0.200 mS cm^{-1}) and the highest average value was for chestnut honey (0.927 mS cm^{-1}). All the analyzed samples complied with the quality criteria of electrical conductivity below 0.8 mS cm^{-1}, the maximum value established by legislation for blossom honey, with the exception of honeydew, chestnut honey, and blends [8,11]. Acacia honey, in most cases, had the lowest amount of ash, and therefore, the electrical conductivity was low. The highest electrical conductivity values for chestnut honey were reported in many studies, within a range between 0.737 and 1.667 mS cm^{-1} (Table 10). The honey samples showed significant differences in terms of ash ($p < 0.01$) and electrical conductivity ($p < 0.001$) when considering the global dataset (Table 3), with the higher differences in conductivity between linden and acacia, and rapeseed and mint, respectively (Table 5).

4.5. Total Phenols Content and Total Flavonoids Content

Honey is known as food and as a natural medicine, used due to its anti-inflammatory, antioxidant, and antibacterial properties. In its composition, there are natural compounds that yield strong antioxidant properties, such as the polyphenolic compounds (phenolic acids, catechins, flavonoids, etc.) [27,71]. These compounds qualitatively and quantitatively varied, and are directly linked with the rich flora, the environment, and the area around the beehive, due to their plant–honey–nectar traceability [13,19,43,44,49,56,72,73]. The chestnut honey sample had the highest average concentration of phenols of 73.2 mg GAE/100 g (Table 3). For the same type of honey from Portugal, similar results in the range of 67.88–73.38 mg GAE/100 g were found by Karabagias et al. [51]. High values of total polyphenols content for chestnut honey from Spain and Georgia ranging from 80.0 to 461.5 mg GAE/100 g were reported by Kharadze et al.; Escuredo et al.; and Rodríguez-Flores et al. [5,67,68]. In acacia honey, total polyphenols were within the range of 13.7–23.1 mg GAE/100 g (Table 3). Similarly, the lowest concentration of total flavonoids (0.87 mg QE/100 g) was determined for acacia honey and the highest content of 4.81 mg QE/100 g in chestnut honey. There was a presence of high amounts of antioxidants in the mint honey of 54.4 mg QE/100 g for polyphenols and 2.87 mg QE/100 g for flavonoids and slightly lower for raspberry honey of 30.9 mg QE/100 g for polyphenols and 2.63 mg QE/100 g for flavonoids. Compared to the results obtained in this study, Pauliuc and Oroian and Pauliuc et al. reported high levels of flavonoids content between 25.36 and 41.35 mg QE/100 g for raspberry honey [8,50].

As summarized in Table 10, studies on antioxidants (TPC, TFC) showed their variable contents in various monofloral honeys. These findings were confirmed by the results of the ANOVA, which indicated significant differences ($p < 0.001$) in the concentrations of both classes of antioxidant compounds in the global dataset (Table 3). For linden honey, TPC ranged between 6.62 and 71.49 mg GAE/100 g, for acacia honey between 1 and 142.61 mg GAE/100 g, for sunflower honey within the range of 20–132.5 mg GAE/100 g, for mint honey between 23.7 and 119.42 mg GAE/100 g, for raspberry honey in the range of 14.48–109.1 mg GAE/100 g, and the highest concentrations were noticed for chestnut honey, between 35.41 and 461.5 mg GAE/100 g. Similarly, different values of TFC were reported: in the range from 0.26 to 6.98 mg QE/100 g for linden honey, from 0.14 to 7.6 mg QE/100 g for acacia honey, and from 1.21 to 22.8 mg QE/100 g for sunflower honey. The highest TFC values were in the range of 2.38 to 41.35 mg QE/100 g in raspberry honey (Table 10). Important factors with a high possibility of influence on the variability in the dataset could be climatic and soil conditions (with large seasonal and yearly variations), as well as the quality of pollen from plants that have different botanical origins and are found in different geographical regions. Different analytical methodologies used to determine these compounds or analytical equipment as possible sources of variation should also be considered [15,51].

However, in all analyzed honey samples, antioxidant compounds were found in variable amounts, which once again proves and confirms the anti-inflammatory, antioxidant, and antibacterial properties of the studied honey samples.

4.6. Mineral Elements' Content

Carbohydrates and water are the main quantitative components of honey, but other important substances such as vitamins and minerals were found in small amounts. Most of the honey components come from plants, but minerals derive from soil following the path of soil–plant–nectar–honey or beehive products [74]. The composition and amount of minerals depend on the environment (soil, water, air), their availability, climatic conditions, botanical origin, and the procedure of harvesting and storage [68,74–76].

The minerals in honeys are highly bioaccessible and play a positive role in human nutrition, prevention of illness, and healing the body. Macro-mineral elements (Ca, K, Na) and trace minerals (Cu, Zn, Mg, Fe) are important in biological systems. Potassium and sodium are electrolytes, and they maintain fluid balance in the body and help the heart and muscle functions. Na also has an essential role in kidney function, the maintenance of optimum blood pressure, and nerve functions. Calcium is involved in mineral homeostasis and physiological performance, has many roles (important for bones, for healing fracture, is indicated in prophylaxis for osteoporosis, and confers a protective role in the musculoskeletal, nervous, and cardiac systems), and acts as a cofactor for several enzymes. Calcium and phosphorus confer a protective role, making honey less cariogenic. Magnesium is a key mineral in honey, which plays an indispensable function in muscle contraction and in transmission of electrical impulses between neurons and contributes to the growth and support of bones and acts as a cofactor for many enzymes, most of which are involved in antioxidant reactions. Mg deficiency contributes to aging and age-related disorders [77–79].

The content of potassium ranged from 46 mg kg^{-1} in rapeseed honey to 1507 mg kg^{-1} in linden honey. The mean concentration of K in the seven studied honey types followed the order: C > L > SF > M > RB > A > RP. The concentrations of calcium were within the range of 58.3–167.5 mg kg^{-1}, following the order: L > C > M > SF > RB > A > RP. The content of magnesium was between 24.8 and 330.6 mg kg^{-1}, the highest in chestnut honey, followed in decreasing order by raspberry honey, sunflower honey, linden honey, mint honey, rapeseed honey, and acacia honey. The lowest amount of sodium (94.5 mg kg^{-1}) was in acacia honey, followed in increasing order by rapeseed honey, raspberry honey, linden honey, mint honey, sunflower honey, and chestnut honey (233.3 mg kg^{-1}). For phosphorus, the lowest average content of 34.1 mg kg^{-1} was found for acacia honey, followed in increasing order by linden honey, rapeseed honey, sunflower honey, mint honey, raspberry honey, and chestnut honey (137.2 mg kg^{-1}) (Table 4). The most abundant element in this study was potassium, and the results are confirmed by several other studies (Table 11). The amount of sodium in the studied honeys was higher compared with the results reported in other studies [55,78,80–83]. The higher concentrations of sodium found in the honey samples collected in this study could be explained by the presence of plants in soils rich in sodium and salts in certain areas in Romania [84]. In the present study, the highest mineral content was determined for potassium, followed by natrium, calcium, and magnesium, and the lowest concentration for phosphorus. Based on the variability in the global dataset, the ANOVA indicated significant differences for Mg and P at the $p < 0.001$ level, and for K at the $p < 0.01$ level (Table 4). No significant differences were observed between the mean values of Ca and Na by testing the variability in the dataset of the seven honey types (Table 4).

In general, the large variations of the studied elements' concentrations were attributed to the composition of the soil, the availability of minerals in the soil, the physiology of the plant, and other factors that can positively or negatively influence the transport of minerals to the nectar. The mineral composition of honey correlates with its color. Dark honeys (such as chestnut honey) contain higher amounts of certain major minerals, such as Ca, K, Mg, and Na, compared with light-colored honeys [79].

Table 11. Mineral elements of the seven floral honeys.

Country	K (mg kg^{-1})	Ca (mg kg^{-1})	Mg (mg kg^{-1})	Na (mg kg^{-1})	P (mg kg^{-1})	Literature Source
			Linden			
Poland	925.2	63.1	28.1	80	-	[75]
Slovenia	1510–2290	48.1–62.5	-	2.9–4.3	-	[80]
Bulgaria	290	46.4	11.5	13.6	-	[85]
Hungary	1027–1883	15.2–67.4	19.8–30.2	5.1–7.4	23.0–42.4	[86]
Croatia	1574.8	387.8	25.5	31.9	-	[87]
Poland	1071.6–2311.6	47.8–102.5	18.9–41.2	-	-	[76]
Romania	955.3	137.9	50.6	123.8	-	[88]
Bulgaria	792	77	21	7.5	49	[55]
Romania	494–735	35.5–76.5	15.7–20.5	22.1–51.1	-	[89]
Hungary	1278	67.9	16.5	9.3	41.5	[81]
Poland	224–528	25.5–48.0	7.3–27	9.2–47.6	35.8–23.8	[82]
			Acacia			
Italy	506	15	5	4.1	-	[90]
Hungary	10–255	17.6–59.6	1.90–15.9	1.60–11.5	27.7–92.3	[86]
Poland	127–196	28.6–69.2	6.5–14	3.8–42.8	71.4–28.6	[82]
Poland	587.2	52.6	24	53.8	-	[75]
Romania	146.7–244.6	1.02–6.9	3.25–6.7	5.06–24.3	-	[74]
Bulgaria	250	46.9	13.1	62.1	-	[85]
Croatia	258.7–360.8	74.4–184.4	16.8–30.7	51.6–168.9	-	[63]
Poland	221.6–431.1	41.7–68.6	16. 7–28.6	-	-	[76]
Romania	553.9	52.9	51.2	171.2	-	[88]
Serbia	188–340	3.8–8.4	3.8–6.8	21–124.3	33–120	[91]
Romania	180.2–252.4	50.1–55.2	11.1–18.1	21.3–24.3	-	[92]
Romania	356–521	5.2–10.2	7.1–18.3	1.9–32.7	-	[89]
Hungary	226.6	12.4	5.2	6.0	24.9	[81]
Bulgaria	126	32	6	8.1	24	[55]
			Rapeseed			
Hungary	103–288	23.7–60.4	13.5–27.6	6.3–16.9	50.9–71.1	[86]
Poland	84.8–494	22.1–62.4	9.5–32.1	7.0–26	35.7–161	[82]
Poland	265.2	48.9	19.2	31.3	-	[75]
Poland	221.6–431.1	41.7–68.6	16.7–28.6	-	-	[76]
Romania	112.6–194.2	87.1–88.6	23.5–23.9	36.1–47.9	-	[92]
Bulgaria	105	46	11	8.5	28	[55]
			Sunflower			
Portugal	276.9	24.9	68.2	87.9	-	[36]
Hungary	245–552	58.2–153	10.2–36.6	4.66–24.5	59.8–144	[86]
Hungary	759	126.4	33.3	13.2	76.3	[81]
Bulgaria	210–260	42.2–56.8	6.9–11	9.5–10.2	-	[85]
Hungary	446.3–790.2	-	24.2–38.7	-	-	[93]
Romania	849.4	163.9	63.8	154.1	-	[88]
Romania	552–574	36.6–60.4	20.4–23.1	24.9–35.2	-	[89]
Romania	234.6–532.1	152.4–200.5	32.6–39.3	47.1–50.7	-	[92]
			Mint			
Tunisia	976.8	221.1	78.1	343.6	59.3	[12]
Spain	200–280	123–148	27.4–36.8	-	-	[18]
Romania	-	1603.5	427.9	-	-	[94]
			Raspberry			
Poland	1104.7	68.8	47.6	48.1	-	[75]
Estonia	125.8–292.7	29.2–53.9	12.1–20.9	4.8–9.7	-	[83]

Table 11. *Cont.*

Country	K (mg kg^{-1})	Ca (mg kg^{-1})	Mg (mg kg^{-1})	Na (mg kg^{-1})	P (mg kg^{-1})	Literature Source
			Chestnut			
Italy	3875	119	49	11.9	-	[90]
Italy	706–714	54–55.9	48.9–49.9	7.5–8.2	143	[82]
Italy	3250–5280	60–130	-	60–90	-	[95]
Slovenia	3670–5520	117–183	-	7.1–9.0	-	[80]
Spain	1615–3770	68–476	30–402	11–84	48–315	[68]
Croatia	2824.4	486.7	59.1	35.8	-	[87]
Hungary	2136–2281	51.6–59.7	25.4–31.7	10.8–18.3	66.4–84.7	[86]
Hungary	1815.8	153	45.4	20.9	79	[81]
Bulgaria	1628	66	16	9.55	32	[55]
Turkey	2524–5125	320.24–463.10	32.05–67.10	28.3–52.0	56.20–71.02	[69]

4.7. FTIR Spectra

The FTIR spectral fingerprints of the seven types of analyzed honeys are shown in Figures 3 and 4. Intense signals were identified in the spectral ranges of 3400–3200 cm^{-1} and 1700–800 cm^{-1}. Many studies carried out on honey have shown that the obtained spectra can be studied by dividing them into spectral regions depending on the vibration of the functional groups:

D0—3500–3100 cm^{-1}, assigned to: O–H stretching (carboxylic acids) and NH$_3$ stretching (free amino acids).

D1—3000–2800 cm^{-1}, assigned to: C–H stretching (carbohydrates).

D2—1700–1600 cm^{-1}, assigned to: O–H stretching/bending (water), C = O stretching (mainly from carbohydrates), and N–H bending of amide I (mainly proteins).

D3—1540–1175 cm^{-1}, assigned to: O–H stretching/bending, C–O stretching (carbohydrates), C–H stretching (carbohydrates), and C = O stretching of ketones.

D4—1175–940 cm^{-1}, assigned to: C–O, C–C stretching (carbohydrates), and ring vibrations (mainly from carbohydrates).

D5—940–700 cm^{-1}, assigned to: the anomeric region of carbohydrates, C–H bending (mainly from carbohydrates), and ring vibrations (mainly from carbohydrates) [32,33,96,97].

The analyzed honey samples of seven floral origins come from different geographical zones, and the physicochemical analysis results obtained in this study confirmed the differences in FTIR spectra (Table 6). The differences obtained for the FTIR spectra observed in Figure 4 are probably related to the amount of carboxylic acids, groups of polyphenols, types of carbohydrates, or other functional groups.

The spectral region D0 of 3100–3500 cm^{-1} corresponds to O–H of carbohydrates, O–H stretching from water, and N–H stretching vibration (amide A band) of the peptide and proteins and polyphenols [75]. The wavenumber was between 3369 cm^{-1} for chestnut honey and 3416 cm^{-1} for linden honey, higher values compared to those reported by Anjos et al. of 3279.64 cm^{-1}, Sabri and See of 3276.79 cm^{-1}, Pauliuc et al. of 3297 cm^{-1}, or by Svečnjak et al. of 3284 cm^{-1} [30,31,34,58].

In the spectral region between 3000 and 2800 cm^{-1}, the presence of bands between 2922 and 2935 cm^{-1} corresponds to stretching vibrations of the C–H bonds of the chemical structure of the carbohydrates [29,33]. Similar values between 2932 and 2960 cm^{-1} were also reported [31,34,98]. In Figure 4, next to the D1 spectral region, a peak can be observed that records values from 2115 to 2121 cm^{-1}, which can be assigned to C = C conjugated and C $\equiv$ C. Honey spectra recorded peaks between 1636 and 1647 cm^{-1}. It is known that the domain represented by the values from 1700 to 1600 cm^{-1} is responsible for C = O stretching (mainly from carbohydrates), O–H stretching/bending (water), and N–H bending of amide I (mainly proteins). The appearance of peaks could also be attributed to stretching band of carbonyl groups C = O and C = C related to phenolic molecules [29,32,33,58,97]. Comparable wavenumber values were found in studies performed by Anjos et al. of 1646.56 cm^{-1}, Sabri and See of 1638.19 cm^{-1}, Pauluic et al. of 1640 cm^{-1}, and by Svečnjak et al. of 1645 cm^{-1} [30,31,34,58].

In Table 6, the wavenumber values are presented for the spectral region of 1200–1500 cm^{-1}. The bands at 1450 and 1454 cm^{-1} are attributed to bending vibration of O–CH and C–C–H in the carbohydrate structure [31,98].

Characteristic of O–H bending vibration of the C–OH group is the presence of a peak in the spectral range of 1340–1350 cm^{-1}. For 1255–1259 cm^{-1}, the overlapping peaks are due to N–H deformation and C–N stretching vibrations from amide II and C−N amide III bands [31,75,98]. Between 1175 and 940 cm^{-1}, the presence of peaks is associated with C–H in carbohydrates and/or C–O and C–C in carbohydrates [31,34,75,96,98]. The band from 1050 to 970 cm^{-1} is responsible for the C–O stretching vibrations of the C–OH group or for the C–C stretch in the carbohydrate structure, ring vibrations (mainly from carbohydrates) [33,34,98]. The spectral region between 940 and 700 cm^{-1} is assigned to the anomeric region of carbohydrates, C–H bending (mainly from carbohydrates), and ring vibrations (mainly from carbohydrates) specific for honey samples [33,34,75,98].

4.8. Correlation and Multivariate Statistical Analyses

The correlations between the investigated quality parameters (Pearson's coefficients for mean values of each parameter corresponding to each type of honey) are shown in Table 6. Significant correlations at $p < 0.001$ were observed for the refractive index with moisture (r = -0.99), solid substances (r = 0.99), and specific gravity (r = 0.99). These parameters were significantly correlated ($p < 0.01$) with Mg (r = 0.91), P (r = 0.91), and total phenolic content (r = 0.79, $p < 0.05$). Color was significantly correlated ($p < 0.001$) with total flavonoid content (r = 0.95), and with free acidity (r = 0.82), total phenolic content (r = 0.78), and two mineral elements, Mg (r = 0.84) and P (r = 0.86), at the $p < 0.05$ level. Similar correlations have also been reported for other honey samples [36].

As concerns the pH, a strong positive correlation at $p < 0.001$ was observed with K (r = 0.95) and at $p < 0.01$ with Ca (r = 0.93) and Na (r = 0.90), as well as the expected correlations with ash (r = 0.93) and electrical conductivity (r = -0.92).

Significant strong correlations were observed for ash with electrical conductivity (r = 0.97), and all mineral cations (K—r = 0.98, Na—r = 0.95, Ca—r = 0.85, and Mg—r = 0.77). Similar findings were also reported by other authors, who observed strong positive correlation of electrical conductivity with potassium [36,99,100]. Among the mineral elements, P was significantly correlated only with Mg (0.98).

Many studies on honey reported correlations between the studied honey parameters: Strong correlation (Pearson's coefficient of r > 0.8) between color and antioxidant compounds (TPC, TFC) for honey samples from Brazil, Sultanate of Oman, Romania, Bangladesh, Serbia, Croatia, Turkey, and Alger [26–28,43,45,69,101–103]. Positive correlations between color and TPC (Pearson's coefficients of 0.8 > r > 0.6) were found by Kavanagh et al. in Irish honey of r = +0.6 and by Aazza et al. in Portuguese honey of r = +0.685 [36,104]. In this research, the highest content of phenolic compounds was found in chestnut honey, which was the sample with the highest ash content, and at the same time, with a darker color (amber). Rapeseed honey (RP3) of dark color (light amber) also had a high content of ash and antioxidants. The high content of polyphenolic compounds, pollen, pigments (carotenoids and flavonoid), and minerals present in honey can contribute to the appearance of a dark color of the honey [8,49,73]. In this study, comparable values of phenols concentrations in the acacia honey sample (A3) and rapeseed honey sample (RP3) were observed, but the flavonoid content was 3.3 times higher in the RP3 sample compared to sample A3, confirming the conclusions of several studies of a strong positive correlation between color and flavonoid content. Strong correlations (Pearson's coefficient r > 0.8) between total phenol and total flavonoid contents are presented in some studies on honey samples from Italy, Serbia, and Algeria [43,60,103]. Kolayli et al.; in a study on chestnut honey, noticed that the ash content and the value of electrical conductivity increased with the pollen content from the studied chestnut honey samples [69].

Lanjwani and Channa reported good correlations between Na and K (r > 0.7) and K and Ca (r > 0.879), and moderate correlations (r = 0.5–0.7) between Na and Ca and Mg for honey samples from Pakistan [6].

Principal component analysis was used to obtain an overview of the honey data and to achieve a better resolution of contributions from different parameters. Via Varimax normalized factor rotation, three factors that explain 86% of the total variance in the dataset were chosen. The loadings and explained variances are presented in Table 8.

Principal component 1 (PC1) is characterized by the refractive index, moisture, solid substances, and specific gravity as parameters with significant loadings, which can be attributed to some fundamental characteristics of the product that contribute to ensuring its stability. The graphical representation of corresponding scores (Figure 5) indicated that PC1 can be interpreted as a component distinguishing the chestnut honey from the other honey types. PC1 did not clearly differentiate the botanical and geographical origins of the investigated honey types.

PC2, which explains 25.15% of the variance in the dataset, includes pH, ash, electrical conductivity, K, and Na as dominant parameters, suggesting a honey origin factor, as mineral content depends on the botanical and geographical origins of honey [36,74,105]. The scores for PC2 (Figure 5a) did not highlight distinct groupings of the samples according to the geographical origin.

PC3, with high loadings of color and total flavonoid content, followed by free acidity, P, and total phenolic content, could be considered a color/antioxidant factor. The corresponding scores outlined three groups of samples: one group with RP1, RP2, M1, and L2, one group that contained only L1, and one group of all the other samples (Figure 5b).

The hierarchical dendrogram obtained for physicochemical parameters determined in honey samples collected in 2017 is shown in Figure 6. K was clearly differentiated from the other parameters. One cluster contained the mineral elements Na and Ca. Free acidity and total phenols were grouped in a cluster, and electrical conductivity and ash were grouped in another distinct cluster.

5. Conclusions

Some of the investigated quality parameters such as humidity, acidity, and pH are useful when the honey is stored for a longer time, while others (sugar content, electrical conductivity, antioxidant compounds, minerals, etc.) could differentiate the quality of honey for its therapeutic use. FTIR spectroscopy can be easily implemented to determine the composition of honey at mainly a qualitative level and to confirm the authenticity of the honey and its possible adulteration. In all analyzed honey samples, significant amounts of antioxidants were found, that once again proved and confirmed the anti-inflammatory, antioxidant, and antibacterial properties of the studied honey types. The positive correlation between the honey's color and its phenolic and flavonoid content indicates that the higher the content of antioxidant compounds, the darker the honey becomes. The mineral elements' concentrations showed the abundance of potassium and sodium.

Many factors, such as improper management in raising and caring for bees, the presence of xenobiotics (pesticides, heavy metals, etc.), pests, specific diseases of bees, improper food given to bees, and finally, collecting honey from the hive and its subsequent handling, can influence the composition and quality of honey. Hence, every year the quality of honey is different, and therefore a systematic quality control of this product is needed.

Romania has a diverse flora, that allows producing a wide variety of honey types. Most of the samples analyzed in this study were collected from Iasi County, an area in eastern Romania of tradition in linden honey. The great variability of the results on honeys from this relatively small area, especially the mineral elements, in accordance with previous studies carried out on the soils of this area, supports the need of monitoring these honey quality parameters and their continuous updating.

In Romania, and elsewhere, honey is being used as food but also as an adjuvant for various diseases. Today, honey is recommended in different diseases as a dietary supplement. To consume the appropriate dose of honey for each health condition, it is necessary to know the level of antioxidants and the amount of important major elements, such as K, Ca, Na, P, and Mg. Studies shows high differences of these components in the same type of honey. The results reported in this study indicate that studies should not be limited to studying the quality of honey in only one year and from one area because of the very large differences between mineral elements in honey samples collected from very close areas.

In the association of mineral elements' concentration in honey with the particularities of the soil mineral composition, the geographical and floral origins confirm the influence of environmental components (soil, water, air, biota) on the quality of honey. However, for a more complete quality assessment, determination of more mineral elements, including toxic ones, in honey, plants, and soil is still needed.

The obtained data and correlations between the studied parameters of honey are complementary with other studies and highlight the variability of honey composition in time and space, and consequently the need to monitor its quality.

Author Contributions: Conceptualization, I.M.P., A.A., S.-M.C.-M. and D.S.; methodology, I.M.P. and A.A.; software, A.A., S.-M.C.-M. and A.P.; validation, I.M.P., A.A., S.-M.C.-M. and A.P.; investigation, I.M.P., A.A. and D.S.; data curation, A.A., S.-M.C.-M. and D.S.; writing—original draft preparation, A.A. and S.-M.C.-M.; writing—review and editing, I.M.P., A.A.,S.-M.C.-M. and D.S.; supervision, I.M.P. and A.A. All authors have read and agreed to the published version of the manuscript.

Funding: This research received no external funding.

Data Availability Statement: The data present in this study are available upon request from the corresponding author.

Acknowledgments: Acknowledgment is given by S.-M.C.-M. to infrastructure support from the Operational Program Competitiveness 2014–2020, Axis 1, under POC/448/1/1 Research infrastructure projects for public R&D institutions/Sections F 2018, through the Research Center with Integrated Techniques for Atmospheric Aerosol Investigation in Romania (RECENT AIR) project, under grant agreement MySMIS No. 127324. A.A. and S.-M.C.-M. are grateful to CERNESIM center for AAS analyses.

Conflicts of Interest: The authors declare no conflict of interest.

References

1. Tanda, A.S. Why insect pollinators important in crop improvement? *Indian J. Entomol.* **2021**, *84*, 223–236. [CrossRef]
2. Pauliuc, D.; Oroian, M. Organic Acids and Physico-Chemical Parameters of Romanian Sunflower Honey. *Food Environ. Saf. J.* **2020**, *19*, 148–155.
3. Oddo, L.P.; Bogdanov, S. Determination of honey botanical origin: Problems and issues. *Apidologie* **2004**, *35*, 2001–2002.
4. da Silva, P.M.; Gauche, C.; Gonzaga, L.V.; Costa, A.C.O.; Fett, R. Honey: Chemical composition, stability and authenticity. *Food Chem.* **2016**, *196*, 309–323. [CrossRef]
5. Escuredo, O.; Rodríguez-Flores, M.S.; Rojo-Martínez, S.; Seijo, M.C. Contribution to the Chromatic Characterization of Unifloral Honeys from Galicia (NW Spain). *Foods* **2019**, *8*, 233. [CrossRef]
6. Lanjwani, M.F.; Channa, F.A. Minerals content in different types of local and branded honey in Sindh, Pakistan. *Heliyon* **2019**, *5*, e02042. [CrossRef]
7. Mădaş, N.M.; Mărghitaş, L.A.; Dezmirean, D.S.; Bonta, V.; Bobiş, O.; Fauconnier, M.-L.; Francis, F.; Haubruge, E.; Nguyen, K.B. Volatile profile and physico-chemical analysis of acacia honey for geographical origin and nutritional value determination. *Foods* **2019**, *8*, 445. [CrossRef]
8. Pauliuc, D.; Dranca, F.; Oroian, M. Antioxidant Activity, Total Phenolic Content, Individual Phenolics and Physicochemical Parameters Suitability for Romanian Honey Authentication. *Foods* **2020**, *9*, 306. [CrossRef]
9. Soares, S.; Amaral, J.S.; Oliveira, M.B.P.; Mafra, I. A comprehensive review on the main honey authentication issues: Production and origin. *Compr. Rev. Food Sci. Food Saf.* **2017**, *16*, 1072–1100. [CrossRef]
10. Sereia, M.J.; Março, P.H.; Perdoncini, M.R.G.; Parpinelli, R.S.; de Lima, E.G.; Anjo, F.A. Chapter 9: Techniques for the Evaluationof Physicochemical Quality and Bioactive Compounds in Honey. In *Honey Analysis*; INTECH: London, UK, 2017; pp. 194–214.
11. European Commission. Council Directive 2001/110/CE concerning honey. *Off. J. Eur. Comm.* **2002**, *L10*, 47–52.

12. Boussaid, A.; Chouaibi, M.; Rezig, L.; Hellal, R.; Donsì, F.; Ferrari, G.; Hamdi, S. Physicochemical and bioactive properties of six honey samples from various floral origins from Tunisia. *Arab. J. Chem.* **2018**, *11*, 265–274. [CrossRef]

13. Halouzka, R.; Tarkowski, P.; Zeljković, S.C. Characterisation of phenolics and other quality parameters of different types of honey. *Czech J. Food Sci.* **2016**, *34*, 244–253. [CrossRef]

14. Milek, M.; Bocian, A.; Kleczyńska, E.; Sowa, P.; Dżugan, M. The Comparison of Physicochemical Parameters, Antioxidant Activity and Proteins for the Raw Local Polish Honeys and Imported Honey Blends. *Molecules* **2021**, *26*, 2423. [CrossRef]

15. Pauliuc, D.; Dranca, F.; Ropciuc, S.; Oroian, M. Advanced Characterization of Monofloral Honeys from Romania. *Agriculture* **2022**, *12*, 526. [CrossRef]

16. Sánchez-Martín, V.; Morales, P.; González-Porto, A.V.; Iriondo-DeHond, A.; López-Parra, M.B.; Del Castillo, M.D.; Hospital, X.F.; Fernández, M.; Hierro, E.; Haza, A.I. Enhancement of the Antioxidant Capacity of Thyme and Chestnut Honey by Addition of Bee Products. *Foods* **2022**, *11*, 3118. [CrossRef] [PubMed]

17. Ratiu, I.A.; Al-Suod, H.; Bukowska, M.; Ligor, M.; Buszewski, B. Correlation study of honey regarding their physicochemical properties and sugars and cyclitols content. *Molecules* **2020**, *25*, 34. [CrossRef]

18. Terrab, A.; Díez, M.J.; Heredia, F.J. Palynological, physico-chemical and colour characterization of Moroccan honeys: III. Other unifloral honey types. *Int. J. Food Sci. Technol.* **2003**, *38*, 395–402. [CrossRef]

19. Gośliński, M.; Nowak, D.; Szwengiel, A. Multidimensional Comparative Analysis of Bioactive Phenolic Compounds of Honeys of Various Origin. *Antioxidants* **2021**, *10*, 530. [CrossRef]

20. Tomczyk, M.; Tarapatskyy, M.; Dżugan, M. The influence of geographical origin on honey composition studied by Polish and Slovak honeys. *Czech J. Food Sci.* **2019**, *37*, 232–238. [CrossRef]

21. Ahmida, N.H.; Elagori, M.; Agha, A.; Elwerfali, S.; Ahmida, M.H. Physicochemical, heavy metals and phenolic compounds analysis of Libyan honey samples collected from Benghazi during 2009–2010. *Food Nutr. Sci.* **2013**, *4*, 33–40. [CrossRef]

22. El Sohaimy, S.A.; Masry, S.H.D.; Shehata, M.G. Physicochemical characteristics of honey from different origins. *Ann. Agric. Sci.* **2015**, *60*, 279–287. [CrossRef]

23. Majewska, E.; Druzynska, B.; Wołosiak, R. Determination of the botanical origin of honeybee honeys based on the analysis oftheir selected physicochemical parameters coupled with chemometric assays. *Food Sci. Biotechnol.* **2019**, *28*, 1307–1314. [CrossRef] [PubMed]

24. Krishnasree, V.; Ukkuru, P.M. Quality Analysis of Bee Honeys. *Int. J. Curr. Microbiol. Appl. Sci* **2017**, *6*, 626–636. [CrossRef]

25. Sant'Ana, L.D.; Ferreira, A.B.; Lorenzon, M.C.A.; Berbara, R.L.L.; Castro, R.N. Correlation of total phenolic and flavonoid contents of Brazilian honeys with color and antioxidant capacity. *Int. J. Food Prop.* **2014**, *17*, 65–76. [CrossRef]

26. Pontis, J.A.; Costa, L.A.M.A.D.; Silva, S.J.R.D.; Flach, A. Color, phenolic and flavonoid content, and antioxidant activity of honey from Roraima, Brazil. *Food Sci. Technol.* **2014**, *34*, 69–73. [CrossRef]

27. Al-Farsi, M.; Al-Amri, A.; Al-Hadhrami, A.; Al-Belushi, S. Color, flavonoids, phenolics and antioxidants of Omani honey. *Heliyon* **2018**, *4*, e00874. [CrossRef]

28. Cimpoiu, C.; Hosu, A.; Miclaus, V.; Ioncas, A. Determination of the floral origin of some Romanian honeys on the basis of physical and biochemical properties. *Spectrochim. Acta Part A* **2013**, *100*, 149–154. [CrossRef]

29. Anguebes, F.; Pat, L.; Ali, B.; Guerrero, A.; Cordova, A.V.; Abatal, M.; Garduza, J.P. Application of Multivariable Analysisand FTIR-ATR Spectroscopy to the Prediction of Properties in Campeche Honey. *J. Anal. Methods Chem.* **2016**, *2016*, 5427526. [CrossRef]

30. Svečnjak, L.; Bubalo, D.; Baranović, G.; Novosel, H. Optimization of FTIR-ATR spectroscopy for botanical authentication of unifloral honey types and melissopalynological data prediction. *Eur. Food Res. Technol.* **2015**, *240*, 1101–1115. [CrossRef]

31. Sabri, N.F.B.M.; See, H.H. Classification of honey using fourier transform infrared spectroscopy and chemometrics. *EProceedings Chem.* **2016**, *1*, 22–26.

32. Sahlan, M.; Karwita, S.; Gozan, M.; Hermansyah, H.; Yohda, M.; Yoo, Y.J.; Pratami, D.K. Identification and classification of honey's authenticity by attenuated total reflectance Fourier-transform infrared spectroscopy and chemometric method. *Vet. World* **2019**, *12*, 1304–1310. [CrossRef]

33. Gok, S.; Severcan, M.; Goormaghtigh, E.; Kandemir, I.; Severcan, F. Differentiation of Anatolian honey samples from different botanical origins by ATR-FTIR spectroscopy using multivariate analysis. *Food Chem.* **2015**, *170*, 234–240. [CrossRef]

34. Anjos, O.; Campos, M.G.; Ruiz, P.C.; Antunes, P. Application of FTIR-ATR spectroscopy to the quantification of sugar in honey. *Food Chem.* **2015**, *169*, 218–223. [CrossRef]

35. Kasprzyk, I.; Depciuch, J.; Grabek-Lejko, D.; Parlinska-Wojtan, M. FTIR-ATR spectroscopy of pollen and honey as a tool for unifloral honey authentication. The case study of rape honey. *Food Control* **2018**, *84*, 33–40. [CrossRef]

36. Aazza, S.; Lyoussi, B.; Antunes, D.; Miguel, M.G. Physicochemical Characterization and Antioxidant Activity of CommercialPortuguese Honeys. *J. Food Sci.* **2013**, *78*, 1159–1165. [CrossRef] [PubMed]

37. Romanian Standards Association. SR (Romanian Standard) 784-3:2009: Honey Bee. Part 3: Analytical Methods. Available online: https://e-standard.eu/en/standard/174480 (accessed on 16 April 2018).

38. Rebiai, A.; Lanez, T. Comparative study of honey collected from different flora of Algeria. *J. Fundam. Appl. Sci.* **2014**, *6*, 48–55. [CrossRef]

39. Bogdanov, S. 2009: Harmonized Methods of the International Honey Commission. Available online: https://www.ihc-platform.net/ihcmethods2009.pdf (accessed on 30 May 2018).

40. Bobis, O.; Marghitas, L.; Rindt, I.K.; Niculae, M.; Dezmirean, D. Honeydew honey: Correlations between chemical composition, antioxidant capacity and antibacterial effect. *Sci. Pap. Anim. Sci. Biotechnol.* **2008**, *41*, 271–277.
41. Johnbull, O.A.; Frances, A.; Isaac, A.O.; Cynthia, A.C. Analysis of mineral content and some heavy metals of honey samples produced and marketed in Imo state, Nigeria: A prerequisite to fomulating a pharmaceutical dosage form. *World J. Pharm. Pharm. Sci.* **2020**, *9*, 130–138.
42. Bodor, Z.; Benedek, C.; Urbin, Á.; Szabó, D.; Sipos, L. Colour of honey: Can we trust the Pfund scale?—An alternative graphical tool covering the whole visible spectra. *LWT—Food Sci. Technol.* **2021**, *149*, 111859. [CrossRef]
43. Živković, J.; Sunarić, S.; Stanković, N.; Mihajilov-Krstev, T.; Spasić, A. Total phenolic and flavonoid contents, antioxidant and antibacterial activities of selected honeys against human pathogenic bacteria. *Acta Pol. Pharm.—Drug Res.* **2019**, *76*, 671–681. [CrossRef]
44. Becerril-Sánchez, A.L.; Quintero-Salazar, B.; Dublán-García, O.; Escalona-Buendía, H.B. Phenolic Compounds in Honey and Their Relationship with Antioxidant Activity, Botanical Origin, and Color. *Antioxidants* **2021**, *10*, 1700. [CrossRef] [PubMed]
45. Flanjak, I.; Kenjerić, D.; Bubalo, D.; Primorac, L. Characterisation of Selected Croatian Honey Types Based on the Combination of Antioxidant Capacity, Quality Parameters, and Chemometrics. *Eur. Food Res. Technol.* **2016**, *242*, 467–475. [CrossRef]
46. Chiș, A.M.; Purcărea, C. Storage effect on antioxidant capacitiesof some monofloral honey. *Stud. Univ. "Vasile Goldiș" Life Sci. Ser.* **2017**, *27*, 91–97.
47. Mărghitaș, L.A.; Dezmirean, D.; Moise, A.; Bobiș, O.; Laslo, L.; Bogdanov, S. Physico-chemical and bioactive properties of different floral origin honeys from Romania. *Food Chem.* **2009**, *112*, 863–867.
48. Chirsanova, A.; Capcanari, T.; Boistean, A. Quality Assessment of Honey in Three Different Geographical Areas from Republic of Moldova. *Food Nutr. Sci.* **2021**, *12*, 962–977. [CrossRef]
49. Smetanska, I.; Alharthi, S.S.; Selim, K.A. Physicochemical, antioxidant capacity and color analysis of six honeys from different origin. *J. King Saud Univ.—Sci.* **2021**, *33*, 101447. [CrossRef]
50. Pauliuc, D.; Oroian, M. Organic Acids and Physico-Chemical Parameters of Romanian Raspberry Honey. *Food Environ. Saf. J.* **2019**, *18*, 298–307.
51. Karabagias, I.K.; Maia, M.; Karabagias, V.K.; Gatzias, I.; Badeka, A.V. Characterization of eucalyptus, chestnut and heather honeys from Portugal using multi-parameter analysis and chemo- calculus. *Foods* **2018**, *7*, 194. [CrossRef]
52. Oddo, L.P.; Piro, R.; Bruneau, É.; Guyot-Declerck, C.; Ivanov, T.; Piskulová, J.; Flamini, C.; Lheritier, J.; Morlot, M.; Russmann, H.; et al. Main European unifloral honeys: Descriptive sheets. *Apidologie* **2004**, *35* (Suppl. 1), S38–S81. [CrossRef]
53. Popescu, R.; Geana, E.I.; Dinca, O.R.; Sandru, C.; Costinel, D.; Ionete, R.E. Characterization of the Quality and Floral Origin of Romanian Honey. *Anal. Lett.* **2015**, *49*, 411–422. [CrossRef]
54. Stihi, C.; Chelarescu, E.D.; Duliu, O.G.; Toma, L.G. Characterization of Romanian honey using physico-chemical parameters and the elemental content determined by analytical techniques. *Rom. Rep. Phys.* **2016**, *68*, 362–369.
55. Atanassova, J.; Yurukova, L.; Lazarova, M. Pollen and inorganic characteristics of Bulgarian unifloral honeys. *Czech J. Food Sci.* **2012**, *30*, 520–526. [CrossRef]
56. Di Marco, G.; Gismondi, A.; Panzanella, L.; Canuti, L.; Impei, S.; Leonardi, D.; Canini, A. Botanical influence on phenolic profile and antioxidant level of Italian honeys. *J. Food Sci. Technol.* **2018**, *55*, 4042–4050. [CrossRef] [PubMed]
57. Sakač, M.; Jovanov, P.; Marić, A.; Četojević-Simin, D.; Novaković, A.; Plavšić, D.; Škrobot, D.; Kovač, R. Antioxidative, Antibacterial and Antiproliferative Properties of Honey Types from the Western Balkans. *Antioxidants* **2022**, *11*, 1120. [CrossRef]
58. Pauliuc, D.; Ciursa, P.; Ropciuc, S.; Dranca, F.; Oroian, M. Physicochemical parameters prediction and authentication of different monofloral honeys based on FTIR spectra. *J. Food Compos. Anal.* **2021**, *102*, 104021. [CrossRef]
59. Alzahrani, H.A.; Alsabehi, R.; Boukraâ, L.; Abdellah, F.; Bellik, Y.; Bakhotmah, B.A. Antibacterial and Antioxidant Potency of Floral Honeys from Different Botanical and Geographical Origins. *Molecules* **2012**, *17*, 10540–10549. [CrossRef]
60. Attanzio, A.; Tesoriere, L.; Allegra, M.; Livrea, M.A. Monofloral honeys by Sicilian black honeybee (*Apis mellifera* ssp. *sicula*) havehigh reducing power and antioxidant capacity. *Heliyon* **2016**, *2*, e00193. [CrossRef]
61. Milosavljević, S.; Jadranin, M.; Mladenović, M.; Tešević, V.; Menković, N.; Mutavdžić, D.; Krstić, G. Physicochemical parameters as indicators of the authenticity of monofloral honey from the territory of the Republic of Serbia. *Maced. J. Chem. Chem. Eng.* **2021**, *40*, 57–67. [CrossRef]
62. Akgün, N.; Çelik, Ö.F.; Kelebekli, L. Physicochemical properties, total phenolic content, and antioxidant activity of chestnut, rhododendron, acacia and multifloral honey. *J. Food Meas. Charact.* **2021**, *15*, 3501–3508. [CrossRef]
63. Uršulin-Trstenjak, N.; Puntarić, D.; Levanić, D.; Gvozdić, V.; Pavlek, Ž.; Puntarić, A.; Puntarić, E.; Puntarić, I.; Vidosavljević, D.; Lasić, D.; et al. Pollen, Physicochemical, and Mineral Analysis of Croatian Acacia Honey Samples: Applicability for Identification of Botanical and Geographical Origin. *J. Food Qual.* **2017**, *2017*, 8538693. [CrossRef]
64. Chiș, A.M.; Purcărea, C. The phisico-chemical caracterisation of sun flower honey from bihor county. *An. Univ. Din Oradea Fasc. Ecotoxicol. Zooteh. Și Tehnol. De Ind. Aliment.* **2015**, *XIV B*, 107–114.
65. Živkov Baloš, M.M.; Jakšić, S.M.; Popov, N.S.; Polaček, V.A. Characterization of Serbian sunflower honeys by their physicochemical characteristics. *Food Feed. Res.* **2021**, *48*, 1–8. [CrossRef]
66. Chiș, A.; Purcărea, C. Quality of chestnut honey modified by thermal treatment. *Studia Univ. Vasile Goldis Arad, Seria Stiintele Vieti.* **2011**, *21*, 573–579.

67. Kharadze, M.; Abashidze, N.; Djaparidze, I.; Vanidz, M.; Kalandia, A. Antioxidant Activity of Chestnut Honey Produced in Western Georgia. *Bull. Georgian Natl. Acad. Sci.* **2018**, *12*, 145–151.
68. Rodríguez-Flores, S.; Escuredo, O.; Seijo, M.C. Characterization and antioxidant capacity of sweet chestnut honey produced in North-West Spain. *J. Apic. Sci.* **2016**, *60*, 19–30. [CrossRef]
69. Kolayli, S.; Can, Z.; Yildiz, O.; Sahin, H.; Karaoglu, S.A. A comparative study of the antihyaluronidase, antiurease, antioxidant, antimicrobial and physicochemical properties of different unifloral degrees of chestnut (*Castanea sativa* Mill.) honeys. *J. Enzyme Inhib. Med. Chem.* **2016**, *31* (Suppl. 3), 96–104. [CrossRef]
70. Pita-Calvo, C.; Vazquez, M. Differences between honeydew and blossom honeys: A review. *Trends Food Sci. Technol.* **2017**, *59*, 79–87. [CrossRef]
71. Ciulu, M.; Spano, N.; Pilo, M.I.; Sanna, G. Recent Advances in the Analysis of Phenolic Compounds in Unifloral Honeys. *Molecules* **2016**, *21*, 451. [CrossRef]
72. Gomes, T.; Feás, X.; Iglesias, A.; Estevinho, L.M. Study of organic honey from the northeast of Portugal. *Molecules* **2011**, *16*, 5374–5386. [CrossRef]
73. Bertoncelj, J.; Doberšek, U.; Jamnik, M.; Golob, T. Evaluation of the phenolic content, antioxidant activity and colour of Slovenian honey. *Food Chem.* **2007**, *105*, 822–828. [CrossRef]
74. Mărghitaș, L.A.; Dezmirean, D.S.; Pocol, C.B.; Ilea, M.; Bobiș, O.; Gergen, I. The development of a biochemical profile of acacia honey by identifying biochemical determinants of its quality. *Not. Bot. Hort. Agrobot. Cluj* **2010**, *38*, 84–90.
75. Kędzierska-Matysek, M.; Matwijczuk, A.; Florek, M.; Barłowska, J.; Wolanciuk, A.; Matwijczuk, A.; Chruściel, E.; Walkowiak, R.; Karcz, D.; Gładyszewska, B. "Application of FTIR spectroscopy for analysis of the quality of honey". *BIO Web Conf.* **2018**, *10*, 02008. [CrossRef]
76. Dżugan, M.; Zaguła, G.; Wesołowska, M.; Sowa, P.; Puchalski, C.Z. Levels of toxic and essential metals in varietal honeys from Podkarpacie. *J. Elem.* **2017**, *22*, 1039–1048. [CrossRef]
77. Kamaruzzaman, M.A.; Chin, K.-Y.; Ramli, E.S.M. A Review of Potential Beneficial Effects of Honey on Bone Health. *Evid. Based Complement. Altern. Med.* **2019**, *2019*, 8543618. [CrossRef] [PubMed]
78. Alqari, A.S.; Owayss, A.A.; Mahmoud, A.A.; Hannan, M.A. Mineral content and physical properties of local and imported honeys in Saudi Arabia. *J. Saudi Chem. Soc.* **2014**, *18*, 618–625. [CrossRef]
79. Solayman, M.; Islam, A.; Paul, S.; Ali, Y.; Khalil, I.; Alam, N.; Gan, S.H. Physicochemical properties, minerals, trace elements, and heavy metals in honey of different origins: A comprehensive review. *Compr. Rev. Food Sci. Food Saf.* **2016**, *15*, 219–233. [CrossRef]
80. Kropf, U.; Stibilj, V.; Jaćimović, J.R.; Bertoncelj, J.; Golob, T.; Korošec, M. Elemental Composition of Different Slovenian Honeys Using k0-Instrumental Neutron Activation Analysis. *J. AOAC Int.* **2017**, *100*, 871–880. [CrossRef]
81. Bodó, A.; Radványi, L.; Kőszegi, T.; Csepregi, R.; Nagy, D.U.; Farkas, Á.; Kocsis, M. Quality Evaluation of Light- and Dark-Colored Hungarian Honeys, Focusing on Botanical Origin, Antioxidant Capacity and Mineral Content. *Molecules* **2021**, *26*, 2825. [CrossRef]
82. Grembecka, M.; Szefer, P. Evaluation of honeys and bee products quality based on their mineral composition using multivariate techniques. *Environ. Monit. Assess.* **2013**, *185*, 4033–4047. [CrossRef]
83. Kivima, E.; Seiman, A.; Pall, R.; Sarapuu, E.; Martverk, K.; Laos, K. Characterization of Estonian honeys by botanical origin. *Proc. Estonian Acad. Sci.* **2014**, *63*, 183. [CrossRef]
84. Tesileanu, R.; Fedorca, M. Solurile Afectate de Săruri din România: Concepte de Bază și Stare Actuală. 2016. Available online: https://www.researchgate.net/publication/294855962_Solurile_afectate_de_saruri_din_Romania_concepte_de_baza_si_stare_actuala (accessed on 4 February 2019).
85. Nikolova, K.; Gentscheva, G.; Ivanova, E. Survey of the mineral content and some physicochemical parameters of Bulgarian bee honeys. *Bulgarian Chem. Commun.* **2013**, *45*, 244–249.
86. Czipa, N.; Kovács, B. Electrical conductivity of hungarian honeys. *J. Food Phys.* **2014**, *27*, 13–20.
87. Bilandžić, N.; Gačić, M.; Đokić, M.; Sedak, M.; Šipušić, Đ.I.; Končurat, A.; Gajger, I.T. Major and trace elements levels in multifloral and unifloral honeys in Croatia. *J. Food Compos. Anal.* **2014**, *33*, 132–138. [CrossRef]
88. Oroian, M.; Amariei, S.; Leahu, A.; Gheorghe, G. Multi-Element Composition of Honey as a Suitable Tool for Its Authenticity Analysis Pol. *J. Food Nutr. Sci.* **2015**, *65*, 93–100.
89. Pătruică, S.; Hărmănescu, M.; Jivan, A.; Cioabă, C.; Simiz, E.; Călamar, C.D. Researches concerning of the mineral content of acacia honey derived on differents country. *Sci. Pap. Anim. Sci. Biotechnol.* **2009**, *42*, 178–181.
90. Bontempo, L.; Camin, F.; Ziller, L.; Perini, M.; Nicolini, G.; Larcher, R. Isotopic and elemental composition of selected types of Italian honey. *Meas. J. Int. Meas. Confed.* **2017**, *98*, 283–289. [CrossRef]
91. Popov-Raljic, J.; Arsic, N.; Zlatkovic, B.; Basarin, B.; Mladenovic, M.; Lalicic, J.; Ivkov, M.; Popov, V. Evaluation of color, mineral substances and sensory uniqueness of meadow and acacia honey from Serbia. *Rom. Biotechnol. Lett.* **2015**, *20*, 10784–10799.
92. Tudoreanu, L.; Codreanu, M.; Crivineanu, V.; Goran, G. The Quality of Romanian Honey Varieties -Mineral Content and Textural Properties. *Bull. USAVM Cluj-Napoca Vet. Med.* **2012**, *69*, 452–458. [CrossRef]
93. Ördög, A.; Tari, I.; Bátori, Z.; Poór, P. Mineral content analysis of unifloral honeys from the Hungarian great plain. *J. Elem.* **2017**, *22*, 271–281.
94. Scripcă, L.A.; Amariei, S. The Influence of Chemical Contaminants on the Physicochemical Properties of Unifloral andMul-tifloral Honey from the North-East Region of Romania. *Foods* **2021**, *10*, 1039. [CrossRef]

95. Di Rosa, A.R.; Leone, F.; Cheli, F.; Chiofalo, V. Novel approach for the characterisation of Sicilian honeys based on the cor-relation of physico-chemical parameters and artificial senses. *Ital. J. Anim. Sci.* **2019**, *18*, 389–397. [CrossRef]
96. David, M.; Hategan, A.R.; Berghian-Grosan, C.; Magdas, D.A. The Development of Honey Recognition Models Based on the Association between ATR-IR Spectroscopy and Advanced Statistical Tools. *Int. J. Mol. Sci.* **2022**, *23*, 9977. [CrossRef] [PubMed]
97. Mendes, E.; Duarte, N. Mid-Infrared Spectroscopy as a Valuable Tool to Tackle Food Analysis: A Literature Review on Coffee, Dairies, Honey, Olive Oil and Wine. *Foods* **2021**, *10*, 477. [CrossRef] [PubMed]
98. Matwijczuk, A.; Budziak-Wieczorek, I.; Czernel, G.; Karcz, D.; Baránska, A.; Jedlínska, A.; Samborska, K. Classification of Honey Powder Composition by FTIR Spectroscopy Coupled with Chemometric Analysis. *Molecules* **2022**, *27*, 3800. [CrossRef] [PubMed]
99. Guler, A.; Bakan, A.; Nisbet, C.; Yavuz, O. Determination of important biochemical properties of honey to discriminate pure and adulterated honey with sucrose (*Saccharum offcinarum* L.) syrup. *Food Chem.* **2007**, *105*, 1119–1125. [CrossRef]
100. Spirić, D.; Ćirić, J.; Dorđević, V.; Nikolić, D.; Janković, S.; Nikolić, A.; Petrović, Z.; Katanić, N.; Teodorović, V. Toxic and essential element concentrations in diferent honey types. *Int. J. Environ. Anal. Chem.* **2019**, *99*, 474–485. [CrossRef]
101. Almeida, A.M.M.; Oliveira, M.B.S.; Costa, J.G.; Valentim, I.B.; Goulart, M.O.F. Antioxidant Capacity, Physicochemical and Floral Characterization of Honeys from the NorthEast of Brazil. *Rev. Virtual Química* **2016**, *8*, 57–77. [CrossRef]
102. Moniruzzaman, M.; Yung, A.C.; Azlan, S.A.B.M.; Sulaiman, S.A.; Rao, P.V.; Hawlader, M.N.I.; Gan, S.H. Identification of phenolic acids and flavonoids in monofloral honey from Bangladesh by high performance liquid chromatography: Determination of antioxidant capacity. *BioMed Res. Int.* **2014**, *2014*, 737490. [CrossRef]
103. Otmani, A.; Amessis-Ouchemoukh, N.; Birinci, C.; Yahiaoui, S.; Kolayli, S.; Rodríguez-Flores, M.S.; Escuredo, O.; Seijo, M.C.; Ouchemoukh, S. Phenolic compounds and antioxidant and antibacterial activities of Algerian honeys. *Food Biosci.* **2021**, *42*, 101070. [CrossRef]
104. Kavanagh, S.; Gunnoo, J.; Marques, P.T.; Stout, J.C.; White, B. Physicochemical Properties and Phenolic Content of Honey from Different Floral Origins and from Rural versus Urban Landscapes. *Food Chem.* **2019**, *272*, 66–75. [CrossRef]
105. Alvarez-Suarez, J.M.; Tulipani, S.; Romandini, S.; Bertoli, E.; Battino, M. Contribution of honey in nutrition and human health: A review. *Med. J. Nutr. Metab.* **2010**, *3*, 15–23. [CrossRef]

 *agriculture*

Article

Analysis of Phylogeny and Genetic Diversity of Endangered Romanian Grey Steppe Cattle Breed, a Reservoir of Valuable Genes to Preserve Biodiversity

Madalina-Alexandra Davidescu [1], Daniel Simeanu [1,*], Dragos-Lucian Gorgan [2,*], Mitica Ciorpac [3] and Steofil Creanga [1]

[1] Department of Control, Expertise and Services, Faculty of Food and Animal Sciences, "Ion Ionescu de la Brad" University of Life Sciences, 700489 Iasi, Romania
[2] Department of Biology, Faculty of Biology, Alexandru Ioan Cuza University of Iasi, 700506 Iasi, Romania
[3] Advanced Research and Development Center for Experimental Medicine (CEMEX), Grigore T. Popa University of Medicine and Pharmacy of Iasi, 700115 Iasi, Romania
* Correspondence: dsimeanu@uaiasi.ro (D.S.); lucian.gorgan@uaic.ro (D.-L.G.)

Abstract: Since 2000, the Food and Agriculture Organization of the United Nations (FAO) has been drawing attention to the increasing numerical decline of Podolian cattle, which include the Romanian Grey Steppe. Currently, this breed is endangered, numbering under 100 heads across the territory of the entire country. Due to its qualities of rusticity, adaptability, and increased resistance to diseases and severe climate conditions, the Grey Steppe is considered a valuable genetic reserve for improving livestock production. This study aimed to quantify the genetic diversity of a population of 32 cattle from the area of N-E Moldova through the analysis of two mitochondrial markers, cytochrome b and the d-loop, which have been proven to be relevant to studies of genetic diversity and phylogeny. The results obtained based on the statistical analysis of the data using nucleotide sequence analysis software (DnaSP, SeaView, MegaX, PopArt, etc.) demonstrated that the breed belonged to the ancestral P'QT haplogroup, with direct descent from *Bos taurus primigenius*. Within this haplogroup, five cattle were identified, which could be used in the selection of crosses, with the aim of preserving valuable genetic resources for the improvement of other cattle breeds and the protection of biodiversity.

Keywords: animal production; genetic diversity; grey cattle; mitochondrial DNA; Podolian cattle

Citation: Davidescu, M.-A.; Simeanu, D.; Gorgan, D.-L.; Ciorpac, M.; Creanga, S. Analysis of Phylogeny and Genetic Diversity of Endangered Romanian Grey Steppe Cattle Breed, a Reservoir of Valuable Genes to Preserve Biodiversity. *Agriculture* **2022**, *12*, 2059. https://doi.org/10.3390/agriculture12122059

Academic Editor: Ligang Wang

Received: 26 October 2022
Accepted: 28 November 2022
Published: 30 November 2022

Publisher's Note: MDPI stays neutral with regard to jurisdictional claims in published maps and institutional affiliations.

1. Introduction

Podolian cattle breeds are considered a form of socio-cultural heritage and a valuable genetic resource due to their high tolerance to extremely harsh environmental climatic conditions, resistance to disease and external parasites, and quick recovery after illnesses, which are important aspects of reproduction and conservation programs for improving livestock production [1–3]. Many historical sources attribute the term Podolian to a shared ancestral origin in Podolia (the modern western Ukraine). Alternative hypotheses have been proposed: Podolian cattle may have spread from the eastern steppe southward into Anatolia and westward into the Balkans and Italy in historical times (3rd–5th century AD) with Eastern European Barbarian people; other authors have suggested a more ancient migration (3 kya BP) from the Near East to Central Italy via the Mediterranean Sea, with a contribution from local wild aurochs via secondary local domestication [3]. Romanian Grey Steppe cattle belong to the category of Podolian breeds, which have been threatened with extinction since 2000, according to the Food and Agriculture Organization (FAO) [4,5]. The common origin of this cattle and other Podolian breeds (Iskar Grey, Bulgarian Grey, Istrian, Slavonian Podolian, Katerini, Hungarian Grey, Maremmana, Podolica, Turkish Grey, etc.) is the wild ancestor *Bos taurus primigenius*, which was declared be extinct around the 16th century [2,6]. This cattle breed was formed over many centuries, under the exclusive influence of natural environmental conditions that imprinted special qualities of hardiness, resistance, and adaptability to severe climatic factors [7–10]. The external peculiarities of the Grey Steppe breed led to the distinction of this breed from the others based on the craniological type and, more precisely, the characteristic shape of the bicolor horns, which are white at the base with a characteristic black dot. At birth, calves are yellow-reddish in color, and after a period of about 2–3 months, they change to shades of grey [11,12]. In 2018, the number of specimens of the Grey Steppe breed decreased below 150 heads [13]. Isolated specimens of this breed can also be found in households in north-eastern Moldova, in the counties of Iasi, Neamt, and Pardina (Danube Delta), currently totaling about 100 heads [14,15]. Presently, within the Research and Development Station for Cattle Breeding Dancu, Iasi, Romania, there is a population of cattle of this breed that is part of a national genetic conservation program, representing the biological material used in this research. The genetic diversity of Podolian breeds is an important asset for countries where agriculture is still an important economic sector [16].

Numerous studies [17–21] focusing on the origin of cattle based on mitochondrial genome analysis have determined that all taurine breeds share a common ancestor in the Near East, around 10 ka ago, during the Neolithic transition. The discovery of uncommon mitochondrial DNA (mtDNA) haplogroups such as P, Q, and R has sparked debate about the potential occurrence of independent or secondary domestication episodes. Recent research on the mitochondrial genome of *Bos taurus* has revealed that macro-haplogroup T is composed of two sister clades, T1'2'3 and T5, with the former including the originally identified haplogroups T1, T2, and T3, and T4, a grouping within T3. The Fertile Crescent is where all T haplogroups most likely originated and underwent domestication. From there, they spread with the movement of domestic *Bos taurus* herds. However, the mtDNA of contemporary taurine breeds is not entirely represented by haplogroups T1–T5. Complete mtDNA sequence analyses have revealed that a small subset of these individuals belongs to three other uncommon haplogroups (P, Q, and R). The mtDNA of haplogroups P and R most likely come from populations of wild aurochsen in Europe, whereas haplogroup Q is most likely of Near Eastern origin.

This research aimed to quantify the genetic diversity of the Grey Steppe cattle population using mitochondrial markers (cytochrome b and the d-loop) that are relevant to studies of genetic diversity, phylogeny, and molecular phylogeography. The main objective was to preserve valuable genetic resources for the improvement of other cattle breeds and the protection of biodiversity. The results of this study demonstrated that the Romanian Grey is a direct descendant of the *Bos taurus primigenius*, which is valuable information for the efforts to conserve the genetic resources of this endangered cattle breed.

2. Materials and Methods

2.1. The Biological Material Studied

In Romania, the Grey Steppe represents one of the oldest autochthonous breeds, adapted to the pedoclimatic conditions of the country. Along with other breeds (Podolica Italiana, Hungarian Grey, Bulgarian Grey, Istrian cattle, Katerini, Turkish Grey, Ukrainian Grey, etc.), it is included in the Podolian group [21]. This breed developed in a natural environment, both in summer and in winter, which granted it exceptional qualities that aew necessary for the improvement of other cattle. In the past, the breed was used intensively for traction, due to its special resistance characteristics, which also ensured an average milk production, noted for its high-fat content (approximately 4.71% in the fifth lactation). With the disappearance of this breed, a series of special qualities such as adaptability, resistance to disease, and high fat content in milk would disappear. The maintenance of various conformation features of the breed, such as the large lyre-shaped horns, gray color, and small waist that are characteristic of the primitive type, is an equally important goal (Figure 1) [22].

Figure 1. The Grey Steppe cattle within the Research and Development Station for Cattle Breeding, Iasi, Romania (original photograph).

2.2. Blood Samples

The first step in achieving the objectives of this research was the collection of blood samples from 32 females of the Grey Steppe cattle breed (Table 1), a population located within the Research and Development Station for Cattle Breeding in Romania, located in northern Moldova. The blood samples were collected by jugular vein puncture using Vacutainer tubes with EDTA (ethylene-diamine-tetra-acetic acid) to prevent clotting, with a capacity of 2 mL.

Table 1. Information on Grey Steppe population analyzed.

Current No.	Cattle Identification No.	Age of Cattle (Months)	Sex	Blood Sample Identification Code [1]
1.	RO242000109988	288	female	blood-ss01
2.	RO242000110002	242	female	blood-ss02
3.	RO245000109998	228	female	blood-ss03
4.	RO241000109812	202	female	blood-ss04
5.	RO247000109811	193	female	blood-ss05
6.	RO243000109786	187	female	blood-ss06
7.	RO243000109800	186	female	blood-ss07
8.	RO242000110205	185	female	blood-ss08
9.	RO243000109723	185	female	blood-ss09
10.	RO241000110120	184	female	blood-ss10
11.	RO241000110274	178	female	blood-ss11
12.	RO242000120579	175	female	blood-ss12
13.	RO243000120587	175	female	blood-ss13
14.	RO242000120656	171	female	blood-ss14
15.	RO243000120678	171	female	blood-ss15
16.	RO243000174164	152	female	blood-ss16
17.	RO242000218328	144	female	blood-ss17
18.	RO243000218587	136	female	blood-ss18
19.	RO143000108432	128	female	blood-ss19
20.	RO248001058693	111	female	blood-ss20
21.	RO507000112314	109	female	blood-ss21
22.	RO506001166289	97	female	blood-ss22
23.	RO504001166308	96	female	blood-ss23
24.	RO508003831422	69	female	blood-ss24
25.	RO502003831482	68	female	blood-ss25
26.	RO504005500784	61	female	blood-ss26
27.	RO501007300790	39	female	blood-ss27
28.	RO508007300858	37	female	blood-ss28

Table 1. *Cont.*

Current No.	Cattle Identification No.	Age of Cattle (Months)	Sex	Blood Sample Identification Code [1]
29.	RO507007405358	29	female	blood-ss29
30.	RO504007405388	28	female	blood-ss30
31.	RO507007448267	27	female	blood-ss31
32.	RO500007448406	22	female	blood-ss32

[1]—"ss" is the abbreviation of the name of the cattle breed in the Romanian language (Sura Stepa).

The blood samples were stored in the freezer under optimal conditions, at a temperature of $-20\,^{\circ}$C until the DNA extraction step.

2.3. Extraction and Quantification of Total DNA from Blood Samples

DNA extraction from blood samples was performed via the automated method with Maxwell™ 16 and 16 MDx instruments, using a special kit provided by the Promega distributor: Maxwell 16 LEV Blood DNA Kit (code-AS1290) [23], containing 50 LEV purification pistons, 50 elution tubes, 20 mL lysis buffer, two 1 mL proteinase K solutions, and 20 mL elution buffer. The DNA samples were quantified using the Nanodrop ASP-3700 spectrophotometer (ACTGene Inc., New Jersey, USA), and the optical density was measured at the absorption rates of A260 nm and A280 nm to calculate the DNA yield.

2.4. Primer, Amplification, Sequencing

The amplification of mtDNA, cytochrome b, and d-loop (mitochondrial DNA control-region sequences) was carried out through PCR analysis (polymerase chain reaction), which is a genomic technique applied for rapidly amplifying millions to billions of copies of a specific segment of DNA using amplification primers. In this research, the PCR amplification of the cytochrome b and d-loop was performed using two pairs of primers, forward and reverse (Table 2), specifically designed on the Bovine Reference Sequence (BRS; GenBank V00654) [24]. The total size of the complete mitochondrial genome of *Bos taurus* is 16,341 bp (Figure 2a). The cytochrome b gene, which has a size of 1140 bp, and the mitochondrial d-loop control region, which has a length of 910 bp (Figure 2b) [25–28], were amplified by the PCR technique.

Table 2. Characteristics of the primer pairs used to amplify the gene sequences by PCR.

Primer Set	Primer Specificity	Sequenced (5′-3′)	Content G [1] + C [2] (%)	GenBank Accession no./Position in Genome [23]	Length of PCR Product (bp)
BCYT	cytochrome b	Forward: TTCTTACATGGAATCTAACCATGA	33.3	V00654.1 14,443–14,466	1140 cytochrome b
	cytochrome b	Reverse: GGGAGGTTAGTTGTTCTCCTTCTC	50.0	V00654.1 473–497	
BRS	d-loop	Forward: CCTAAGACTCAAGGAAGAAACTGC	45.8	V00654.1 15,718–15,741	910 d-loop
	d-loop	Reverse: CAGTGAGAATGCCCTCTAGGTT	50.0	V00654.1 496–517	

[1] "G"—guanine nitrogen base. [2] "C"—cytosine nitrogen base.

The isolated and purified DNA samples were amplified by the PCR technique. The final volume of the PCR was 25.5 μL, including 2 μL DNA samples; 12.5 μL GoTaq® Green Master Mix (Promega, Madison, WI, USA); 1.5 μL of each primer (forward and reverse); and 7.5 μL nuclease-free water. The reaction mixture was supplemented with 0.5 μL $MgCl_2$. In the case of the amplification of cytochrome b fragments, the temperature of primer alignment was 62 $^{\circ}$C, and in the case of the amplification of d-loop fragments, the temperature of primer alignment was 60 °C. The PCR program for the amplification reaction of cytochrome b and d-loop fragments included 35 amplification cycles under the conditions represented in Table 3.

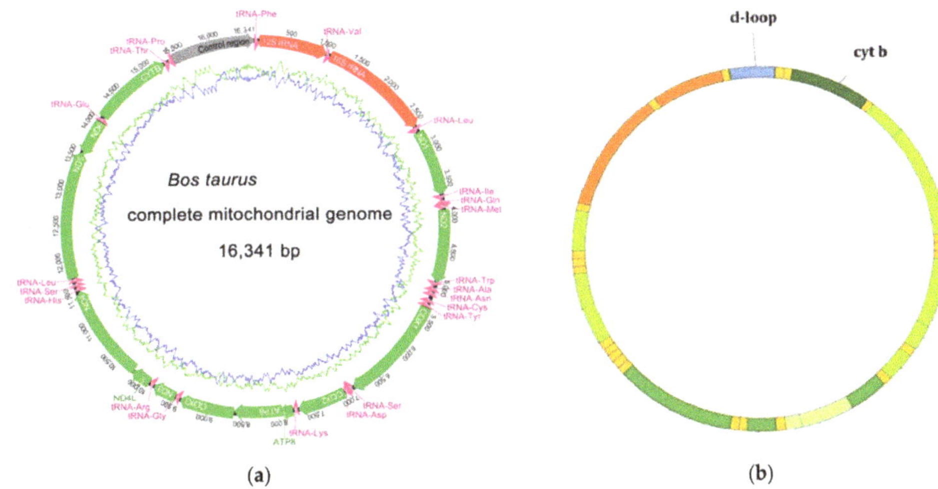

(a) (b)

Figure 2. The complete mitochondrial genome of *Bos taurus*: (**a**) complete mitochondrial genome of *Bos taurus* (16,340 bp); (**b**) mitochondrial markers analyzed—cytochrome b (1140 bp) and d-loop-(910 bp) [29].

Table 3. PCR program: cytochrome b and d-loop sequence amplification.

Stages *	Cytochrome b Amplification Conditions		d-Loop Amplification Conditions	
	Temperature	**Time**	**Temperature**	**Time**
1. Denaturing	94 °C	2 min	94 °C	2 min
2. Annealing	62 °C	30 s	60 °C	30 s
3. Extending	72 °C	5 min	72 °C	5 min

* ×35 amplification cycles.

The cytochrome b and d-loop region were successfully amplified; then, the PCR fragments were sent to Macrogen Europe Sequencing Laboratory, Amsterdam, The Netherlands for sequencing. To determine the nucleotide sequence of cytochrome b and d-loop, Sanger sequencing was applied using the primers detailed in Table 2.

2.5. Data Analysis

Data analysis was carried out using a series of programs (Table 4). The primary processing of the sequences in the form of fluorograms and their correction was carried out using the DNABaser program (sequence assembly software). To identify the phyloge-netic relationships of the Grey Steppe cattle breed with other Podolian cattle breeds, we aligned both sequences of the complete mitochondrial genome of *Bos taurus* and the individual cytochrome b genes and mitochondrial control regions (d-loop) specific to the Podolian cattle breeds (Grey Steppe, access no. HM596474.1; Bulgarian Grey, access no. KF373019.1; Istrian, access no. MZ901619.1; Slavonian Podolian, access no. MZ901634.1; Katerini, access no. DQ518336.1; Hungarian Grey, access no. GQ129207.1; Podolica, access no. EU177843.1; and Turkish Grey, access no. EF126309.1) from the NCBI-GenBank [23]. For each individual, forward and reverse sequences were concatenated with the resulting sequences from the two primers, and the two strands were cut, resulting in a unique sequence. The alignment of all cytochrome b and d-loop gene sequences from the analyzed individuals was carried out using ClustalW [30] with the MegaAlignment module. Sequence comparison and phylogenetic tree tracing was carried out using the MEGA X program (Molecular Evolutionary Genetics Analysis, Center for Evolutionary Medicine and Informatics, Tempe, AZ, USA).

Table 4. List of programs used for data analysis.

Data Analysis	Program Used
Amplicon sequencing ➢ *Sanger sequencing*	CEQ™ 8000 BECKMAN COULTER Life Sciences macrogen EUROPE
Alignment of chromatograms and correction of raw sequences ➢ *DNA Baser*	DNA BASER sequence assembly software
Sequence alignment ➢ *Mega X*	MEGA Molecular Evolutionary Genetics Analysis
Calculation of the optimal substitution model ➢ *jModelTest*	jModelTest
Haplotype network construction ➢ *PopArt*	PopART
Construction of phylogenetic trees ➢ *SeaView/PhyML*	SEA- PhyML
Nucleotide sequence diversity analysis ➢ *DnaSP*	DnaSP

The analyzed sequence diversity was estimated using LaunchDnaSP version 4.50.3 software [31]. PopART software was used for the median-joining network analysis. The neighbor-joining tree was built using the Kimura 2-parameter model with the following parameters: 1000 bootstrapping replicates, a gamma distribution (+G) with five rate categories, and evolutionary invariability (+I).

3. Results

3.1. Validation of Amplification of PCR Products in Agarose Gel Electrophoresis

The PCR amplicons were validated in 1% agarose gel with 0.5xTBE buffer, after which migration was performed at a voltage of 100 volts for 30 min (Figure 3). A molecular weight marker of 100 base pairs was used to estimate the size of the amplified fragments. The amplification of the gene sequences of interest was successfully achieved, with the primers having high specificity for these mitochondrial markers. The length of the PCR products was approximately 1140 bp for the cytochrome b sequences and approximately 910 bp for the d-loop sequences, in agreement with the values found for the complete sequence of the *Bos taurus* mitochondrial genome (GeneBank accession number: V00654).

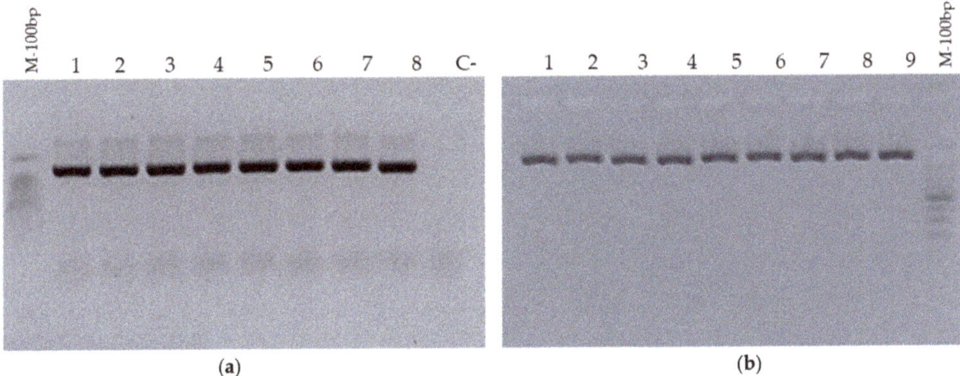

(a)   (b)

Figure 3. Amplification of mitochondrial markers: (**a**) Amplification of cytochrome b sequence (M, 100 bp marker; C-, negative control; 1–8: sample numbers). (**b**) Amplification of d-loop sequence (M, 100 bp marker; C-, negative control; 1–9: sample numbers).

3.2. The Proportion of Nitrogenous Bases in the Nucleotide Sequences of Grey Steppe for Cytochrome b and d-Loop Mitochondrial Markers

The complete sequence of the mtDNA cytochrome b (1140 bp) [32] was obtained for all 32 samples. The base composition was as follows: 31.2% adenine, 25.1% thymine, 30.2% cytosine, and 13.4% guanine. The base composition for the mtDNA d-loop (910 bp) [33] was as follows: 32.8% adenine, 29.0% thymine, 24.3% cytosine, and 13.9% guanine (Figure 4).

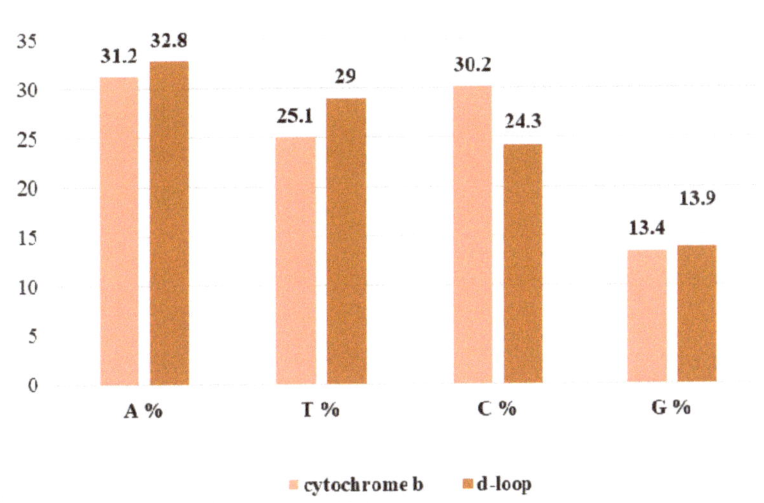

Figure 4. The frequency of nitrogenous bases (A: adenine; T: thymine; C: cytosine; and G: guanine) for the sequences of cytochrome b and d-loop.

3.3. The Specificity Coefficient

Based on the frequencies of the four nitrogenous bases, the specificity coefficient was calculated, the value of which is given by the A+T/C+G ratio and shows the differences between the individuals of a species. This coefficient is calculated in molecular phylogeny studies to observe differences in nucleotide composition. Normally, it has values in the range of 1.2–1.5, characteristic of most animal species. If we refer to the nitrogenous bases with the highest frequency, two types of DNA can be distinguished, namely, AT-type DNA (when A+T > C+G) and GC-type DNA (when A+T < C+G). In the case of the cytochrome b nucleotide sequences, the nitrogenous bases A and T predominated; therefore, all the individuals analyzed had characteristic AT-type DNA (A+T > C+G). The same conclusion was also reached in the case of the nucleotide sequences of the mitochondrial d-loop control region. The specificity coefficient for the two concatenated nucleotide sequences had an average value of 1.45, which fell within the range of values specific to animal organisms.

3.4. Dynamics of the Rate Evolution of Cytochrome b and d-Loop Mitochondrial Markers

Regarding the genetic variability at the d-loop region, after aligning and trimming the 32 sequences, the haplotypes and variable nucleotide positions were assessed in relation to the reference sequence V00654 from GenBank. The T3/T4 subclade haplogroup had the highest frequency, as predicted for European cattle breeds. The diversity of haplotypes was 0.908 ± 0.005. The haplotypes observed in Romanian Grey Steppe were the same as those seen in Hungarian Grey and Podolian cattle (Hungarian Grey, Bulgarian Grey, Ukrainian Grey, Istrian, and Slavonian Syrmian).

Based on the gene sequences of the cytochrome b and d-loop mitochondrial markers, the demographic and spatial expansion of the Grey Steppe population was analyzed, using the Mismatch distribution calculation model, which shows the distribution of the number of differences identified between pairs of haplotypes. As a rule, this distribution can be unimodal (specific to populations that have undergone either a recent demographic expansion or a spatial expansion characterized by a high degree of migration) or multimodal (in the case of populations in demographic equilibrium). In the case of the Grey Steppe, a multimodal distribution resulted from the analysis of both the cytochrome b nucleotide sequences and the d-loop mitochondrial control region (Figure 5a,b).

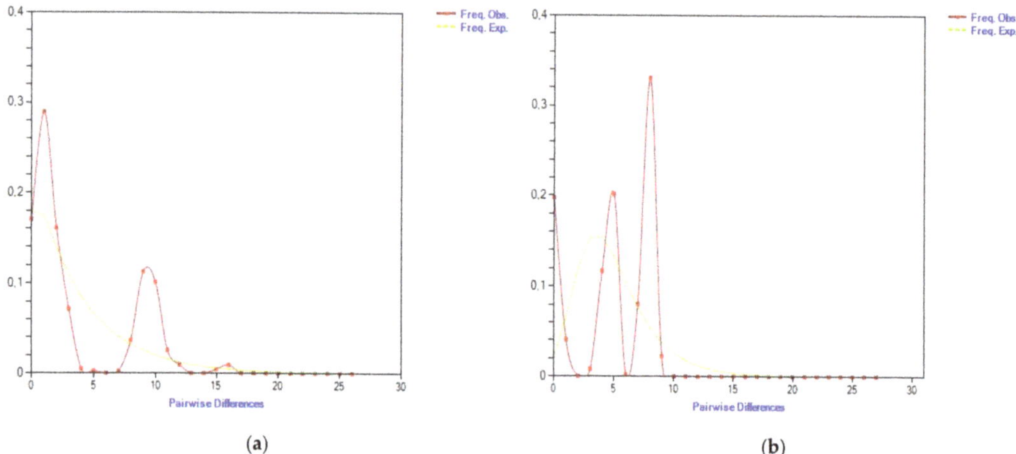

Figure 5. Demographic expansion of Grey Steppe population: (**a**) cytochrome b; (**b**) d-loop.

The alignment of all cytochrome b and d-loop gene sequences was performed by the ClustalW method using the MegAlign module, which is based on joining the forward and reverse sequences for each individual by aligning them with the primer sequences, cutting the two chains, and joining them into a single chain. After alignment, for each dataset (cytochrome b sequences and d-loop sequences, respectively), the optimal substitution model was checked, using the jModelTest program [34] based on the Akaike Information Algorithm Criterion [35] (AIC). The best-fit models of nucleotide substitution were chosen using jModelTest. This program employs five distinct model selection procedures, including hierarchical and dynamical likelihood ratio tests (hLRT and dLRT), Akaike and

Bayesian information criteria (AIC and BIC), and a decision theory method (DT). It also provides model-selection uncertainty estimates; parameter importance estimates; and model-averaged parameter estimates, including model-averaged tree topologies. Additionally, jModelTest provides novel tree optimization algorithms and model-averaged phylogenetic trees (determining both topology and branch length). The optimal substitution model for each analysis is shown in Table 5. Following the analysis, 22 variable sites (1.9%) were observed, of which 18 were informative sites (1.6%). In the case of d-loop sequences, the number of variable sites was 18 (2.0%), and the number of informative sites was 13 (1.4%).

Table 5. The optimal substitution model calculated in the jModelTest program (cytochrome b and d-loop).

		Cytochrome b				d-Loop	
Model		TrN+I [6]				TPM3uf	
Partition		010020				012012	
-lnL [1]		1708.1951				1354.6687	
K [2]		64				67	
freqA [3]	0.3125	R(a)	1.0000	0.3278		R(a)	23.9836
freqC [3]	0.3021	R(b)	4.4591	0.2440		R(b)	200.0000
freqG [3]	0.1352	R(c)	1.0000	0.1379		R(c)	0.0586
		cytochrome b				d-loop	
freqT [3]	0.2501	R(d)	1.0000	0.2902		R(d)	23.9836
ti/tv [4]	-	R(e)	0.5005	-		R(e)	200.0000
	-	R(f)	1.0000	-		R(f)	0.0586
p-inv [5]	0.8680	gamma	-			gamma	-

[1] -lnL: negative log likelihood. [2] K: number of estimated parameters. [3] freqA, freqC, freqG, freqT: frequency of nitrogen base A, C, G, and T, respectively. [4] ti/tv: transition/trasversion ratio. [5] p-inv: proportion of invariable sites. [6] I: invariable sites

The optimal substitution model was chosen from a confidence interval of 100%, with eight candidate models for both cytochrome b TrN+I6 (with an estimated mean log-likelihood value of 1708.1951) and d-loop TPM3uf (with an estimated mean log-likelihood value of 1354.6687). The values for the nitrogenous base frequencies were 31.2% for A, 25.1% for T, 13.4% for G, and 30.2% for C (cytochrome b sequence). A<->G and T<->C type transversions had values of 4.4591 and 0.5005, respectively (Figure 6a), their ratio being 0.755 (Table 6).

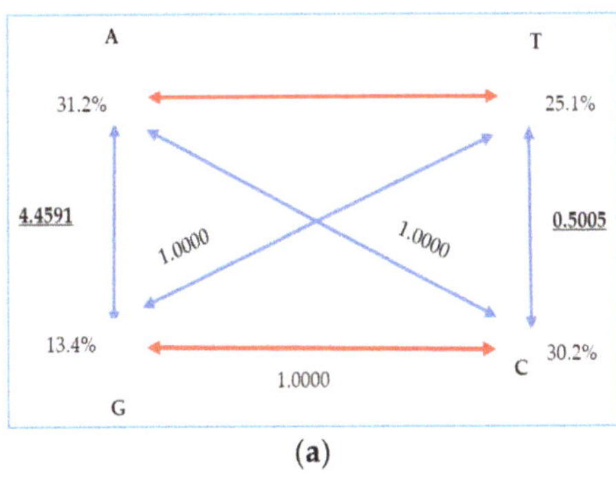

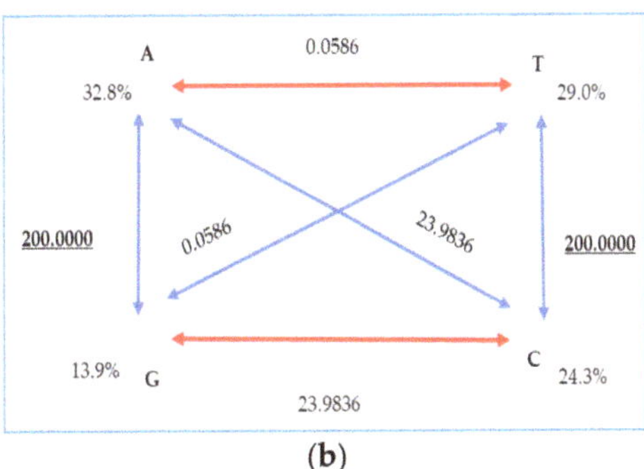

Figure 6. Substitution rates for each nucleotide, calculated in jModelTest: (**a**) cytochrome b gene sequences; (**b**) d-loop gene sequences.

Table 6. Number of sites and Ti/Tv ratio for cytochrome b and d-loop mitochondrial markers specific to the Grey Steppe population.

Sequence	No. of Informative Sites	%	No. of Variable Sites	%	Ti/Tv [1]
cytochrome b	18	1.6	22	1.9	0.755
d-loop	13	1.4	18	2.0	5.107

[1] ti/tv: transition/trasversion ratio.

In the case of d-loop gene sequences, the frequencies of the nitrogenous bases were as follows: 32.8% A, 29.0% T, 24.3% C, and 13.9% G. The Ti/Tv ratio showed a value of 5.107 (Table 6 and Figure 6b).

3.5. Haplotype Frequency Assessment by Analysis of Cytochrome b Gene Sequences and Mitochondrial d-Loop Control Region

By analyzing and interpreting the gene sequences of the two mitochondrial markers, four haplotypes with different frequencies were identified (T1, T2, T3/T4, and P′QT), into which the 32 individuals under study were classified (Figure 7).

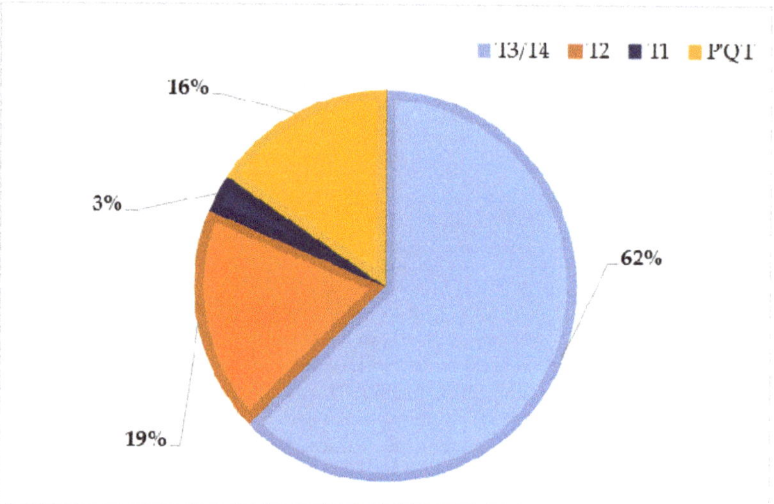

Figure 7. The frequency of haplotypes in Grey Steppe cattle population analyzed.

The highest haplotype frequency was represented by the T3/T4 haplotype, which was identified in 20 analyzed individuals (62%): SS_02; SS_03; SS_04; SS_06; SS_07; SS_08; SS_09; SS_10; SS_12; SS_13; SS_14; SS_15; SS_16; SS_17; SS_20; SS_21; SS_23; SS_24; SS_29; and SS_31. Six individuals fell into the T2 haplotype, representing 19%: SS_01; SS_05; SS_18; SS_19; SS_28; and SS_33. The haplotype P′QT was identified in five individuals, with a frequency of 16%: SS_22; SS_25; SS_26; SS_30; and SS_34. (This haplotype was also identified in the analysis of the nucleotide sequences from taurine specimens with direct descent from *Bos taurus primigenius*.) A single individual, SS_11, was included in the T1 haplotype, which had the lowest frequency at only 3% (Table 7).

Table 7. Identified haplotypes and representative individuals for each haplotype in Grey Steppe population.

Haplotypes Identified	Representative Individuals	Total Individuals/Haplotype
T3/T4	SS_02; SS_03; SS_04; SS_06; SS_07; SS_08; SS_09; SS_10; SS_12; SS_13; SS_14; SS_15; SS_16; SS_17; SS_20; SS_21; SS_23; SS_24; SS_29; SS_31	20
T2	SS_01; SS_05; SS_18; SS_19; SS_28; SS_33	6
T1	SS_11	1
P′QT	SS_22; SS_25; SS_26; SS_30; SS_34	5

3.6. Haplotype Network Analysis and Phylogenetic Tree Construction

The analysis of the nucleotide sequences using the Network program resulted in the haplotype network graphically represented in Figure 8a,b. Four major haplotypes (T1, T2, T3, and P′QT) were identified, with specific connection networks. Each haplotype corresponded to a certain number of individuals (Table 7).

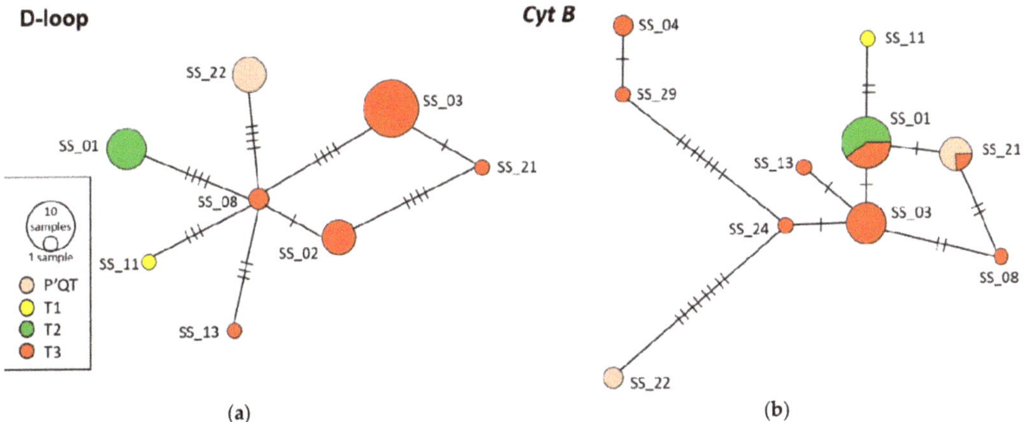

Figure 8. The network haplotypes of Grey Steppe cattle population: (**a**) d-loop; (**b**) cytochrome b.

Through the analysis of the distribution of the four haplotypes, it was found that the T3 haplotype had the highest weight, being identified in 20 individuals out of the 32 that were analyzed.

The identification of the P′QT haplogroup following the analysis of the nucleotide sequences of the Grey Steppe cattle breed indicated that this haplotype is of the ancestral type, being specific to *Bos taurus primigenius*, from which this breed evolved. Within this haplogroup, five individuals were identified (SS_22; SS_25; SS_26; SS_30; and SS_34), representing 16% of the specimens analyzed. The genetic distance between haplotypes T1, T2, and T3 varied between one and four sites. The presence of T-derived haplotypes suggested the demographic expansion of this population (Figure 9).

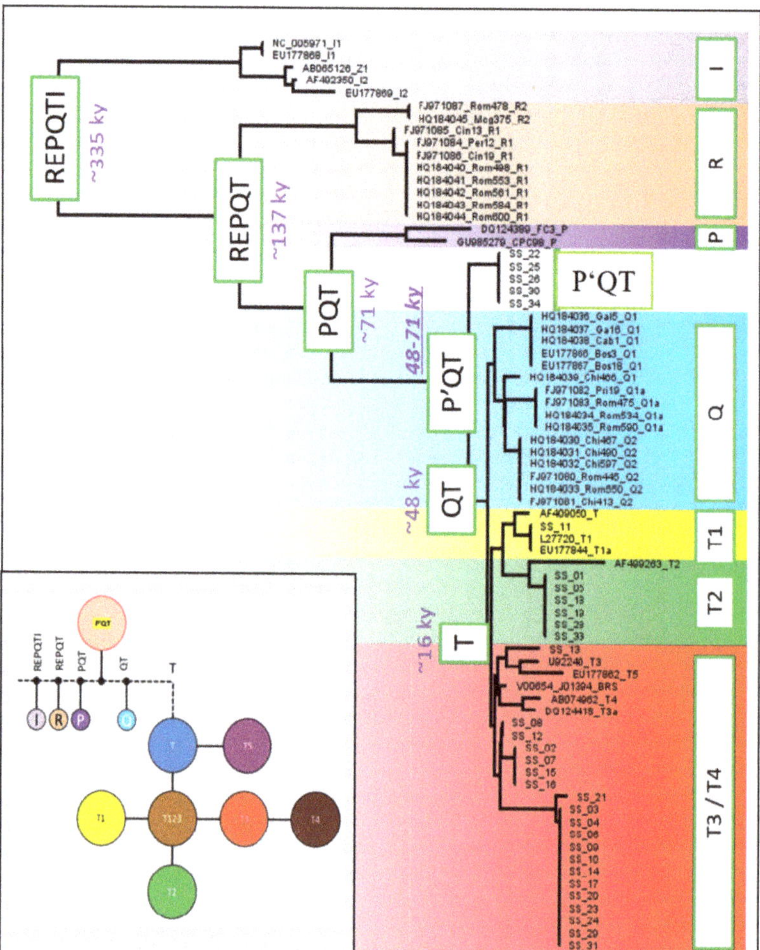

Figure 9. Phylogenetic tree obtained from the data analysis of the cytochrome b gene and the estimation of divergence time.

4. Discussion

To date, only a few studies have been conducted on the genetic composition of Podolian cattle breeds. Most were confined to a few breeds and focused on the nuclear genome. This study reported, for the first time, the genetic diversity and phylogenetic characteristics of indigenous Romanian Grey cattle based on the sequencing of mtDNA markers. The analysis of genetic diversity and phylogeny plays an essential role in the genetic improvement and selection of cattle breeds for sustainable breeding and management programs in many countries. The type and number of haplotypes constitutes a key indicator in the maternal line of a cattle breed's genetic diversity. The numerous studies of mtDNA in the *Bovinae* species [2,19,36–38] have demonstrated the existence of five major haplotypes specific to the genus *Bos taurus*, the wild ancestor of domesticated cattle (T1, T2, T3, and T4), as well as two haplotypes for *Bos indicus* (I1 and I2). Recent research [39–43] has shown that almost all taurines belong to macro-haplogroup T, and the estimated divergence time is ~16,000 years, indicating a narrow bottleneck in the evolutionary history of taurines in the *Bos taurus* genus. Macro-haplogroup T is divided into two sister subclades (T5 and T1/T2/T3), the predominant subclade being T1/T2/T3. Over time, T4 was integrated into T3 [44]. Haplogroup P was another prominent haplogroup in European aurochs, but it has not been found in current European cattle. However, from a dataset of over 3000 haplotypes, haplogroup P has only been detected in three modern breeds: Asian cattle, Korean cattle, and Chinese cattle [45]. This haplogroup is thought to be a relic of wild auroch introgression into the early domesticated bovine gene pool [45,46].

Other studies [47,48], aimed at analyzing the nucleotide sequences of the d-loop specific to an endangered Italian taurine breed, led to the identification of another haplogroup belonging to this breed, namely, haplogroup Q. Haplogroup Q was also identified following the phylogenetic analysis of ancestral European cattle, with descent from *Bos taurus primigenius*. A study by Achilli et al. [19], regarding the origin of taurines based on mitochondrial genome analysis, demonstrated that not all taurines in Europe belong to haplogroup T. In this study, we analyzed the gene sequences of the mtDNA from 26 European cattle breeds (22 from Italy and 4 from other regions of Europe). Most breeds fell into macro-haplogroup T and its subclades. Of the analyzed breeds, 1.4% were representative of haplogroups P and Q, which are specific to ancestral cattle from northern and central Europe with *Bos taurus primigenius* as their common ancestor. Haplogroup Q is phylogenetically close to macro-haplogroup T. Studies in the literature show that haplogroup Q and haplogroup T subclades are implicated in the same domestication event in the Fertile Crescent [49–51].

It has been suggested that haplogroup P belongs to breeds domesticated in the Near East, which then moved following the migration of humans. Another recent article by Senczuk et al. [21] contributed several findings regarding the origin and evolutionary history of the Podolian cattle breeds. Haplogroup P is characteristic of ancestral taurines of central and northern Europe and a few modern taurine breeds. In the haplogroup Q line, Italian, Egyptian, and Neolithic European cattle breeds can be distinguished, and in the R line, Italian cattle breeds can be distinguished from the Podolian Grey group. Compared to modern European cattle breeds, cattle from the "Podolian Grey" group, which also includes the Grey Steppe breed that was the subject of this research, present many ancestral characteristics (long horns, longevity, and adaptability to climatic and environmental conditions). The evolutionary history of European cattle is dominated by introgression events with both ancestral cattle and other breeds. According to the latest findings, the introgression between Indic and European taurines dates back ~4200 years [21,52,53].

Other studies on Podolian breeds found 13 haplotypes in 5 Istrian cattle and 5 haplotypes in 47 Slavonian Syrmian Podolian cattle [54], as well as 7 haplotypes in 39 Bulgarian Grey cattle [55]. Significantly lower levels of genetic diversity were observed in Serbian Podolian (0.709; $n = 11$) and Ukrainian Grey cattle (0.000; $n = 8$) [56], which could be explained by the lower sample size or the influence of human breeding activities (inbreeding, population bottlenecks). Moreover, modern cattle populations and wild cattle have high haplotype diversity, with the predominance of haplogroups T1, T2, and T3 [19,24].

Through molecular investigations, we found that the cytochrome b gene was the mitochondrial marker capable of evaluating the phylogenetic relationships most accurately, showing particular relevance for highlighting genetic differences, while the d-loop molecular marker presented the best topological support.

5. Conclusions

The identification of the P′QT haplogroup following the analysis of the mitochondrial marker nucleotide sequences of the Grey Steppe cattle breed indicated that this haplogroup is of an ancestral type, being specific to the wild *Bos taurus primigenius*, from which this breed evolved. The present research might help to save the endangered Romanian Grey Steppe. The findings from a small population of endangered Romanian Grey cattle demonstrate the presence of significant genetic variety and underline the importance of the breed in the formation of overall genetic biodiversity in cattle. Saving the Romanian Grey breed from extinction will assist both Romanian agriculture and global genetic heritage. The results of this study demonstrated that the Romanian Grey is a direct descendant of *Bos taurus primigenius*, representing a valuable tool for efforts to conserve the genetic resources of this endangered cattle breed.

Author Contributions: Conceptualization, M.-A.D.; methodology, M.-A.D., M.C. and D.-L.G.; software, M.-A.D. and M.C.; validation, M.-A.D. and D.S.; formal analysis, M.-A.D., S.C., D.-L.G. and M.C.; investigation, M.-A.D., S.C. and D.-L.G.; data curation, D.S.; writing—original draft preparation, M.-A.D.; writing—review and editing, M.-A.D., M.C. and D.S.; supervision, M.-A.D. All authors have read and agreed to the published version of the manuscript.

Funding: This research received no external funding.

Institutional Review Board Statement: In this study, blood samples were collected from 32 cows, in accordance with the EU directive UE 2016/679 on "general data protection regulation". Samples were collected through the puncture of the jugular vein, and sampling procedures were in accordance with the EU Directive 2010/63/EU "on the protection of animals used for scientific purposes". Animals were not subjected to experimental factors.

Data Availability Statement: The data present in this study are available on request from the first author and corresponding authors.

Conflicts of Interest: The authors declare no conflict of interest.

References

1. Groeneveld, L.F.; Lenstra, J.A.; Eding, H.; Toro, M.A.; Scherf, B.; Pilling, D. Genetic diversity in farm animals—A review. *Anim. Genet.* **2010**, *11*, 6–31. [CrossRef]
2. Pariset, L.; Mariotti, M.; Nardone, A.; Soysal, M.I.; Ozkan, E.; Williams, J.L.; Dunner, S.; Leveziel, H.; Maroti-Agots, A.; Bodo, I.; et al. Relationships between Podolic cattle breeds assessed by single nucleotide polymorphisms (SNPs) genotyping. *J. Anim. Breed. Genet.* **2010**, *127*, 481–488. [CrossRef]
3. Ilie, D.E.; Cean, A.; Cziszter, L.T.; Gavojdian, D.; Ivan, A.; Kusza, S. Microsatellite and Mitochondrial DNA study of native Eastern European cattle populations: The case of the Romanian Grey. *PLoS ONE* **2015**, *10*, e0138736. [CrossRef]
4. Scherf, B.D. *World Watch List for Domestic Animal Diversity*, 3rd ed.; FAO: Rome, Italy, 2000. Available online: http://www.fao.org/docrep/009/x8750e/x8750e00.htm (accessed on 23 May 2022).
5. Creangă, Ș.; Dascălu, D.L.; Ariton, A.M. *Conservation of Grey Steppe Cattle Breed from Romania in Order to Ensure the Biodiversity of the Genetic Resources of the Animal Populations*; Pim: Iași, Romania, 2018; pp. 15–23, ISBN 20978-606-13-4638-7.
6. Teneva, A.; Todorovska, E.; Tyufekchiev, N.; Kozelov, L.; Atanassov, A.; Foteva, S.; Ralcheva, S.; Zlatarev, S. Molecular characterization of Bulgarian livestock genetic resources: Genetic diversity in Bulgarian grey cattle as revealed by microsatellite markers. *Biotechnol. Anim. Husb.* **2005**, *21*, 35–42. [CrossRef]
7. Creangă, Ș.; Maciuc, V.; Leonte, C. Genetical conservation for Romanian Grey Steppe cattle breed from SCDCB Dancu, Iasi county. *Sci. Pap. Vet. Med. Ser.* **2008**, *51*, 702–706.
8. Georgescu, S.E.; Manea, M.A.; Zaulet, M.; Costache, M. Genetic diversity among Romanian cattle breeds with a special focus on the Romanian Grey Steppe Breed. *Rom. Biotechnol. Lett.* **2008**, *14*, 4194–4200.
9. Creangă, Ș.; Chelmu, S.S.; Maciuc, V.; Bâlteanu, V.A. Molecular characterization of Grey Steppe cattle breed for its genetic biopreservation. *Rom. Biotechnol. Lett.* **2013**, *18*, 8893–8900.
10. Maciuc, V.; Radu-Rusu, R.M. Assessment of Grey Steppe cattle genetic and phenotypic traits as valuable resources in preserving biodiversity. *Environ. Eng. Manag. J.* **2018**, *17*, 2741–2748.
11. Maciuc, V. *Cattle Breeding Management*; Alfa: Iași, Romania, 2006; pp. 76–78, ISBN 973-8953-18-9.
12. Creangă, Ș.; Maciuc, V. *Grey Steppe Cattle Breed from Romania*; Alfa: Iași, Romania, 2010; pp. 47–58, ISBN 978-606-540-034-4.
13. Dascălu, D.L.; Creangă, Ș.; Borș, S.I.; Doliș, M.G.; Simeanu, D.; Donosă, R. The Romanian Grey Steppe characterisation—The breed evolution in the last century from the numerical and morphological point of view. *Danub. Anim. Genet. Resour.* **2017**, *2*, 70–75.
14. Davidescu, M.A.; Ciorpac, M.; Madescu, B.M.; Porosnicu, I.; Creangă, Ș. Analysis of the Genetic diversity of endangered cattle breeds based on studies of genetic markers. *Sci. Pap. Anim. Sci. Biotechnol.* **2021**, *54*, 60–63.
15. Davidescu, M.A.; Grădinaru, A.C.; Creangă, Ș. Endangered romanian cattle breeds–between traditional breeding and genetic conservation. *Sci. Pap.-Anim. Sci. Ser. Sci. Pap.-Anim. Husb. Ser.* **2021**, *75*, 66–75.
16. Grădinaru, A.C.; Petrescu-Mag, I.V.; Oroian, F.C.; Balint, C.; Oltean, I. Milk Protein Polymorphism Characterization: A Modern Tool for Sustainable Conservation of Endangered Romanian Cattle Breeds in the Context of Traditional Breeding. *Sustainability* **2018**, *10*, 534. [CrossRef]
17. Bollongino, R.; Edwards, C.J.; Alt, K.W.; Burger, J.; Bradley, D.G. Early history of European domestic cattle as revealed by ancient DNA. *Biol. Lett.* **2006**, *2*, 155–159. [CrossRef] [PubMed]
18. Bollongino, R.; Elsner, J.; Vigne, J.D.; Burger, J. Y-SNPs do not indicate hybridisation between European aurochs and domestic cattle. *PLoS ONE* **2008**, *3*, e3418. [CrossRef]

19. Achilli, A.; Bonfiglio, S.; Olivieri, A.; Malusà, A.; Pala, M. The multifaceted origin of taurine cattle reflected by the mitochondrial genome. *PLoS ONE* **2009**, *4*, e5753. [CrossRef] [PubMed]

20. Sinding, M.H.S.; Gilbert, M.T.P. The draft genome of extinct European Aurochs and its implications for de-extinction. *Open Quat.* **2016**, *2*, 7. [CrossRef]

21. Senczuk, G.; Mastrangelo, S.; Ajmone-Marsan, P.; Becskei, Z.; Colangelo, P.; Colli, L.; Ferretti, L.; Karsli, T.; Lancioni, H.; Lasagna, E.; et al. On the origin and diversification of Podolian cattle breeds: Testing scenarios of European colonization using genome-wide SNP data. *Genet. Sel.* **2021**, *53*, 48. [CrossRef]

22. Dascălu, D.L. Conservation of the Romanian Grey Steppe Breed. Ph.D. Thesis, Iasi University of Life Sciences, Iași, Romania, 2014.

23. PCR Master Mix—Promega Corporation. PCR Master Mix. Available online: https://www.promega.ro/products/pcr/taq-polymerase/master-mix-pcr/?catNum=M7502 (accessed on 12 May 2021).

24. GenBank Overview-NCBI. Bos taurus Complete Mitochondrial Genome. Available online: https://www.ncbi.nlm.nih.gov/nuccore/V00654.1 (accessed on 15 September 2021).

25. Seroussi, E.; Yakobson, E. Bovine mtDNA D-loop haplotypes exceed mutations in number despite reduced recombination: An effective alternative for identity control. *Animal* **2010**, *4*, 1818–1822. [CrossRef]

26. Wang, X.; Zhang, Y.Q.; He, D.C.; Yang, X.M.; Li, B.; Wang, D.C.; Guang, J.; Xu, F.; Li, J.Y.; Gao, X.; et al. The complete mitochondrial genome of *Bos taurus* coreanae (Korean native cattle). *Mitochondrial DNA A DNA Mapp. Seq. Anal.* **2016**, *27*, 120–121. [CrossRef]

27. Liu, S.J.; Lv, J.Z.; Tan, Z.Y.; Ge, X.Y. The complete mitochondrial genome of Uruguayan native cattle (*Bos taurus*). *Mitochondrial DNA B Resour.* **2020**, *5*, 443–444. [CrossRef]

28. Arbizu, C.I.; Ferro-Mauricio, R.D.; Chávez-Galarza, J.C.; Vásquez, H.V.; Maicelo, J.L.; Poemape, C.; Gonzales, J.; Quilcate, C.; Corredor, F.-A. The Complete Mitochondrial Genome of a Neglected Breed, the Peruvian Creole Cattle (*Bos taurus*), and Its Phylogenetic Analysis. *Data* **2022**, *7*, 76. [CrossRef]

29. Guo, X.; Pei, J.; Xiong, L.; Bao, P.; Zhu, Y.; Wangdui, B.; Wu, X.; Chu, M.; Yan, P.; Ding, X. The complete mitochondrial genome of Shigaste humped cattle (*Bos taurus*). *Conserv. Genet. Resour.* **2017**, *10*, 789–790. [CrossRef]

30. Thompson, J.D.; Higgins, D.G.; Gibson, T.J. CLUSTAL W: Improving the sensitivity of progressive multiple sequence alignment through sequence weighting, position-specific gap penalties and weight matrix choice. *Nucleic Acids Res.* **1994**, *22*, 4673–4680. [CrossRef] [PubMed]

31. Rozas, J.; Sánchez-DelBarrio, J.C.; Messeguer, X.; Rozas, R. DnaSP, DNA polymorphism analyses by the coalescent and other methods. *Bioinformatics* **2003**, *19*, 2496–2497. [CrossRef] [PubMed]

32. Tarekegn, G.M.; Ji, X.Y.; Bai, X.; Liu, B.; Zhang, W.; Birungi, J.; Djikeng, A.; Tesfaye, K. Variations in mitochondrial cytochrome b region among Ethiopian indigenous cattle populations assert *Bos taurus* maternal origin and historical dynamics. *Asian-Australas. J. Anim. Sci.* **2018**, *31*, 1393–1400. [CrossRef]

33. Yang, S.L.; Lin, R.Y.; Xu, L.X.; Cheng, L. Analysis of polymorphisms of mitochondrial DNA D-loop and Mc1R gene in Chinese Wuchuan Black cattle. *J. Appl. Anim. Res.* **2014**, *42*, 487–491. [CrossRef]

34. Darriba, D.; Taboada, G.L.; Doallo, R.; Posada, D. jModelTest 2: More models, new heuristics and parallel computing. *Nature Methods* **2012**, *9*, 772. [CrossRef]

35. Akaike, H. Anew look at the statistical model identification. *IEEE Trans. Autom. Control* **1974**, *19*, 716–723. [CrossRef]

36. Achilli, A.; Olivieri, A.; Pellecchia, M.; Uboldi, C.; Colli, L. Mitochondrial genomes of extinct aurochs survive in domestic cattle. *Curr. Biol.* **2008**, *18*, 157–158. [CrossRef]

37. Bonfiglio, S.; Ginja, C.; De Gaetano, A.; Achilli, A.; Olivieri, A.; Colli, L.; Tesfaye, K.; Agha, S.H.; Gama, L.T.; Cattonaro, F.; et al. Origin and spread of *Bos taurus*: New clues from mitochondrial genomes belonging to haplogroup T1. *PLoS ONE* **2012**, *7*, e38601. [CrossRef]

38. Di Lorenzo, P.; Lancioni, H.; Ceccobelli, S.; Colli, L.; Cardinali, I.; Karsli, T.; Capodiferro, M.R.; Sahin, E.; Ferretti, L.; Marsan, P.A.; et al. Mitochondrial DNA variants of Podolian cattle breeds testify for a dual maternal origin. *PLoS ONE* **2018**, *13*, e0192567. [CrossRef]

39. Olivieri, A.; Gandini, F.; Achilli, A.; Fichera, A.; Rizzi, E.; Bonfiglio, S.; Battaglia, V.; Brandini, S.; De-Gaetano, A.; El-Beltagi, A.; et al. Mitogenomes fromEgyptian Cattle Breeds: New Clues on the Origin of Haplogroup Q and the Early Spread of *Bos taurus* from the Near East. *PLoS ONE* **2015**, *10*, e0141170. [CrossRef] [PubMed]

40. Park, S.D.E.; Magee, D.A.; McGettigan, P.A.; Teasdale, M.D.; Edwards, C.J.; Lohan, A.J.; Murphy, A.; Braud, M.; Donoghue, M.T.; Liu, Y.; et al. Genome sequencing of the extinct Eurasian wild aurochs, *Bos primigenius*, illuminates the phylogeography and evolution of cattle. *Genome Biol.* **2015**, *16*, 234. [CrossRef]

41. Cubric-Curik, V.; Novosel, D.; Brajkovic, V.; Rota Stabelli, O.; Krebs, S.; Sölkner, J.; Šalamon, D.; Ristov, S.; Berger, B.; Trivizaki, S.; et al. Large-scale mitogenome sequencing reveals consecutive expansions of domestic taurine cattle and supports sporadic aurochs introgression. *Evol. Appl.* **2022**, *15*, 663–678. [CrossRef] [PubMed]

42. Dorji, J.; Vander-Jagt, C.; Chamberlain, A.; Cocks, B.; MacLeod, I.; Daetwyler, H. Cattle Maternal Diversity Inferred from 1,883 Taurine and Indicine Mitogenomes. *Res. Sq.* **2021**. [CrossRef]

43. Dorji, J.; Vander Jagt, C.J.; Chamberlain, A.J.; Cocks, B.G.; MacLeod, I.M.; Daetwyler, H.D. Recovery of mitogenomes from whole genome sequences to infer maternal diversity in 1883 modern taurine and indicine cattle. *Sci. Rep.* **2022**, *12*, 5582. [CrossRef]

44. Liu, H.; Zhai, J.; Wu, H.; Wang, J.; Zhang, S.; Li, J.; Niu, Z.; Shen, C.; Zhang, K.; Liu, Z.; et al. Diversity of Mitochondrial DNA Haplogroups and Their Association with Bovine Antral Follicle Count. *Animals* **2022**, *12*, 2350. [CrossRef]

45. Edwards, J.; Bollongino, R.; Scheu, A.; Chamberlain, A.; Tresset, A.; Vigne, J.D.; Baird, J.F.; Larson, G.; Ho, S.Y.; Heupink, T.H. Mitochondrial DNA analysis shows a Near Eastern Neolithic origin for domestic cattle and no indication of domestication of European aurochs. *Proc. Biol. Sci.* **2007**, *274*, 1377–1385. [CrossRef]
46. Stock, F.; Edwards, C.J.; Bollongino, R.; Finlay, E.K.; Burger, J.; Bradley, D.G. Cytochrome b sequences of ancient cattle and wild ox support phylogenetic complexity in the ancient and modern bovine populations. *Anim. Genet.* **2009**, *40*, 694–700. [CrossRef]
47. Magee, D.A.; Hugh, D.E.; Ceiridwen, J.E. Interrogation of modern and ancient genomes reveals the complex domestic history of cattle. *Anim. Front.* **2014**, *4*, 7–22. [CrossRef]
48. Maretto, F.; Ramljak, J.; Sbarra, F.; Penasa, M.; Mantovani, R.; Ivankovic, A.; Bittantea, G. Genetic relationship among Italian and Croatian Podolian cattle breeds assessed by microsatellite markers. *Livest. Sci.* **2012**, *150*, 256–264. [CrossRef]
49. Felius, M.; Beerling, M.-L.; Buchanan, D.S.; Theunissen, B.; Koolmees, P.A.; Lenstra, J.A. On the History of Cattle Genetic Resources. *Diversity* **2014**, *6*, 705–750. [CrossRef]
50. Pitt, D.; Sevane, N.; Nicolazzi, E.L.; Hugh, D.E.; Park, S.D.E.; Colli, L.; Martinez, R.; Bruford, M.W.; Orozco-terWenge, P. Domestication of cattle: Two or three events? *Evol. Appl.* **2019**, *12*, 123–136. [CrossRef] [PubMed]
51. Verdugo, M.P.; Mullin, V.E.; Scheu, A.; Mattiangeli, V.; Daly, K.G.; Maisano Delser, P.; Hare, A.J.; Burger, J.; Collins, M.J.; Kehati, R.; et al. Ancient cattle genomics, origins, and rapid turnover in the Fertile Crescent. *Science* **2019**, *365*, 173–176. [CrossRef]
52. Loftus, R.T.; MacHugh, D.E.; Bradley, D.G.; Sharp, P.M.; Cunningham, E.P. Evidence for two independent domestications of cattle. *Proc. Natl. Acad. Sci. USA* **1994**, *91*, 2757–2761. [CrossRef]
53. Upadhyay, M.R.; Chen, W.; Lenstra, J.A.; Goderie, C.R.; MacHugh, D.E.; Park, S.D.; Magee, D.A.; Matassino, D.; Ciani, F.; Megens, H.J.; et al. European Cattle Genetic Diversity Consortium & RPMA Crooijmans Genetic origin, admixture and population history of aurochs (*Bos primigenius*) and primitive European cattle. *Heredity* **2017**, *118*, 169–176. [CrossRef]
54. Ivankovic, A.; Paprika, S.; Ramljak, J.; Dovc, P.; Konjacic, M. Mitochondrial DNA-based genetic evaluation of autochthonous cattle breeds in Croatia. *Czech J. Anim. Sci.* **2014**, *59*, 519–528. [CrossRef]
55. Hristov, P.; Teofanova, D.; Neov, B.; Radoslavov, G. Haplotype diversity in autochthonous Balkan cattle breeds. *Anim. Genet.* **2014**, *46*, 92–94. [CrossRef]
56. Kantanen, J.; Edwards, C.J.; Bradley, D.G.; Viinalass, H.; Thessler, S.; Ivanova, Z.; Kiselyova, T.; Ćinkulov, M.; Popov, R.; Stojanović, S.; et al. Maternal and paternal genealogy of Eurasian taurine cattle (*Bos taurus*). *Heredity* **2009**, *103*, 404–415. [CrossRef]

MDPI

St. Alban-Anlage 66

4052 Basel

Switzerland

Tel. +41 61 683 77 34

Fax +41 61 302 89 18

www.mdpi.com

Agriculture Editorial Office

E-mail: agriculture@mdpi.com

www.mdpi.com/journal/agriculture